17 50

FROM ENGINEERING SCIENCE TO BIG SCIENCE

The NACA and NASA Collier Trophy
Research Project Winners

NASA SP-4219

FROM ENGINEERING SCIENCE TO BIG SCIENCE

The NACA and NASA Collier Trophy Research Project Winners

Edited by Pamela E. Mack

The NASA History Series

National Aeronautics and Space Administration
NASA Office of Policy and Plans
NASA History Office
Washington, D.C.   1998

Library of Congress Cataloguing-in-Publication Data

From Engineering Science to Big Science: the NACA and NASA Collier Trophy Research Project Winners/edited by Pamela E. Mack.
p. cm.—(NASA SP: 4219) (The NASA history series)

    Includes bibliographical references and indexes.
    1. Collier Trophy.
    2. Aerospace Engineering—United States—History.
    3. Airplanes—United States—Design and Construction—History.
    4. Aeronautics—Research—United States—History.

I. Mack, Pamela Etter.	II. Series.	III. Series: The NASA History series.
TL537.F76	1997	97-27899
629.1'0973—dc21		CIP

For sale by the U.S. Government Printing Office
Superintendent of Documents, Mail Stop: SSOP, Washington, DC 20402-9328
ISBN 0-16-049640-3

Table of Contents

Introduction xi
Pamela E. Mack, Editor

Chapter 1   "Engineering Science and the Development of the NACA 1
            Low-Drag Engine Cowling," James R. Hansen

Chapter 2   "Lew Rodert, Epistemological Liaison, and Thermal 29
            De-Icing at Ames," Glenn E. Bugos

Chapter 3   "Research in Supersonic Flight and the Breaking 59
            of the Sound Barrier," John D. Anderson, Jr.

Chapter 4   "The Transonic Wind Tunnel and the NACA 91
            Technical Culture," Steven T. Corneliussen

Chapter 5   "The Whitcomb Area Rule: NACA Aerodynamics Research 135
            and Innovation," Lane E. Wallace

Chapter 6   "The X-15 Hypersonic Flight Research Program: 149
            Politics and Permutations at NASA," W.D. Kay

Chapter 7   "The Collier as Commemoration: The Project Mercury Astronauts 165
            and the Collier Trophy," Jannelle Warren-Findley

Chapter 8   "Managing America to the Moon: 193
            A Coalition Analysis," W. Henry Lambright

Chapter 9   "The Human Touch: The History of the 213
            Skylab Program," Donald C. Elder

Chapter 10 "LANDSAT and the Rise of Earth Resources Monitoring," 235
            Pamela E. Mack

Chapter 11 "Voyager: The Grand Tour of Big Science," Andrew J. Butrica 251

Chapter 12 "The Space Shuttle's First Flight: STS-1," Henry C. Dethloff 277

Chapter 13 "More Favored than the Birds: The Manned 299
            Maneuvering Unit in Space," Anne Millbrooke

| Chapter 14 | "The Advanced Turboprop Project: Radical Innovation in a Conservative Environment," Mark D. Bowles and Virginia P. Dawson | 321 |

| Chapter 15 | "Return to Flight: Richard H. Truly and the Recovery from the *Challenger* Accident," John M. Logsdon | 345 |

| Chapter 16 | "The Hubble Space Telescope Servicing Mission," Joseph N. Tatarewicz | 365 |

About the Authors ... 397

Index ... 401

The NASA History Series ... 425

Acknowledgments

Whenever historians take on a project of historical investigation such as this, they stand squarely on the shoulders of earlier investigators and incur a good many intellectual debts. We must acknowledge the assistance of several individuals who aided in the preparation of this study of aerospace research and development projects that have received the Collier Trophy over the years. First, Lee D. Saegesser, NASA Archivist between 1967 and 1997, was instrumental in obtaining documents used in the preparation of the work; Stephen J. Garber, assistant historian in the NASA History Office, critiqued the text; M. Louise Alstork, edited the work and prepared the index; and Nadine J. Andreassen helped with proofreading and compilation. Second, the history representatives at the various NASA Centers provided much needed assistance: Virginia Butler, Kevin Coleman, Deborah G. Douglas, Michael Q. Hooks, J.D. Hunley, Keith Koehler, William A. Larsen, Richard Layman, Elaine Liston, Dan Pappas, Jane Riddle, and Mike Wright. Third, the staffs of the NASA Headquarters Library and the Scientific and Technical Information Program provided assistance in locating materials; and archivists at various presidential libraries, the National Archives and Records Administration, the National Air and Space Museum, and in other research centers aided with research efforts. Fourth, the NASA Headquarters Printing and Design Office developed the layout and handled printing for this volume. Specifically, we wish to acknowledge the work of Janie E. Penn, Lillian Gipson, Patricia Talbert, and Kimberly Jenkins for their editorial and design work. In addition, Michael Crnkovic, Stanley Artis, and Jeffrey Thompson saw the book through the publication process. Thanks are do them all.

Several individuals read portions of the manuscript or talked with me about the project, in the process helping me more than they could ever know. These include Roger E. Bilstein, Michael L. Ciancone, Tom D. Crouch, Dwayne A. Day, David H. DeVorkin, Deborah G. Douglas, Andrew Dunar, Linda Neumann Ezell, Charles J. Gross, R. Cargill Hall, Richard P. Hallion, Gregg Herken, Norriss S. Hetherington, Robin Higham, Francis T. Hoban, Karl Hufbauer, Sylvia K. Kraemer, John Howard E. McCurdy, John E. Naugle, Allan E. Needell, David H. Onkst, Craig B. Waff, Stephen P. Waring, and Ray A. Williamson. All of these people would disagree with some of the areas chosen for emphasis, with many of the conclusions, and with a few of the themes, but such is both the boon and the bane of historical inquiry.

The Robert J. Collier Trophy is awarded annually, since 1911, for the greatest achievement in aeronautics and astronautics in America, with respect to improving the performance, efficiency, or safety of air or space vehicles, the value of which has been thoroughly demonstrated by actual use during the preceding year. Formerly called the Aero Club of America Trophy, it was renamed the Robert J. Collier Trophy in 1922. The National Aeronautics and Space Administration and its predecessor agency, the National Advisory Committee for Aeronautics (NACA), have received the trophy twenty times since 1929.

Introduction

by Pamela E. Mack

For many scientists and science and technology policy analysts, the cancellation of the Superconducting Supercollider project in 1994 served as a symbol of a fundamental change in public and congressional attitudes towards Federal funding for large science and technology projects. At minimum, government funded big science and big technology were not likely to continue to grow at the pace that characterized the Cold War era. Politicians in the United States seemed to have turned against funding very expensive research and development projects without clear, practical goals, probably because they believed such projects tended to take on a life of their own and require more and more funding. In the eyes of most policymakers, funding for innovation in science and technology could no longer easily be justified by the promise of great benefits from the new technology, both because such promises were viewed skeptically and because policymakers believed that budgetary pressures precluded even worthwhile new programs unless they directly saved money for the government.

Even before the trend started to turn, historians of science and technology had made important steps in understanding the development of big science and big technology in a number of different institutional settings, and the changing current climate can give new perspective. Scholarly interest in "big science" arose out of the perception of scientists in the 1950s and 1960s that the experience of doing science had changed in a fundamental way (at least in some fields) because of the increasing prevalence of expensive instruments and large externally funded research projects. Engineers did not experience a parallel shift of similar intensity; they already had experience with large government-funded projects (such as dams). But, at least in some fields, engineers working on large-scale, government-funded research and development did experience a shift to a particular new kind of big technology. For example, at the National Aeronautics and Space Administration this "big technology" involved large projects with a high political profile, quite different from the systematic research into fundamental design parameters that characterized the "engineering science" approach typical of the National Advisory Committee for Aeronautics before the war.[1]

Most historians studying big science and technology have focused either on basic science (particularly high-energy physics) or on military research and development.[2] Obviously, the National Aeronautics and Space Administration (NASA) and its predecessor organization, the National Advisory Committee for Aeronautics (NACA), provide another important example. The leaders of NASA during the Apollo program realized that they were pioneers in large program management as well as in space travel, but there has been little integration of the larger background to that story or systematic attention to the role of large project management in the issues NASA has faced since Apollo.

The NACA and NASA provide an opportunity to study changes in the pattern of major research and development projects over a significant span of time in a government context quite different from the Department of Defense. The chapters of this book discuss a series of case studies of notable technological projects carried out at least in part by the NACA and NASA. The case studies chosen are those projects that won the National

---

1. I have distinguished between "big science" and "big technology," but NASA uses the term "big science" to include both.
2. For references to the big science literature, see below; for a discussion of how the study of large technology fits into broader historiographical trends in history of technology, see John M. Staudenmaier, "Recent Trends in the History of Technology," *American Historical Review* 95 (June 1990): 715–26.

Aeronautic Association's (NAA) Collier Trophy for "the greatest achievement in aviation in America, the value of which has been thoroughly demonstrated by use during the preceding year." Looking back on the whole series of projects we can examine both what successes were seen as important at various times, and how the goals and organization of these notable projects changed over time.[3]

The Collier Trophy provides a way of selecting a series of case studies of projects that can be compared over a fairly long span of time. This volume covers projects that received their awards from 1929 to 1994. From the point of view of scholars who have studied government support for science and technology, this span of years covers three important periods. The period after World War I saw limited experimentation with the role of the government in supporting research most importantly in the form of engineering science. The period during and after World War II saw an explosion in the government role in science and technology, with another burst after Sputnik. Finally, a reevaluation of science and technology as public goods started from one side of the political spectrum in the late 1960s and took on new momentum from the other side in the 1980s. From the point of view of the rise of big science and technology, the projects in this book take us through a period when budgets, the number of people and organizations involved, and bureaucracy dramatically increased for most NACA and NASA projects. Not all the later projects covered in this book were large by the standards of their own time, but even the smaller ones, such as the Manned Maneuvering Unit (chapter 13) or the Fuel-efficient Turboprop (chapter 14), took form in an environment of political and bureaucratic pressures that had developed in NASA because of its role as a big-technology agency.

The series of case studies included here present some of the most successful projects in the history of the NACA and NASA. Each illuminates the development and limitations of big technology at these agencies as an example of the larger phenomenon of the development of engineering science and big science. The work of Walter Vincenti and James Hansen has made aeronautical engineering in general and the NACA in particular the standard example of engineering science.[4] While historians have used high-energy physics as the standard example of big science, NASA has some claim to the role of standard example for big technology (using patterns that to a considerable extent were set by the NACA). Apollo-era NASA Administrator James Webb certainly sought to make that claim by writing a book on *Space Age Management: The Large-Scale Approach,* and the idea had enough public resonance to turn the phrase—"If we can send a man to the moon why can't we. . . ?"—into a cliché.[5] Apollo did not provide the model for the future that Webb had hoped, but NASA continued to grapple in a very public way with the problems of conducting large-scale technology-development projects that required support from diverse interest

---

3. These projects do not represent simply a collection of success stories. While some were major triumphs for the NACA or NASA others did not live up to their initial promise, represented responses to major failures, or earned their awards more for public appeal than for technological achievement. Other projects, such as the Viking Mars landing, might have deserved the Collier Trophy more than some included here—the Collier Trophy provides an interesting sample, not a list of the NACA and the NASA's most successful projects. The most that can be said of all these projects is that they gained the praise of the aerospace community; within the context of the time and that community they represent successes.

4. See chapters 1, 3, and 4 and Walter G. Vincenti, *What Engineers Know and How They Know It: Analytical Studies in the History of Aeronautical History* (Baltimore, MD: Johns Hopkins University Press, 1990).

5. James E. Webb, *Space Age Management: The Large-Scale Approach* (New York, NY: McGraw Hill, 1969). See also Leonard Sayles and Margaret Chandler, *Managing Large Systems: Organizations for the Future* (New York, NY: Harper and Row, 1971). The "If they can send a man to the Moon" cliché eventually evolved into a joke; in Philadelphia in the late 1970s a business called Hong Kong Custom Tailors advertised with the line: "If they can send a man to the Moon why can't they make a suit to fit me?"

groups.⁶ The case studies in this book illuminate some of the key issues of big science and big technology, including the role of politics, the management of large enterprises, the relationship between basic research and research and development for practical ends, and the declining role of the individual leader or inventor.

The Collier Trophy

The Collier Trophy is the most prestigious award for aerospace achievement in the United States, and the recipients of the trophy have long been proud of the recognition the Collier Trophy brought their activities. While the projects covered in this volume would deserve study whether or not they had won the Collier Trophy, a volume focused on the winners of a particular award should give some attention to the history and character of that award. In fact, the history of the Collier Trophy and its parent organization, the National Aeronautic Association, provide a unique perspective on prizes for scientific and technological achievement.

The United States has had and still has a number of aviation and aerospace organizations, ranging from booster groups to professional societies. The National Aeronautic Association fits somewhere in the middle of that range. In turn, its prize is shaped by the composition of the committee that awards it and by a series of rules, in particular that the prize be given for an achievement in the preceding year. While the Nobel Prize is usually given for an accomplishment whose significance has been proven by years of experience, the Collier Trophy represents an almost concurrent evaluation of an achievement (like the Pulitzer Prize, it sometimes lacks the wisdom of hindsight).

In its early years, the National Aeronautic Association and its predecessor organization, the Aero Club of America, sought to foster American aviation in all its forms, and therefore both served as a booster club and advocated an increasingly professional approach to aviation.⁷ The Aero Club of America was formed by members of the Automobile Club of America in 1905, just two years after the first successful flight by the Wright brothers. The model of the Automobile Club led the Aero Club into such activities as training and licensing pilots and lobbying the Federal government to give more attention to military aviation during the build-up to the United States' entry into World War I.⁸ As aviation expanded during the War and the club suffered from divisiveness, it tended to lose its central role. Its members responded by negotiating a series of mergers with other clubs, starting with a merger with the American Flying Club in 1920.⁹ In 1922, a merger with the National Air Association (NAA) led to a new name, the National Aeronautic Association, and new bylaws that emphasized promoting aviation and lobbying for uniform federal regulation of the aviation industry.¹⁰

---

6. For a specific discussion of the failure of attempts to apply Apollo or Department of Defense models to social problems, see Bruce L.R. Smith, *American Science Policy Since World War II* (Washington, DC: The Brookings Institution, 1990), pp. 76-77. For an introduction to the social construction analysis of the role of interest groups in technological change, see Wiebe E. Bijker, Thomas P. Hughes, and Trevor Pinch, *The Social Construction of Technological Systems: New Directions in the Sociology and History of Technology* (Cambridge, MA: MIT Press, 1987).

7. William Kroger, "For Greatest Achievement: The Story Behind American Aviation's Most Prized Award," *National Aeronautics*, December 1944, pp. 15, 18, 26.

8. Bill Robie, *For the Greatest Achievement: A History of the Aero Club of America and the National Aeronautic Association* (Washington, DC: Smithsonian Institution Press, 1993), p. ix. The federal government took over pilot licensing from the club (at the club's urging) in 1926.

9. *Ibid.*, p. 100.

10. *Ibid.*, p. 103–05. The leadership of the new organization was dominated by industrialists, though none were at the time primarily employed in the aviation industry.

The Collier Trophy had been established by the Aero Club of America in 1911. The club had a new president at that time, Robert J. Collier. Collier had inherited the *Collier's* publishing enterprise and fortune in 1909 and also belonged to the community of upper-class men interested in expensive, dangerous sports, such as big game hunting, yacht racing, and polo. Many members of that community saw aviation as the next frontier for sport, and Collier owned two Wright biplanes by 1911, though he did not learn to fly until the spring of 1912.[11] Interested in improving aviation and in promoting safety, Robert Collier decided to sponsor a trophy, not for another airplane race, but for "the greatest achievement in aviation in America, the value of which has been thoroughly demonstrated by use during the preceding year."[12] Collier used his political connections to give the trophy prestige, in particular arranging for it to be presented by the President of the United States (an arrangement that has continued to the present). In its early years, the trophy was usually awarded to inventors for specific technologies such as hydroplanes developed by Glenn Curtiss (in 1911 and 1912), an automatic stabilizing device invented by Orville Wright (1913), and a gyroscopic control invented by Elmer and Lawrence Sperry (1914).[13]

Because of changes in the parent organization, in the 1920s and 1930s the Collier Trophy came to be awarded more often to organizations rather than individual inventors. The U.S. Air Mail system won the trophy in 1922 and 1923 for its safety record and for night flying, the Army Air Service won in 1924 for the first flight around the world, and the Aeronautics Branch of the Department of Commerce won in 1928 for the development of airways and air navigation. The NACA won its first Collier Trophy in 1929 for developing principles for the design of improved engine cowlings. This new pattern of awards reflected the merger of the Aero Club of America into a new organization, the National Aeronautic Association, which put a much greater emphasis on promoting government sponsorship and regulation of aviation. Between the formation of the National Aeronautic Association in 1922 and 1944, eleven Collier trophies listed government agencies or corporations as the first or only recipient, and four more listed organizations along with a key individual.[14] Orville Wright objected to this pattern in a 1944 letter that called for a return to the pattern of awarding the trophy to individuals for specific inventions.[15]

---

11. Collier was involved in promoting aviation before learning to fly himself. In the spring of 1911 he loaned a plane to the Army for maneuvers, then in the fall of 1911 he staged a large air meet on his estate, at which his planes were piloted by flyers from the Wright School. Kroger, "For Greatest Achievement," p. 18.

12. Robie, *For the Greatest Achievement*, p. 83, quoting from the Bulletin of the Aero Club of America, 1912. The trophy was originally named Aero Club of America Trophy. It was called the Collier Trophy from 1922, when the Aero Club became the National Aeronautic Association, but the name was not officially changed until 1944. The bronze trophy was the work of sculptor Ernest Wise Keyser, a former student of Augustus Saint-Gaudens who had also sculpted the memorial of Robert Collier's father. The resulting trophy represents the triumph of man over natural forces; it weighs 525 pounds and is now on permanent display at the National Air and Space Museum. Kroger, "For Greatest Achievement," p. 18.

13. Sperry won two Collier Trophies, in 1914 and 1916. In the eyes of his biographer, these did not compare in importance to the prize he won for his aircraft improvements in an international Competition for Safety in Aeroplanes held in France in 1914 or to the John Fritz Medal awarded to Sperry by the leading engineering societies of America in 1926. Thomas P. Hughes, *Elmer Sperry: Inventor and Engineer* (Baltimore, MD: Johns Hopkins University Press, 1971), pp. 199–200 and 307–08. In part, this is because aviation was not Sperry's first priority, but it also reflects the limited prestige of the Collier Trophy at a time when the United States had fallen behind other countries in aviation.

14. For a complete list of Collier Trophy winners see Robie, *For the Greatest Achievement*, pp. 229–36.

15. Wright's letter is quoted at length by Alex Roland, *Model Research: The National Advisory Committee for Aeronautics, 1915–1948*, 2 Vols. (Washington, DC: NASA SP–4103, 1985), 1:351 (note 36). Wright comments that "An examination of the list of recipients since that time will reveal that after the N.A.A. came into possession of it the awards have been mostly to U.S. government bureaus and to manufacturing companies instead of to individuals. This, no doubt, is due to the fact that individuals have more modestly [sic] than bureaus and corporations, and that individuals do not have the 'brass' to seek the award, while bureaus and companies have no lack in that respect." Roland comments that "Wright was seventy-two when he wrote that letter, just four years from death, but he was not senile and he was not a bitter old man. He was simply the patriarch of aviation, free to call a spade a spade."

However, Wright's protest could not reverse the declining role of the individual inventor or redefine the trophy. The Collier Trophy was an award for achievement, not for invention, and could be given to pilots or organizations as readily as to inventors.

The NAA appointed a new committee each year to select the Collier Trophy winner. The President of the Association nominated the members of the selection committee, often including previous winners. The nine members of the 1943 committee give a sense of the interests involved: Grover Loenig, advisor on aircraft of the War Production Board (WPB) (chair), Dr. George W. Lewis, Director of Aeronautical Research for the NACA; William R. Enyart, President of the NAA (ex officio); Gill Robb Wilson, aviation editor of the *New York Herald Tribune*; Major Lester D. Gardner, chairman of the council of the Institute of Aeronautical Sciences; Roger Wolfe Kahn, a famous private pilot; Laurence P. Sharples, chairman of the board of the Aircraft Owners and Pilots Association; William P. MacCracken, Jr., general counsel, and William P. Redding, the treasurer of the NAA.[16] By the 1990s, the selection committee had grown to thirty to forty members, but continued to represent leaders of all facets of the aerospace industry.[17] The varying types of projects receiving awards covered in this volume suggests that the character of the selection committee tended to vary somewhat on the basis of the interests of the NAA President and the Association. The Collier Trophy should therefore be understood as a reflection of attitudes and priorities in the community of aviation enthusiasts and those employed in aerospace-related work in industry and government. It did not have as much built-in protection from bias and short-term fads as the Nobel Prize, but those involved in the Collier award process valued very highly the prestige of the trophy and sought to preserve that prestige by choosing appropriate awardees. The trophy had little to back up its significance except for its long history and the tradition that it was awarded by the President of the United States; its importance rested on the luster of the winners.[18]

The NACA, NASA, and Government Research

The projects whose stories are told in this book provide a series of case studies of changes in the research and development process in a government setting over the period from the 1920s to the 1990s. They fit into a story of increasing government support for science and technology through one particular government agency, which like all organizations and people has been shaped by its own unique history. A brief survey of that history provides important background for any attempt to draw broader conclusions.

The National Advisory Committee for Aeronautics (NACA) helped set the precedent for government funding of research and development in twentieth century America, a precedent that represented a very significant change from nineteenth century assumptions. Even in the nineteenth century the Federal government had provided support for research

---

16. Kroger, "For Greatest Achievement," p. 18. There is no particular significance to 1943; it is simply the only published list of selection committee members that I have encountered. The award that year went to Captain Luis DeFlorez of the U.S. Navy Reserve for his contribution to the safe and rapid training of combat pilots and crews.

17. The current practice is that the President of the National Aeronautic Association extends about sixty invitations to participate in the selection committee, and the committee is composed of whatever number accept the invitation. Most of those invited are members of the Association; presidents of the Air Clubs affiliated with the National Aeronautic Association are automatically invited. The members of the selection committee meet in person, and after discussing the recommendations and entries vote by secret ballot on that year's award. Telephone interview with Jill Baucom, Administrative Assistant, National Aeronautic Association, December 15, 1995.

18. A large monetary award or a more important sponsoring organization would have given the trophy a more objective source of prestige. Instead, the Collier Trophy maintained its status as the "most prized of all aviation honors in the United States" by tradition alone (the quote is from Robie, *For the Greatest Achievement*, p. x).

in certain key areas where a consensus could be reached about how to serve the public good, such as the Coast Survey and the Department of Agriculture. But a constitutional principle that such functions belonged to the states except in times of national emergency continued after the Civil War, and *laissez faire* economic theories actually led to an even more limited definition of the proper role of government in areas that might be considered competition with industry.[19] The new government sponsorship of research and development that had its roots in World War I represented a significant change in the role of the state, and a change that met significant resistance.

The NACA was only one of a number of organizations created as a result of lobbying by scientists and engineers for a new government role in research and development in World War I.[20] The NACA got off to an early start; President Wilson signed the Naval Appropriations Bill that created the National Advisory Committee for Aeronautics in March 1915. The scientists, engineers, and enthusiasts who had lobbied for the bill for more than four years wanted government funding of aeronautical research to allow the United States to catch up with rapid developments in Europe, where the possibilities of the Wright brothers' invention had sparked more interest than in the United States. The legislation did not pass until the outbreak of war provided an additional push, and the bill did nothing more than create an advisory committee and provide it with a small appropriation. The NACA then set out to invent its own role.[21] In its first few years, the new Committee played a significant role in the wartime coordination of industry and used some of its small budget to sponsor research at private institutions. Its leaders made the building of a new laboratory their highest priority, despite considerable opposition.[22] The laboratory at Langley Field, in Virginia, established the NACA as a Federal research agency despite its title as an advisory committee. After the war ended, debates over the role of the Federal government in supporting and regulating aviation created considerable uncertainty about the future of the NACA. In the end, other aviation related functions—regulation and the sponsorship of infrastructure—were assigned to the Department of Commerce, leaving the Committee with research as its central role.[23]

At the Langley Memorial Aeronautical Laboratory, dedicated in June 1920, NACA scientists and engineers set out to establish the place of the Federal government in peacetime aviation research. The laboratory provided fairly up-to-date facilities: a wind tunnel, an engine-dynamometer laboratory, and a general research laboratory building. A series of conflicts between personnel at the laboratory and the NACA Headquarters in Washington, DC, tended to dominate the concerns of the leadership, but technical personnel had the equipment they needed to do worthwhile research.[24] The laboratory developed a focus on aeronautical principles in order to take advantage of its wind tunnel facilities and to avoid competition with the military services (which wanted to maintain control of testing and setting specifications for new aircraft designs for military missions), the National Bureau of Standards, and industry (which had facilities for engine research).[25] The NACA found a niche not only in its choice of research program but also in how it approached research problems: "The strength of the NACA seems to be that it had the luxury of pursuing incrementally over a long period of time

---

19. For a thorough survey, see A. Hunter Dupree, *Science in the Federal Government: A History of Policies and Activities to 1940* (Cambridge, MA: Belknap Press of Harvard University Press, 1957).
20. For a survey of the impact of World War I on science, see Daniel J. Kevles, *The Physicists: The History of a Scientific Community in Modern America* (New York, NY: Alfred A. Knopf, 1978), pp. 102–54.
21. Roland, *Model Research*, 1:24–25.
22. During the war years the NACA spent more than half its total budget on building its laboratory rather than on immediate war–related projects. Roland, *Model Research*, 1:30–31, 46.
23. *Ibid.*, ch. 3.
24. *Ibid.*, 1: 80–87.
25. *Ibid.*, 1: 87–89.

answers to problems that were of great interest to the commercial and military worlds."[26] In other words, the NACA could pursue engineering science: systematic investigation of the parameters needed for engineering design. The leaders of the NACA initially thought that the Committee had to establish its reputation by scientific (not engineering) achievement, and hired Max Munk from Germany because of his theoretical reputation.[27] The necessity of practical results to justify Federal funding, and the dominant role of engineers on the NACA main committee gradually reversed that attitude, establishing the relationship between theoretical and practical research as a central tension within the laboratory and for the agency as a whole.[28] Chapters 2, 3, and 4 of this volume show the central role of these issues in the NACA in the years before and even continuing during and after World War II.

As aviation technology became more complex in the late interwar period, the NACA found itself sponsoring not only research on components and design parameters, but large-scale research and development projects. World War II brought a return to more practical concerns (see chapter 2, for example), but with the greater emphasis on government-funded technology characteristic of the war years, it also provided the NACA with broader experience in large development programs and some push to take bigger risks.[29] Perhaps most notably, members of the aviation community saw supersonic flight as the next step, but making that step required both theoretical research (chapter 5), wind tunnel testing (chapter 4), and actual building of experimental aircraft (chapters 3 and 6). Those experimental aircraft were no longer prototypes of new military aircraft, but were designed solely for research purposes. The NACA therefore found itself in the business of contracting with industry for the design and manufacture of radically new vehicles. The X-15 project in particular (chapter 6) differed little in scale and scope from space projects of a few years later. The increasing sophistication of the technological challenges chosen by the NACA was leading the agency toward a project organization typical of big technology even before funding became politicized.

NACA leaders felt some uncertainty about this transition from a role that centered on basic research and problem solving to one centered on taking responsibility for large projects, and they did not push to take a major role in space research in the period before Sputnik.[30] The agency's budget had not grown with its role; it depended on partnerships with the Department of Defense for the funding of large projects, such as the X-15. While the NACA did not initially move to seize the new opportunities opened by the launch of Sputnik in October 1957, those opportunities proved significant and a unique confluence of circumstances soon thrust the NASA into the center of the Sputnik response.[31] If the American people demanded that the United States meet aggressively the challenges of the Soviet Union, President Dwight D. Eisenhower at least hoped to keep that effort out of the hands of the Department of Defense, whose mission he wanted to keep aimed at national security.[32] He assigned the problem of what to do about space to his science advisor, and Killian immediately turned to the NACA as a possible alternative to Department of Defense

---

26. Roger D. Launius, private communication to author, May 29, 1996.
27. Munk had earn two Ph.D. degrees from the University of Göttingen, one in engineering and another in physics.
28. Roland, *Model Research*, 1:89–98.
29. In particular, the NACA was criticized because the United States lost the wartime race to develop a jet aircraft. See Edward W. Constant II, *The Origins of the Turbojet Revolution* (Baltimore, MD: Johns Hopkins University Press, 1980).
30. Roland, *Model Research*, ch. 12, particularly 1:288.
31. *Ibid.*, 1: 290. Roland reports that the subject of Sputnik did not arise at the NACA annual meeting held less than two weeks later.
32. See Walter A. McDougall, . . . *the Heavens and the Earth: A Political History of the Space Age* (New York, NY: Basic Books, 1985), ch. 6, for a discussion of Eisenhower's motivations.

control of the space program. Meanwhile, by December 1957, NACA leaders began trying to define and lobby for a new role for the NACA in space research.[33]

In February 1958, the President's Science Advisory Committee recommended that the NACA be expanded into a new civilian space agency. NACA leaders found themselves and their vision of how a research agency should operate given second place in the new organization. This marginalization started when Eisenhower assigned an executive branch team to write the necessary legislation. The Bureau of the Budget had long wanted to reorganize the NACA's committee structure, and insisted that the new agency be organized hierarchically with an administrator appointed by the President.[34] NACA leaders had assumed that the new agency would continue a traditional NACA pattern by seeking research assignments and funding for cooperative projects from the military services, but Congress wanted space projects to be defined by NASA, not the Department of Defense.[35] Eisenhower and his advisors had similar interests and insisted on a division of space activities between NASA and the Department of Defense instead of cooperative projects on the model of the X-15. The President's Science Advisor, James R. Killian, Jr., finally stepped into a deadlocked discussion in which the NACA and the Department of Defense's Advanced Research Projects Agency were trying to divide the space program. Killian insisted that all space activities without a clear military mission be assigned solely to the new National Aeronautics and Space Administration.[36] This decision committed the new agency to a focus on large projects rather than research into basic principles, suddenly completing a transition that had been in its early stages with projects like the X-15.

Not all of the new agency took on the new style, but most of the attention of its leadership and the public went to the space race in the 1960s. Hesitantly under the Eisenhower administration, and then with a surge of confidence after Kennedy's decision to go to the Moon, NASA leaders shaped the agency towards the pursuit of large research and development projects whose justification lay as much in national prestige and a belief that space was the new frontier as in specific scientific and practical objectives.[37] The Mercury project (chapter 7) represented a mix of old and new constraints and opportunities, but the decision to go to the Moon gave NASA a few years of high priority, generous funding, and public support (chapter 8). The flush years of the early 1960s depended on congruence between the space program and perceptions of national needs; they did not represent support for a space program for its own sake. That congruence made possible the success of Apollo under the inspired leadership of NASA Administrator James E. Webb, a fine manager and a master of the delicate maneuvering necessary to exert leadership from an administrative position.[38] But it left NASA in an unstable position, identified with a relatively short-lived national agenda item rather than with a permanent mission.[39]

---

33. Roland, *Model Research*, 1:291–93. Roland reports that within the NACA "opinion was divided, roughly along generational lines, between the young men who wanted the NACA to campaign for a broad new role in space and the old hands who preferred a more cautious expansion of the NACA's current activities" (p. 292).

34. *Ibid.*, 1:294–95.

35. McDougall, *Heavens and the Earth*, argues that this sentiment was strongest in the House of Representatives, and that Senator Lyndon Johnson supported the Pentagon's claim while publicly arguing for the peaceful uses of space (p. 173).

36. Roland, *Model Research*, 1:296–99. See also James R. Killian, Jr., *Sputnik, Scientists, and Eisenhower: A Memoir of the First Special Assistant to the President for Science and Technology* (Cambridge, MA: MIT Press, 1977).

37. See for example, W. Henry Lambright, *Powering Apollo: James E. Webb of NASA* (Baltimore, MD: Johns Hopkins University Press, 1995) and McDougall, *Heavens and the Earth*.

38. Lambright, *Powering Apollo*, pp. 8–9. Lambright's introduction is a wonderful explanation of the fundamental issues of leadership that arise in executive agencies, which in theory are supposed only to carry out policies set by the President.

39. Lambright argues that the congressional consensus in support of Apollo lasted "barely two years," *Ibid.*, p. 9.

The Apollo period gave the agency a sense of momentum, but by the late 1960s public and political support had shifted and the agency found it could not get the funding it needed to sustain that momentum.[40] Public support declined once it was clear that the U.S. would win the race to the Moon and, in addition, the Vietnam War led both to a budget crunch and to the development of a new left-wing critique of science and technology.[41] In this increasingly hostile environment, NASA leaders struggled to maintain the agency's tradition of large projects to put people in space and to adapt to the new realities of maintaining a program without a national consensus about the importance of the space race. In the 1970s, the space agency and its supporters tried a variety of approaches to rebuild the public support that had made so much possible in the Apollo program. Skylab (chapter 9) represented an effort to prove both the value of human beings in space and to hang an expansive space program on the hook of science. Landsat (chapter 10) sought to bring the benefits of the space program back to Earth, an effort that did not get adequate support either in NASA or in the rest of government, but which looked enough like the wave of the future to get a Collier Trophy in 1974. The Space Shuttle (chapter 12) became identified with a new vision of routine, relatively economical access to space, a promise which the vehicle could never quite meet.[42] The goal of the agency through most of this period was to find a way to continue space exploration in an era of diminishing funding.

The old models became increasingly problematic in the 1980s. Starting with President Jimmy Carter's efforts to cut back big government, NASA leaders found themselves under pressure to commercialize or privatize more operations. At first these pressures had little effect; an emphasis on the routine operation of the Space Shuttle as a "space truck" perhaps represented a new way of thinking for the agency, but the shuttle accident made it clear that the vehicle could not fully fill that role. Both the *Challenger* accident and the problems of the Hubble Space Telescope led to significant criticisms of NASA management, and to changes in management structure to address the pressing problems that had been identified (chapters 15 and 16). These immediate changes fed into a push for broader changes; starting in the early 1990s the leaders of NASA began to explore alternatives to the big science model. NASA Administrator Daniel S. Goldin's call for a "faster, better, quicker" way of doing business involved not only criticism of the old large-project model but also an attempt to develop an alternative.

Big Science, Big Technology

The changes that took place in the NACA and the NASA form part of a larger pattern that historians call the rise of big science. The case studies covered in this book give a sample of projects over the key period for the development of big science. They do not represent classic cases—the classic case for big science is usually high-energy physics—but they widen our understanding of how government support and increasing project size affected the research and development community well beyond the borders of physics. These cases show both the strengths and the limitations of the "big science" approach; in fact NASA may be one of the first agencies where people have begun to be aware of the limits of bigger and bigger projects and to explore alternatives.

---

40. The momentum of technological development is necessarily more a matter of institutions than of any inherent line of development of the technology itself. See Thomas P. Hughes, *Networks of Power: Electrification in Western Society, 1880–1930* (Baltimore, MD: Johns Hopkins University Press, 1983), ch. VI.

41. Bruce L.R. Smith argues that a broad consensus in support of federal funding of research and development disintegrated in the second half of the 1960s under criticism from both the left and the right. See Smith, *American Science Policy Since World War II*, ch. 4, particularly pp. 75–76.

42. The classic critique of the space shuttle for not living up to the exaggerated promises that had been used to gain approval for the project is Alex Roland, "The Shuttle, Triumph or Turkey?" *Discover* 6 (November 1985): 29–49.

Ever since Derek J. De Solla Price published *Little Science, Big Science* in 1963, historians have used various concepts of big science as one basis for trying to understand how the practice and character of science have changed in the twentieth century.[43] Our understanding of big science has developed significantly in more than thirty years, and, in addition, we have begun to explore qualitative as well as quantitative effects of scale on technology as well as on science. Some historians of technology object to lumping big technology with big science, arguing that big technology has its own independent history, with close ties to big business. But in the case of government support for research, the confusion between science and technology starts not in the minds of historians writing about the projects but in the minds of the policymakers and scientists who shaped and advised these projects. NASA leaders regularly referred to the agency's success in the conduct of "big science" even when the projects involved aimed at technological rather than scientific ends, and in many NACA and NASA projects, technological and scientific ends were irrevocably intermixed. The NACA and NASA research projects stories told in this book show some of the complexities of this relationship between science and technology.

Looking at science first, the simplest argument makes World War II a turning point in the rise of big science. The development of large telescopes and a few other large scientific instruments before World War II trained some leaders of the scientific community in administration of large scientific projects. They, in turn, put their experience to use in a series of very successful weapons-development projects during the war. By the end of the war, the military services had come to believe that they needed to continue to support basic scientific research, and significant progress had been made towards a consensus that the Federal government should support large research projects for civilian purposes. Scientists who had been involved in wartime projects hoped for continued government funding, and while they lobbied for civilian funding agencies such as the AEC and NSF, they also worked out a compromise of interests with the military services to get funding from the Department of Defense on terms that most scientists found agreeable.[44] Once new funding mechanisms had been worked out and the start of the Cold War had restored a sense of urgency, government funding for scientific research moved into another growth phase. This gave a significant number of scientists (at least in certain fields) an opportunity to work on a new scale, managing large budgets and tackling scientific problems with expensive instruments and teams of investigators who might all be listed as co-authors on a single scientific paper. These changes affected not just the conduct of scientific research on certain questions, but also what questions scientists asked; some fields of science came to focus on questions that could only be answered with big instruments. These changes in science transformed universities; they became dependent on Federal grants and contracts as the major sources of research funding for basic science.[45]

More detailed studies of post-World War II science and technology have revealed a more complex picture. Even in physics, big science represented a choice of styles and organizational approaches, not an inevitable response to particular discoveries in high-energy physics.[46] Other fields of science felt the effects of big science less, and small science

---

43. For the early history of the term "big science," which actually dates back to the late 1950s, see James H. Capshew and Karen A. Rader, "Big Science, Price to the Present," *Osiris* (second series) 7 (1992): 4–18.

44. See, for example, Daniel Greenberg, *The Politics of Pure Science* (New York, NY: New American Library, 1967) and Kevles, *The Physicists*, ch. XXII.

45. For the university side of the story good places to start are Stuart W. Leslie, *The Cold War and American Science: The Military–Industrial–Academic Complex at MIT and Stanford* (New York, NY: Columbia University Press, 1993) and Ronald L. Geiger, "Science, Universities, and National Defense, 1945–1970," *Osiris* (second series) 7 (1992): 26–48.

46. Peter Galison and Bruce Hevly, eds., *Big Science, The Growth of Large Scale Research* (Stanford, CA: Stanford University Press, 1992), pp. 3–8.

attitudes survived even in some areas where big instruments were used.[47] Big projects not dependent on a single instrument took on different characteristics than those organized around one piece of hardware, and the degree to which the research was focused on science or technology and was goal-oriented or curiosity-driven made a tremendous difference in the character of big projects. NASA could not be compared directly to a federal physics laboratory; the scale may have been similar but the mix of goals was different.[48] But most kinds of big science and technology shared certain common themes involving the relationship between science and technology and the problems of public relations, administration, and funding.[49]

In the case of the NACA and NASA we can identify certain characteristics of big science and big technology that form clear, though by no means uniform, trends. First, NACA and then NASA became increasingly caught in a web of bureaucratic and political obligations. The kinds of popular projects that might become Collier Trophy winners had to provide political or bureaucratic capital to the agency or its supporters, not just research results. Without that note, they stood little chance of being recognized for "outstanding achievement." Second, larger projects required more complex formal organization to keep control of the details. Individual leaders and innovators became less important, and the planning process became more important. Third, research and development projects became more complicated in fundamental ways over this period. More and more different kinds of expertise went into a single project, and the developers of technology were often no longer in close communication with the users. Fourth, the experience of researchers and the approaches they took to their research changed as projects grew larger and more bureaucratic. Fifth, attitudes towards funding research changed, though not just in one direction. Before World War II the emphasis was on practical results, while after the war basic research became more acceptable. A shift away from willingness to support basic research for its own sake occurred around the time of Apollo, with a new emphasis on cost-benefit calculations but also more willingness to fund projects on the basis of popular support.

Any government agency must cultivate bureaucratic and political support in order to survive, but as projects got larger and more expensive (or budgets got tighter) that process shaped more and more of what the NACA and NASA did. The NACA had served its constituencies carefully (mostly by providing practical results) to maintain political support, and the very creation of NASA served political ends at least as much as science and technology. NASA did very well in the 1960s because a growing emphasis on the space race expanded the agency's political and popular support, but that support put the agency into the Washington power game to a greater extent than the NACA had usually experienced. This trend accelerated with the end of Apollo, because the winning of the race to the Moon brought not a reduction in political pressures, but a more complex web of constituencies as NASA leaders sought to cobble together enough support to continue a large-scale space program. In the 1970s and 1980s, NASA had to play bureaucratic politics and look for new ways to serve political agendas in order to maintain a program on anything like the scale established for Apollo.

---

47. For some examples of other fields of science see Arnold Thackray, ed., *Science After '40, Osiris* (second series) 7 (1992). For a discussion of how big science was not inevitable even in high-energy physics see John Krige, "The Installation of High-Energy Accelerators in Britain After the War: Big Equipment but not "Big Science," in Michelangelo DeMaria, Maria Grillia, and Fabio Sebastiani, eds., *Proceedings of the International Conference on the Restructuring of Physical Science in Europe and the United States, 1945–1960* (Singapore: World Scientific, 1988).

48. For a comparison between high-energy physics and space programs (not only in the U.S. but also in other countries) see John Krige, ed., *Choosing Big Technologies* (Geneva, Switzerland: Harwood Academic Publishers, 1993).

49. Capshew and Rader "Big Science" provide one useful thematic introduction; I take my themes partly from their discussion of Alvin Weinberg's warning that the three diseases of big science are "journalitis, moneyitis, and administratitis" (p. 5).

NASA's human space flight programs also represented significant challenges in the development of large-scale management.[50] Apollo was an overwhelmingly large and complex program, but the Space Shuttle introduced further challenges by requiring both technological innovation and routine, long-term management control. Not all NASA projects in this period were large (see chapter 14 on the fuel-efficient turboprop), but the space program became increasingly identified with large, spectacular projects that got public attention. The agency and its contractors became accustomed to a technological style that they sometimes called big science, though it had more to do with technology than science. In particular, they preferred programs to build one or two large satellites, or a large platform carrying many sensors, over projects that would launch many small satellites, each carrying one or two sensors (probably somewhat less capable than those a large platform could support). Even when astronauts were not involved, these relatively large and complex programs required many layers of management, paperwork, and checks and counterchecks, to control a system that was too complex for a small group of people to keep track of and which needed extremely careful risk management because of the public embarrassment of large failures.[51] Individual leadership was harder to exert on projects of this scale, and the planning process tended to become an increasingly political negotiation.

Large, involved projects dependent on outside political and bureaucratic support also became fundamentally more complex because they had to serve many masters. Researchers in space science complained particularly vocally about this change, because they assumed that space science projects should be conducted in whatever manner would best serve the interests of scientists. A project like the Hubble Space Telescope servicing mission served the scientists using the instrument but, in addition, NASA achieved important political ends through its success (chapter 16). The inevitable conflicts of interest sometimes irked the science community, especially as Congress set the agenda for space science in such missions as the Grand Tour (chapter 11). Projects with practical goals raised even more fundamental problems, particularly for an agency as focused on research and development for its own sake as NASA. In the cases of Landsat (chapter 10), and the fuel-efficient turboprop (chapter 14), NASA successfully developed technology to do the job, only to find that the intended users were not as interested as had been predicted. In the first case, the problem lay in part in NASA's technology transfer efforts, but in the second case changes in economic parameters and issues relating to public opinion kept the new technology from being put to effective use.

Within these projects, the experiences of scientists and engineers had also undergone a fundamental change. The individual inventor had almost disappeared from view, though individuals might still invent small parts of large, complex systems.[52] Teamwork and the ability to provide intellectual leadership while not having control over the entire project became critical skills. Government funding made possible projects that would probably never have received funding in a corporate research and development laboratory because the total cost was too high or the payoff too uncertain or too far in the future. Pressure for quick results, while very real, could be less intense than in other settings.

50. Lambright, *Powering Apollo*, is a good place to start for this issue.
51. In the 1990s a new generation of advocates for small satellites developed this critique of what they perceived as a NASA culture of bureaucratic control and large–scale programs. The impact of this challenge to the old way of doing things is not yet clear, but it has already had some impact on the congressional committees that oversee NASA's budget and on the leadership of the agency itself. For a good example of the critique, see John R. London III, *LEO [Low–Earth Orbit] On the Cheap: Methods for Achieving Drastic Reductions in Space Launch Costs*, Research Report No. AU–ARI–93–8 (Maxwell Air Force Base, AL: Air University Press, October 1994).
52. The decline of the individual inventor and the rise of complex systems in the corporate world has been laid out by Thomas P. Hughes in *Networks of Power and in American Genesis: A Century of Invention and Technological Enthusiasm* (New York, NY: Simon and Schuster, 1989).

Finally, big science and technology shaped and were shaped by changing attitudes towards the relative roles of basic and applied research. In the period before World War II, government funding required practical justifications, but the NACA found a niche for less goal-driven research by pioneering work in engineering science, exploring some of the fundamental parameters of flight. World War II made the Federal government more willing to fund basic research in some fields, particularly in physics where exploratory research had proved its military value most clearly. NASA certainly funded more basic science than the NACA because its mission included space science, but that scientific research formed a relatively small part of a large agency. When NASA took in the national goal of putting people in space it committed itself to a vision that was not centered on basic research.[53] Outside factors reinforced this tendency. Starting in the late 1960s, critics of the space program, some of them in the Executive Office, began to demand cost-benefit analyses for at least some space missions (see chapter 10). This change resulted from the space program's lower political priority and from a larger trend towards demanding tighter justification for government sponsored research. In particular, the Mansfield amendment in 1970 prohibited the Department of Defense from funding basic research with no military purposes.[54] However, by the 1990s the trend had split: Congress seemed to favor projects that were unabashedly basic science (at least if they were not too expensive) or those that would clearly save the government money (though then the question arose of why private industry couldn't do the job).[55] The tension between basic and applied research and between research and development and routine operations was complicated by a constantly shifting environment.

We can also see in the projects covered in the last few chapters of this volume the beginning of a challenge to the big science model. The return to flight of the Space Shuttle (chapter 15) and the Hubble Servicing Mission (chapter 16) represent successful recoveries from failures caused by management problems, not just inevitable bad luck. The failures showed some of the limits of big science, and the recovery efforts involved at least in part attempts to change the big science style of operation (for example, the role of individual leadership in the shuttle case, chapter 15). Since the early 1990s NASA has met significant criticism not just for bureaucracy, but for assuming that large projects are the best way of achieving any end. Studies of such concerns as lowering the cost of launch vehicles have concluded that "to achieve this goal, it will be necessary to bring about major cultural changes within the aerospace community."[56] That particular study pointed out that cheaper systems are not necessarily smaller, but changes such as mass production and a greater tolerance for failure represent major changes to the big science, big technology approach. A new NASA Administrator, Daniel S. Goldin, appointed in April 1992, established as one of his initiatives "A shift away from the pursuit of big science and engineering programs toward 'faster, better, and cheaper' ones."[57] It is too soon to know whether this represents the beginning of the end of the dominance of big science, but its values are certainly being questioned in a new way within NASA.

---

53. Many scientists have criticized NASA's emphasis on putting people in space as being a waste of money from a scientific point of view. For examples of the perspective of scientists, see Homer E. Newell, *Beyond the Atmosphere: Early Years of Space Science* (Washington, DC: NASA SP–4211, 1980) and Steven G. Brush, "Nickel for Your Thoughts: Urey and the Origin of the Moon," *Science 21* (1982): 891–98, as well as ch. 11 and 16 and their references.

54. For an analysis of the Mansfield amendment as part of a larger trend, see Smith, *American Science Policy Since World War II*, pp. 81–82.

55. Committee on Earth Studies, Space Studies Board, National Research Council, *Earth Observations from Space: History, Promise, and Reality* (Washington, DC: National Academy of Sciences, 1995), pp. 102–103.

56. London, *LEO on the Cheap*, p. 149.

57. NASA Federal Laboratory Review, "Executive Summary," located at *http://www.hq.nasa.gov/office/fed–lab/exec.html*, March 20, 1996.

Chapter 1

Engineering Science and the Development of the NACA Low–Drag Engine Cowling

by James R. Hansen

The agency that preceded NASA, the National Advisory Committee for Aeronautics (NACA), won its first of five Collier Trophies in 1929, and did so basically for advancing a counterintuitive idea. The idea, which flew in the face of a conventional wisdom about proper aircraft design, ventured the following: *covering up*—not leaving open to the air— the cylinders of an air-cooled radial engine could not only dramatically reduce aerodynamic drag but actually *improve* engine cooling. The immediate product of this startling engineering insight was the NACA's development of a low-drag engine "cowling," the winner of the 1929 Collier Trophy.

Put simply, the NACA cowling was a metal shroud for a radial air-cooled engine. However, the purpose of the shroud involved much more than hiding an ugly engine or keeping the rain out; rather, its main function was to *cool* a hot engine. This is what ran so contrary to what throughout the 1920s had been the practical solution to the problem of air-cooling an engine, that was, exposing the red–hot engine cylinders to an outside rush of cooling air. Besides improving the cooling of the engine, the NACA cowling—designed as it was to be a *streamlined* shroud—also worked to reduce drag. This allowed an airplane to fly faster and farther on less fuel, a significant technological accomplishment in the late 1920s, and one that deserved to win the National Aeronautic Association's (NAA's) award for the year's greatest achievement in American aviation.[1]

Deserving the Collier Trophy is not to say, however, that the NACA's low-drag engine cowling was everything that it was cracked up to be. In the years following the Collier Trophy, American aviation journalists generally exaggerated the significance of the cowling, and NACA publicists claimed more credit for the aircraft industry's adoption of the cowling than the government research organization deserved. Almost everyone outside the aircraft industry itself failed to appreciate the true character of the NACA's cowling work and credited *science* rather than *engineering* as its source, an all-too-common mistake made in modern American society. Partly as a result of this misapprehension, spokesmen for aviation progress—most of them rabid technological enthusiasts—did not know enough to explain that the cowling was *not* really an *invention* in the classic sense, for different crude cowlings were already available and in limited use around the world. Nor did they know enough to make clear that every cowling had to be custom fitted: that the cowling was not a magical tin shape that could be applied generically to just any airplane (at least not with great success), because the effectiveness of the cowl depended significantly upon the *shape* of the airplane behind it. If the NACA engineers at Langley Memorial Aeronautical Laboratory (LMAL), who were responsible for developing the original prize-winning cowling, had tested it with certain other aircraft of the era, such as a Bellanca or

---

1. For an excellent technical summary of how cowlings function, past and present, see Peter Garrison, "Cowlings," *Flying* 113 (February 1986): 58–61.

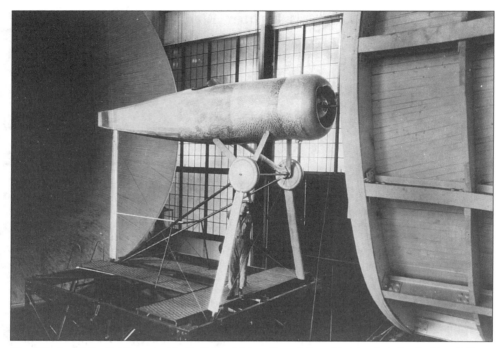

One of the four Collier Trophies received by the National Aeronautics and Space Administration's Langley Research Center, Hampton, Va., was in 1929 for the development of the cowling for radial air cooled engines. By the end of September 1928, tests of cowling No. 10 in the Propeller Research Tunnel shown here demonstrated a dramatic reduction in drag. (NASA Photo 87-H-1250)

Stinson, rather than with the Curtiss Hawk AT-5A and Lockheed Air Express that flew with it so successfully, the NACA cowling would not have performed nearly so well.[2]

But these things about the NACA cowling were never well understood outside of the aeronautical engineering community, and they were certainly not communicated very successfully to the broader aviation public at the time. In the era from Lindbergh to the New Deal, the United States' aviation publicists—devout believers in a "winged gospel" and in an airplane symbolic of the boundless promise of the American future—did not understand the technology well enough to see any advantage in making practical qualifications about the engineering of cowlings.[3] Perhaps some of them realized that the people who built airplanes already had the good sense to understand the subtleties of the NACA research program: that the cowling was not so much an invention or new standard piece of equipment as it was a *process* or method, with every airplane and engine con-

---

  2. For a concise history of the NACA cowling program at NACA Langley, see Ch. 5: "The Cowling Program: Experimental Impasse and Beyond," in James R. Hansen, *Engineer in Charge: A History of the Langley Aeronautical Laboratory, 1917–1958* (Washington, DC: NASA SP-4305, 1987), pp. 123–39. See also Alex Roland, *Model Research: The National Advisory Committee for Aeronautics, 1915–1958*, 2 Vols. (Washington, DC: NASA SP-4103, 1985), 1:111–13.

  3. See Joseph J. Corn, *The Winged Gospel: America's Romance with Aviation, 1900–1950* (New York, NY: Oxford University Press, 1983).

figuration requiring a special, customized cowling for optimum results.⁴ Perhaps some considered the distinctions too technical for the wider aviation public to understand.

More likely, they were as misled as the rest of American society by a heroic theory of invention in which a few great geniuses like Thomas Edison and the Wright brothers, not industrial teamwork—and certainly not government bureaucracy—deserved most of the credit for technological progress. If it was *not* heroic invention, then the NACA cowling was not really original; it constituted "mere development" and did not deserve to win a prestigious national award like the Collier Trophy.⁵ Better that the award be presented to an individual genius, just as the Collier Trophy itself had been won ten of the last fourteen times since the inaugural award to Glenn H. Curtiss for development of the "hydroaeroplane," or flying boat, in 1911.⁶ But the fact that the National Aeronautic Association's judges had awarded the Collier to the NACA in 1929 was proof enough of heroic invention. Thus, with heroic inventors in mind, those explaining the significance of the

The NACA received the Collier Trophy in 1929 for developing a cowling to fit over the engine which increased the speed of the test aircraft from 118 to 137 miles per hour, an increase of sixteen percent. The cowling was later adapted to other aircraft. This photo shows NACA mechanics installing, in 1928, a cowling for testing. (NASA Photo 90-H-189)

---

4. I wish to thank my colleague in the history department at Auburn University, Stephen L. McFarland, for contributing valuable insights into my understanding of the NACA cowling as a process rather than an invention.

5. For a critique of the heroic theory of invention, see George Basalla, *The Evolution of Technology* (New York, NY: Cambridge University Press, 1988), pp. 21, 26, 59–60. See also Ch. 2: "Emerging Technology and the Mystery of Creativity," in John M. Staudenmaier, *Technology's Storytellers: Reweaving the Human Fabric* (Cambridge, MA: MIT Press, 1985), especially pp. 40–45.

6. The first winner of the Collier Trophy, 1911, was Glenn H. Curtiss, for the "hydroaeroplane." Other Collier "individual" winners before 1929 included: Orville Wright, for developing the automatic stabilizer (1913); Elmer A. Sperry, for gyroscopic control (1914) and the drift indicator (1916); Grover Loening, for the aerial yacht (1921); Sylvanus Albert Reed, for developing the metal propeller (1925); and Charles W. Lawrance, for his radial air–cooled engine (1928). By the late 1920s, the Collier Trophy was recognized as the most prized of all aeronautical honors to be accorded in the United States; the winner received the award from the president of the United States. On the history of the Collier Trophy, see Frederick J. Neely, "The Robert J. Collier Trophy: Its Origin and Purpose," *Pegasus* (December 1950): 1–16, and Bill Robie, *For the Greatest Achievement: A History of the Aero Club of America and the National Aeronautic Association* (Washington, DC: Smithsonian Institution Press, 1993).

The Curtiss Hawk used in NACA Tests, in November 1928, before (above) and after (below) installation of the cowling. (NACA Photo 3018)

(NACA Photo 3019)

NACA cowling did so in close accordance with popular expectations, however naive, about where valuable new technology came from and how it moved from conception to practical reality.[7]

As the following essay intends to show, the technological process represented in the NACA's cowling investigation was of a particular type that has often proved fundamental to progress not only in aviation but in all engineering fields. It was not the path of inspired genius the public had come to want, but neither was it mere development. Rather, the NACA cowling was something more fundamental and harder to identify, let alone comprehend. It was the fruitful product at a government laboratory of what historians of technology have come to call *engineering science:* a solid combination of physical understanding, intuition (and counterintuition), systematic experimentation, and applied mathematics.[8] As such, the NACA cowling evolved during the 1930s into the mature type of basic technological achievement that has been extremely hard for the non-technical American public to understand and appreciate for what it is, but which must be explained, understood, and appreciated in a democratic society if basic applied research is to be supported and adequately funded.

Who Asked the Question?

As most successful research programs do, the NACA cowling investigation started with a question: "Is it possible to extend a cowling outward over the exposed cylinders of a radial air-cooled engine without interfering too much with the cooling?" It is significant for NACA history that the question, which brought the breakthrough counterintuitive answer, was asked at the NACA's first annual manufacturers' conference, which was held at Langley Memorial Aeronautical Laboratory on May 24, 1926. This event became the NACA's "rite of spring." A combined technical meeting and public relations extravaganza, the annual conference gave the NACA research staff an opportunity to ascertain the problems deemed most vital by the aircraft industry so that it could incorporate them as far as possible into its research programs. At the same time, the conference gave the staff a chance to publicize its recent accomplishments before individuals who rarely had the time to read the NACA's published technical reports but who needed, and wanted, to know what the NACA was doing. The conference also gave the research staff at Langley a chance to bang a big drum before congressmen and other public officials who "had neither the time nor the qualifications to read the technical reports" but who played critical roles in the appropriations of government money. The event started in 1926 as a modest and relaxed one-day affair, but it soon grew into an elaborately staged pageant that took weeks of preparation by the NACA staffs both at Langley and in Washington. By 1936, the spectacle lasted two days, the first day for executives of the aircraft industries and government officials, the second "for personnel of the government agencies using aircraft, representatives of engineering societies, and members of professional schools." In 1926, only forty-six attended the conference; ten years later, more than 300 people were attending each *session*, including aviation writers who reported fully on the laboratory's presentations in newspapers and journals.[9]

---

7. Other individual winners were Glenn Curtiss, again, in 1912; Orville Wright in 1913; Elmer A. Sperry, in 1914; W. Sterling Burgess in 1915; Elmer A. Sperry, again, in 1916; Grover Loening in 1921; Sylvanus Albert Reed in 1925; Major E. L. Hoffman in 1926; and Charles L. Lawrance in 1927.

8. For a rich historical treatment of the role of engineering science in American aeronautics, see Walter G. Vincenti, *What Engineers Know and How They Know It: Analytical Studies from Aeronautical History* (Baltimore, MD: Johns Hopkins University Press, 1990).

9. For a discussion of the NACA's annual aircraft engineering conferences, see Hansen, *Engineer in Charge*, pp. 148–58, and Roland, *Model Research*, 1:111–13.

The identity of the person who asked the pivotal question about engine cowlings is uncertain, but the subject is worth some speculation because of what it says about the aviation community and its process of discovery in the late 1920s. No one attending the conference ever went on record about who first asked the question about cowlings, and those who lived long enough to be interviewed by historians (and remember the question being asked) do not remember who it was that did the asking. One likely candidate is Charles W. Lawrance, who by 1926 was part of the Wright Aeronautical Corporation in Paterson, New Jersey. In the early 1920s, Lawrance had built his own small engine company around a pioneering air-cooled radial engine known as the Whirlwind J-1. The Navy loved the engine, but Lawrance's company nevertheless struggled to remain solvent and could not avoid a buy-out by the huge Wright company. With the resources of the Wright Corporation behind him, Lawrance kept improving his engine and, by 1927, had a nine-cylinder, 220–HP Whirlwind J-5 in mass production. This outstanding radial air-cooled engine powered Lindbergh across the Atlantic in 1927, Sir Charles Kingsford-Smith across the Pacific in 1928, U.S. Army pilots Hegenberger and Maitland from Oakland to Hawaii in 1927, and Commander Richard E. Byrd over the South Pole in 1929. So impressive was the engine's performance, which was highly publicized because of these benchmark flights—especially Lindbergh's—that the NAA awarded Lawrance its Collier Trophy for 1927 in recognition of his marvelous engine.[10] Given the fact a Sperry Messenger airplane equipped with an air-cooled Lawrance engine was demonstrated in a Langley wind tunnel at the NACA conference's morning session in May 1926, one might imagine that Lawrance asked the question about cowlings, but there is no real evidence he did.

Perhaps an even more likely candidate was Captain Holden C. ("Dick") Richardson, an officer in the Navy's Bureau of Aeronautics and one of the original members of the NACA's main committee (from 1915–1917). Richardson, who had completed a master's degree in engineering from the Massachusetts Institute of Technology (class of 1907), was one of the Navy's leading aircraft designers. Having "honed his skills in the fields of hydrodynamics and aerodynamics" at the Philadelphia and Washington navy yards (at the latter working with Captains David W. Taylor and Washington I. Chambers on the wind tunnel in the experimental model basin), flying boats became his expertise.[11] Along with Dr. Jerome C. Hunsaker (a future NACA chairman, 1941–1956) and Captain George C. Westervelt, Richardson was one of the designers of the Navy's famous NC-4 (NC for Navy-Curtiss) flying boats, a 25,000-pound aircraft that successfully flew the Atlantic in 1919. In the mid-1920s, as head of the design section of the Navy's Bureau of Aeronautics' (BuAer's) material division, he was one of the Navy leaders working hardest to bring about the design of metal flying boats, notably the PN class, which were originally equipped with liquid-cooled Packard engines. Various problems with the heavy engines prompted the Navy in 1927 to move to air-cooled engines (two 525-HP Wright R-1750 Cyclone radials) for the PN-10, the first of the Navy's all-metal seaplanes.[12] At the time of the NACA's first manufacturers' conference in May 1926, which Richardson attended, this conversion to the radial was still being pondered. Thus, the subject of this engine and its potential for further improvements—aerodynamic and otherwise—through an advanced cowling was high on the list of Richardson's concerns.

---

    10.   On the early in–flight achievements of the Wright Whirlwind engine, see Terry Gwynn–Jones, "Farther: The Quest for Distance," in the Smithsonian Institution's *Milestones of Aviation* (New York, NY: Hugh Lauter Levin Associates, Inc., 1989), p. 54.

    11.   For information on the naval career of Holden C. Richardson, see William F. Trimble, *Wings for the Navy: A History of the Naval Aircraft Factory, 1917–1956* (Annapolis, MD.: Naval Institute Press, 1990) and William F. Trimble, *Admiral William A. Moffett: Architect of Naval Aviation* (Washington, DC: Smithsonian Institution Press, 1994).

    12.   Trimble, *Wings for the Navy*, pp. 97–99.

Therefore, it would not be at all surprising if the cowling question came from Dick Richardson, an aircraft designer totally absorbed in the unique problems of naval aviation. Without a doubt, one of the more urgent questions facing the designers of naval aircraft in the 1920s was how to reduce the drag of radial engines without degrading their cooling. During the early 1920s, the navy had decided that the lighter air-cooled engine, with its short crankshafts and crankcases and no radiators, offered a more practical solution to most of its aircraft power-plant problems than did the heavier liquid-cooled engine with its water jacket, radiator, and gallons of coolant favored by the army. The jarring confrontations of naval aircraft with arresting gear on aircraft carriers resulted in too many cooling system maintenance problems at sea, including loose joints, leaks, and cracked radiators. However, subsequent experience also made it clear to the Bureau of Aeronautics (established under the direction of Admiral William A. Moffett in 1921) that existing air-cooled designs wasted considerable power. The finned cylinders of the radial engine, projected into the external airstream, caused high drag. Navy engineers attempted to reduce this drag by putting a propeller spinner (a rounded cover) over the hub and covering the crankcase and inner portions of the cylinders with a metal jacket, but this left the outer ends of the cylinders jutting into the airstream.[13]

With this persistent design problem in mind, it would have been very sensible for Captain Richardson to ask at the NACA conference whether the research staff at Langley could determine how much a cowling could be extended outward over the cylinders of the radial engine in order to reduce drag without excessive interference with cooling. The answer promised significant advantages for all sorts of aircraft, especially shipboard fighters, as well as the Navy's PN-10 flying boats.

I Didn't Want People to Expect too Much

The immediate circumstances prompting the cowling question in May 1926 was a demonstration in Langley's new Propeller Research Tunnel, a monster facility whose kinks were still being worked out in May 1926 and whose routine operation was still almost a year away. During the morning session of the conference, as part of a tour of various Langley facilities, the NACA turned on the big tunnel so everyone could witness its operation. Mounted on the test balance in the wind stream was a small Sperry Messenger airplane, with its radial air-cooled Lawrance engine running. The Propeller Research Tunnel, or PRT as it came to be known, was only the NACA's third wind tunnel, the largest one built. The PRT was in fact the largest tunnel built to that time anywhere in the world. Designed to accommodate a full-scale propeller, the throat of the PRT was a spacious twenty feet in diameter. This was four times the size of the largest wind tunnel at Langley, and it meant that the PRT structure required sixty-four times the volume of any tunnel built there before. Furthermore, for full-scale tests of propellers to be practical, the tunnel's airflow had to reach at least 100 MPH, and to achieve that it took 2000 HP—ten times the power it took to drive NACA Wind Tunnel No. 1 (operational June 1920) and eight times what it took to drive the NACA's second wind tunnel, the revolutionary Variable-Density Tunnel (or VDT, operational October 1922). Both the VDT and PRT were conceived by Dr. Max M. Munk, the NACA's brilliant German import. As neither the city of Hampton nor the nearby Newport News generating plants were large enough to supply the necessary electricity to power the PRT, the NACA had obtained two surplus

---

13. See Herschel H. Smith, *Aircraft Piston Engines: From the Manly Balzer to the Continental Tiara* (New York, NY: McGraw–Hill, 1981), pp. 97–113.

1,000-HP diesel submarine engines from its friends in the U.S. Navy. Thus, any demonstration of this huge beast of a machine made a powerful impression.[14]

What made the PRT demonstration even more exciting was the fact that the NACA, by May 1926, had not yet been able to get the tunnel's diesel engines running properly. To get the big submarine engines to turn over, a blast of compressed air had to be used, a minor explosion that startled the uninitiated. For the morning visitors, the Langley engineers ran the tunnel on the compressed air for about a minute, with the little Sperry Messenger airplane up in the test section with its engine running also. The demonstration was not only memorable—very noisy and a little scary—but also, as the NACA found out that afternoon, question provoking. Whether it was Charles Lawrance, Captain Richardson, or someone else who asked the critical question about cowlings early in the afternoon session, we do know from the historical record that several other people immediately spoke up to second the interest. By the end of the afternoon, it was clear to the NACA that airplane designers were rather desperate to know more about the potential of engine cowlings, that they considered it the job of the government laboratory to provide the basic information, and that the PRT might be just the right place to make a systematic experimental study. The inaugural NACA conference thus served its purpose well and set the stage for positive NACA-industry-military services interaction for years to come.

The NACA's Washington office (it was hardly ever called "Headquarters" until after World War II) responded immediately by authorizing Langley to conduct a free-flight investigation of the effects of various forms of cowling on the performance and engine operation of a Wright Apache (borrowed from the navy) and prepare a systematic program of cowling tests in the PRT, a facility that made it possible for the first time anywhere to test full-size propellers and other aircraft components in a wind tunnel.[15]

The organizing thinker and team leader of the NACA's original cowling program at Langley was Fred E. Weick, one of the most remarkable aeronautical engineers in the history of American aeronautics.[16] Born near Chicago in 1899, Weick (pronounced Wyke) developed an avid interest in aviation by the age of twelve, going to air meets at nearby Cicero Field and engaging in model airplane competitions. Upon graduation from the University of Illinois in 1922, he began his professional career as a draftsman with the original U.S. Air Mail Service. After a short stay with the Yackey Aircraft Company (during which time he worked in a converted beer hall in Maywood, Illinois, transforming war-

---

14. For the design details of the Propeller Research Tunnel, see "The Twenty-Foot Propeller Research Tunnel of the National Advisory Committee for Aeronautics," *Technical Report (TR) 300*, 1928, by Fred E. Weick and Donald H. Wood. For their history, see Hansen, *Engineer in Charge*, pp. 87–90 and pp. 144–45, and *The Wind Tunnels of NASA* (Washington, DC: NASA SP–440, 1981), pp. 5–6, by Donald D. Baals and William R. Corliss.

15. NACA research authorization (RA) no. 172, "Effect of Various Forms of Cowling on Performance and Engine Operation of Air-Cooled Pursuit Airplane," approved by the Executive Committee, 30 June 1926; RA 215, "Effect of Cooling and Fuselage Shape on the Resistance and Cooling Characteristics of Air-Cooled Engines," approved June 22, 1927. The NACA research authorization files are in the Langley Historical Archives (LHA), Floyd L. Thompson Technical Library, NASA Langley Research Center, Hampton, VA.

The Navy lent the Apache aircraft to NACA Langley in the summer of 1926, but soon recalled it. Though the recall forced the laboratory to suspend cowling work on the Apache and its Whirlwind engine, RA 172 was kept open until 1932. Langley carried out most of its later cowling tests under RA 215.

16. Fred E. Weick and James R. Hansen, *From the Ground Up: The Autobiography of an Aeronautical Engineer* (Washington, DC: Smithsonian Institution Press, 1988). Over the years Weick made many significant contributions to the advancement of aeronautical technology, including development of the steerable tricycle landing gear, the conventional gear used today—even for the Space Shuttle. His most widely recognized achievement, the Ercoupe, has been the favorite airplane of thousands of private flyers since the first production model of it came out in 1940. And his revolutionary Ag-1 and Piper Pawnee set lifesaving standards of lasting benefit to both the agricultural airplane and general aviation industries. His autobiography tells his entire life story in fascinating detail, from his pioneering work with the U.S. Air Mail Service in the early 1920s, through his Navy and NACA years, to his many years in manufacturing for ERCO and Piper.

Fred E. Weick, head of the Propeller Research Tunnel section, 1925–1929. (NASA photo)

surplus Breguet fourteen biplanes into "Yackey Transports"), he started a job with the U.S. Navy Bureau of Aeronautics in Washington, D.C., where, within a matter of months, the NACA's director for research, George W. Lewis (1882–1948), personally recruited him for important work to be done at Langley, some 120 miles to the southeast. (The NACA's Washington office was located in an adjacent wing of the Navy building, thus facilitating close relations between the NACA and the Navy.) Weick arrived at Langley in November 1925 just in time to take over the design and construction of the new Propeller Research Tunnel, the job Lewis had specifically picked him to do.[17]

In the weeks following the May 1926 conference, Weick and a small team of engineers and technicians laid out a program for the cowling tests that was tailor-made for the capabilities of Langley's big new tunnel. The primary method Weick chose to employ was something just becoming known to engineers as *experimental parameter variation*, which has since been defined as "the procedure of repeatedly determining the performance of some material, process, or device while systematically varying the parameters that define the object or its conditions of operation."[18] Although just being fully articulated in the 1920s, the method itself was ancient. Greek military engineers had varied the parameters of full-scale machines to find the most effective dimensions for their catapults hundreds of years before the time of Christ.[19] During the Industrial Revolution, engineers had used the method to explore the performance of new construction materials and steam engines.[20] The success of the first powered airplane in 1903 followed application of the fundamentals of the method used by the Wright brothers while testing airfoils in their homemade wind tunnel.[21] Over the centuries, many different types of engineers used parameter variation precisely because it permitted solution of a complex problem without a complete understanding of all aspects of the problem. When a complex research problem needed practical solution, and hypotheses were more scattershot than pinpoint because complex understanding was still a distant goal, the technique systematized the pragmatic researcher's only real choice for a course of action: a combination of brainwork, guesswork, and trial and error. By observing the effects of slight changes made one at a time in planned, orderly sequence, an engineer like Fred Weick could add progressively to his knowledge about the actual performance of whatever was being investigated. Seeking effects now and saving causes for later, he could use what he did know, circumvent what he did not know, and discover what would work.

For Weick, the advantages of using such a proven method, though intuitively clear and logical, were a rather recent revelation. While at BuAer in 1924 he learned, from propeller work carried out by William F. Durand and Everett P. Lesley at Stanford University, what he called "the advantages of using a systematic series of independent variables in experimental research."[22] (Even earlier, as a senior engineering student at the University of Illinois, he had based a paper on variable-pitch propellers on data from the Durand-Lesley propeller tests in the Stanford wind tunnel.)[23] So it was a method that had proven immensely practical to him in his own work, which gave him confidence to try it again.

---

17. Weick and Hansen, *From the Ground Up*, pp. 49–59.
18. Walter G. Vincenti, "The Air-Propeller Tests of W. F. Durand and E. P. Lesley: A Case Study in Technological Methodology," *Technology and Culture* 20 (October 1979): 743–44.
19. Barton C. Hacker, "Greek Catapults and Catapult Technology: Science, Technology, and War in the Ancient World," *Technology and Culture* 9 (January 1968): 34–50.
20. For references, see Vincenti, "Air-Propeller Tests," pp. 714–15.
21. See Peter Jakab, *Visions of a Flying Machine: The Wright Brothers and the Process of Invention* (Washington, DC: Smithsonian Institution Press, 1990), pp. 138–52.
22. Weick and Hansen, *From the Ground Up*, p. 60.
23. Vincenti, "Air-Propeller Tests," p. 740.

Recognizing that he should extend the cowling investigation well beyond the range of immediate interest, Weick pinpointed the extremes. Obviously, one extreme was a bare engine with no cowling at all; everyone who knew anything about aerodynamics assumed that it would have maximum cooling, but maximum drag as well. The value of the other extreme—enclosing the engine completely—no one had anticipated because that form seemed to exclude all possibility of air cooling. For smooth flow around the exterior of the cowl, Weick modeled an engine nacelle on the best available airship form, with the idea of bringing in cooling air at the center of the nose. Then the amount of cowling was systematically varied from one extreme to the other until he had produced ten different cowling shapes, ready for testing in the PRT.[24] "After I had completed the outline of a tentative cowling test program," Weick remembered in his autobiography (published in 1988, when Weick was 89), "the NACA sent it to the military air services and to various manufacturers that had shown interest at the May 1926 conference, and it was approved by all of them. Fortunately, getting their okay took some time, because the propeller research tunnel was at this point in no sense ready to operate."[25] The PRT was not ready for actual testing until early 1927, at which time the systematic experiments began.

The first round of tests in the PRT initiated a process of cowling development that lasted at Langley for more than a decade, into the late 1930s. With the process came significant design refinement and a far deeper understanding of all the beneficial things properly cowled engines could do for an airplane in flight. Most importantly, from the viewpoint of expanding engineering knowledge, the process eventually resulted in a far better understanding of *how* cowlings do what they do. In retrospect, the process was divided into four stages: (1) 1926 to 1929, definition of the cowling's parameters, a stage which ended with the NACA's public announcement of a successful low-drag design that won the Collier Trophy; (2) 1929 to 1931, an important series of engine placement and free-flight cowling tests that resulted in a strong identification throughout the NACA with the empirical method; (3) 1931 to 1934, when the laboratory began by outlining a new three-pronged experimental attack on cowling and cooling problems, but ended in an impasse when that attack stalled; and (4) 1934 to 1936 and beyond, when a more analytical approach to cowling research began to emerge out of this stalemate to answer some of the basic questions that the empirical approach of the preceding three stages had left unanswered. Experimental parameter variation led to results in each of the first three stages; practical use was made of observed performance effects. By the fourth and final stage, it was time to search beneath the effects for causes. It was time to go after that distant goal of complex understanding. By the start of World War II, which in some respects saw the final, culminating evolution of the propeller-driven airplane, this ultimate goal had been largely achieved.[26]

In 1927, Weick's team at Langley stood at square one. According to Weick:

> *The goal that we had set for ourselves was a cowled engine that would be cooled as well as one with no cowling whatsoever. This program proceeded easily enough until the complete cowling, covering the entire engine, was first tried. At this point, some of the cylinder temperatures proved to be much too high. After several modifications to the cooling air inlet and exit forms, and the use of internal guide vanes or baffles, we finally obtained satisfactory cooling with a complete cowling.*

24. Weick and Hansen, *From the Ground Up*, p. 60.
25. *Ibid.*, pp. 60–61.
26. See *Quest for Performance: The Evolution of Modern Aircraft* (Washington, DC: NASA SP-468, 1985), by Laurence K. Loftin, Jr. Loftin was an aeronautical engineer who worked at Langley from 1944 until his retirement from NASA in 1971. Much of my analysis of the four stages of Langley's cowling work that follows in this essay is based on Weick's autobiographical account. See also "The N.A.C.A. Cowling," *Aviation* 25 (November 17, 1928): 1556–57 and 1586–90, by Fred E. Weick; and "Notes on the Design of the N.A.C.A. Cowling," *Aviation* 27 (September 21, 1929): 636–38, by William H. McAvoy.

Donald H. Wood, a 1920 graduate in mechanical engineering from Rensselaer Polytechnic Institute who had been working at Langley since 1924, was in charge of the actual operation of the testing, and the first of these modifications was made while Weick was away on a vacation. When Weick returned to work, it was obvious to him that "the boys were on to something, and from that time on we all worked very hard on the program."[27]

The airplane that the engineers worked with in the PRT was a Wright Apache, a small airplane, which was equipped with a J-5 Whirlwind air-cooled engine. They measured the cooling effectiveness of each of the ten cowlings, investigating their different effects on propulsive efficiency. Each experimental shape underwent numerous, systematically planned variations. With the help of Elliott G. Reid (a 1923 master's graduate in aeronautical engineering from the University of Michigan), the head of Langley's atmospheric wind tunnel ("NACA No. 1") who had been studying the effects of Handley-Page wing slots, Weick designed a cowl that brought outside air in and around the engine via a slot at the center of the nose. The potential of a complete cowl then began to look more enticing. The researchers had to modify the cooling air inlet several times, and install guide vanes or baffles to control the air in its passage for a more efficient heat transfer. They also had to design an exit slot that released the air at a slightly higher velocity and lower pressure than it entered the cowling with, but they finally obtained satisfactory cooling with a complete cowl, which they called "No. 10." This cowling covered the engine entirely and used slots and baffles to direct air over the hottest portions of the cylinders and crankcase.

To everyone's surprise, the No. 10 cowling reduced drag by a factor of almost three. As Weick remembered, "The results of this first portion of cowling tests were so remarkable that we decided to make them known to industry at once. In November 1928, I wrote up Technical Note 301, 'Drag and Cooling with Various Forms of Cowling for a Whirlwind Engine in a Cabin Fuselage,' which the NACA published immediately." The summary of the report was as follows:

> *The National Advisory Committee for Aeronautics has undertaken an investigation in the 20-foot Propeller Research Tunnel at Langley Field on the cowling of radial air-cooled engines. A portion of the investigation has been completed in which several forms and degrees of cowling were tested on a Wright Whirlwind J-5 engine mounted in the nose of a cabin fuselage. The cowlings varied from the one extreme of an entirely exposed engine to the other in which the engine was entirely enclosed. Cooling tests were made and each cowling modified if necessary until the engine cooled approximately as satisfactorily as when it was entirely exposed. Drag tests were then made with each form of cowling and the effect of the cowling on the propulsive efficiency determined with a metal propeller. The propulsive efficiency was found to be practically the same with all forms of cowling. The drag of the cabin fuselage with uncowled engine was found to be more than three times as great as the drag of the fuselage with the engine removed and nose rounded. The conventional forms of cowling in which at least the tops of the cylinder heads and valve gear are exposed, reduced the drag somewhat, but the cowling entirely covering the engine reduced it 2.6 times as much as the best conventional one. The decrease in drag due to the use of spinners proved to be almost negligible.*

In concluding the summary, Weick argued that use of the form completely covering the engine was "entirely practical" under service conditions, but warned that "it must be carefully designed to cool properly."[28] In conjunction with the appearance of this report, the

---

27. Weick and Hansen, *From the Ground Up*, p. 66.
28. *NACA Technical Note 301*, quoted in Weick and Hansen, *From the Ground Up*, p. 66. See also Weick's, "Drag and Cooling with Various Forms of Cowling for a 'Whirlwind' Radial Air-Cooled Engine, I," *NACA TR 313*, 1929, and "II," *TR 314*, 1929.

NACA's Washington office announced to the press that aircraft manufacturers could install the NACA's low-drag cowling as an airplane's standard equipment for about $25 and that the possible annual savings from industry's use of the invention was in excess of $5 million—more than the total of all NACA appropriations through 1928.[29]

With the initial round of wind tunnel investigations completed, Langley borrowed a Curtiss Hawk AT-5A airplane from the Army Air Service, that was already fitted with a Wright Whirlwind J-5 engine, and applied cowling No. 10 for flight research. "These tests showed that the airplane's speed increased from 118 to 137 miles per hour with the new cowling, an increase of nineteen MPH," Weick wrote in his autobiography. "The results of the instrumented flight tests had a little scatter, and we could have been justified in claiming that the increase in speed was twenty MPH instead of 19, but I wanted to be conservative. I didn't want people to expect too much from this cowling, so we called it 19."[30]

Godsend

But the lid on the cowling breakthrough was about to be lifted. On February 4–5, 1929, Frank Hawks, who was already famous for his barnstorming and stunt flying, established a new Los Angeles to New York nonstop record (eighteen hours, thirteen minutes) flying a Lockheed Air Express equipped with a NACA low-drag cowling that increased the aircraft's maximum speed from 157 to 177 miles per hour. The day after the feat, the Committee received the following telegram:

COOLING CAREFULLY CHECKED AND OK. RECORD IMPOSSIBLE WITHOUT NEW COWLING. ALL CREDIT DUE NACA FOR PAINSTAKING AND ACCURATE RESEARCH. [signed] GERRY VULTEE. LOCKHEED AIRCRAFT CO.[31]

In the following months, as the NACA reported in its annual report to the President of the United States at the end of 1929, "all the high–speed records in this country in the past year were made with airplanes powered with radial air-cooled engines using the N.A.C.A. type cowling."[32] Amid a burst of publicity—some of it exaggerated—about the benefits of the NACA cowling, the National Aeronautic Association announced in January 1930 that the NACA had won the Collier Trophy for the greatest achievement in American aviation in 1929.

The NAA presented the award to the NACA at a brief ceremony on the grounds of the White House on June 3, 1930, "before a small but distinguished gathering of aeronautical authorities."[33] President Herbert Hoover presented the trophy to Dr. Joseph S. Ames, the NACA chairman (1927–1939). Significantly, none of the speakers said anything

---

29. Regarding the NACA's public announcement of the cowling, see George W. Lewis's, "Cowling and Cooling of Radial Air-Cooled Engines," transcript of speech before the Society of Automotive Engineers, Detroit, April 10, 1929, Accession 61 A 195 (Box 25), Records of the National Advisory Committee for Aeronautics, National Archives and Records Administration, Washington, DC.

30. Weick and Hansen's *From the Ground Up*, p. 67. See also Thomas Carroll, "Flight Tests of No. 10 Cowling," in E. P. Warner and S. Paul Johnston, *Aviation Handbook* (New York, 1931), p. 145.

31. Telegram dated February 6, 1929, NACA Langley Correspondence Files, Code A176–11, Langley Central Files (LCF), NASA Langley Research Center, Hampton, VA.

32. *Fifteenth Annual Report of the National Advisory Committee for Aeronautics,* 1929 (Washington, DC: U.S. Government Printing Office, 1930), p. 63.

33. *Ibid.*, p. 2.

Dr. Joseph S. Ames, Director of Research for the NACA, was awarded the Collier Trophy, in 1930, for NACA's work developing the low-drag cowling. President Herbert Hoover is making the award. (NASA Photo 90-4348)

to qualify the significance of the design breakthrough or to focus the attention on engineering rather than on science—in fact just the opposite:

> *Senator Hiram Bingham, president of the National Aeronautic Association, opened the ceremony by explaining the history and status of the Collier trophy and read the award citation. President Hoover, in presenting the trophy to Dr. Joseph S. Ames, chairman of the National Advisory Committee for Aeronautics, commended the committee on the scientific* [author's emphasis] *research which had developed the cowling. Doctor Ames, in accepting the trophy on behalf of the committee, said in part: "A scientist receives his reward from his own work in believing that he has added to human knowledge; but he is always gratified when his work is recognized as good by those competent to judge."* [34]

One would hope that Ames, an accomplished physics professor at (and later president of) the Johns Hopkins University, understood that the NACA cowling was producing solid, but not fantastic, results and that there was no magic in the tin shape. As a member of the NACA Main Committee since the NACA's establishment in 1915, he certainly should have

---

34. *Ibid.*, p. 3.

known enough about the research process at Langley to appreciate the systematic character of the laboratory work that made the breakthrough possible. He should also have known that the genuine achievement of the NACA cowling was part of an experimental process more natural to engineering than to any of the sciences per se; that the cowling certainly was not the product of inspired genius; and that there was still a lot of work to be done to make any great use of it, mostly by industry. But if Dr. Ames knew these things, he did not announce them at the White House; and why should he have done so? The NACA was still a fledgling agency uncertain of its political support; Wall Street had just crashed months before; and the Hoover administration's support for on-going aeronautical research and development (R&D) was so tenuous that the NACA was going to need all the boosterism it could get just to survive. (In December 1932, as part of his plan to reduce expenditures and increase efficiency in government by eliminating or consolidating unnecessary or overlapping Federal offices, Hoover signed an executive order to abolish the NACA—something that he had recommended doing in the mid-1920s when serving as secretary of commerce. The election of Franklin D. Roosevelt cancelled President Hoover's mergers and left the NACA intact.)[35]

The 1929 Collier Trophy thus seemed a godsend to the NACA; certainly Ames and the other leaders of the NACA saw it that way. (It is more than coincidental that John F. Victory, the executive secretary of the NACA, was serving as treasurer of the National Aeronautic Association in the year that the NACA first won the Collier. No NACA official had served on the NAA executive committee before 1929.) The pleasant recognition not only justified the funding levels the NACA had gotten in 1929 and 1930—$836,700 and $1.3 million, respectively, which seems modest but was in fact nearly $300,000 more than it had ever received—but was also timely support for the NACA's request for more money (the FY 1931 appropriation would turn out to be $1.36 million) to continue construction of a large, new, full-scale wind tunnel at Langley, one even larger than the PRT. It was not the time to be dirtying the water with complex thoughts about the authentic nature of engineering breakthroughs; rather, it was the time to give the aviation public what it wanted. Great science. Heroic thoughts to match the feat of Lindbergh. Magical technology. Tin shapes that produced miraculous results. That is the sort of "right stuff" that "flew" with the aviation public in the 1930s, as it still does today. The "honest stuff" about the details of the NACA research program was too down-to-earth and technically complicated. Better just to call all of your achievements "science."

After all, in 1930, no one yet was absolutely sure whether the NACA was an organization for science or for engineering. Congress had created the NACA in 1915 "to supervise and direct the scientific and technical problems of flight with a view to their practical solution."[36] The leaders of America's embryonic aviation establishment, however, had been in sharp disagreement over how to interpret this mandate. Some had felt that the NACA should remain small and continue to serve as merely an advisory body, devoted to pure scientific research. (With qualifications, Dr. Ames had tended to support this view.) Others had argued that the NACA should grow larger and combine basic research with engineering and technology development. This second group, led by the NACA's ambitious director of research George Lewis (M.S. in mechanical engineering, Cornell University, 1910), wanted the NACA to attack the most pressing problems obstructing the immediate progress of American aviation, particularly those that were vexing the fledgling military air services and aircraft manufacturing and operating industries.[37]

---

35. Hansen, *Engineer in Charge*, p. 145.
36. The full text of the law establishing the NACA in 1915 (Public Law 271, 63rd Congress, approved March 3, 1915) is reprinted in *Engineer in Charge*, p. 399, appendix A.
37. James R. Hansen, "George W. Lewis and the Management of Aeronautical Research," in William M. Leary, ed., *Aviation's Golden Age: Portraits from the 1920s and 1930s* (Iowa City, IA: University of Iowa Press, 1989), pp. 93–112.

Under Lewis's careful direction (he served as director of research from 1919 to 1947), the NACA moved slowly but surely along the second course. By the mid-1920s, engineers, not scientists, were in charge at Langley, and the keystone of the NACA's charter rested securely in their notion of "practical solutions." Over the next twenty years, the NACA conducted research into basic aerodynamic, structural, and propulsion problems whose solutions led to the design of safer, faster, higher-flying, and generally more versatile and dependable aircraft. With these aircraft, the United States became a world power in commercial aviation and Allied victory in World War II was assured. In the opinion of many experts, the NACA did "at least as much for aeronautical progress as any organization in the world."[38]

Engineering or Science?

Much of the credit for this impressive record rests with the NACA's *engineering* approach to the technological problems. Scientific principles undergirded aeronautical development, of course, and basic discoveries in the physics of airflows definitely played a major role in focusing the effort. But it was engineering research and development that really brought the progress. When Langley laboratory started flight testing in 1919 (the first LMAL wind tunnel did not begin operating until June 1920), frail wooden biplanes covered with fabric, braced by wires, powered by heavy water-cooled engines, and driven by hand-carved wooden propellers still ruled the airways. The principles of aeronautical engineering had yet to be fully discovered, and only a few programs at major schools like MIT and the University of Michigan existed to find them and teach them to students. The design of aircraft remained a largely intuitive and empirical practice requiring bold speculation and daring, in both a financial and technological sense.

In terms of engineering, there were still a number of bothersome and potentially dangerous unknowns. As evidenced in the question asked of the NACA at the 1926 conference, no one knew for sure how to reduce engine drag without degrading cooling. But there were so many of these questions still needing to be asked. No one knew with certainty how to shape wings to increase lift or to diminish the effects of turbulence. No one knew how and when flaps, ailerons, and other control surfaces worked best. No one knew if it was even worthwhile to retract landing gears (according to various pundits, the added weight and complexity of a retractable undercarriage would not be worth the saving in air resistance). Substantial increases in aerodynamic efficiency might follow on the heels of correct answers to just a few of these technical concerns, but no one knew exactly how, or even whether to try, to get at them.

It was, therefore, unfortunate—and tremendously misleading to the aviation public—for Dr. Ames, at the White House ceremony, to commend the NACA on the "scientific research which had developed the cowling," for it was not science, but engineering—and not scientists, but engineers like Fred Weick and his PRT team—who actually deserved the credit. Engineering deserved the credit not only for the NACA cowling but for most of the design revolution then beginning to take place in American aeronautics. Ames's acceptance speech was thus like congratulating the Wright brothers for being scientists rather than engineers, thereby missing the essential points of what they had actually achieved and how they achieved it. Of course, the Wrights *had been* portrayed all too often as scientists. In this sense, Ames's attribution

---

38. C. G. Gray, "Dr. G. W. Lewis," *The Aeroplane*, August 27, 1948.

for the cowling was in keeping with the American tradition of co-opting engineering achievements for science.

The failure to distinguish between scientific and engineering achievement haunted the NACA throughout its history, but never more so than in the early 1930s. The most outspoken critic of the NACA at that time, Frank Tichenor, the editor of the journal *Aero Digest*, mislabeled the NACA cowling "a development rather than an original work" and misjudged it as being far less effective than the Townend ring, a rival cowling concept developed simultaneously by Hubert C. Townend at the British National Physical Laboratory.[39] Tichener did so largely because he took the NACA at its own words about being a *scientific* organization and because he failed to appreciate that aviation progress during the era really depended on engineering being in charge, as it was at Langley laboratory, not science. In his regular monthly column, "Air—Hot and Otherwise," Tichenor attacked the NACA in late 1930 and early 1931. In the February 1931 issue, he stated the gist of his criticism:

> *In these columns in December, I reviewed the conditions prevailing in the National Advisory Committee for Aeronautics which prevent it from functioning in a manner useful to the best interests of the industry it purports to serve. . . . The importance of a wise and honest expenditure of public funds appropriated specifically for scientific* [author's emphasis] *research and not for a cheap substitute for it, is generally recognized.*

In his column, subtitled "The NACA Counters," Tichenor then took on a "defender of NACA management," Dr. Edward P. Warner, editor of the rival trade journal *Aviation* and a long-time member of the NACA's Committee on Aerodynamics and Committee on Materials for Aircraft (Warner had served temporarily in 1920 as Langley laboratory's chief physicist), who had prepared a response to Tichenor's December 1930 column "Why the NACA?"[40] In his editorial response, published in *Aviation* in January 1931, Warner "skirted the definition of 'scientific research'"[41] and by inference, seemed to concede (as Langley chief of aerodynamics Elton W. Miller also did in an unpublished response he prepared for the NACA Washington Office, which Warner received before writing his own

---

39. Frank Tichenor, "Air—Hot and Otherwise," *Aero Digest* (February 1931): 24. The history of the NACA cowling-Townend ring rivalry has yet to be written. In the beginning, neither the British NPL nor the American NACA appear to have been aware of the other's cowling work. The NPL published the results of its ring research just before the NACA's cowling reports appeared. To impress American manufacturers with the value of its cowling, the NACA did place its design into competition with the Townend ring. George Lewis told Glenn L. Martin, for example, that Martin's B-10 bomber would not only fly significantly faster than its present maximum speed of 195 miles per hour, but would also land slower and more safely, if the engine's Townend ring were replaced by the NACA No. 10 cowl. Pratt & Whitney, the builder of the engine for the airplane, was contractually committed to using the ring. Martin eventually adopted the NACA cowling for the B-10, increasing the airplane's maximum speed by 30 MPH to 225 and also reducing its landing speed significantly. In 1933 and 1934, the army purchased more than 100 B-10s, rescuing Martin from the worst of the Depression. What the cowling did for the B-10's performance may well have been why Martin won the production contract and why Boeing's B-9, in competition with the Martin aircraft, lost. The B-9 used the Townend ring. See Lloyd S. Jones, *U.S. Bombers, 1928 to 1980s*, 3rd ed. (Falbrook, CA: Aero Books, 1981), pp. 30–32. The overall competitive situation fed the fire of the transatlantic dispute and resulted in a long series of patent suits. For NACA Langley's reaction to, and role in, the patent dispute, see Elton W. Miller to LMAL engineer-in-charge, "Criticism of Committee's Attitude with Reference to Townend Ring Cowling," March 3, 1931, File A176–11, LCF; George W. Lewis to LMAL, "NACA Cowling and Claim of Townend Patent," August 12, 1931, *ibid.*; "Report of Meeting between Representatives of NACA and of the Army and Navy to Discuss the Cowling Patent Situation," June 21, 1932, *ibid.* On the Townend ring specifically, see H. C. H. Townend, "The Townend Ring," *Journal of the Royal Aeronautical Society* 34 (October 1930): 813–48. For a contemporary analysis of cowling development, see J. D. North, "Engine Cowling," *Journal of the Royal Aeronautical Society* 38 (July 1934): 566–612.

40. Tichenor, "Why the NACA," *Aero Digest* (December 1930): 47ff; "The N.A.C.A. Counters," *Aero Digest*, January 1931, pp. 50ff; Edward P. Warner, editorial, *Aviation* 30 (January 1931): 3–4.

41. Roland, *Model Research*, 1:133.

rejoinder)[42] that very little NACA work "could be classified as fundamental, according to general acceptance of the term." Still, the NACA research program was scientific, as it involved (in Miller's words) "accumulated and accepted knowledge, systematized and formulated with reference to the discovery of general truths on the operation of general laws." Like Miller, Warner argued that Tichenor was looking at aeronautical R&D at Langley laboratory (a place Tichenor apparently had never visited) in the wrong way: just because research at Langley had a practical object, it did not mean that it was not scientific.[43]

But Tichenor did not grasp the point, largely because he saw an all-too-dramatic dropoff from science to whatever else came, in his view, *below* it. (NACA leaders believed that Tichenor's anti-NACA columns were in fact being fueled—and perhaps even drafted— by *Aero Digest* consultant, Dr. Max Munk, the eccentric German aerodynamicist who had conceptualized the VDT and PRT at Langley but who had been forced to resign as LMAL chief of aerodynamics in early 1927 after a revolt of all the sections heads in the aerodynamics division against his autocratic style of supervision. Elton Miller was Munk's successor and had played a major part in the revolt.)[44] If it was not science at the NACA, then for the *Aero Digest* editor (and for the disgruntled Dr. Munk, who really should have known better), it was "a cheap substitute." There was nothing in between, and certainly nothing on par, with science.

Thus, Tichenor took Warner's response—which did not make a terribly clear case for the requirements of an engineering approach to basic applied research but tried instead only to claim the values of science for the NACA—and he turned them against the government organization. (Warner had earned a master's degree in physics at MIT in 1919 and, following his brief hiatus at LMAL, taught in the school's pioneering aeronautical engineering program into the mid-1920s, when he became a consultant in Washington, DC, to the President's Aircraft Board, better known as the Morrow Board, after its chairman Dwight Morrow.)[45] Responding to Warner, Tichenor wrote:

> *It almost looks as though the defender of the N.A.C.A. management in his own heart agrees with us; and although he finds it expedient to depreciate our criticism, he writes as though he himself would like to see reform effected. He does not call attention to one successful research, nor one scientific advancement which can be credited to the N.A.C.A. . . . Nor does he suggest that such advances can be expected in the future. . . .Our principal criticism, the absence of scientific research, is tacitly admitted. Such research, he contends, is the proper sphere of universities, not of the N.A.C.A.*

Tichenor bolstered his case with references to the NACA's own language, its own executive policy decisions, and to the NACA charter itself:

> *Now, we have not, merely as the result of our own judgment, specified scientific research as the task of the N.A.C.A.; we quoted this as the NACA's task from the Committee's own annual reports. The defender of the N.A.C.A. cannot logically ignore this point altogether, as he does, for it is the most important consideration, the keynote of the*

---

42. As Roland points out in *Model Research* (1:356, n. 18), Elton W. Miller's comments appeared in a 19 December 1930 memorandum to LMAL engineer-in-charge Henry Reid, who then forwarded it to George Lewis in the NACA Washington office as "Comments on the Article in the December 1930 Issue of *Aero Digest*, Entitled 'Why the N.A.C.A.?'" dated January 2, 1931, Accession 55 A 312, Records of the NACA, National Archives.
43. Elton W, Miller memorandum, December 19, 1930, cited in *ibid.*
44. See Roland, *Model Research*, 1:132–35.
45. On E.P. Warner's career in aeronautics, see Roger E. Bilstein, "Edward Pearson Warner and the New Air Age," in Leary, ed., *Aviation's Golden Age*, pp. 113–26.

N.A.C.A.'s shortcomings. This is not a question of opinion only; rather, it is far more a question of keeping faith, of loyalty to duties defined by the supervising body of the N.A.C.A. The policy of conducting scientific research was adopted ten years ago by the presiding [Main] Committee, made up of the foremost experts of the country. In all annual reports since then, it has been recorded as the accepted policy of this body. It has been pleaded for in hearings before Congressional committees. It has formed the basis for public appropriations.

Tichenor then asked the key question, one much more insightful than the *Aero Digest* editor ever realized at the time: "Does the defender of the N.A.C.A. mean to imply that there is one policy for obtaining appropriations and for general advertising and publicity purposes and quite another one for the actual service and activity within the walls of the N.A.C.A.?"[46]

The answer, honestly, was, yes; there were two *practices*, if not policies. Not that the NACA was consciously involved in any deception; it was just that the NACA as an organization was not yet self-conscious enough in 1930 about the value of engineering at its research laboratory to extricate itself from the public relations dilemma. The American people expected scientific achievement and did not really understand engineering. The NACA charter said it was the job of the NACA "to supervise and direct the scientific study of the problems of flight with a view to their practical solution;" Tichenor thus thought he was calling the NACA to task when he asked, "If money is appropriated for scientific research, can we consider it of no consequence that those funds are spent for something else?"; while Warner thought the NACA research staff was doing exactly what it was supposed to do in seeking practical solutions, no matter exactly what one called it. In Tichenor's purist opinion, "Either there is scientific research or there is not," and Congress in 1915 had "decreed that the N.A.C.A. should conduct scientific research." In the NACA's more utilitarian view, "Research need not necessarily be aimless to be scientific."[47]

The two sides were talking past one another. What Tichenor needed to understand, and what the NACA itself needed to grasp more fully and communicate far better and more often to the aviation public, was that a methodologically sophisticated approach to solving technological problems, later to be called *engineering science*, was developing in the American engineering profession in the first decades of the twentieth century—and that it, not pure science, held the key to unlocking aviation progress and igniting the airplane design revolution of the 1930s. The fact that engineering had come to dominate the character of the work at NACA Langley was not something to bemoan and condemn, as Tichenor was doing; it was something to praise, explain, and fully exploit.

Because Tichenor did not understand the many advantages of engineering science, he dismissed the NACA cowling work as cut-and-try development. With the actual invention of the cowling, the editor charged, "the N.A.C.A. had nothing whatsoever to do." Nevertheless, according to Tichenor, the NACA was claiming that, "had it not been for the NACA," the industry would not be adopting it. He wrote:

The industry is alleged to be so timid that the information about improvements available is not sufficient to induce it to adopt them; the industry needs the guiding hand of the N.A.C.A.; the industry does not trust and has no confidence in its own speed tests made by its own pilots. The implication is that, instead, it waits until the N.A.C.A.

---

46. Tichenor, "The N.A.C.A. Counters," p. 50.
47. *Ibid.*, pp. 50 and 122; Elton W. Miller memorandum to LMAL engineer-in-charge, attached to "Comments on the Article in the December 1930 Issue of *Aero Digest*, Entitled 'Why the N.A.C.A.?'," dated January 2, 1931, Accession 55 A 312, National Archives.

> *measures in pounds and ounces the diminishment of the drag in consequence of some improvement and then computes the increase in the speed. The industry, it is seriously alleged, has more confidence in such computed speed gain than in speed directly observed. How grotesque! We really have cause to admire the courage of one who advances such opinions.*[48]

Edward P. Warner, in turn, reassured the NACA privately that Tichenor's indictment was without force in the aircraft industry. On January 5, 1931 he wrote to George Lewis: "One thing you never need to worry about in any year is the worth-whileness of the work that you are guiding. I have never overheard so much comment on anything that appeared in *Aero Digest* as on Frank Tichenor's attack on the Committee, and the comment has been about ninety-eight percent unfavorable—and I have already been receiving congratulations."[49]

By the time this debate broke out, NACA Langley's cowling program had already evolved into a distinct second stage, one still rooted in the engineering approach to solving the outstanding technological problems. In Fred Weick's formulation, "The second part of the cowling program covered tests with several forms of cowling, including individual fairings behind and individual hoods over the cylinders, and a smaller version of the new complete cowling, all mounted in a smaller, open-cockpit fuselage. We also performed drag tests with a conventional engine nacelle and with a nacelle having the new complete design."[50] Though the individual fairings and hoods proved ineffective in reducing drag, Weick and his colleagues found that the reduction with the complete cowling over that with the conventional cowling was in fact over twice as great as with the larger cabin fuselage. Data from the Curtiss Hawk AT-5A flight tests confirmed this conclusion.[51]

In early 1929, Langley's flight research division mounted NACA low-drag cowlings on the engines of a Fokker trimotor. Although Weick did not supervise these tests, he followed their results closely.

> *The comparative speed trials proved extremely disappointing. Separate tests on the individual nacelles showed that cowling the Fokker's nose engine gave approximately the improved performance we expected. Cowling the wing nacelles, however, gave no improvement in performance at all. This was strange, because the wind-tunnel tests had already demonstrated convincingly that one could obtain much greater improvement with a cowled nacelle than with a cowled engine in front of a large fuselage. Some of us started to wonder how the position of the nacelle with respect to the wing might affect drag.*[52]

This was a critical design issue, especially for multi-engine aircraft, as big commercial and military aircraft were bound to be. In the case of the Fokker (as well as the Ford) trimotor, the original design location of the wing engines was slightly below the surface of the wing. As the air flowed back between the wing and nacelle, and the distance between them increased toward the rear of the nacelle, the expansion required was too great for the air to flow over the contour smoothly. The LMAL flight research group, in association with the PRT team, tried fairing-in this space, but achieved only a small improvement.[53]

---

48. Tichenor, "The NACA Counters," p. 124.
49. Quoted in Roland's *Model Research*, 1:356, n. 22.
50. *From the Ground Up*, p. 67, by Weick and Hansen.
51. "The Effect on Airplane Performance of the Factors That Must Be Considered in Applying Low-Drag Cowling to Radial Engines," *TR 414*, 1932, by Wiliam H. McAvoy, Oscar W. Schey, and Alfred W. Young.
52. Weick and Hansen, *From the Ground Up*, p. 72.
53. *Ibid.*

Nevertheless, the lab's systematic, empirical approach soon yielded its dividend. With the help of his assistants, Weick laid out a series of model tests in the PRT with NACA-cowled nacelles placed in twenty-one different positions with respect to the wing above it, below it, and within its leading edge. "Where it appeared pertinent, extra fairing was put between them," Weick recalled.[54] The resulting data on the effect of the nacelle on the lift, drag, and propulsive efficiency of the big Fokker trimotor made it clear that the optimum location of the nacelle was directly in line with the wing, and with the propeller fairly well ahead. Although their primary emphasis was on drag and improved cooling, the tests at Langley also confirmed that a cowling No. 10 of the radial engine, if situated in the optimum position, could in some cases actually increase the lift of the airplane's wing.[55] "Without the complete cowling," Weick and the others learned, "the radial engine in this position spoiled the maximum-lift coefficient of the wing. With the cowling, and the smooth airflow that resulted from it, the maximum-lift coefficient was actually increased."[56] In transmitting this important information confidentially to the army, navy, and industry, the NACA helped build a several-months lead for American aircraft designers over rival European companies. After 1932, nearly all American transport and bombing airplanes—including the Douglas DC-3, Boeing B-17, and many other famous aircraft of the era that followed—employed radial wing-mounted engines with the NACA-cowled nacelles located approximately in what Weick and his associates had identified as the optimum position.

Weick and his colleagues remained extremely proud of this contribution for the rest of their lives. In his autobiography, Fred wrote: "This combination, according to some historians, was one of the important advances that enabled airliners to become financially self-supporting, that is, without the need for government subsidy."[57] As such, it fulfilled the NACA's public mandate, put another feather in the cap of the still fledgling government research organization, and demonstrated again, for better reasons than even the original ones, that the NACA's winning of the Collier Trophy in 1929 was well deserved.

The cowling was winning so much respect in the late 1920s and early 1930s that the NACA came to identify itself more and more with the systematic experimental approach that had been the basis of the successful cowling research. In 1930, the head of the Langley aerodynamics division, Elton W. Miller (B.S. in mechanical engineering from George Washington University, class of '08) reported to engineer-in-charge Henry J. E. Reid (B.S. in electrical engineering from Worcester Polytechnic Institute, class of '19) that "an effort is being made throughout the Laboratory to conduct every investigation in as thorough and systematic a manner" as the cowling program.[58] The following year, George Lewis told Reid to hang, in his office or along the corridor of the LMAL administration building, a copy of the following quotation from a speech by President Hoover in praise of Thomas Edison:

> *Scientific discovery and its practical applications are the products of long and arduous research. Discovery and invention do not spring full-blown from the brains of men. The labor of a host of men, great laboratories, long, patient, scientific experiments build up the structure of knowledge, not stone by stone, but particle by particle. This adding of fact to fact some day brings forth a revolutionary discovery, an illuminating hypothesis, a great generalization of practical invention.*[59]

---

54. *Ibid.*
55. Donald H. Wood, "Tests of Nacelle–Propeller Combinations in Various Positions with References to Wings, I—Thick Wing—NACA Cowled Nacelle—Tractor Propeller," *TR 436*, 1932.
56. Weick and Hansen, *From the Ground Up*, pp. 72–73.
57. *Ibid.*, p. 73.
58. Elton W. Miller to LMAL engineer–in–charge, December 19, 1930, file A176–11, LCF.
59. As quoted in Roland's *Model Research*, 1:105.

Although this quotation fell short of the whole truth about how progress was made in science and technology, it was closer to the realities of the cowling achievement than was the myth of heroic invention; Lewis's request for it to be displayed at Langley indicates that some NACA leaders certainly possessed a more mature understanding of the nature of technological change than they were willing to grant for, or explain to, the public at large. Clearly the pattern of work behind the cowling—the NACA's greatest public success to date—contributed to a clearer sense of institutional identity and mission, even if the agency as a whole was not doing much to enhance the public's understanding of the technological process at work.

Experimental Impasse

However, given what was to take place during the third stage of cowling research at Langley, from 1931 to 1934, one cannot be too sure even whether this clearer identity for the NACA was an altogether good thing—that is, whether Langley's confidence in systematic parameter variation would continue to signify technological momentum or turn into technological inertia.

A distinct third stage of cowling research began at Langley when many more aircraft manufacturers decided to adopt the NACA design as standard high-performance equipment. A few companies did rather well with their applications of the NACA No. 10 cowling, especially those that put a series of adjustable flaps around the circumference of the metal jacket in the hope of better regulating the release of used air. (Those that tried to encourage more cooling flow by employing larger exit openings failed, however, sometimes to the point of nullifying the external drag advantage.) With the development of twin-row engines such as the Pratt & Whitney R-1830 of 1933–34—with one row of cylinders behind the other—whole new problems arose.[60] This situation challenged Langley to obtain more trustworthy data on the general aerodynamic properties of the proven NACA design. Practical results had been obtained from experimental parameter variation, and they had been used profitably. Now it was time for a clearer understanding of them, so that still more results could eventually be achieved.

Three major branches of the laboratory became involved in the ambitious program. The power plants division worked to improve the efficiency of radial-engine cooling by varying such engine parameters as pitch, width, thickness, and shape of the fins. The 7 x 10-foot wind tunnel section, using small models, sought the best possible cowling arrangement for necessary cooling with minimum drag by streamlining the front and rear openings, changing the size of the nacelle, and altering the camber of the cowling's leading edge. The PRT team was then to verify the results of the tests made by the other two groups. Full scale propeller-cowling-nacelle units were to be tested under conditions of taxiing, takeoff, and level flight.[61] Don Wood was now the head of the PRT section. In April 1929, Fred Weick took a position with the Hamilton Aero Manufacturing Company in

---

60. John V. Becker, *The High–Speed Frontier: Case Studies of Four NACA Programs*, 1920–1950 (Washington, DC: NASA SP-445, 1980), pp. 140–41.

61. *Nineteenth Annual Report of the National Advisory Committee for Aeronautics* (Washington, DC: U.S. Government Printing Office, 1934), p. 10; Arnold E. Biermann and Benjamin Pinkel, "Heat Transfer from Finned Metal Cylinders in an Air Stream," *TR 488*, 1934; Donald H. Wood, "Tests of Nacelle-Propeller Combinations in Various Positions with Reference to Wings, II—Thick Wing—Various Radial-Engine Cowlings—Tractor Propeller," *TR 436*, 1932; *ibid.*, "III—Clark Y Wing—Various Radial–Engine Cowlings—Tractor Propeller," *TR 462*, 1933; James G. McHugh, *ibid.*, "IV—Thick Wings—Various Radial-Engine Cowlings—Tandem Propellers," *TR 505*, 1934; E. Floyd Valentine, *ibid.*, "V—Clark Y Biplane Cellule—NACA Cowled Nacelle—Tractor Propeller," *TR 506*, 1934; Donald H. Wood and Carlton Bioletti, ibid., "VI—Wings and Nacelles with Pusher Propeller," *TR 507*, 1934.

Milwaukee, Wisconsin, a subsidiary of the United Aircraft and Transport Corporation. He returned to Langley in less than a year as assistant chief of the LMAL aerodynamics division, a position from which he could work with any of the wind tunnels as well as the flight section. In this capacity, Weick stayed in touch with the cowling program but it did not monopolize his time and energies as before.[62]

Though the first two parts of the program advanced without much difficulty, the PRT tests under Don Wood—the final and most important part—ran into major problems soon after starting in 1933: the 100-mile-per-hour tunnel could simulate only the climb speeds of the cowled engine being used (a borrowed Pratt & Whitney Wasp); the obsolete shell-type baffles employed to deflect cooling air toward the hottest parts of the engine were too loose for the NACA researchers to work with effectively;[63] and, more importantly, certain anomalies that no one at the lab could explain plagued the cowling drag measurements. Together these problems contributed to a growing "maze of contradictory data" about cowlings. Despite five years of NACA experimentation and three years of general industrial flight test experience, American aeronautical engineers felt a "general suspicion" that there was "something mysterious or unpredictable determining the efficiency of engine cowling."[64]

To move beyond this experimental impasse, Langley's cowling research needed some analytical help. It was eventually provided by the head of the laboratory's small Physical Research Division, Theodore Theodorsen (Dr. Ing., Universitetet I Trondheim, '22). A Norwegian-born engineer-physicist with a trigger mind and tremendous power of concentration, Dr. Theodorsen had already seen, in Langley's pattern of airfoil testing in the variable-density tunnel (VDT), the need for experimental routine to be fertilized with a stronger dose of theory. In the curious introduction to his seminal 1931 report on the "Theory of Wing Sections of Arbitrary Shape"—curious at least in an NACA report for stating a bold personal opinion and implicitly taking part of the parent organization to task—Theodorsen had asserted that

> *a science can develop on a purely empirical basis for only a certain time. Theory is a process of systematic arrangement and simplification of known facts. As long as the facts are few and obvious no theory is necessary, but when they become many and less simple theory is needed. Although the experimenting itself may require little effort, it is, however, often exceedingly difficult to analyse the results of even simple experiments. There exists, therefore, always a tendency to produce more test results than can be digested by theory or applied by industry.*

What Theodorsen believed the NACA needed in order for it to move beyond the impasse now blocking the progress of its experimental cowling program was more attention to the "pencil-and-paper" work that could lead to a complete mathematical and physical understanding of the basic internal and external aerodynamics of the different cowling

---

62. Weick and Hansen, *From the Ground Up*, pp. 85–114.

63. To direct cooling air around the hot engine cylinders, LMAL engineers had tried a number of different deflectors. One of the conceptually more refined ones tried in the early 1930s was the loosely fitting "shell" baffle. (At about this same time Pratt & Whitney and Vought were finding this type of baffle inferior to the tightly fitting pressure baffle.) Though LMAL tested the double-row R-1830 engine installation with a pressure baffle system in the Full-Scale Tunnel in 1934 (*TR 550*, "Cooling Characteristics of a 2–Row Radial Engine," 1935, by Oscar W. Schey and Vernon G. Rollin), thorough investigation came only in 1936 (*TN 630*, "Energy Loss, Velocity Distribution, and Temperature Distribution for a Baffled Cylinder Model," 1937, by Maurice J. Brevoort). For analysis of the comparative value of the two types of baffles, see Becker, *High–Speed Frontier*, pp. 141–43.

64. Rex Beisel, "The Cowling and Cooling of Radial Air-Cooled Engines," *Society of Automotive Engineering Journal* 34 (May 1934): 159.

shapes.⁶⁵ And what this meant in terms of the history of Langley's method of cowling research was a turning away from experimental parameter variation, and toward that distant goal of complex understanding.

Theodorsen first perceived new cream to be skimmed off the top of the old cowling and cooling investigation while serving on the LMAL editorial committee that reviewed the draft report on the tests of the full-scale propeller-cowling-nacelle units in the PRT. After pointing to the blunt afterbody of the nacelle as the probable source of the anomalies that had been observed in the drag data, he suggested to his colleagues that the stalled cowling program could be completed as planned (and his resolution of the drag anomalies verified) by a new, more comprehensive and analytical full-scale investigation. Its aim, underscored Theodorsen, would be both to improve basic understanding of the obscure cooling mechanisms of the cowled engine and to put the understanding of the relationship between internal flow and drag on a more rational basis. The provocative suggestion was adopted; engineer-in-charge Henry Reid transferred most of the cowling work and many of its key personnel to Theodorsen's division.⁶⁶

The PRT team had previously focused almost entirely on the net effect of the cowling on drag and engine temperatures. What Theodorsen now proposed was to investigate the fundamental flow involved. In part, the approach of Theodorsen's new cowling research team still followed that of experimental variation. The Wasp engine having proved inadequate as part of the test bed, they built a full-scale wind tunnel model with a dummy engine, which had one cylinder heated electrically. Numerous combinations of more than a dozen nose shapes, about a dozen skirts, six propellers, two sizes of nacelles, and various spinners were tested. But hoping to produce a detailed handbook by which designers could better understand the actual functioning of the NACA cowl, they also included extensive measurements of pressure in both the external and internal flows.

Langley's revised cowling program thus remained primarily experimental, but it now also allowed quantitative analysis and computation of these flow pressures. This quantitative analysis, which had been lacking in the PRT's previous work, eventually produced some new NACA cowling designs, but more importantly it provided solid answers to virtually all the remaining questions about the fundamental principles of the cowling and cooling of radial engines.⁶⁷ It demonstrated conclusively that the early NACA designs had been "quite haphazard and often aerodynamically poor," and had cooled the engine successfully only by a crude excess of internal flow and internal drag—a conclusion that engineers in the aircraft industry, notably at Vought, had already arrived at on their own, on behalf of Pratt & Whitney and its R-1830 engine.⁶⁸ Designers of future cowlings, like airfoil designers, would have to be much more sensitive to such subtleties as the ideal angle of the cowling's leading edge attack on the local airflow. The fourth stage of cowling work at Langley even demonstrated as fact something that everyone had unconsciously assumed to be physically impossible when the cowling research began in 1926: a proper engine cowling could, by making the enclosed baffled engine act in essence as a ducted radiator for cooling, lower operating temperatures more than could full exposure of cylinders in the airstream. With this counterintuitive reality confirmed, the national aeronautical establishment could now begin to focus on more specific,

---

65. Theodore Theodorsen, "Theory of Wing Sections," *TR 411*, printed in the NACA annual report of 1932, p. 29.
66. Theodorsen to LMAL engineer-in-charge, June 28, 1935, file R1600–1, LCF; telephone interview, James G. McHugh, Hampton, VA, with author, June 13, 1983.
67. Theodorsen, Maurice J. Brevoort, George Stickle, and Melvin Gough, "Full-Scale Tests of a New Type NACA Nose-Slot Cowling," *TR 595*, 1937; Theodorsen, Brevoort, and Stickle, "Full-Scale Tests of NACA Cowlings," *TR 592*, 1937, and *TR 662*, 1939.
68. Becker's *High-Speed Frontier*, pp. 142–43.

higher-speed applications of cowlings, work that would prove essential to the design of military aircraft used by the United States and her allies in World War II.

Demystifying the Cowling

The history of the cowling research from 1926 to 1936 celebrates the victory of the NACA's winning the National Aeronautic Association's prestigious Collier Trophy for 1929, but it illustrates a more fundamental point about applied basic research. No matter how practical or otherwise advantageous any one research method may be, it always has some disadvantages. Systematic parameter variation had enabled the researchers at Langley to delineate a cowling that significantly reduced the drag of a radial engine without degrading its cooling, but because initial success came rather quickly and easily, they did not have to understand exactly why the cowling worked. When questions and doubts arose, and data seemed contradictory and mysterious, the original empirical method was unable to proceed. Only then did Theodorsen design the research program whose goal was an understanding that went far beyond the mere collection of overall performance data on a variety of promising but arbitrary shapes. The cowlings that resulted from the Theodorsen program did not beat the earlier shapes as regards external drag (which is only a weak function of cowl shape), but with the tight baffles, small exit areas, and low internal drag made possible by the NACA's new criteria of understanding, the total drag of Theodorsen's shapes was dramatically less.

Three-quarters of a century after the initial cowling breakthrough, historians of aeronautics still tend to treat the NACA cowling as a magical piece of tin wrapped around an engine, and they still tend to misinterpret the NACA for its failure to be *scientific*. As a result, they fail not only to appreciate the systematic character of the laboratory work that made the initial design breakthrough possible, but also to pick up on the later work by Theodorsen and engineering groups in the aircraft industry that made the important final breakthrough in understanding possible. The success of the cowling was not due to magic. Nor was it the result of simple cut-and-try or advanced theory demonstrating its ultimate superiority over empiricism. Rather, the cowling was the product of fruitful engineering science.

Ultimate success in research is never inevitable, however. Without the help of Theodorsen or someone else with comparable analytical and mathematical talents, the cowling research at Langley might have remained indefinitely at the point of impasse. Much of the responsibility for misunderstanding the true achievement of the NACA cowling program belongs to the NACA, whose leaders and publicists of the late 1920s and early 1930s, in seeking to gain respect and additional funding for the honestly meritorious operations (and future wind-tunnel building projects) of their struggling research agency, exaggerated the mysterious wonders of the NACA cowling and continued to stress the scientific character of all NACA research when they should have been advancing a more utilitarian view of basic research methodology—and of technological progress. In doing so, they condoned the miscasting of the cowling as a heroic invention—which, in some key respects, represented it as something *less* than it was.

With its winning of the Collier Trophy for 1929, the NACA missed an excellent opportunity to explain to the aviation public, which was growing ever larger and generally more informed during the post-Lindbergh era, what successful applied research done by the government was really all about. Even if the NACA had provided brilliant explanations, of course, the public might not have cared to listen. But for the general technological literacy of the country, it would have been worth the try. And at the very least, the NACA would not have left itself so open to criticism from Frank Tichenor and other critics, as well as later historians, for overselling what really *did* amount to one of the most significant *types* of accomplishments within the NACA's capability.

The original counterintuition that won the NACA its first Collier Trophy was remarkable enough to merit winning the award, because it laid open to public view the many potential advantages of a low-drag engine cowling. But that strange opening idea, which was hard enough for the public to understand, represented only the *first* step in a much more complicated "learning for design" process. Beyond the conceptual breakthrough there was much more to be done by American engineers before truly remarkable results in aircraft performance could be achieved. The NACA's Langley laboratory in Virginia, where a culture of "the engineer in charge" took hold in the 1930s, still had to carry out a rigorous experimental program and analysis. It was then up to the aircraft industry, not the NACA itself—which, after all, was not in the business of *designing* aircraft—to incorporate the cowling development into the larger revolution just taking wing in 1929. In just a few years this revolution would lead to such advanced airplanes as the Douglas DC-3 and Boeing B-17, with cantilever wings, retractable landing gear, efficiently cowled radial engines, controllable-pitch propellers, and all-metal, stressed-skin construction. Without its integration into this larger technological development, moving from the various shapes of ungainly wooden biplanes to sleek metal monoplanes, the singular existence of a low-drag NACA cowling would have been almost meaningless.

Engineering science is not easy for the layperson to understand. Partly for this reason, back in the early 1930s, the NACA had outspoken critics. Some of the criticisms were valid. The NACA's publicists did exaggerate the cowling's significance and took too much credit for the aircraft industry's adoption of the cowling. They could have done a far better job of explaining what *really* had been accomplished and how important it all was: that is, how systematic research was moving things along nicely and how Langley's Propeller Research Tunnel, a modestly-priced and brand new public facility was already paying off in spades by permitting a team of engineers to work in a wind tunnel with full-scale airplanes. Better experimental equipment was leading to more comprehensive and more useful data. The aircraft industry was benefitting from the government's help—and was very thankful for it. It was that simple.

This is what the NACA could have said, and perhaps should have said, to the aviation public rather than leave most people with the impression that a magical piece of equipment had been invented and that science was responsible for it. Like the engineering of cowlings itself, which was work honestly done and honestly explained in NACA's technical reports by talented engineers like Fred Weick, more accurate public expressions out of the NACA's Washington office, although requiring much more understanding from those who both articulated and received them, could perhaps have served the cause of the NACA better. They could have done so by explaining to the paying public how basic applied research gets done in a laboratory setting and how painstaking research fuels technical progress.

As hyperbole and myth, NACA statements from which people inferred a heroic invention of the cowling seem, indeed, to have had some short-term political value. But one can wonder if such exaggerations have, in the long run, made it harder to justify public funding for slow-but-sure technological endeavors. Granted, it might have been chancy public relations for the NACA, especially in the middle of the Great Depression, to take the high road and distinguish its research from pure science and heroic invention; it very well could have backfired. But in historical perspective, a more honest and fully informative approach by the NACA to the importance of its basic activity seems worth the risk. The cowled engines of American airplanes probably would not have performed any better, but the public context for government R&D may have matured a bit—and in the long run, led to a more informed public, wiser political decisions, and more logical next steps.

Chapter 2

Lew Rodert, Epistemological Liaison, and Thermal De-Icing at Ames

by Glenn E. Bugos

A paradox in aircraft icing research took the National Advisory Committee for Aeronautics (NACA) further into actual aircraft design than it had ever before ventured. To gather data on new de-icing equipment under natural icing conditions, and do so safely, NACA needed an aircraft already invulnerable to the dangers of icing. So Lewis A. Rodert, leader of NACA icing research from 1936 to 1945, built his own de-icing system on two aircraft—first a small Lockheed 12A and next a Curtiss C-46 transport that would become flying laboratories for further research. "Seldom before," wrote Edwin Hartman, NACA's representative in southern California and Rodert's liaison to aircraft manufacturers, "had NACA's research work been carried so far into the hardware stage or so far in achieving a complete and satisfying solution to a major operational problem."[1]

Yet when Rodert received his Collier Trophy in December 1947, the practicality of his innovation had hardly been established. As evidence of practicality, the press release noted only that his specially-modified C-46 flew through the weather that grounded other aircraft. Manufacturers had begun building similar de-icing systems, though few followed Rodert's suggestions. Still, despite the narrow practicality of Rodert's work, he was indeed largely responsible for getting industry off its duff. The Collier Trophy, given annually in recognition of outstanding achievement in aeronautics, testified to the peculiar and fruitful synergism of his personality with the NACA advisory committee form of research.

Rodert was a short, intense man, just forty years old when he won the award. Born in Kansas City and raised on a farm in Kansas, Rodert studied at the Kansas City Junior College before transferring and graduating with a Bachelor's degree in 1930 from the University of Minnesota.[2] He instructed in aeronautical engineering at Duluth Junior College in Minneapolis before moving briefly to Curtiss Aeroplane & Motor Company in Buffalo, New York. He joined NACA's Langley laboratory in 1936 to do de-icing work, transferred to the new Ames laboratory in California in 1941, quit briefly to join industry in 1946, then returned to NACA as chief of the flight research branch for the new Cleveland laboratory. The Flight Safety Foundation cited Rodert in 1953 for his "aircraft fire prevention research work" while at the Cleveland center, and his alma mater gave him the 1954 University of Minnesota Outstanding Achievement Medal. In 1956 Rodert joined Lockheed in Burbank, California, as a special assistant on research management, then quickly disappeared from the aviation scene. Former co-workers passed rumors of his decline into mental illness.

Rodert put everyone on edge with his show-me attitude. Rodert encountered many philosophies of de-icing, and accepted none easily. In the aeronautical research community—rife with epistemological insecurities, where unequivocal proofs were the most exasperating part of any researcher's daily life—work moved forward because peers conferred upon each other the initial benefit of the doubt. Rodert broke that unspoken rule by calling

---

1. Edwin P. Hartman, *Adventures in Research: A History of Ames Research Center, 1940–1965* (Washington, DC: NASA SP-4302, 1970), p. 77.
2. "Rodert, Lewis August," *Who's Who in World Aviation* (Washington, DC: American Aviation Publications, Inc., 1958), p. 376.

Lewis A. Rodert, then Chief of the Flight Research Branch at the NACA Lewis laboratory, was awarded the Collier Trophy for 1946 for his pioneering work in the development and practical application of a thermal ice prevention system for aircraft. (NASA Photo)

everything into question, especially the widespread belief that de-icing was a complex and intractable problem. He did so because of his wartime ethos of urgency, his farm-boy abruptness, his distrust of mathematical obfuscation, his own predilection for trial-and-error engineering, and his power over the NACA testbed aircraft. Nor did Rodert shy from making his own problematical pronouncements—he was especially quick in proclaiming the BTUs required to de-ice a plane—then working like hell to prove himself right. People had opinions about Rodert, both good and bad, and expressing these opinions caused everyone to think more precisely about their own de-icing work.

Rodert was no organization man. He was a poor manager. He did, however, expertly exploit the most fundamental structure of the NACA research organization—its system of nested advisory committees. Committee business allowed him to visit with virtually everybody—manufacturers, airlines, and military pilots—to hash out the details of thermal de-icing. Furthermore, Rodert worked oblivious to the rarefied distinctions between basic and applied research that then gripped so many NACA officials, and that today guides so much historical analysis of the NACA. Rodert judged everything simply on how well it kept ice off an aircraft in flight. This study of Rodert's work, therefore, focuses on his role as epistemological liaison—on the practical work involved in establishing certainty for himself, and amongst the many groups mobilized to defeat the icing menace.

Defining an Approach

Following a joint Army-Navy request, in 1928 NACA researchers initiated a small-scale investigation of aircraft icing, then a big mystery as well as a big cause of aircraft crashes. First the NACA surveyed air mail and airline pilots on which aircraft were most likely to ice, and collected reports on crashes attributed to icing. They built a small six-inch refrigerated wind tunnel, the first icing research tunnel in the world, and watched how ice formed on an airfoil. And they installed a free-flight icing rig under the shoulder-mounted wing of an old Fairchild F-17 cabin monoplane. There they mounted a thermometer and a small but visible wing section, on which they sprayed water as the aircraft passed through freezing air.

NACA pilot William H. McAvoy, by just watching this wing section as ice formed, confirmed some suppositions about icing. Ice did indeed form "mushroom" shapes projecting forward of the leading edge, rather than smooth sheets coating the airfoil. Pilots should expect, McAvoy continued, that ice also formed on fast-turning propellers with mushroom projections. Ice that hardened far back on the wing posed no problems because it adhered poorly and slipped off easily. McAvoy also collected anti-icing pastes from the airlines—greases and oils, and water soluble compounds like glycerin, honey,

Karo syrup, and soap—but discovered these actually induced icing by trapping ice crystals until huge hunks formed. From his window-side survey of the state of the art in aircraft de-icing, McAvoy had established a way of studying icing—flight tests to frame questions about the impact of ice on aircraft performance.

NACA theoreticians Theodore Theodorsen and William C. Clay directed the tunnel experiments as part of a broader research program on turbulent airflow. By mounting an electrically-heated, brass wing section in the tunnel, and impregnating it with thermocouples, they showed that heat transfer between an airfoil and its atmosphere varied directly with airspeed and closely followed the pressure distribution of air along the airfoil. Local transmission of heat was high along the leading edge, diminishing to zero by the thirty percent chord.[3] With this tunnel set up, Theodorsen and Clay also tried out some ideas on thermal de-icing—that is, applying heat to melt ice as it formed. McAvoy also tested thermal de-icing on NACA's free-flight apparatus. NACA shops built a small metal airfoil, of four-foot chord and two-feet span, and mounted it under the Fairchild. Once ice formed, the pilots turned on a small boiler in the engine exhaust manifold and measured how much steam was required to keep ice from forming or to melt ice once it had.

As early as 1931 NACA had established the principle of thermal de-icing as strongly as doubts about its practicability. Theodorsen and Clay concluded that steam heat might de-ice wood-composite wings but the system would be "excessively heavy," especially if designed to de-ice all the struts and support wires that then held together such wings.[4] The best system, they suggested, would use waste heat from the exhaust stream, but this would likely await development of new all-metal monoplane aircraft. "The recommendation for the guidance of those who must encounter [icing] conditions," concluded McAvoy, "appears to lie entirely along the lines of their avoidance."[5]

On the last night of 1934, an aircraft slammed into an Adirondacks mountain killing its passenger and crew of four. The weather remained cold so that a crash inspector, curious that the aircraft had not burned, found the carburetors completely choked with ice. The engines likely just suffocated and stopped, leaving the pilot no way to de-ice and restart it. Publicity prompted the Commerce Department to investigate and discover that, during 1934, twenty-six planeloads of passengers had been forced down by carburetor icing.[6] Some of the most disastrous crashes in aviation history had been attributed to icing, and airline executives widely believed that their industry would never boom until they erased this element of danger.

Pennsylvania-Central Airlines resurveyed its route system for winter flying conditions, raising some minimum ceilings and adjusting ranges. American Airlines improved their runways for winter operations, Northwest added staff for better flight and weather planning, TWA prohibited its pilots from landing when icing conditions prevailed below 1,000 feet, and United Airlines started paying their pilots a base salary in addition to flight pay so they would have no disincentive to cancel flights in bad weather.[7] This winter, wrote an airline executive in December 1937, "is the best opportunity the industry has ever had to

---

3. Imagine a chord line running straight backwards from the leading to the trailing edge of a wing, with a total distance expressed as 100 percent to account for taper along a wingspan. A thirty percent chord measurement is a point 3/10ths of this distance backwards from the wing's leading edge. The higher the chord number, the farther backwards it is.

4. Theodore Theodorsen and William C. Clay, "Ice Prevention on Aircraft by Means of Engine Exhaust Heat and a Technical Study of Heat Transmission From a Clark Y Airfoil," *NACA Technical Report No. 403* (1931): 3.

5. Thomas Carroll and William H. McAvoy, "The Formation of Ice Upon Exposed Parts of an Airplane in Flight," *NACA Technical Note No. 293* (July 1928): 10.

6. George W. Gray, *Frontiers of Flight: The Story of NACA Research* (New York: Alfred A. Knopf, 1948), p. 320.

7. J.A. Browne, Meteorologist-In-Charge, "Ice Accretion Within the Convective Layer," TWA Meteorological Department, Technical Note No. 4 (June 1940), in Stanford Libraries.

demonstrate to the public that air transportation is more than reasonably safe." Their strategy: "cooperate with the weather in a big way."[8]

But airline operators and manufacturers ultimately wanted to defeat the icing menace, not cooperate with it. Aircraft already rivaled the steamship and train for speed and economy; but it lacked regularity. Radio navigation aids had brought aircraft to the brink of being all-weather conveyances, until the temperature dropped. Lacking a technological fix to icing these airlines cancelled or delayed flights—an estimated one-tenth of all flights—at first sight of icing clouds. Icing became a consuming challenge, to both airline economics and engineer pride. Lewis A. Rodert joined the NACA Langley Memorial Aeronautical Laboratory (LMAL) in September 1936, and teamed with Alun R. Jones to re-invigorate NACA's icing research with youth, stubbornness, and a fresh perspective on icing problems.

Ice caused aircraft to crash by adding weight and preventing the pilot from climbing above the icing clouds, so that the aircraft gradually lost altitude and slammed into the ground. That was how most people understood the danger of icing. Rodert and Jones started their studies by showing that icing seldom enveloped the aircraft with weight, but rather icing incapacitated small but crucial parts. As McAvoy had proved with his photos of mushroom-shaped ice projections, and as Rodert and Jones confirmed, ice accreted along the wing and tail leading edges disturbing lift and adding drag. Ice clogged the interstices of rudders and ailerons, preventing control and inducing buffeting. It changed the aerodynamic profile of the propeller, causing it to vibrate and exert less thrust per horsepower. It coated windshields, so the pilot flew blind. Ice made antenna wires oscillate and snap, and generated static that rendered useless most radio communication and navigation. It distorted pitot shapes, so that pilots got erroneous airspeed readings. And it clogged carburetors, suffocating the engine. Frequently, the pilot lost each of these systems—engine, wings, control surfaces, indicators, radio, sight—within minutes. With their lives at stake, pilots of ice-hindered aircraft had little time for the careful observations NACA researchers promised to make.

Using a DC-3 Mainliner loaned by United Airlines, in September 1937 Rodert and Jones glued sponge rubber to the leading edge of the wing, simulating ice formations, and showed how a small layer of ice had a big impact on lift, drag, and stalling.[9] NACA headquarters authorized construction of a larger icing tunnel at Langley. LMAL technicians insulated the tunnel with a crude layer of kapok pulled from surplus Navy life preservers, and added an open tank of ethylene glycol cooled by dry ice as refrigeration. This tunnel worked well enough for Rodert to further chart the impact of ice on aerodynamic efficiency, and to prove that a full size wing section could be de-iced with exhaust heat.[10] But Rodert lost patience with tunnel research as he learned that tunnel ice bore little relation to the natural ice he hoped to defeat.

The B.F. Goodrich Rubber Company ran a small icing tunnel in Akron, where they verified the pneumatic de-icer they had introduced in 1930. The pneumatic de-icer was a strip of rubberized cloth holding inflatable rubber tubes that attached to the leading edge of a wing or tail. When the pilot unexpectedly encountered icing, he shot compressed air into the strip, cracking the ice so that the wind stream swept it off. It worked well enough to become standard equipment on large transports by the late 1930s, but never well

---

8. T. Park Hay, "Operators Project Safety Program for Winter Operations," *Aero Digest* 31 (December 1937): 24-25.

9. Lewis A. Rodert and Alun R. Jones, "Profile Drag Investigations of an Airplane Wing Equipped with Rubber Inflatable De-Icer," *NACA Advanced Confidential Report* (December 1939).

10. After some perfunctory studies in June 1938, NACA easily converted the tunnel into a pressurized two-dimensional low-turbulence tunnel for studies of the shift from laminar to turbulent flow along an airfoil–the use for which NACA most likely intended it. James R. Hansen, *Engineer in Charge: A History of the Langley Aeronautical Laboratory, 1917-1958* (Washington, DC: NASA SP-4305, 1986), pp. 110-11.

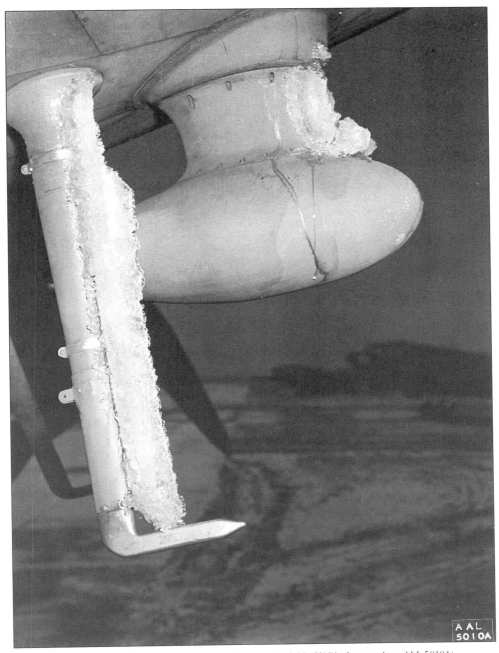

Ice jutting forward on the radio antenna and airspeed pitot mast of a C-46. (NASA photo no. Ames AAL-5010A).

enough that aircraft could deliberately fly into icing conditions. Rodert and Jones held the tenet that nothing restrict where aircraft could fly.

Rodert and Jones also claimed the rubber boots were in no way fail-safe. Pilots already knew they were not very clean—they ballooned with changes in air pressure or returned wrinkled on the smooth airfoil contour after inflating. In carefully controlled test flights Rodert discovered pneumatic de-icers worked in really very limited conditions. They seldom cracked ice cleanly, and the jagged edges more quickly accumulated lumps of ice. Furthermore, the pneumatic de-icer attached to the wing at ten percent chord, with strips that protruded into the airstream that further accumulated ice at the place most likely to disrupt lift. If a de-icer failed—and a bullet hole through one shoe would destroy pressure in the whole system—profile drag could increase 458 percent over an unprotected wing, putting the aircraft in greater peril.[11] B.F Goodrich failed to see danger in this, contended Rodert, because the ice created in their tunnel bore little relation to natural ice. Goodrich sprayed water in big drops, which created a smooth coating of glaze ice. Natural icing was more likely to be opaque, crystalline rime ice, created when very small supercooled droplets ran into a crystallizing structure like a wing. Any tunnel that verified the utility of the pneumatic de-icer caused Rodert to doubt the entire enterprise.

So Rodert and Jones kept their research in free flight as often as possible, and worked on thermal de-icing to replace the pneumatic boot. They built a more elaborate icing installation between the double wings of a Martin XBM-1 dive bomber loaned to NACA by the Navy. But rather than using a heavy steam boiler, Rodert and Jones diverted hot exhaust directly from the engine into the model section. NACA Engineer-Test Pilots William H. McAvoy and Lawrence A. Clousing flew the XBM-1 into cold air, turned on the water spray, and a camera recorded how quickly the ice melted away. By early 1938 Rodert and Jones were convinced thermal de-icing held great promise. Confirming their optimism were reports, leaked through Naval Intelligence from London, that the Germans had added heat de-icing systems to two production aircraft, the Junkers Ju.88 and Dornier 217E.[12] The Germans had first studied thermal de-icing in late 1920s, as had NACA, but had accelerated their research under the Nazi regime. With war on the horizon, and airlines still agitating about the icing menace, Rodert and Jones thought it high time to prototype a complete thermal de-icing system and test it in real clouds.

The Lockheed 12A

NACA headquarters, anticipating funding for icing studies, allowed the Langley Flight Research Branch to buy a twin-engine, all-metal Lockheed 12A light transport. Rodert got dibs on converting it into what NACA researchers traditionally built so well—a sophisticated and dedicated testing facility, but in the form of a flying laboratory. The 12A would easily accept a "hot wing:" the wing outer panels held no fuel tanks, detached easily at the nacelles, and the engine exhaust stacks were close to the wing leading edge. Most important, the 12A was built by a company interested in staying on the forefront of icing

---

11. "Ice Off The Wings," *Business Week* (March 16, 1940): 21; Rodert to chief of the LMAL aerodynamics division, June 24, 1940; File AF1-15a; Box 66; Central Files, 1939-1957; Records of NACA Ames Aeronautical Laboratory, Record Group 255; National Archives–Pacific Sierra Region, San Bruno, CA. My thanks to Kathleen O'Connor, NARA-San Bruno, for her help in making these records available. [Hereafter, citations to Ames records are abbreviated, so the above citation would follow this formula: RG255/Central/66/AF1-15a).

12. Royal Aircraft Establishment, "Report No. E.A. 14/10 Enemy Aircraft: Junkers Ju.88, entitled Description of Main Plane De-Icing System," December 1940 (RG255/Central/101/AF19-10). Other, though vague, reports had already appeared in aviation periodicals.

research. Lockheed vice president and chief engineer Hall L. Hibbard assigned the 12A modifications high priority.[13]

Rodert and Jones started with Lockheed blueprints to sketch a hot wing. They added a butterfly valve in the engine exhaust stack to divert hot gas (at 1500°F) into a four inch diameter tube, running close to the leading edge but insulated from the wing structure, and exhausting out the end of the wing tube. To cool the tube and improve heat transfer, an intake scoop sent fresh air around the tube, then through holes in the spar web into the wing structure, and exhausting out louvers at the aileron hinges. They repeatedly calculated wing strength, since heat weakened metal structures, especially one modified with new tubes and holes. By August 1939 the designs were ready, and NACA went looking for a sponsor.

The Navy BuAer (Bureau of Aeronautics) was so enthusiastic about the idea that they asked Rodert to make the modifications on a Navy production aircraft. Navy PBY patrol boats anchored off the Aleutian Islands had special icing problems. Waterplanes were not easily covered with protective tarps, so thick ice formed on them overnight. Splash during taxiing added more sheet ice. The Navy needed a de-icer with enough punch to knock this thick glaze ice completely off the wing, and Rodert's design promised to do so. But Rodert had the 12A blueprints ready to go; switching aircraft would deter him from test flights the coming winter. Further, NACA had no facilities for modifying seaplanes. So BuAer sent a draftsman from its San Diego depot to Langley in September 1939, and NACA engineers helped him modify their 12A blueprints to fit a Consolidated PBY-2 Catalina patrol boat. BuAer hired Rodert and Jones to draft specifications for the PBY-2, especially the heat transfer calculations that helped Consolidated define the thermal performance of the system. When the PBY-2 was ready for testing the following summer, BuAer offered Rodert a job. But Rodert stuck with the NACA, and tied his lot with its patrons in the U.S. Army Air Corps (USAAC).

The USAAC signed a job order for the 12A wings in November 1939. Major C.M. Cummings of the Equipment Branch at Wright Field had helped Rodert at several crucial stages, and supported his project without change. The United States, in any type of war, was vulnerable to two avenues of attack—by air over Alaska or Newfoundland—both with severe ice storms. Germany, Rodert later wrote, "has aircraft which can fly in almost any kind of weather, irrespective of icing conditions. There cannot be a possible defense against such aircraft without similar or superior equipment."[14] American aircraft must be able to fly through any clouds; indeed pilots will likely seek protective cover in them. For $25,000, the AAC bought new wings and a modified windshield from Lockheed, and loaned them to NACA for research. While Lockheed fabricated the wings, Rodert and his Flight Research Branch prepared for a move westward.

NACA had already begun construction on the new Ames Aeronautical Laboratory adjacent to the Navy's Moffett Field on the flat bay lands near Sunnyvale, California. Compared to the humid air over the Virginia Tidewater, the cold Sierra Mountain air mixing with the warm, moist air rising off the San Francisco Bay made excellent icing conditions. Furthermore, Rodert had freed his research from wind tunnels, and by July 1940 the well-equipped shops and hangars at Ames were ready for his group. Test pilots McAvoy and Clousing ferried out an old North American O-47 they would use until the 12A was ready. Alun Jones rejoined them in January 1942, along with Carr Neel, an engineer who became increasingly involved in the work. Since the icing research was the first project at the new Ames laboratory, engineer-in-chief Smith DeFrance lent constant aid to his Ames Flight Research Branch.

---

13. Clarence L. Johnson, Lockheed Aircraft Corporation, "Wing Loading, Icing and Associated Aspects of Modern Transport Design," *Journal of the Aeronautical Sciences* 8 (December 1940): 43–54.

14. Rodert to LMAL Engineer-in-Charge, "Memo: Progress of ice research on Lockheed 12A Airplane," May 27, 1940 (RG255/Central/66/AF1-15a).

McAvoy picked up the 12A with hot wings from Lockheed's Burbank plant on January 22, 1941. Back at Ames they flew it enough to be sure the heat did nothing to weaken the wing. Then they went hunting for ice.

During March and April 1941, McAvoy, Clousing, and Rodert took up the 12A almost everyday, scanning the horizon for ever more severe icing. United Airlines had compiled atmospheric data to help its pilots avoid icing on their routes; the Ames group used this data to seek out the ice.[15] The Weather Bureau office at the Oakland Airport confirmed that they would find the best icing flying westward from Sacramento to Donner summit in the Sierras. George W. Lewis, director of aeronautical research at NACA Washington headquarters, had recommended that Rodert attach a two-foot long, unheated strut above the right wing. Thus, in one photo they could contrast the clean hot wing with the icing on the unprotected "tell-tale" strut. Lewis was delighted a few months later when he received his copy of the first report out of Ames—Rodert, McAvoy and Clousing's "Preliminary Report on Flight Tests"—"So I am going to celebrate by taking a copy over to Dr. Ames."[16]

While the icing over California was regular, that spring it was hardly severe. To secure ever more dramatic photographs, the group ventured the 12A further north and east. On March 20, 1941, while flying through cumulous clouds over Superior, Michigan, at 9,000 to 11,000 feet, with air temperature at twenty-six to thirty degrees, they got pictures of three inches of ice on the strut while the wing below, on only half heat, was clean. Icing on the few unprotected parts turned so severe on a flight between Minneapolis and Fargo that the 12A slowed thirty-five mph from just the added drag. While flying northward along the Pacific coast, Rodert reported: "The airplane was struck by an electrical charge which melted the trailing edge of one propeller blade and the edges of the airplane struc-

The Lockheed 12A ice research airplane at Ames. (NASA photo no. Ames ALL-1166).

---

15. R.L. McBrien, "Icing Problems Attendant to the Operation of Transport Aircraft," *Aviation* 4 (May 1941): 138.
16. Lewis to AAL, April 5, 1941; in File 50-14D Deicing Problems; Box 247; General Correspondence (Numeric File); Records of NACA, Record Group 255; National Archives, College Park, MD. [hereafter abbreviated WDC: RG255: collection name: file name]

ture at several points."[17] The lightning strike grounded the 12A during a week of excellent icing conditions, but proved they were indeed flying into severe conditions. (McAvoy would win the 1943 Octave Chanute Award of the Institute of the Aeronautical Sciences, and Clousing the 1947 Award, for their test flying in severe icing conditions.)[18]

The drama, the photographs, the urgency all helped Rodert protect and expand his program from a number of competitors. B.F. Goodrich was working hard to improve their pneumatic de-icers. At the 6,288 foot summit of Mt. Washington in New Hampshire, Goodrich mounted a test wing like a weather vane so it stayed in constant wind. There a design team tried out new de-icers with hundreds of smaller, self-sealing inflatable tubes, snap-action distributor valves, flexible camouflage sprays, water-repellent rubbers, and non-adhesive sprays like the silicone *Icex*.[19] As a result, pneumatic de-icers remained in wide use—and the long wing span of the Douglas C-54 transport was the widest ever—during and well after World War II.

Others preferred new chemical de-icers. Chemicals worked in two ways. Alcohol-based fluids lowered freezing temperatures. Other slick, oil-based fluids, exuded from wing leading edges or sprayed on before take-off, prevented ice crystals from adhering to the wing surface. The British especially advocated chemical de-icing. They claimed Americans like Rodert were misled about the war dangers of icing by inaccurate reporting of early Royal Air Force raids over Germany. As far as the RAF was concerned, chemicals sprayed easily onto any aircraft, lasted for a complete mission, and kept off North Atlantic ice. The Royal Aircraft Establishment at Farnsborough was perfecting a Dunlop strip which leaked a steady stream of chemical along the wing during longer flights. Chemists at the Naval Research Laboratory, looking for quick relevance on U.S. entry into the war, concocted similar anti-icing pastes and fluids. Since Rodert had the only aircraft known to withstand icing, they regularly asked him to try out new fluid recipes. It was highly likely icing conditions over the North Atlantic differed from those over North America, Rodert concluded, but all fluids tested poorly. Perhaps the British realized this too, because they increasingly cancelled icing-bound flights out of distrust of their equipment. For the first three years of the war, in a period of otherwise exceptional technical cooperation, British and American icing researchers kept their distance. Farnsborough transferred the two-engine Bristol bomber they used for icing research to Ottawa in April 1941, and for most of the war the Allies communicated only through the National Research Council of Canada.

The Ames group reported some important discoveries in the spring of 1941 that confirmed the value of thermal de-icing.[20] Most important, the heat required in free flight was much less than indicated in wind tunnel tests. A seventy-degree rise over the ambient dry-air temperature at 200 mph was enough to weaken the bond between the ice crystals and the wing (though a 100° F rise had a safer margin). Furthermore, heat concentrated on the leading edge—less than ten percent chord—was enough to protect the trailing parts of the wing. Thus, exhaust heat never weakened the wing structure.

NACA also reported how much heat would damage the structure. Lockheed had designed another "cellular" wing, which passed exhaust gas through large chambers directly on the leading edge with no additional cooling air. Lockheed volunteered to rig the wing with 107 thermocouples, far more than specified, to get information on how evenly it transmitted heat. When flying the cellular wing near Ames in July 1941, McAvoy had applied only partial heat when expansion at the leading edge caused buckling aft of

---

17. Gray, *Frontiers of Flight*, p. 312.
18. "File: McAvoy, Wm. A." (WDC: RG255: Biography File: Box 25).
19. "De-Icing Test Rig," *Aero Digest* 51 (November 1, 1945): 88+; Dwight L. Loughborough, Howard E. Greene, and Paul A. Roush, "A Study of Wing De-Icer Performance on Mount Washington," *Aeronautical Engineering Review* 7 (September 1948): 41-50.
20. Lewis A. Rodert and Lawrence A. Clousing, "A Flight Investigation of the Thermal Properties of an Exhaust Heated Wing De-Icing System on a Lockheed 12-A Airplane," *NACA Advanced Restricted Report* (June 1941).

the rear shear beam, threatening destruction of the aircraft.[21] Ames quickly replaced it with the exhaust tube wing, having just learned the upper limits of wing heating. This information was directly useful to the firms that designed and built aircraft—whom NACA referred to by the venereal term "the manufacturers"—and they requested a great many copies of Rodert and Clousing's flight test reports.

To fly into ice clouds and survive, the Ames group necessarily became expert on the impact of ice on the total aircraft. "I am surprised to find," noted Engineer-in-Chief Smith DeFrance, "that there are so many details which have not been anticipated before the de-icing tests were started."[22] Frosting prevented photographs out cabin windows; Clousing and McAvoy found they needed better instruction on flying blind; electrically-heated pitots looked clean even when ice in the throat skewed pressure readings; exhaust gas corroded the aluminum alloy at the wing tip; and the radio broke regularly. Rodert persuaded United Airlines to install in the 12A a radio they had specially adapted for ice flying. He asked the Massachusetts Institute of Technology (MIT) to design electric-resistance heating for the twenty-five foot long antenna wire that stretched between the cabin and the tail. And he asked the Naval Research Laboratory and the Air Corps labs at Wright Field to design loop antennas that would not collect static as they encountered precipitation. Any sharp corner or gadget protruding into the airstream, Rodert constantly reminded manufacturers, was an invitation both to icing and static electricity.

In less than a year of flight testing on the 12A, and early experience with the Navy PBY-2, thermal de-icing looked promising. Manufacturers kept pressing Rodert for more details on the 12A installation, which Rodert preferred to deliver in person rather than through reports. Rodert knew manufacturers could improve upon his 12A design—especially in reducing weight by better integrating the tube into the wing structure—and thought being vague about details might prompt them to innovate. Rodert instead claimed expertise in flight testing. The NACA Special Subcommittee on De-icing Problems, which served as Rodert's peer review group, and from which he often sought advice on how best to report data, encouraged this division of labor.

The Subcommittee did not actually convene until April 1941. Rodert was not initially a member, though its charge was to "help in keeping the research organization in touch with the practical problems that require attack by research."[23] Early committee meetings would have likely exasperated Rodert: just a bunch of guys sitting around talking about icing. They freely dispensed fragments of experience, ill-formed ideas, and random observations, and passed resolutions on which isolated aircraft parts most needed Rodert's attention.[24] They collected and amended dozens of letters:

> My dear Doctor: The industry is yelling to beat the band for a windshield that they can see through in rain and ice. Is there anyway you can expedite your activity on your improved windshield?[25]

But the committee gave a free hand to NACA's research bureaucracy, and it gave a free hand to Rodert, to integrate and prioritize these requests.

---

21. "Memo: Progress on Wing And Propeller De-Icing," Comdr. D.W. Tomlinson, U.S.N.R., Chairman, NACA Technical Subcommittee on De-Icing Problems, March 28, 1942 (RG255/Central/104/AF19-10a).
22. DeFrance to Lewis, February 19, 1942 (RG255/Central/66/AF1-15a).
23. Edward Warner, NACA, to Frank R. Collbohm, Douglas Aircraft Corporation, March 5, 1941 (WDC: RG255: General Correspondence [Numeric File]: Box 247: File 50-14D).
24. "Minutes of Meeting of Special Subcommittee on Deicing Problems, Committee on Aerodynamics," April 15, 1941 (WDC: RG255: General Correspondence [Numeric File]: Box 247: File 50-14B "Deicing Problems, Minutes").
25. Edgar S. Garrell, president of Air Transport Association of America, to G.W. Lewis, June 18, 1940 (WDC: RG255: General Correspondence [Numeric File]: Box 247: File 50-13D "Deicing Conference").

But Rodert found allies among the committee chairman. He had met J.W. Tomlinson in 1939 when Tomlinson was on the NACA aerodynamics committee and vice president of engineering for Transcontinental & Western Airlines of Kansas City. Tomlinson had seen the Ju.88 on a trip to Germany and, even though he had a predisposition toward the rubber de-icers used on his fleet, he understood what Rodert was working toward. And Tomlinson kept writing Rodert letters of introduction and beating the bushes for icing tests. Tomlinson was called to active status with the Air Primary Training Command in April 1942. His last act as chairman was to meet with Disney Studios to have them make an educational film to "effectively register" the icing issue in the minds of young servicemen.

Karl O. Larson became subcommittee chairman in 1942 and shifted its aegis from the NACA Committee on Aerodynamics to the Committee on Operational Problems. Yet Larson supported Rodert's desire to just make and verify ice-invulnerable aircraft, and not approach icing as an operational problem. Larson was chief engineer for Northwest Airlines which, like all airlines during the war, had subordinated passenger travel to military transport. Northwest's biggest military contract came from the Air Corps Ferry Command to run the "Alaskan airway" between Minneapolis and Fairbanks. Flight experience taught Larson that the route was a natural and reliable icing laboratory. He assembled at the Minneapolis municipal airfield, near Northwest's headquarters, the equipment and technicians needed to keep aircraft flying through ice clouds.

Rodert, Clousing, and McAvoy had already talked of setting up flight test operations in the north, central states. They wanted a new base with reliable blasts of arctic air, light traffic, and no mountains for when they flew blind, and freezing air at ground level so they could photograph ice on the aircraft underside after it landed. Both Clousing and Rodert knew Minnesota—Rodert from his years at the University of Minnesota—and knew Minneapolis offered all that.

Larson convened an NACA Committee for the Winter Flight Laboratory in June 1942, which proposed that the Air Corps give Northwest a $55,000 contract to provide NACA with an office and access to Northwest facilities and personnel.[26] Northwest managed operations and maintenance, while NACA directed a cooperative research project. The Ice Research Project opened in November 1942, and that winter hosted more than ten visitors per week in addition to the seven pilots and seventy-five mechanics on duty. Airlines and manufacturers were invited to send engineers with new equipment to test. The Weather Bureau sent a meteorologist to collect data and develop hypotheses on which atmospheric conditions caused icing. The Air Corps remained hands-off, to avoid duplicating operations at its existing Cold Weather Test Station at Ladd Field in Fairbanks, and sent only pilots from Wright Field, Ohio. Their task, however, was crucial: to fly thirteen aircraft with new de-icing equipment, including the first aircraft de-iced by heated air.

Heated Air De-Icing

Rodert had formed some negative opinions of heated air—that is, chemically normal air as opposed to burnt exhaust gas with its attendant carbon gases and water and gas vapors. While trying to complete thermal de-icing of the 12A in late 1939, without resorting to convoluted ducting, Rodert had canvassed industry for a heater to put remotely in the tail. Stewart-Warner sold a gasoline-burning heater, for automobiles, that put out 8,500 BTUs per hour. Rodert asked if they might upgrade it to put out 75,000 BTUs, with less weight and very cold air intake. Stewart-Warner proposed linking ten burners together,

---

26. Karl O. Larson, "Proposal for the Establishment of the Winter Flight Laboratory," June 1942 (RG255/Central/113/AM16-15).

but could not get it to Rodert in time. So that winter he put a pneumatic de-icer on the 12A tail. He tried again the following summer, starting with a gasoline heater Curtiss-Wright used for cabin heating. It too proved weak, so Rodert built a long exhaust duct to prove the concept of thermal de-icing in the tail.

Rodert had better luck using heated air to de-ice the 12A windshield. The Pittsburgh Plate Glass (PPG) Co. helped Ames find a laminated safety glass that conducted heat well, and mount double-panes with a ¼ inch gap through which heated air flowed. Putting exhaust heat into the windshield was unsafe—seepage would dump toxic gas into the pilot's face, and Rodert wanted to hinge the inside pane so the pilot could move it out of his line of sight in warm weather. Rodert found that air diverted from the cabin heat exchanger was warm enough to keep the windshield free of ice, yet cooler than the critical temperature of the plastic binders. As early as November 1941 Rodert flatly contradicted Boeing's public thinking that much higher heat was required, and pronounced that, at an airspeed of 150 mph, only 1,000 BTUs per square foot per hour was needed to keep any windshield at 50° F, and thus free from ice. United Air Lines liked the PPG windshield well enough to retrofit it onto all its DC-3s.[27]

Manufacturers were simply afraid of exhaust gas. A bullet hole or weakened seam could poison the cabin (though Rodert designed airflow to exhaust out the wing). A failed engine would send raw, explosive gas vapors into the wing tube or gasoline leaking from a wing tank might ignite against the hot tube (though Rodert claimed the wing got no hotter than if left parked in a tropical sun.) Exhaust gas corroded aluminum and manufacturers refused to take the weight penalty of using stainless steel, as Rodert had done on the 12A. In addition to the dangers of exhaust gas, de-icing the entire aircraft with heated air held some advantages. Manufacturers could couple heated air ducting more neatly with the skin, saving the weight and strength penalties of the exhaust tube. Heated air could be vented out small holes on the wing surface with minimal drag. And a steady source of heated air could provide the cabin comfort all aircraft then lacked. The problem, however, was finding a steady source of heated air.

Rodert turned his full attention to heated air in September 1941, after learning the Glenn L. Martin Company would use a cabin heater to de-ice the wings of a B-26. Since manufacturers accepted only heated air de-icing, Rodert planned to stay one step ahead of them. He toured plants in January 1942 and, after telling manufacturers de-icing required less heat than previously thought, now he had to tell them their heat exchangers were too weak. To prove this point, in April 1942 Ames again modified the 12A wings—putting corrugated ducting on the right wing and sheet ducting with baffles on the left—to concentrate heated air on a narrower chord of the leading edge. Ames craftsmen built a cast aluminum heat exchanger that transferred heat from the exhaust stream into fresh air flowing to the wings. They also built a variety of heat warning and dump valve controls.

To take advantage of this expertise and to "relieve industry of the design and development work," the Army Air Forces (AAF) asked Ames to build a complete heated air de-icing system to retrofit into the Consolidated B-24D Liberator.[28] The B-24D was a high-wing, four-engine heavy bomber which would have a long production run. The system would include hot wings and tail, an electrically-heated antenna, an alcohol-based windshield wiper, an anti-static system for the wings and antenna, and a carbon monoxide indicator for the

---

27. R.L. McBrien, "An Aircraft Double Windshield–Its Development and Use," *SAE Journal* 51 (October 1943): 350-55.

28. Alun R. Jones and Lewis A. Rodert, "Development of Thermal Ice-Prevention Equipment for the B-24D Airplane," Confidential Memorandum Report for the Material Center, USAAF, September 11, 1942 (RG255/Central/104/AF19-10K) 2.

cabin.²⁹ Heated air would exhaust through half-inch holes along the top wing surface, and then travel backwards with the boundary layer. This satisfied AAF specifications that the wings got a 70° F temperature rise over the forward 20 percent of chord and a 20° F rise back to 75 percent chord. Engineers for the AAF Materiel Command approved Ames' blueprints, and in May 1942 Ames acquired B-24D No. 111678 (soon redesignated the XB-24F-CO). The Ames erection shop procured all materials, metals and fasteners, built the wing tubing, and installed it into the aircraft. Consolidated sent senior engineer Howard F. Schmidt and several draftsmen to Ames, who completed production drawings as the work progressed. As early as June 1942, the B-24D did well in test flights around the Bay area. Rodert declared he had standardized a work outline for retrofitting de-icing into existing aircraft, and was willing to take on more. Then problems arose with the heat exchangers.

Ames had bought exchangers from two exhaust systems specialists—AiResearch Manufacturing Company of Los Angeles and Solar Aircraft Company of San Diego. They were stock designs, scaled up for greater output than ever achieved in an aircraft. When they failed, Ames commissioned other firms to submit prototypes—AiResearch offered a different hollow-finned exchanger, Hanlon & Wilson Company sent a pin-type exchanger, and Stewart-Warner Corporation offered a multiple-fin type exchanger that delivered the required BTUs but buckled under the blast and heat of the exhaust stream. Once word got out of Rodert's quest for an exchanger for a mass-produced bomber, Ames was swamped with prototypes. The Ames erection shop designed a few themselves, applying their new expertise in brazing compounds, metal conduction, and pressure drops.

Rodert's entire plan hinged on getting a workable heat exchanger, and he was confident he could find one. The German Ju.88, after all, had used heat exchangers—a series of four along a single exhaust stream—and Rodert heard reports that the Germans had also put similar exchangers on the Ju.52, Ju.188, Ju.388, and the four-engine Ju.290 search bomber. Rodert considered the Ju.88 "a splendid de-icing system" and got Wright Field to send him sections of the Ju.88 exchanger, now on the scrap heap, so he could look for some secret the drawings didn't convey.³⁰ Rodert also wrote to Martin, asking for exchangers Ames could not duplicate from blueprints. It was common, Rodert discovered, for an exchanger's actual and predicted performance to differ as great as four times. Ames made a flying test bed out of its C-47 and, in their desperate search for a workable exchanger, Ames pilots carried aloft thirty-two different designs during the summer of 1942. Once trial and error indicated which exchangers promised results, Jones or Neel drove a batch across the Bay to the Berkeley laboratory of L.M.K. Boelter, where Ames bought analytical insight.

Boelter, a professor of mechanical engineering and associate dean of engineering at the University of California, was the sort of teacher who kept perpetual office hours. As a student, Jones had worked with Boelter on an earlier NACA contract seeking advice on placing thermocouples to study heat transfer along the wing surface. Boelter read widely—even translating articles on heat exchanger theory from Italian and German—and was fascinated with the process of perfecting equations to predict real-world performance of heat transfer systems. Boelter also understood the challenge of measuring tiny drops of airborne water from his tests of evaporative cooling towers. So Jones learned much from his free-ranging conversations with Boelter, though their mission at hand was perfecting airborne heat exchangers.

Ames asked Boelter to expand his group that summer of 1942 to run bench tests on all promising heat exchangers. Boelter's goal was to measure static pressure drops and

---

29. "Liberator's New Thermal Anti-Icer," *Aero Digest* 43 (November 1943): 26–27. In a telephone interview (December 7, 1995) Alun R. Jones claims that Ames never actually exhausted heated air along the Liberator wing.

30. Rodert to DeFrance, November 26, 1941 (WDC: RG255: General Records Relating to Ames Research, 1938–1952: Box 48: File-61 "Icing Research 1939-41"); and DeFrance to NACA, "Memo: De-Icing Installation of JU-88 Airplane at Wright Field," November 29, 1941 (RG255/Central/104/AF19-10).

A cutaway display model of a wing leading edge, with a corrugated inner skin to direct heated air. (NASA photo no. Ames A-10679).

rates of heat transfer, devise a theory of exchanger performance, perfect an equation of design parameters so that predicted values approached measured performance, and ultimately offer to a single number for ranking exchanger performance.[31] Rodert and his AAF patrons knew any number would be riddled with error, but hoped that Boelter could simply standardize the errors—in thermocouple placement, pressure drop and conductivity measures, and BTU output—so that it would still help in comparative rankings. Boelter's work on aircraft heat exchangers was widely praised, as the sort of analytical work Ames should have done on all facets of its icing research.[32]

---

31. This was Boelter's L/D ratio of unit thermal conductance, where L was the length of the heat transfer surface and D was the hydraulic diameter of the ventilating and exhaust pipes. See L.M.K. Boelter, R.C. Martinelli, F.E. Romie, and E.H. Morrin, "An Investigation of Aircraft Heaters: XVIII–A Design Manual for Exhaust Gas and Air Heat Exchangers," *NACA Advanced Restricted Report* WR W-95 (August 1945); and file "University of California," (RG255/Central/107/AF22-20).

32. Boelter spent his entire career with the University of California. He ran a test station for the California Division of Motor Vehicles to verify designs of headlights, built a heat-power laboratory to improve the efficiency of internal combustion engines, and worked on ways to diffuse heat so citrus orchards would not frost. In 1944 he moved to the Los Angeles campus to establish a school with a "unified engineering curriculum" to train young men working in the aircraft industries. He wrote widely on engineering education, in which he encouraged laboratory precision, teaching students about the scope of an engineering problem, and integrating mathematical analysis directly into their work. On Boelter and his contemporaneous work, see Llewellyn Michael Kraus Boelter, *Heat Transfer Notes* (Berkeley, CA: University of California, 1946); *Reprints* (Bindery date 1944) [308xB669co, Bancroft Library, University of California, Berkeley]; and Harold A. Johnson, ed., *Heat Transfer, Thermodynamics, and Education: Boelter Anniversary Volume* (New York: McGraw Hill, 1964), pp. vii–viii.

The British began following progress in heat exchangers, and softening their allegiance to chemicals, under the guise of better flame suppression. British bombers lit up at night because flames shot from their exhaust stacks as the hot exhaust ignited the fresh air. By moving heat into the wing tube, and thus cooling the exhaust stream below 1300° F, a heat exchanger prevented this re-ignition and torching. The U.S. Navy confirmed the prospects of flame dampening, by noting that its PBYs could fly only 200 feet over an aircraft carrier, at part throttle, without being detected. The Royal Aircraft Establishment (RAE) representative to the Ice Research Project, J.K. Hardy, began following Rodert's work, offering a good dose of skepticism that helped NACA refine its reporting.

General Electric's supercharger engineering department, which built turbochargers powered by the exhaust stream, invited themselves to standardize exhaust instrumentation. Their concern—shared by the Army Air Forces—was that putting a heat exchanger in the path of an exhaust stream pulsating at seventeen cycles per second might back up the flow of gases through the engine and impede engine performance. So as the summer dragged on and the BTU output of the exchangers steadily improved, Rodert turned his attention to ram pressures at the air intake scoop and pressure drops on the wing side of the exchanger.

By September 1942, the Ames group had approved five exchangers rated around 300,000 BTUs per hour that did not greatly diminish the range and speed of the B-24D. They weighed only thirty pounds, occupied a cylindrical space eight inches in diameter and twenty-two inches long. The complete de-icing system weighed an acceptable 300 pounds, less than 1.5 percent of the total gross weight of the aircraft. Pneumatic de-icers protecting only the wing and tail leading edges, weighed in at 230 pounds. As soon as the B-24F was out the door and on its way to Minneapolis—following a brief inspection stop at the Consolidated Plant—the Army Air Forces delivered to Ames a Boeing B-17F Flying Fortress.

Ames drew from their work on the B-24F to quickly retrofit de-icing equipment onto the larger B-17F. They started with the same heat exchangers, then modified those that buckled under the greater heat blast. Unsure of which exchangers would least impact range and speed, the B-17F carried an older exchanger designed for cabin warming in one nacelle and a proposed production exchanger—bought from McQuay, Inc., the Trane Company, and AAF engineers at Wright Field—in each of its other three nacelles. The Ames group tested pressure distribution around the exchangers well into the fall of 1943. They installed additional thermocouples, and tried out some valves to adjust heat flows from the four engines around the wings. By January 1943, Ames and visiting Boeing draftsmen had prepared corrected B-17F production drawings, and the aircraft was ready for icing tests in Minneapolis.

That same month, Ames outlined "preliminary design considerations" for the most complete de-icing system yet, for a Curtiss-Wright C-46 Commando transport. The Army Air Forces, impressed with the plans, delivered to Ames C-46 No. 41-12293 in March 1943, once the Ames group returned from Minneapolis. As Rodert and Jones struggled to write up the B-17F and B-24D test results that manufacturers clamored for, they turned their attention to the C-46.[33] The C-46 was then America's largest transport, much bigger than the B-17 and B-24, with a stressed wing that required more careful revisions and a long series of mock-ups. Ames built and tested two wing inner skins—with baffles on the right wing and corrugation on the left. Because the C-46 was to be an all-weather aircraft, Ames had to protect the propellers, windshields, antennas, carburetors and

---

33. Lewis A. Rodert and Alun R. Jones, "Development of Thermal Ice Prevention Equipment for the B-17F Airplane," Advanced Restricted Report 3124, WR A-51 (August 1943); Lewis A. Rodert and Alun R. Jones, "Development of Thermal Ice Prevention Equipment for the B-24D Airplane," Advanced Confidential Report, WR A-35 (February 1943).

other parts vulnerable to icing. The wider radius of the C-46 propeller, especially, demanded a new approach to de-icing.

Ames had closely followed innovations in these other parts, but now Rodert had to make specific recommendations. Rodert's committee especially urged him to move forward: "The consensus of the subcommittee is that the thermal method of aircraft de-icing has been proved to be sound."[34] Rodert should now help pilots follow the one rule boldfaced in every manual on de-icing: "You must maintain your airspeed."[35]

Propellers and Carburetors

Rodert's work with propellers, as with wings, started with proof that de-icing was crucial, then showing how it was easier than previously thought. By stopping and feathering propeller blades in flight, Rodert, Clousing and McAvoy discovered how propeller icing usually started with a thin pencil of ice formed at the aerodynamic dead-center of the leading edge. Rodert's position that this pencil was a necessary precursor to de-icing proved controversial. A slight temperature rise weakened its attachment enough that centrifugal force spun it off, whereas a great amount of heat was needed to prevent it forming. Yet manufacturers claimed the pencil induced vibration as it unbalanced the propeller, and became a flying missile when spun off.

Chemicals also weakened the pencil adhesion, and their use dominated propeller de-icing. Hamilton Standard offered viscous *Icelac*, the British their Mark F9 Kilfrost paste, and the Naval Research Laboratory their P-85 paste—which absorbed ice crystals on a tacky, glycerin-like surface before sloughing off the propeller.[36] Slick lacquers—like one developed by MIT—kept ice crystals from adhering to the propeller surface. Or a steady stream of alcohol expelled from a slinger ring at a propeller hub and directed along a slot-

The Curtiss C-46 flying ice-research laboratory at the Ice Research Base. (NASA photo no. Ames ALL-3895).

---

34. "Minutes of Meeting of Special Subcommittee on Deicing Problems, Committee on Aerodynamics," May 13, 1942, p. 14 (WDC: RG255: General Correspondence [Numeric File]: Box 247: File 50-14B "Deicing Problems, Minutes").
35. *Ice Formation on Aircraft: Aerology Series No. 1* (Washington, DC: Training Division, Bureau of Aeronautics, U.S. Navy, 1942).
36. "Propeller Icing," *Scientific American* 172 (April 1945): 215.

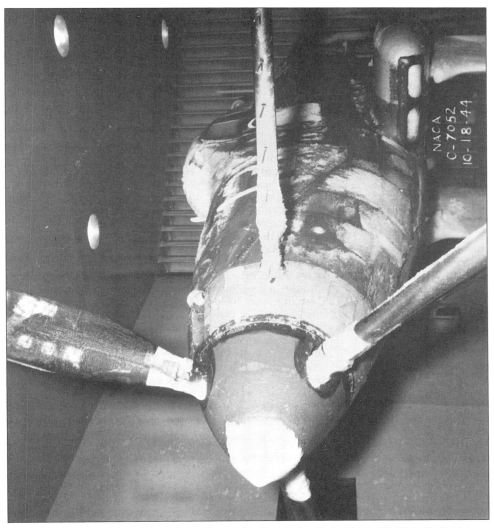

An nacelle assembly in a Lewis Laboratory tunnel test showing icing on the propeller, October 18, 1944. (NASA photo no. NACA C-7052).

ted rubber panel, cooled the icing temperature.[37] None of the chemicals, however, worked longer than an hour. The lacquers pitted and eroded; the pastes sloughed off the faster propellers too quickly; the alcohol tanks depleted if used prophylactically. A three-blade propeller used three quarts of alcohol per hour, and manufacturers hesitated to put reserve tanks of highly-flammable fluids near engine nacelles. To improve de-icer fluid

---

37. A good description of slinger rings is David Gregg, "Carburetor and Propeller Anti-Icers," *Aviation* 40 (March 1941): 42–43+.

economy, Monsanto tried to develop trimethyl phosphate—used in automotive antifreeze—as a universal de-icing fluid for all parts of the aircraft. Still, the slotted surface that directed fluids over a propeller disturbed its aerodynamic efficiency.

So during 1942, Rodert turned his attention to thermo-electric blade shoes—hard, neoprene strips imbedded with high-resistance wires and built into the leading edge of the propeller blade. The group assembled at Minneapolis that winter—especially engineers from Ames, Wright Field, the National Research Council of Canada, and the Hamilton Standard Propellers Division of the United Aircraft Company—verified the proper size, span, and heat output of the shoes. (Goodrich sent engineers to test a proprietary shoe but, in order to protect their trade secrets, kept on the outskirts of NACA-led studies.) As with heated air de-icing, the biggest problem was adequate power. They had to match the shoe with a generator built into the propeller hub—too big a generator drained engine power, too small left the shoe underheated. In an April 1943 report, Rodert offered no theory of how to determine the right quantity of heat, but suggested some empirical rules of thumb: an optimum shoe span over twenty percent chord, along ninety percent of blade radius, and a hub-generator putting out 2.5 watts per square inch. Generating the five kW needed for complete protection of the B-17 sapped twenty-eight horsepower from the four propellers, and added 120 total pounds. The AAF committed to thermo-electric boots for its medium-sized bombers, but had Ames keep working on a better de-icer for the C-46.

The larger radius of C-46 blades made it impractical to heat a boot that long with existing hub-generators. Since the larger blades were hollow, Ames and Curtiss-Wright engineers proposed pumping heated air into the hollow blades, circulating it through baffles to better transfer heat to the surface, and ejecting it out the tips. Though the exhaust tips imposed no special drag, these engineers failed to devise a method for getting enough hot air into the propeller core. (Researchers at Cleveland experimented with burning fuel inside the core to generate heat.) NACA also tried internal electrical heating, running the resistance wires along the inside surface. In the end, the C-46 left for Minneapolis with external thermoelectric boots and a promise of smaller and lighter hub-generators, which soon followed.

Rodert likewise had to recommend a system for de-icing the carburetor. Three types of ice can silence an engine. Impact ice forms around the air intake or ducting to the carburetor as supercooled moisture hits a crystallizing surface. Throttling ice encrusts the interior surface of the carburetor, when moisture-laden air expands rapidly. Fuel-evaporation ice clogs the passageways to the cylinders, when vaporizing gasoline robs heat from air in the carburetor. Since throttling and fuel-evaporation ice forms whenever the air holds moisture, regardless of temperature, research into carburetor icing proved quite complicated.

As an interim precaution, the Civil Aeronautics Authority (CAA) specified that passenger aircraft have pipes to return hot exhaust into the air entering the carburetor, which could melt all three types of ice. The Ames group were satisfied that hot air return on their 12A would keep their carburetor invulnerable. But hot air pipes had weight, and because the hotter air burned less efficiently than cooler, denser air, pilots used it only when they suspected icing. In July 1940 the Engineering and Maintenance Committee of the Air Transport Association of America passed a resolution urging NACA to expedite research into carburetor de-icing.

So the NACA convened a Special Subcommittee on Induction System Icing under the Committee on Power Plants (it remained separate from the Subcommittee on De-icing Problems). Rodert was not a committee member, but they asked him to tour engine testing tunnels—at Wright Field, Wright Aeronautical, Goodrich, Pratt & Whitney, and the Naval Aircraft Factory—and find one for the induction tests. Rodert was most impressed with the carburetor test box at the Naval Aircraft Factory. But it was booked doing expedited production testing, as was every facility save the old altitude chamber of the National Bureau of Standards (NBS).

An air-heated propeller designed by Curtiss and installed on the NACA C-46. The heated air exhusted out the orifice at the tip of each blade. (NASA photo no. Ames A-8646).

In September 1940, the Subcommittee reluctantly agreed to fund at $25,000 a year a research program led by Dr. Leo B. Kimball of the NBS. The Subcommittee—after throwing out all sorts of speculation—suggested Kimball start by building a window into his test stand so he could pass quick judgment on two existing de-icing systems. Then he should begin deliberate study of icing instrumentation, a temperature and pressure survey of the carburetor, and a process of fundamental research.

Kimball spent several months meticulously constructing a test stand for a Wright engine in his laboratory, trying to simulate rain and altitude. He studied the changing chemical composition of exhaust gases as octane and air combusted at various temperatures before starting on tests of alcohol injection, the first part of his stated task. When NACA pressed Kimball, in June 1941, to release some useful results, he looked through the observation window of his test stand and mimicked some rules of thumb Rodert had offered long ago: avoid any protuberances into the airflow, like bolt heads, and keep air flowing smoothly through cross sections that are geometrically similar.[38] He then returned to his calibrations of measurements on icing, moisture, temperature, and throttle openings. In March 1942, Kimball was ready to shoot hot air into his test-stand carburetor, the second part of his research program.

By then Rodert, who had largely solved problems of wing and windshield icing, began seeing carburetor de-icing as the reverse salient to making an ice-impervious aircraft. Whenever asked to comment on Kimball's progress, Rodert iterated that Kimball's strategy should be more like his:

> *I believe that it is better to employ trial-and-error methods in the search for a solution than to devote too much energy to analyzing the causes and effects of the many factors involved in the icing phenomenon. When an apparently satisfactory solution has been found, research leading to a complete understanding of the fundamentals may be required to perfect it. Such work is easily defined, because we then know what we are after.*[39]

Rodert praised the more directed research program pursued by the United Aircraft Corporation to improve its Pratt & Whitney engines.[40] And he was encouraged when simple anti-icing tests were added to the Army-Navy specifications for carburetors so the excellent NAF tunnel could begin collecting data on induction icing. In February 1943, Rodert convinced NACA headquarters to move Kimball's test stand to Minneapolis for studies of the XB-17 engine induction system, and to build him a cowled engine test stand that he could tow by car through the clouds along the Sierras. In the meantime, Rodert's group determined that, as rules of thumb for the C-46, they would avoid alcohol sprays, try resistance heating on carburetor parts, and otherwise keep the intake stream at 90° F for all its research aircraft. Carburetor icing delayed none of the test flights in Minneapolis.

Yet in Rodert's haste to devise design rules of thumb using cut-and-try methods, he neglected more theoretical analysis of icing conditions and heat transfer. This approach did not go unchallenged. By the summer of 1944 the de-icing community would be rife with disagreements over how to specify a workable system and who should enjoy the flexibility to improve upon it.

---

38. Kimball submitted weekly progress reports (RG255/Central/102/AF-19-10a).
39. Rodert to Engineer-in-Charge, "Memo: Prelimary report on Icing Tests of Aircraft Engine Induction Systems," by Leo B. Kimball, July 28, 1942 (RG255/Central/101/AF19-10).
40. Victor J. Skoglund, "Icing of Carburetor Air Induction Systems of Airplanes and Engines," *Journal of the Aeronautical Sciences* 8 (October 1941): 437–62.

Pushed into Theory

On a tour of southern California manufacturers in April 1943, Rodert was outraged to find that none were actually building thermal de-icing into production aircraft. After two years of expedited work, during which Rodert thought of little else, Ames had designed and proven de-icing for the B-24 and B-17, and had consulted on many more installations. AAF pilots had already flown the XB-24F over 200 hours in expedited service tests. Consolidated was already installing thermal de-icing systems into three Navy production aircraft: PBY-5 Catalinas, PB2Y-3 Coronados, and the PB4Y-1, the world's fastest flying boat.[41] Consolidated had developed soft tooling for the B-24 retrofit, including a dimpled inner skin they found easier to fabricate. Consolidated was even installing the new heat exchangers, but for cabin heating only. Even after public pronouncements that production B-24s sporting thermal de-icing would soon change the face of air battle, the AAF was still retrofitting pneumatic de-icers on B-24s as they left the plant. When Rodert asked why, Consolidated blamed "red tape:" they were confused by conflicting specifications from the AAF Materiel Command, and thus had not prepared final production specifications for approval.[42]

If Rodert had thorns in his side, they were AAF Lt. Myron Tribus and Douglas Aircraft Company. Douglas was one of the first manufacturers to design hot wings, by adopting Rodert's 12A design to their XA-26 light bomber. But Douglas never liked exhaust gas or heat exchangers. Instead Douglas adapted a gasoline-burning cabin heater from its DC-3 and scaled it up for wing de-icing. Gasoline burners would be lighter, removable, less vulnerable to gun fire, and independent of engine failure. Burner temperature was more easily controlled, so excess heat would never weaken the wing structure. Further, Douglas wanted no exchangers blocking its ejector-type stack, which turned exhaust gas into jet thrust. But Rodert's rules of thumb offered conflicting advice on how to calculate thermal requirements of gas burners. Rodert ran his early flights conservatively and got high numbers, and then had to convince manufacturers they could use less heat. Without clearer calculations, by April 1941, Douglas still considered thermal heating "too experimental."[43]

Design politics within Douglas further encouraged them to pass blame to Rodert. Douglas charged the entire weight of thermal de-icing to its equipment group, which also bought pressurizers and air conditioners. They, in turn, wished to charge much of this weight back to the wing and structures group by emphasizing the role of inner skin in transmitting heat. The structures engineers refused any responsibility for the system, however, until the thermodynamics group specified more exactly what thermal stress de-icing imposed on the leading edge. And the thermodynamics group, because they specialized in heat transfer theory, wanted NACA to provide some kind of theory rather than just empirical design rules.[44]

Myron Tribus came to Rodert's attention in June 1942 when, as an undergraduate in Boelter's mechanical engineering laboratory, he clued in Douglas to another inconsistency in Rodert's heat transfer calculations. Tribus entered the Army Air Force in September 1942, and was sent to Wright Field to prepare specifications for de-icing systems. Wright Field engineers had been enthusiastic and compliant customers of Rodert's work. Rodert reported in November 1941 that Wright Field had agreed that all plans for de-icing "will be referred to me for approval until the Air Corps has developed a group of experienced

---

41. "Development of Exhaust Gas De-Icer Revealed by Consolidated Vultee," *Western Flying* 23 (September 1943): 120; "New Techniques Applied to Anti-Icing Problem," *Aero Digest* 43 (September 1943): 225–27.

42. Rodert to DeFrance, "Memo: Production airplanes employing thermal ice-prevention equipment," June 15, 1943; DeFrance to NACA, June 18, 1943 (RG255/Central/101/AF19-20).

43. E.H. Heinemann, chief engineer, Douglas Aircraft Company to E.P. Hartman, NACA, February 28, 1941; Heinemann to DeFrance, April 3, 1941; DeFrance to Douglas Aircraft, August 13, 1941 (RG255/Central/93/AF4-10).

44. As reported later in Alun R. Jones to Engineer-in-Charge, June 21, 1945 (RG255/Central/104/AF19-20).

men in this field."[45] Wright Field had virtually plagiarized the first draft of its specifications from BuAer's specification SR-105, issued in December 1941, which in turn borrowed Rodert's rules of thumb on heating.[46]

When Tribus arrived at Wright Field he declared the specifications too inflexible. Surely Rodert's wing design of ten to fifteen percent chord heating to 100° F rise and exchanger design of 1000 BTUs per foot per hour, while good rules of thumb, would not fit all airfoils, wing structures, cruising speeds, or types of clouds. Air energy loss through tortuous ducting in the B-24 already made Tribus and Consolidated question Rodert's 1,000 BTU figure. Tribus knew Rodert's system worked, so instead he held up the uncertainty that the system might be lighter. Tribus wanted the specifications to state the temperature criteria needed to prevent icing—"in air actually containing water droplets"—then lend manufacturers flexibility in designing the wing ducting and heating systems. To do so, he wanted NACA to provide better data on the meteorological conditions for icing and better heat transfer calculations for wet air, like that in clouds.

Rodert, on the other hand, was in constant, personal contact with the manufacturers precisely because most wanted explicit design advice. Rodert had successfully designed new equipment of notable simplicity and good margins of error using the simpler dry-air calculations. He considered Tribus' preoccupation with wet-air both irrelevant, since weakening the ice adhesion bond was a sufficient concern, and too complex, since all air turned turbulent in the presence of water. Rodert further knew that Tribus harped on weight issues because he had a right to, and not because Rodert's design was too heavy. Many engineers in industry considered Rodert a peer, though quirky, and enjoyed mulling over aircraft design with him. This face-to-face contact, and not just the practical orientation, most distinguished Rodert's approach from NACA's traditional mode of encountering manufacturers—which used the NACA committee structure as a filter.[47] As Rodert expanded his program he told his bosses: "We hope that the NACA policy of permitting a close coordination of our work with the needs of the [AAF] Material Center, Bureau of Aeronautics, airline operators and manufacturers will continue."[48]

But Tribus considered such liaison his prerogative, and had access to the AAF job orders that Ames depended upon. Throughout 1943, DeFrance accepted more work on calculating wet-air heat transfers even though Rodert, with so many practical problems still to solve, considered such number-crunching an annoyance. This was good news to Douglas, which continued to complain about discrepancies in Rodert's calculations as a strategy for gaining greater freedom to design their gas-burning system. During 1943 Douglas had many planes with different types of de-icing equipment: C-74, SB2D-1, XTB2D-1, C-47, and C-54. Donald Douglas Jr. had invited Rodert to visit his plants every summer, but usually engaged Rodert on issues peripheral to his work on the hot wing—issues like how best to heat big aircraft cabins, how constant speed propellers could distinguish between friction drag from icing and normal drag from pitch change, and what to do about melted ice flowing backwards and clogging the ailerons.

At a public meeting of the American Society of Mechanical Engineers (ASME), Alun Jones spoke up about Douglas still specifying inadequate heat on the prototype DC-6.

---

45. Rodert to Engineer-in-Charge, November 14, 1941 (RG255/Central/104/AF19-20).
46. Army Air Forces Specification No. R-40395, "Anti-Icing Equipment for Aircraft. (Heated Surface Type) General Specification For:" April 21, 1942 (RG255/Central/93/AF5-20).
47. NACA's work on cowlings, another Collier-winning endeavor, also showed how NACA researchers innovated not just in new components, but also new protocols for adapting that component to each aircraft under development; see James R. Hansen, "Engineering Science and the Development of the NACA Low-Drag Engine Cowling," in this volume.
48. Rodert to Engineer-In-Charge, "Memo: Ice Research at AAL," May 8, 1942 (RG255/Central/101/AF19-10).

Douglas had pushed ahead in specifying de-icing equipment based on icing conditions—20,000 feet altitude, 0° F free air temperature, 0.5 grams per cubic meter liquid water content, and 205 mph true airspeed—rather than simple temperature rise. Douglas and Stewart-Warner had announced that they designed a burner that weighed twenty-two pounds, put out 240,000 BTUs per hour, with a tungsten igniter that worked at any altitude. United Air Lines intended to buy the DC-6, and sided with Jones. United pilots had flown Ames' C-46, and wanted their procurement contract for the DC-6 to specify similar performance. Douglas, however, noted that the Ames' reports never specified this performance data, but only design criteria.

So Rodert and Jones asked DeFrance for time to prepare a text on thermal requirements for de-icing using existing dry-air calculations. DeFrance said no. Manufacturers already understood dry air work; to stay on the cutting edge Ames had to move into the more controversial wet air work. When George Lewis of NACA headquarters asked Rodert and Jones to prepare a manual of standardized data on heat exchanger performance, DeFrance protected them, saying their time would be wasted writing manuals for junior engineers. Captain William A. Bennett, Jr., the AAF Materiel Command Liaison Officer to Ames, asked DeFrance to allow him to release preliminary data from Ames exchanger tests. DeFrance agreed, and also asked Boelter to spend the summer revising reports into "The Comparative Performance of Several Exhaust Gas-Air Heat Exchangers." (Rodert congratulated Consolidated Vultee Aircraft Corporation, in June 1944, when they took the initiative of releasing a Thermodynamics Manual summarizing Ames data and design experience.)

DeFrance seemed especially sensitive to what roles NACA could play in directing the industry. He also sensed that Rodert's zeal had raised some hackles. NACA had indeed proved that thermal de-icing held promise, and generated some excellent design rules of thumb. But even the urgency of wartime should not allow Rodert to intrude on the procurement responsibilities of young Lt. Tribus. The AAF had begun giving icing research contracts to more compliant institutions. Nor did urgency allow NACA to come between manufacturers and their customers. DeFrance knew the Ames group would need to shift its focus to more theoretical issues of icing and heat transfer, and encouraged Rodert and Jones to redefine their research agenda before others forced them to.

For example, the AAF announced in April 1943 that it would assume control of the Minneapolis operations, rename it the Ice Research Base (IRB), and expand into a hangar not needed by the Air Transport Command. When the base reopened that September all testing was directed by those engineers in the AAF Engineering Division who needed to standardize acceptance tests, write manuals and technical orders, and approve production drawings of de-icing equipment for twelve new aircraft. Rodert sat out the 1943-44 testing season but kept the XB-24F at Ames until December. DeFrance sent Carr Neel to represent NACA and run the IRB "experimental program," which Tribus had restricted to flight tests in the C-46 "icing lab."[49] Rodert waited until January 1944 to release the C-46 to the Ice Research Base, then had it returned to him the following month.

The C-46 Icing Laboratory

Because Ames had built into the C-46 very complete de-icing equipment, they could fly it into the most severe icing conditions and collect data. For the next two years, Ames pilots would fly the C-46 on a triangle route—from San Francisco northward toward Seattle and inland toward Salt Lake City. Local newspapers often report-

---

49. William A. Bennett, AAF Materiel Command Liaison to DeFrance, September 20, 1943 (RG255/Central/113/AM16-15a).

A laboratory test section of an electrically-heated airfoil for the C-46, in November 1945, with the thermocouples and nichrome electrical heating elements already installed. (NASA photo no. Ames A-9896).

ed on passengers stuck weather-bound at an airport, only to see the C-46 barrel through the clouds to a safe landing. Ames research took on a different hue once centered on this C-46.

Beginning in 1944 they focused on statistical definitions of the meteorological conditions for icing. Rodert had previously dismissed all work on icing indicators, arrogantly expecting his thermal system to work so naturally that pilots had no need to know when they encountered icing. Furthermore, the only practical indicators measured accumulation of ice in order to activate pneumatic de-icers, even though pilots agitated for an indicator that measured the rate of accumulation so they knew when to fly out of icing clouds.[50] Rodert avoided the debate between accumulation versus rate of accumulation indicators to wait for research on measuring more "fundamental" icing conditions: liquid water content of clouds, free air temperature, droplet size, and the distribution of droplet sizes. (Small drops would deflect around the wing by boundary layers; larger drops would slam into the wing.)

Free-air temperature was tough enough to measure; doing so in a cloud of unknown moisture content evoked special ingenuity. Jones directed the work on icing

---

50. "Ice Indicator," *Scientific American* 167 (December 1942): 280; "Ice Indicator Developed," *Aviation* 41 (October 1942): 138.

indicators—both what they indicated and how well—while Weather Bureau meteorologist William Lewis, working with the NACA Subcommittee on Meteorological Problems, suggested hypotheses on which data best portended icing. J.K. Hardy, the British wartime representative and an impartial observer to the Rodert-Tribus dispute, offered to stay with the Ames group through 1947 and use this wet-air data to work up a de-icing theory. Hardy began by calculating the dissipation of heat in wet air from Rodert's dry-air equations, then devised a theory to predict the de-icing performance of the C-46. NACA engineers devised an "optical rainbow recorder" to provide continuous measurement of the water content in clouds, a dew-point recorder and drop size recorders in their search for "further accuracy and simultaneous, continuous and instantaneous recording of all meteorological data."[51] Jones considered it especially challenging work. "The determination of the amount of free water in a cubic foot of cloud through which you are flying at 150 to 200 m.p.h.," noted Jones, "is a problem to be approached with respect."[52]

They began collecting data on icing in other parts of the world, like that encountered by the American-run Chinese National Airways ferrying cargo from India to China over the Himalayan Hump. Icing conditions stretched from 12,000 to 16,000 feet, so a DC-3 could not drop below or climb above it. Ice often formed four inches thick, completely blocking the windshield, and brought down more than 100 transport aircraft flying the Hump during the war, including nine in one day.

Rodert, whose reputation continued to spread, spent more time on ill-defined icing issues—like heat transfer in Navy airships, de-icing aircraft carrier decks, using a static electric field to repulse cloud droplets, protecting the protruding landing lights on the B-17, and frost on cabin windows. And he made one last effort at thermal de-icing the 12A with waste exhaust gas. This time he mixed 15 percent exhaust directly with air, to a temperature of 300° F, and then pumped it through the thin integral skin along the wing leading edge. This avoided the air pressure problems of heat exchangers, minimized maintenance problems and corrosive acids of pure exhaust, and saved the weight of gasoline burners. But it still produced unacceptable levels of condensation.

By the close of the war most manufacturers were set on using thermal de-icing, but with gasoline burners and methods Rodert had earlier bet against. Most postwar transport aircraft—like the Boeing B-50 Stratocruiser, the Douglas DC-6, the Martin 202—carried thermo-electric propellers, single-pane non-electrostatic windshields with very hot air blasts on the outside, and hot wings with gasoline-burning heaters built into the nacelles.[53] Stewart-Warner's *South Wind* heater, now improved, could put out 300,000 BTUs per hour from just 2.8 gallons of engine gasoline. Only Consolidated was using NACA-type heat exchangers, on its C-99 and model 39 cargo aircraft, despite trouble with its air-discharge valves. Still, the airlines considered Rodert a hero for calling manufacturers' results into question and expediting development across the board.

Rodert's ties with the airlines improved as the war came to a close. The chief engineer from Pennsylvania-Central Airlines visited Ames and noted that PCA lost $78,000 in the first quarter of 1944 alone by holding aircraft on the ground. Most airlines still suffered 20 percent downtime during the winter months. He was planning on retro-

---

51. Alun Jones to Engineer-in-Charge, "Memo: Suggested icing research program for Ames Laboratory," August 21, 1946 (RG255/Central/104/AF19-20).
52. Gray, *Frontiers of Flight*, p. 327.
53. "Boeing Thermal Anti-Icing," *Aero Digest* 55 (December 1947): 71–2; "Martin 2-0-2 Performance," *Aero Digest* 54 (May 1947): 59; "Thermal De-Icing on the DC-6," *Aero Digest* 53 (August 1946): 89+; "Propeller Electric De-Icing System For Wide-Range Operating Conditions," *Aviation* 46 (April 1947): 65; "Postwar Sky Giant," San Francisco *Chronicle*, February 26, 1945.

fitting thermal de-icing on all his aircraft and wanted NACA's advice. By 1944, most airlines had specified thermal de-icing equipment on their new aircraft and many were retrofitting it on their old aircraft. After several airlines asked for permission to fly the C-46, Rodert noted "the plans for the future of the airlines may serve as a good guide for expansion of NACA research facilities."[54] DeFrance planned a conference on thermal de-icing for the airlines that, to encourage more open discussion, would exclude the military.

Meanwhile, Rodert's relations with the AAF deteriorated. Rather than himself collect the data and write the specifications he thought so vital, Tribus continued to blame NACA for not doing so: "Designers, in short, are designing heated wings on the basis of very general information derived from experiments, which have been neither analyzed nor correlated." DeFrance retorted: "The Laboratory believes that a rigorous analytical treatment of the icing phenomenon is desirable from an academic viewpoint; however....most questions originate from a reluctance to make the required changes to an existing airplane or to install adequate heating capacity in a new airplane, and that the inquirer is usually seeking an escape from the design requirements shown necessary by our data."[55]

Demobilizing Icing Research

The conflict between Tribus and Rodert came to a head in January 1945, when Tribus publicly presented a paper excerpting NACA reports.[56] Tribus had written the paper back in July 1943, even as he pressed Rodert to hurry the release of the written data that manufacturers clamored for. Rodert's report writing, however, was slowed by new research, the usually slow NACA peer review process, and Army Air Forces classification. Tribus' paper was among the first icing reports downgraded from "confidential" to "restricted." Tribus then manipulated the system to get his paper completely declassified for the January 1945 meeting of the Society of Automotive Engineers, a meeting at which Rodert introduced him, and the paper for which Tribus won the Society's Wright Brothers Medal. Rodert was furious. DeFrance argued Rodert's case, contending that Rodert had dedicated himself to this work only to be robbed of tribute. "Due to the classification imposed on ice-prevention research" by the AAF, DeFrance wrote, whenever Rodert presented papers he "was required to speak only in general terms thereby impairing the quality and value of his paper."[57] Rodert was further incensed that Tribus would claim credit for work—on dry air heating requirements and exchanger design—he routinely belittled. Rodert saw his group suffer a morale decline from laboring so hard in obscurity. Thereafter, NACA more carefully claimed credit for their work on thermal de-icing, and Rodert and Jones found time

---

54. Rodert to Engineer-in-Charge, "Memo: Discussion at Ice Research Project," May 19, 1943 (RG255/Central/104/AF19-20).

55. F.O. Carroll, Chief, Engineering Division, AAF Materiel Command (Tribus' boss) to G.W. Lewis, July 25, 1944; DeFrance to NACA "Subject: Problems associated with the development of thermal anti-icing equipment," August 9, 1944 (RG255/Central/101/AF19-20).

56. Lt. Myron Tribus, "Report on the Development and Application of Heated Wings: Army Air Forces Technical Report No. 4972;" presented at the War Engineering Annual Meeting of the Society of Automotive Engineers, Detroit, January 8, 1945 as "Development and Application of Heated Wings," *SAE Journal* 54 (June 1946): 261-68; see also "Cracking Ice," *Air Forces Magazine* (January 1945).

57. Memo: DeFrance to Lewis, February 28, 1945 (RG255/Central/101/AF19-10). Compare Tribus' report (note 56) with Lewis A. Rodert, "Recent Trends in Airplane Ice Prevention Technique," paper prepared for a meeting of the Society of Automotive Engineers, New York, April 5, 1944 (RG255/Central/113/AM14-20a); *SAE Journal* 52 (December 1944): 586–91.

to finish reports of wartime work.[58] (Tribus, meanwhile, after pressing his seniors so hard for "A Theory of Heat Anti-Icing," returned for a master's degree at war's end rather than entering industry, and remained an academic his entire career.)

Rodert left Ames in September 1945, to cash in on the growing demand for heat exchangers he had created. He joined the Indianapolis-based South Wind Division of Stewart-Warner Corporation, with whom he had worked closely in the past, and remained active with NACA as chair of the de-icing committee.

But within a year he was back with NACA, to head the growing flight research branch at the NACA Aircraft Engine Research Laboratory in Cleveland. As early as January 1940 Rodert had outlined a program of engine and propeller icing research that was later written into the laboratory's agenda.[59] By VJ Day, the Cleveland laboratory had forty-nine people engaged in icing research, compared with thirty-two at Ames. There Rodert had three research aircraft at his disposal. A B-25 was calibrated so that they could switch off one part of the de-icing system, like the propeller, and then measure how icing there affected total aircraft performance. A B-24 had special installations for tests of windshields and antenna—placing them at various angles and measuring both drag and proclivity to icing. These flight tests mirrored many done during the war, except they were done with much greater precision.

Rodert's flight research complemented icing tunnel research under the direction of Wilson H. Hunter, former chief of Goodrich's icing tunnel. NACA built the world's largest refrigeration plant to serve Cleveland's new high-altitude wind tunnel, and the plant had surplus chilling for a smaller six by nine foot icing research tunnel. Winds at velocities up to 320 mph passed through a heat exchanger that cooled it down to minus sixty-five degrees, when a spray bar shot water into the refrigerated airstream. The tunnel had three sections—one for propellers, one for rotary wings, and one for engines, wings, windshields, and antennas. While droplet size in the tunnel was ten times too large to truly simulate natural icing, the Cleveland group made good use of it. They designed an inertial separator for carburetor de-icing, essentially a curve in the intake housing that divided heavier, moisture-laden air from the drier air that then entered the engine. They also analyzed the trajectory of a water-drop around an airfoil, to understand how water intercepted by a heated body was dispersed.

The Civil Aeronautics Board was so enthralled with the continuing improvements to thermal de-icing that in July 1946 they proposed, then shelved, rules requiring it on all transports that might fly into ice. Under the proposed rules hundreds of aircraft already in service would have to be grounded and retrofited at great cost. Airline engineers contended the CAB requirements "would force premature installation of devices that have not been fully proven."[60] Douglas especially claimed they simply could not retrofit thermal ducting into their popular DC-3 and DC-4s, meaning only one-third of the total airline fleet could have hot wings by January 1948, the proposed enforcement date. Yet most manufacturers already built thermal de-icing into their new aircraft—the Douglas DC-6, Martin 202 and 303, Republic Rainbow, Consolidated 240, and Boeing 377 and 417—gambling that their designs would satisfy still-forthcoming CAB certification rules.

In this context, the award of the 1946 Collier Trophy to Rodert probably served several purposes. The award signaled enthusiasm among airline owners and the other representatives of civil aviation on the Collier committee, "that ice has been virtually eliminated

---

58. The first synoptic report was the "Icing Research" summary and Alun R. Jones, "An Investigation of a Thermal Ice-Prevention System for a Twin-Engine Transport Airplane," Technical Report 862, in NACA, *Thirty-Second Annual Report, for 1946* (Washington, DC: U.S. Government Printing Office, 1949), pp. 33–34, 443–79.

59. Rodert to chief of aerodynamics division, LMAL, "Memo: Review of propeller ice research and a proposed ice research unit," January 27, 1940 (WDC: RG255: General Records Relating to Ames Research, 1938–1952: Box 48: File -61 "Icing Research 1939-41").

60. "CAB Shelves Deicing Proposal in Face of Industry Opposition," *Aviation News* 6 (September 23, 1946): 27–28.

as a major menace in air transportation."[61] Rodert was nominated for the award by John F. Victory, the executive secretary of the NACA and honorary secretary of the NAA Collier committee. Victory's careful description of Rodert's work highlighted the unrestricted research flights of the C-46, was vague about Rodert's methods, and merely mentioned its impending use on production aircraft: "Mr. Rodert's contributions involved the determination of the amount of heat required and where it was most needed, and the development of a practical means of conducting the heat to those areas in sufficient quantities without impairing the performance of the aircraft."[62] That is, Rodert was a palatable choice because his work drove the rest of the industry and undergirded de-icing specifications generally, but none specifically.

Rodert won each step in the balloting that year, with nine of eleven votes cast. General Carl Spaatz withdrew his earlier nomination of the USAF 72nd Reconnaissance Squadron (VLR), not wishing to draw too much attention to the new techniques they developed for mapping the polar regions. Committee chairman William Burden nominated Igor Sikorsky, but several other companies also staked claims to recent improvements in helicopters. Jackie Cochran nominated the propeller division of Curtiss-Wright Corporation for the reversible propeller used as a landing brake. Curtiss noted that their "purely American development" of the reversible propeller helped the aircraft carrying the atomic bomb to Hiroshima avoid the problems of maneuvering, overshooting, and aborting take-offs on the shorter forward airfields in the Pacific. Because the propeller allowed braking independent of runway conditions, it made American transit aircraft more all-weather. Technology that helped America's burgeoning air transportation industry in the post war period fit nicely with the Committee's subtle political leanings. Three Curtiss engineers won honorable mention for this work, but the Trophy went to Rodert.[63]

> "President Truman, in presenting the Trophy to Rodert, had a personal interest in this year's winner not only because Rodert is a native Missourian but also because the President's plane, The Independence, a Douglas DC-6, is one of the first production models utilizing the thermal system."[64]

In fact only a few military aircraft carried Rodert's complete de-icing system as he shook Truman's hand in December 1947, but "one or more of these features are to be found on all postwar combat and multi-engine transport aircraft already flying or in the design stage."[65] Rodert let others interpret the significance of his Collier. At the time of the award, NACA was still undergoing an uncertain transition from a wartime to a peacetime research institution. The Trophy forced NACA leaders to confront differing perceptions of NACA's role. Some argued the NACA had lost its pre-war independence; that it too often let the military services set its research agenda; that it was wet-nursing marginal designs from companies that had failed to invest enough in basic research capabilities.[66] Others argued that Rodert's icing research was a model for how NACA should more directly serve the needs of postwar American aviation.

---

61. Robert McLarren, "NACA Research Ends Ice Hazard," *Aviation Week* 47 (December 22, 1947): 24–27.
62. J.F. Victory to William A.M. Burden, "Recommendation for Collier Trophy Award for 1946," October 30, 1947, in 1946 Collier Trophy File: Awards and Trophies Reference Files: Library of the National Air and Space Museum, Washington, DC. My thanks to Paul Silbermann for alerting me to these records and making them available to me.
63. National Aeronautics Association, "Minutes of Collier Trophy Committee Meeting," November 3, 1947, Washington, DC; "Nominations for the 1946 Robert J. Collier Award," (National Air and Space Museum (NASM) Library: Awards and Trophies Reference Files: 1946 Collier Trophy File).
64. "Rodert Wins Collier Trophy," *Aeronautical Engineering Review* 7 (January 1948): 7.
65. "NACA Engineer Gets Collier Trophy for Thermal De-Icing," *Aviation Week* 47 (December 15, 1947): 26.
66. Alex Roland, *Model Research: The National Advisory Committee for Aeronautics, 1915–1958*, (Washington, DC: NASA SP-4103, 1985), 1:194-98; 234–37.

Lew Rodert accepting the Collier Trophy from President Harry S Truman in December 1947. (NASA Photo NACA C-20298)

Rodert himself soon turned from icing research to another operational problem—preventing fires following aircraft crashes. Of the 121 passenger aircraft crashes during 1946, twenty-two involved fires, and sixteen of those fires started after the aircraft hit the ground.[67] The airlines, concerned with any public perception that air travel was unsafe, asked the NACA Committee on Operating Problems to approach this problem as well. A group at the Cleveland laboratory, led by Abe Silverstein, outlined a research program in the "Reduction of Hazards Due to Aircraft Fires" to discover why aircraft flamed after impact. Rodert reprised his role of liaison with manufacturers and all other agencies concerned with aircraft crashes, and secured NACA funding and access to the Ravenna Arsenal in Ohio for full-scale crash tests. The Civil Aeronautics Authority (CAA) crash research facility at Indianapolis competed intensely for this same research funding, especially since the results would be used in CAA design codes for safer aircraft. But the more aggressive NACA program got the backing of the Aircraft Industry Association, representing the manufacturers, and the Air Transport Association, representing the airlines.

NACA's icing research program, meanwhile, had reached a natural termination point. In October 1948, Jones and Lewis returned the meteorological data the CAA would

---

67. Virginia P. Dawson, *Engines and Innovation: Lewis Laboratory and American Propulsion Technology* (Washington, DC: NASA SP-4406, 1985), p. 117.

eventually use for its design specifications.[68] Soon thereafter, DeFrance closed out Ames' work on icing, sold the 12A for scrap, and sent all those still interested in icing research to Cleveland. At Cleveland, Irving Pinkel of the physics section assumed leadership of icing research, and broadened their work into the physics of the icing cloud. They improved the water atomizer of the icing tunnel so it sprayed more natural droplet sizes, and used it to calibrate a simplified pressure-type icing rate meter that became standard equipment on most jet transports. With this new meter, the CAA collected icing data during many regular airline flights and, by the late 1950s considered the icing menace resolved if not exactly solved.

NACA shut its icing tunnel in 1957, and archived its data on icing.[69] Air travel was then done mostly with turbojets, which provided plenty of hot air from the mid-stages of their compressors to heat the wing as it passed through the boundary layer control ducts. And with pressurized cabins, aircraft could cruise well above icing clouds at 18,000. Icing was still a problem near the ground, but modern airports had better de-icing fluids to apply before take-off.

But by the late 1970s, aircraft technology again had evolved so that the icing menace reappeared. More efficient turbofans generated less waste heat, supercritical wing shapes proved tougher to de-ice, some de-icing fluids were eliminated as hazards to runways and watersheds, and deregulation put into service more small aircraft and helicopters that flew low through icing clouds. In 1978, the National Aeronautics and Space Administration reopened its icing tunnel, outlined a research program focusing on flight tests, and secured the cooperation of government agencies, military services, university researchers, manufacturers, and suppliers. Once again they sought to solve the operational problems of aircraft icing—assisted by the knowledge acquired earlier by Lewis A. Rodert, his associates, and rivals—and with rules of thumb evolving, perhaps someday, to a theory of aircraft icing.

---

68. Alun R. Jones and William Lewis, "Recommended Values of Meteorological Factors to be Considered in the Design of Aircraft Ice-Prevention Equipment," NACA TN 1855 (March 1949).

69. On a similar award-winning but "archived" technology, see Mark D. Bowles and Virginia P. Dawson, "The Advanced Turboprop Project: Radical Innovation in a Conservative Environment," in this volume. The rise and fall of fuel prices from the 1970s to the 1980s prompted, then undid, the ATP project, much like the environmental contingencies of wartime flying through all conditions prompted de-icing and turbojet flight in the 1950s rendered it irrelevant.

Chapter 3

Research in Supersonic Flight and the Breaking of the Sound Barrier

by John D. Anderson, Jr.

> "We call the speed range just below and just above the sonic speed—Mach number nearly equal to 1—the transonic range. Dryden [Hugh Dryden, well-known fluid dynamicist and past administrator of the National Advisory Committee for Aeronautics] and I invented the word 'transonic'. We had found that a word was needed to denote the critical speed range of which we were talking. We could not agree whether it should be written with one s or two. Dryden was logical and wanted two s's. I thought it wasn't necessary always to be logical in aeronautics, so I wrote it with one s. I introduced the term in this form in a report to the Air Force. I am not sure whether the general who read it knew what it meant, but his answer contained the word, so it seemed to be officially accepted. . . . I will remember this period (about 1941) when designers were rather frantic because of the unexpected difficulties of transonic flight. They thought the troubles indicated a failure in aerodynamic theory."[1]

The morning of Tuesday, October 14, 1947, dawned bright and beautiful over the Muroc Dry Lake, a large expanse of flat, hard lake bed in the Mojave Desert in California. Beginning at 6:00 a.m., teams of engineers and technicians at the Muroc Army Air Field readied a small rocket-powered airplane for flight. Painted orange, and resembling a 50-caliber machine gun bullet mated to a pair of straight, stubby wings, they carefully installed the Bell X-1 research vehicle in the bomb bay of a four-engine B-29 bomber of World War II vintage. At 10:00 a.m., the B-29 with its soon-to-be historic cargo took off and climbed to an altitude of 20,000 feet. As it passed through 5,000 feet, Captain Charles E. (Chuck) Yeager, a veteran P-51 pilot from the European theater during World War II, struggled into the cockpit of the X-1. This morning Yeager was in pain from two broken ribs incurred during a horseback riding accident the previous weekend. However, not wishing to disrupt the events of the day, Yeager informed no one at Muroc about his condition, except his close friend Captain Jack Ridley, who helped him to squeeze into the X-1 cockpit. At 10:26 a.m., at a speed of 250 miles per hour, the brightly painted X-1 dropped free from the bomb bay of the B-29. Yeager fired his Reaction Motors XLR-11 rocket engine and, powered by 6,000 pounds of thrust, the sleek airplane accelerated and climbed rapidly. Trailing an exhaust jet of shock diamonds from the four convergent-divergent rocket nozzles of the engine, the X-1 soon approached Mach 0.85, the speed beyond which there existed no wind tunnel data on the problems of transonic flight in 1947. Entering this unknown regime, Yeager momentarily shut down two of the four rocket chambers, and carefully tested the controls of the X-1 as the Mach meter in the cockpit registered 0.95 and increased still. Small invisible shockwaves danced back and forth over the top surface of the wings. At an altitude of 40,000 feet, the X-1 finally started to level off, and Yeager fired one of the two shutdown rocket chambers. The Mach meter moved smoothly through 0.98, 0.99, to 1.02. Here, the meter hesitated then jumped to

---

1.   Theodore von Kármán, *Aerodynamics* (Ithaca, NY: Cornell University Press, 1954), p. 116.

1.06. A stronger bow shockwave now formed in the air ahead of the needlelike nose of the X-1 as Yeager reached a velocity of 700 miles per hour, Mach 1.06, at 43,000 feet. The flight was smooth; there was no violent buffeting of the airplane and no loss of control as feared by some engineers. At this moment, Chuck Yeager became the first pilot to fly faster than the speed of sound, and the small but beautiful Bell X-1, became the first successful supersonic airplane in the history of flight.[2]

The Bell X-1. (NASA photo)

---

2. This description of the first supersonic flight is excerpted from John D. Anderson, Jr., *Modern Compressible Flow: With Historical Perspective* (New York, NY: McGraw-Hill Book Co., 1990 2d ed.), pp. 2–4. For a general reference, from Chuck Yeager's point of view, see General Chuck Yeager and Leo Janos, *Yeager: An Autobiography* (New York, NY: Bantam Press, 1985). For a definitive history of the circumstances leading up to and surrounding the development and flight testing of the Bell X-1, see Richard P. Hallion, *Supersonic Flight* (New York, NY: Macmillan, 1972).

As the sonic boom from the X-1 propagated across the California desert, this flight became the most significant milestone in aviation since the Wright brothers' epochal first flight at Kill Devil Hills forty-four years earlier. But in the history of human intellectual accomplishment, this flight was even more significant; it represented the culmination of 260 years of research into the mysteries of high-speed gas dynamics and aerodynamics. In particular, it represented the fruition of twenty-three years of insightful research in high speed aerodynamics carried out by the National Advisory Committee for Aerodynamics (NACA)—research that represented one of the most important stories in the history of aeronautical engineering. The purpose of this chapter is to tell this story. The contribution by the NACA to the Bell X-1 was much more technical than it was administrative. Therefore, this chapter will highlight the history of that technology.

The NACA's work on high-speed aerodynamics described in this chapter is also one of the early examples in the history of aerodynamics where *engineering science* played a deciding role. Beginning in 1919, the NACA embarked on a systematic intellectual quest to obtain the *knowledge* required to eventually *design* proper high-speed airfoil shapes. Historian James R. Hansen, in his chapter on the NACA low-drag engine cowling, in the present book, asks the following question about the cowling work: Was it science, or was it engineering? He comes to the conclusion that it was somewhere in between—that it was an example of engineering science in action at the NACA. In arriving at this conclusion, Hansen draws from the thoughts in Walter Vincenti's book, *What Engineers Know and How They Know It*, where Vincenti clearly makes the following distinction between science and engineering: science is the quest for new knowledge for the sake of enhancing understanding, and engineering is a self-standing body of knowledge (separate from science) for the sake of designing artifacts. For the purpose of the present chapter, I suggest this definition of engineering science: *Engineering science is the search for new scientific knowledge for the explicit purpose of (1) Providing a qualitative understanding which allows the more efficient design of an engineering artifact, and/or (2) Providing a quantitative (predictive) technique, based on science, for the more efficient design of an engineering artifact*. In this chapter we will see that NACA researchers in the 1920s and 1930s were working hard to discover the scientific secrets of high-speed aerodynamics just so they could properly design airfoils for high-speed flight—truly engineering science in action. Also, within the general framework of the historical evolution of aerodynamic thought over the centuries, the NACA's high-speed research program is among the earliest examples of engineering science, although that label had not yet been coined at the time.

The Prehistory of High–Speed Flight: Point and Counterpoint

Most golfers know the following rule of thumb: When you see a flash of lightning in the distance, start counting at a normal rate—one, two, three. . . . For every count of five before you hear the thunder, the lightning bolt struck a mile away. Clearly, sound travels through air at a definite speed, much slower than the speed of light. The standard sea level speed of sound is 1,117 feet per second—in five seconds a sound wave will travel 5,585 feet, slightly more than a mile. This is the basis for the golfer's "count of five" rule of thumb.

The speed of sound is one of the most important quantities in aerodynamics; it is the dividing line between subsonic flight (speeds less than that of sound) and supersonic flight (speeds greater than that of sound). The Mach number is the ratio of the speed of a gas to the speed of sound in that gas. If the Mach number is 0.5, the gas flow velocity is one-half the speed of sound; a Mach number of 2.0 means that the flow velocity is twice that of sound. The physics of a subsonic flow is totally different from that of a supersonic flow—a contrast as striking as that between day and night. This is why the first supersonic flight of the X-1 was so dramatic, and why the precise value of the speed of sound is so important in aerodynamics.

Knowledge of the speed of sound is not a product of twentieth century science. Precisely 260 years before the first supersonic flight of the X-1, Isaac Newton published the first calculation of the speed of sound in air. At that time it was clearly appreciated that sound propagated through air at some finite velocity. Newton knew that artillery tests had already indicated that the speed of sound was approximately 1,140 feet per second. The seventeenth century artillery men were preceding the modern golfer's experience; the tests were performed by standing a known large distance away from a cannon, and noting the time delay between the light flash from the muzzle and the sound of the discharge. In Proposition 50, Book II of his *Principia* (1687), Newton calculated a value of 979 feet per second for the speed of sound in air—fifteen percent lower than the existing artillery data. Undaunted, Newton followed a now familiar ploy of theoreticians; he proceeded to explain away the difference by the existence of solid dust particles and water vapor in the atmosphere. However, in reality Newton had made the incorrect assumption in his analysis that the air temperature inside a sound wave was constant (an isothermal process), which caused him to underpredict the speed of sound. This misconception was corrected more than a century later by the famous French mathematician, Pierre Simon Marquis de Laplace, who properly assumed that a sound wave is adiabatic (no heat loss), not isothermal.[3] Therefore, by the time of the demise of Napoleon, the process and equation for the speed of sound in a gas was fully understood.

This is not to say that the precise value of the speed of sound was totally agreed upon. The debate lasted well into the twentieth century. Indeed, although this event is little known today, the NACA was an arbiter in setting the standard sea level speed of sound. On October 12, 1943, twenty-seven distinguished U.S. leaders in aerodynamics walked through the doorway of NACA Headquarters at 1500 New Hampshire Avenue in Washington, DC. They were attending a meeting of the Committee on Aerodynamics, one of the various adjunct committees set up by the main NACA. Among the experts present were Hugh L. Dryden from the Bureau of Standards, and John Stack, whose career as an aerodynamicist at the NACA Langley Memorial Laboratory was on a meteoric rise. Also present was Theodore von Kármán, director of the Guggenheim Aeronautical Laboratories at Cal Tech, who represented an intellectual pipeline to the seminal aerodynamic research by Ludwig Prandtl at Göttingen University in Germany, where von Kármán had been Prandtl's Ph.D. student before World War I. After subcommittee reports on progress in helicopter aerodynamics, and recent aerodynamic problems in wing flutter and vibration, the matter of speed of sound was brought up as new business by John Stack, who stated that "the problem of establishing a standard speed of sound was raised by an aircraft manufacturer."[4]

Stack reported that the Committee's laboratory staff had surveyed the available information on specific heats of air—thermodynamic information that goes into the calculation of the speed of sound—which led to a calculated value of the speed of sound of

---

3. Pierre Simon Marquis de Laplace, "Sur la vitesse du son dans l'aire et dan l'eau," *Annales de Chimie et de Physique*, 1816.

4. Minutes of the Meeting of Committee on Aerodynamics, October 12, 1943, p. 9. Found by the author in the John Stack files at the NASA Langley Research Center Archives, Langley Research Center, Hampton, VA. Originally marked with security classification Confidential, the minutes have since been declassified. The Langley Archives are kept by Richard T. Layman, who was exceptionally helpful to the author during the course of research for this chapter.

John Stack, Langley Research Center scientist, was presented the Collier Trophy in 1947, awarded for his conception of transonic research airplanes. His research contributed to the X-1 breaking the sound barrier on October 14, 1947. (NASA Photo No. LMAL 48991).

1,116.2 feet per second. Measured values gave weighted means of 1,116.8 to 1,116.16 feet per second. Dryden noted that the specific heats were "not necessarily the same for all conditions" and suggested that the Committee select 1,117 feet per second as a round figure for a standard value of the speed of sound for sea level conditions for aeronautical usage. The outcome of this discussion appeared in the meeting minutes: "After further discussion it was agreed that the recommendation of a standard value for the speed of sound would be left for Dr. Dryden and Mr. Stack to work out jointly." Today, the accepted standard speed of sound depends on which "standard atmosphere" table you look at, ranging from a value of 1,116.4 feet per second in the 1959 ARDC Model atmosphere to 1,116.9 feet per second in the 1954 ICAO Model atmosphere. However, for engineering purposes this is splitting hairs, and Dryden's suggestion of a round value of 1,117 feet per second is still used today for many engineering calculations. Here is a little-known example of how the NACA played a role in the fundamentals of high-speed compressible aerodynamics—even to the mundane extent of providing to industry a "standard" value of the speed of sound.

On October 14, 1947, as the Bell X-1 nudged closer to Mach one, a region of the aerodynamic flow over the wing became locally supersonic. This is because the airflow increases its velocity while moving over the top of the wing, and hence there is always a region of the flow over the wing where the local velocity is larger than the velocity of the airplane itself. As the X-1 accelerated through Mach 0.87, a pocket of locally supersonic flow formed over the top of the wing. This supersonic pocket was terminated on the downstream end by a shockwave oriented almost perpendicular to the flow—called a normal

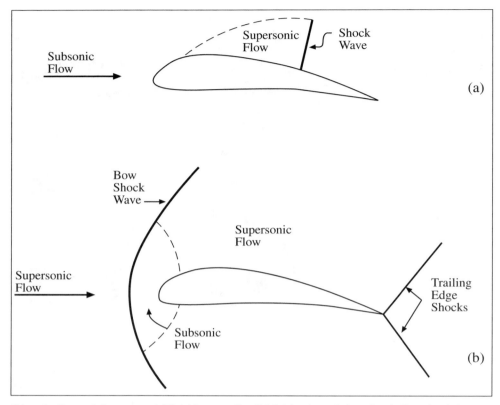

Schematic of transonic flow over an airfoil. (a) Freestream flow slightly below the speed of sound, typically a subsonic freestream Mach number from about 0.8 to 0.999. (b) Freestream flow slightly above the speed of sound, typically a supersonic freestream Mach number from 1.0 to about 1.2.

shock (as shown above). This shock formation was the culprit which made flight through Mach one such a harrowing concern at that time. Finally, when the X–1 accelerated through Mach one to supersonic speeds, another shock wave formed a short distance in front of the nose; this shock, called the bow shock, was curved and more oblique to the flow (As shown above). Shock waves are extremely thin regions—much thinner than the thickness of this page—across which dramatic and almost discontinuous increases in pressure and temperature occur. Shock waves are a fact of life in the aerodynamic flow over transonic and supersonic airplanes.

Knowledge of shock waves is not unique to the twentieth century; their existence was recognized in the early nineteenth century. The German mathematician G. F. Bernhard Riemann first attempted to calculate shock properties in 1858, but he neglected an essential physical feature and hence obtained incorrect results.[5] Twelve years later, William John Rankine, a noted engineering professor at the University of Glasgow, correctly derived the

---

5. A shock wave is, in thermodynamic language, an irreversible process, caused by viscosity and thermal conduction effects inside the shock wave. A measure of the amount of irreversibility is a thermodynamic variable called entropy, which from the Second Law of Thermodynamics always increases in any process involving such irreversibilities. The entropy of a gas always increases as it passes through a shock wave. Unfortunately, Riemann made the incorrect assumption that the entropy remained constant across a shock.

proper equations for the flow across a normal shock wave. Not cognizant of Rankine's work, the French ballistician Pierre Hugoniot rediscovered the normal shock wave equations in 1887. To the present day, the governing equations for flow across a shock wave are called the *Rankine-Hugoniot equations*, in honor of these two men.[6] This work was expanded to include oblique shock waves by the famous German aerodynamicist, Ludwig Prandtl and his student Theodor Meyer at Göttingen University in 1908.[7] Hence, only five years after the first flight by the Wright brothers, the necessary theory for the calculation of shock wave properties in a supersonic flow was in hand, albeit considered a purely academic subject at that time.

The nineteenth century was also a time of experimental work on supersonic flow. Perhaps the most important event was the proof that shock waves were not just a figment of the imagination—they really existed in nature. This proof was given by the physicist-physician-philosopher Ernst Mach in 1887. Mach, while a professor of physics at the University of Prague, took the first photographs of shock waves on a body moving at supersonic speeds. Shock waves are normally invisible to the naked eye. But Mach devised a special optical arrangement (called a shadowgraph) by which he could see and photograph shock waves. In 1887, he presented a paper to the Academy of Sciences in Vienna where he showed a photograph of a bullet moving at supersonic speeds. Using his shadowgraph system, the bow shock and trailing edge shock were made visible (as shown below). This historic photograph allowed scientists, for the first time in history, to actually see a shock wave. The experimental study of shock waves was off and running.

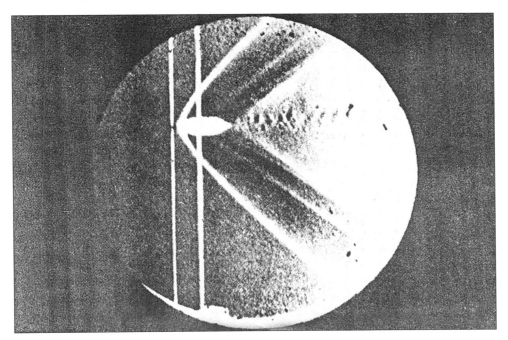

Photograph of a bullet in supersonic flight, published by Ernst Mach in 1887.

---

6.   John D. Anderson, Jr., *Modern Compressible Flow: With Historical Perspective*, (New York, NY: McGraw Hill, 1990), pp. 92–95.
7.   *Ibid.*, pp 140–43.

This prehistory of supersonic flight, both theoretical and experimental, was carried out by basic researchers who were interested in the subject on an academic basis only. The true practical value of this work did not come to fruition until the advent of supersonic flight in the 1940s. However, this is an excellent example of the value of basic research on problems that appear only purely academic at the time. In the 1940s, when basic supersonic flow theory and fundamental understanding of shock waves was suddenly needed due to the advent of high-speed airplanes and rockets, it was there—quietly residing and sleeping in a few dusty books and archive journal articles in the library.

In light of our earlier discussion of engineering science, was this early work on shock waves engineering science? Emphatically no! The researchers involved in this work were after scientific knowledge, and just that. There was no force behind these researchers driving them to design any related engineering artifacts at the time.[8]

Compressibility Problems: The First Inklings (1918–1923)

Airplane aerodynamics, from the time of the Wright Flyer to the beginning of World War II, assumed that changes in air density were negligible as the air flowed over the airplane. This assumption, called *incompressible flow*, was reasonable for the 350 mph or slower flight speeds of airplanes during that era. Theoretically, it was a tremendous advantage to assume constant density, and physically the low-speed aerodynamic flows usually exhibited smooth variations with no sudden changes or surprises. All this changed when flight speeds began to sneak up close to the speed of sound. Aerodynamic theory had to account for changes in the air density in the flow field around the airplane, and physically the flow field sometimes acted erratically, and frequently surprised and greatly challenged aerodynamicists. Aerodynamicists in the 1930s simply threw these phenomena into one pot and called them generically "compressibility problems."

Ironically, the first inklings of compressibility problems occurred during the age of the strut-and-wire biplanes, with flight velocities about as far away from the speed of sound as you can get. It had to do with an airplane part, namely the propeller. Although typical flight speeds of World War I airplanes were less than 125 miles per hour, the tip speeds of propellers, because of their combined rotational and translational motion through the air, were quite large, sometimes exceeding the speed of sound. This fact was appreciated by aeronautical engineers at the time. This drove the British Advisory Committee for Aeronautics to show some interest in compressible flow theory. In 1918 and 1919, G. H. Bryan, working for the Committee at the Royal Aeronautical Establishment, carried out a theoretical analysis of subsonic and supersonic flows over a circular cylinder (a simple geometric shape chosen for convenience). He was able to show that in a subsonic flow the effect of compressibility was to displace adjacent streamlines farther apart. His analysis was cumbersome and complex—a harbinger of things to come—and provided little data of value. But it was evidence of the concern felt by the British over the effects of compressibility on propeller performance.[9]

At the same time, Frank Caldwell and Elisha Fales of the propeller branch of the Army Air Service Engineering Division at McCook Field in Dayton, Ohio, took a purely experi-

---

8. *Report for the Year 1909–10*, Advisory Committee for Aeronautics, England, p. 5.
9. G.H. Bryan, "The Effect of Compressibility on Streamline Motions," R & M No. 555, Technical Report of the Advisory Committee for Aeronautics, Vol. I, Dec. 1918; G.H. Bryan, "The Effect of Compressibility on Streamline Motions, Part II," R & M No. 640, Advisory Committee for Aeronautics, April 1919.

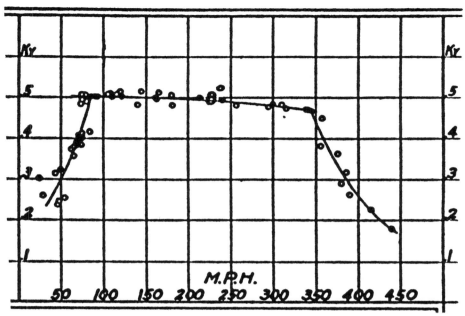

The first data to show the adverse compressibility effects of high-speed flow over an airfoil. Caldwell and Fales, NACA TR 83, 1920. This is a plot of lift coefficient, Ky, versus velocity in miles per hour. The definition used for Ky at that time differed from the modern definition of lift coefficient (usually denoted by C_L today) by a factor of two, i.e., $C_L = 2$ Ky. The large drop in Ky seen at the right of the graph is the adverse effect of compressibility. (The ramp in Ky seen at the left of the graph was not explained by Caldwell and Fales; it is the present author's educated guess that the ramp was a low Reynolds number effect, because of the small size of the airfoil models used, namely a one-inch chord.)

mental approach to the problem. (This was the beginning of a blurred dichotomy between British and American research on compressibility effects. Over the next two decades, the major experimental contributions to understanding compressibility effects were to be made in the United States, principally by the NACA, and the major theoretical contributions were to be made in England.) In 1918, Caldwell and Fales designed and built the first high-speed wind tunnel in the United States—purely to investigate the problems associated with propellers. The tunnel velocity range was from 25 to a stunning 465 miles per hour. It had a length of almost nineteen feet, and the test section was fourteen inches in diameter. This was a big and powerful machine for its day. Six different airfoils, with thickness ratios (ratio of maximum thickness to the chord length) from 0.08 to 0.2, were tested. At the higher speeds, the results showed "a decreased lift coefficient and an increased drag coefficient, so that the lift-drag ratio is enormously decreased." Moreover, the airspeed at which these dramatic departures took place was noted as the "critical speed."[10] Because of its historical significance, some of their data is shown above, reproduced directly from NACA TR 83. Here, the lift coefficient for the airfoil at eight-degree

---

10. The critical Mach number is precisely defined as that freestream Mach number at which sonic flow is first encountered on the surface of a body. The large drag rise due to compressibility effects normally occurs at a freestream Mach number slightly above the critical Mach number; this is called the drag-divergence Mach number. In reality, Caldwell and Fales had reached and exceeded the drag-divergence Mach number in their experiments. But their introduction of the word "critical" in conjunction with this speed was eventually the inspiration for its use in later coining the term "critical Mach number."

angles of attack is plotted versus airstream velocity. Note the dramatic drop in lift coefficient at the "critical speed" of 350 miles per hour—the compressibility effect. This plot, and ones like it for other angles of attack that were published in NACA TR 83, are the first published data in the history of aerodynamics to show the adverse effects of compressibility. Although Caldwell and Fales made an error in the reduction of their data (an understandable error associated with the inexperience of dealing with compressible flow conditions at the early date of 1919) which caused their reported lift and drag coefficients to be about ten percent too low at the higher speeds, this did not compromise the dramatic and important discovery of the large increase in drag and decrease in lift when the airfoil sections were tested above the "critical speed." Moreover, they were the first to show that the "critical speed" for thin airfoils was higher than that for thick airfoils, and hence by making the airfoil section thinner, the adverse compressibility effects can be delayed to higher Mach numbers. This was an important finding, and one which would have a lasting impact on high-speed vehicle design.[11]

It is noteworthy that the fledgling NACA was the government agency which published the results of Caldwell and Fales.[12] The NACA was carrying out its duty as stated in Public Law 271, which created the Committee in 1915, namely "to supervise and direct the scientific study of the problems of flight, with a view to their practical solution, and to determine the problems which should be experimentally attacked, and to discuss their solution and their application to practical questions." Publishing the Caldwell and Fales work is in the latter category—the NACA was already earmarking compressibility effects as a problem "which should be experimentally attacked."

In the chronology of events, the British were next to examine the effects of compressibility on propellers. In 1923, G. P. Douglas and R. McK. Wood, two aerodynamicists at the Royal Aeronautical Establishment, tested model propellers at high rotational speeds in the seven-foot low-speed wind tunnel (100 miles per hour airstream) at the National Physical Laboratory in London.[13] They also carried out flight tests on a DeHaviland D.H. 9A biplane. Their data were the global measurements of the thrust and torque generated by the whole propeller, so the details of the compressibility effects affecting the airfoil sections at the tip of the propeller were somewhat obscured. However, one of their conclusions anticipated the adverse effects of compressibility, namely that "higher tip speeds than at present used will probably involve a serious loss of efficiency."

---

11. This author, upon studying Caldwell and Fales detailed data reduction, has found that, although they recognized that the density of the airflow changed inside the wind tunnel at the higher speeds, their accounting for this in calculating their lift and drag coefficients from their measured lift and drag forces was done incorrectly. They thought they had worked their data reduction so that "density does not enter into the calculation." Rather, they expressed their lift and drag coefficients in terms of the impact pressure—the difference between total and static pressure. This is why they said that "density does not enter into the calculation." But they incorrectly and rather naively used the incompressible Bernoulli equation to replace the velocity-squared term in the definition of lift coefficient with the impact pressure. This resulted in about a ten percent error in the values of their reported lift and drag coefficients at high speeds. For more details, see John D. Anderson, Jr., *The History of Aerodynamics, and its Impact on Flying Machines* (New York, NY: Cambridge University Press, 1997).
12. F.W. Caldwell, and E. Fales, "Wind Tunnel Studies in Aerodynamic Phenomena at High Speed." NACA TR 83, 1920.
13. G.P. Douglas and R. McK. Wood, "The Effects of Tip Speed on Airscrew Performance. An Experimental Investigation of the Performance of an Airscrew Over a Range of Speeds of Revolution from 'Model' Speeds up to Tip Speeds in Excess of the Velocity of Sound in Air," R & M No. 884, Advisory Committee for Aeronautics, 1923.

The Compressibility Burble—NACA's Seminal Research, 1924–1929

Meanwhile, the NACA was forging ahead. During the 1920s, the Committee sponsored a series of fundamental experiments in high-speed aerodynamics at the Bureau of Standards with Lyman J. Briggs and Dr. Hugh L. Dryden. Hugh Dryden was a fresh, young Ph.D. graduate from Johns Hopkins University in physics; he had received his Ph.D. in 1919 at the age of twenty. (Dryden much later was to become the Director of Research for the NACA from 1947 to 1958.) This work progressed in three stages, each one documented in a separate NACA Technical Report, and covered the period from 1924 to 1929. As before, the primary motivation for this research was to understand the compressibility effects at the tips of propellers.

The first stage simply confirmed the trends already observed by Caldwell and Fales four years earlier. Briggs and Dryden, with the help of Lt. Col. G. F. Hull of the Army Ordnance Department, jury-rigged a high-speed wind tunnel by connecting a vertical standpipe thirty inches in diameter and thirty feet high to a large centrifugal compressor at the Lynn Works of the General Electric Company in Massachusetts. At the other end of the pipe was a cylindrical orifice that served as a nozzle 12.24 inches in diameter. With this device "air speeds approaching the speed of sound were obtained."[14] Unlike Caldwell and Fales, Briggs and Dryden used the proper equations for compressible flow to calculate the air velocity. Although not yet in the standard textbooks, these equations were known by Dryden as a result of his Ph.D. studies in physics. (The first engineering textbook in English to focus on compressible flow was not published until 1947.)[15] Rectangular planform models, with a span of 17.2 inches and a chord length of three inches, were placed in the high-speed airstream, and lift, drag and center-of-pressure were measured. The results supported the earlier trends observed by Caldwell and Fales. In particular, Briggs found:[16]

(1) Lift coefficient for a fixed angle of attack decreases very rapidly as the speed increases.
(2) The drag coefficient increases rapidly.
(3) The center-of-pressure moves back towards the trailing edge.
(4) The "critical speed" at which these occur decreases as the angle of attack is increased and the airfoil thickness is increased.

In 1924, the culmination of this work, as well as that which went before, was the waving of a red flag—compressibility effects were nasty, and they markedly degraded airfoil performance. But nobody had any fundamental understanding of the physical features of the flow field which were causing these adverse effects. This was not to come for another decade.

Briggs and Dryden made an important step towards this fundamental understanding in the second stage of their work. Because the Lynn Works compressor was no longer available to them, Briggs and Dryden moved their experimental activity to the Army's Edgewood Arsenal, where they constructed another high-speed wind tunnel, this one much smaller, with an airstream only two inches in diameter. However, by careful design of the small airfoil models, two pressure taps could be placed in each model. Seven iden-

---

14. L.J. Briggs; G.F. Hull; and Hugh L. Dryden, "Aerodynamic Characteristics of Airfoils at High Speeds," NACA TR 207, 1924.
15. Hans W. Liepmann, and Allen E. Puckett, *Introduction to Aerodynamics of a Compressible Fluid* (New York, NY: John Wiley and Sons, 1947).
16. L.J. Briggs and Hugh L. Dryden, "Pressure Distribution Over Airfoils at High Speeds," NACA TR 255, 1926.

tical models were used, each one with different locations of the pressure taps. A total of thirteen pressure tap locations, seven on the upper surface and 6 on the lower surface, were employed (for the reader who is counting, the seventh model had only one tap).

With this technique, Briggs and Dryden measured the pressure distributions over the airfoil at Mach numbers from 0.5 to 1.08. The results were dramatic! Beyond the "critical speed," the pressure distributions over the top of the airfoil exhibited a sudden pressure jump at about one-third to one-half the distance from the leading edge, followed by a rather long plateau towards the trailing edge. Such a pressure plateau was familiar—it was very similar to that which exists over the top surface of an airfoil in low-speed flow when the airfoil stalls at high angle of attack. And it was well known that airfoil stall was caused by the separation of the flow off the top surface of the airfoil. Briggs and Dryden put two-and-two together, and concluded that the adverse effects of compressibility were caused by flow separation over the top surface, even though the airfoil was at low (even zero) angle of attack. To substantiate this, they conducted oil flow tests, wherein a visible, pigmented oil was painted on the model surface, and the model was placed in the high-speed airstream. During the tests, the tell-tale flow separation line formed on the oil pattern. Clearly, beyond the "critical speed," flow separation was occurring on the top surface of the airfoil. The next question was: Why? What was causing the flow to separate? The answer to this question still lay eight years in the future.

Was this work of Briggs and Dryden engineering science? Emphatically yes! Their experiments were designed to obtain basic scientific information about the physics of the high-speed flow over an airfoil, but always for the purpose of learning how to design better airfoil shapes for high-speed flight.

The third stage of the work by Briggs and Dryden was utilitarian, and was in keeping with the stated duty of the NACA to work on the problems of flight "with a view to their practical solution." Towards the end of the 1920s, they carried out a large number of detailed measurements of the aerodynamic properties for 24 different airfoils at Mach numbers from 0.5 to 1.08. The airfoils chosen were those conventionally used by the Army and the Navy for propellers, consisting of the standard family of British-designed RAF airfoils, and the American-designed Clark Y family. These data provided the first definitive measurements on *standard* series of airfoils showing compressibility effects.[17]

It should be noted that theoretical solutions of high-speed compressibility effects in a subsonic flow were virtually non-existent during the 1920s. The only major contribution was that by the famous British aerodynamicist Herman Glauert, who rigorously derived a correction to be applied to the low-speed, incompressible lift coefficient in order to correct it for compressibility effects.[18] This was the first of a series of theoretical rules labeled "compressibility corrections." Because it was known that Ludwig Prandtl in Germany had also derived the same rule a few years earlier, but had not published it, Glauert's result has come down through the decades as the *Prandtl–Glauert Rule.* However, such compressibility corrections are applicable to the variation of lift coefficient with speed *below* the "critical speed," and hence have no way of predicting the lift coefficient in the "compressibility burble."

Throughout this, the primary motivation for all the above work on compressibility effects was for application to airplane propellers. But the focus was about to change, and change dramatically.

---

17. L.J. Briggs and Hugh L. Dryden, "Aerodynamic Characteristics of Twenty–Four Airfoils at High Speeds," NACA TR 319, 1929.
18. H. Glauert, "The Effect of Compressibility on the Lift of an Airfoil," *Journal of the Royal Society* 118 (1927): 113. Also published as R & M No. 1135, Advisory Committee for Aeronautics, September 1927.

John Stack and the NACA Compressible Flow Research—A Breakthrough

In July 1928, a young New Englander, born and raised in Lowell, Massachusetts, began his career with the NACA Langley Memorial Aeronautical Laboratory. Having just graduated from the Massachusetts Institute of Technology with a B.S. degree in aeronautical engineering, John Stack was assigned to the Variable Density Tunnel, the premier wind tunnel in the world at that time. Stack was absolutely dedicated to aeronautical engineering. While in high school, he earned money so that he could take a few hours of flight instruction in a Canuck biplane. He helped out with the maintenance of a Boeing biplane owned by one of his part-time employers. Before he went to college, he had made up his mind to be an aeronautical engineer. However, his father, a carpenter who was also very successful in real estate, wanted his son to study architecture at MIT. Instead, when Stack entered MIT, he enrolled in aeronautical engineering, keeping it a secret from his father for the first year, but with the understanding approval of his mother. Much later, Stack commented: "Then when Dad heard about it, it was too late to protest."[19]

When John Stack first walked into the Langley laboratory that July of 1928, a year's worth of design work had already been done on Langley's first high-speed tunnel, and the facility was already operational with an open throat test section.[20] Success had been achieved by the work of Briggs and Dryden, and the growing importance of high-speed research was perceived by some visionaries. Because of this perception, Joseph S. Ames, President of Johns Hopkins University and the new Chairman of the NACA, in 1927 gave priority to high-speed wind tunnels and research.[21] Eastman Jacobs, who had joined the NACA in 1925 after receiving his B.S. degree in mechanical engineering from the University of California, Berkeley, was the chief designer of the open-throat eleven-inch High Speed Tunnel. (Jacobs would later earn an international reputation for his work on the famous NACA airfoil sections in the 1930s, and for his conception of, and pioneering research on, the NACA laminar flow airfoils just before the beginning of World War II.) An innovative aspect of the eleven-inch High Speed Tunnel was that it was driven from the twenty atmosphere pressure tank of the Langley Variable Density Tunnel. For a change in models in the Variable Density Tunnel, the twenty atmosphere tank which encased the entire tunnel was blown down to one atmosphere; this represented a wasted energy source which the Langley engineers ingeniously realized could be tapped for the eleven-inch High-Speed Tunnel. The 5,200 cubic foot capacity of the high pressure tank allowed about one minute of operation for the tunnel. John Stack was given the responsibility for improving the High-Speed Tunnel by designing a closed throat. This improved facility, shown on the next page, was operational by 1932. It was his participation in the design and development of the eleven-inch High-Speed Tunnel that launched John Stack on his life-long career in high-speed aerodynamics.

While Stack was working on the High-Speed Tunnel, an event occurred in England which made a great impression on him, and which would rapidly refocus the NACA high-speed research program. On Sunday, September 13, 1931, a beautiful, highly streamlined Supermarine S.6B flashed through the clear early afternoon sky at Calshot, near Portsmouth along the southern English coast. Flown by Flt. Lt. John N. Boothman, this exquisite racing

---

19. Lou Davis, "No Time for Soft Talk," *National Aeronautics*, January 1963, pp. 9–12. This is an interesting biographical article written about Stack at the time of his receiving the 1962 Wright Memorial Trophy Award from the National Aeronautic Association.

20. James R. Hansen, *Engineer in Charge: A History of the Langley Aeronautical Laboratory, 1917–1958* (Washington, DC: NASA SP–4305, 1987), p. 446.

21. Donald D. Baals and William R. Corliss, *Wind Tunnels of NASA* (Washington, DC: NASA SP–440, 1981).

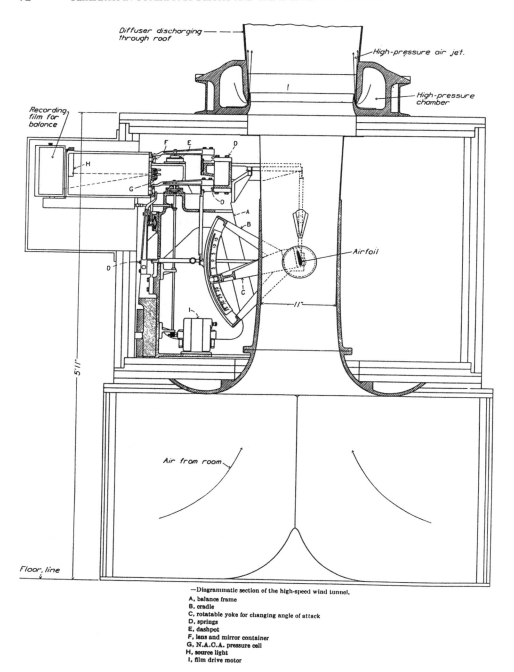

The 11-inch High-Speed Tunnel at NACA Langley; closed-throat modification in 1932.

airplane averaged a speed of 340.1 mph around a long, seven-lap course, winning the coveted Schneider Trophy permanently for Britain. Later that month, on September 29, Flt. Lt. George H. Stainforth set the world's speed record of 401.5 mph in the same S.6B. Looking at this figure, it does not take an aerodynamic expert to appreciate that by 1931 the concept of streamlining in order to reduce drag had taken root. The Supermarine S.6B simply *looked* like it could fly at 400 miles per hour—at Mach 0.53, over half the speed of sound. Suddenly, the aeronautical engineer's concern over compressibility effects on propeller tips, an important but tolerable situation, became an absolutely major concern over compressibility effects on the airplane itself, a problem of showstopper proportions.

Such concern was beginning to dawn on the aircraft industry itself. In 1936, Lockheed's Kelly Johnson began early design studies for the P-38, which was the first airplane to encounter major, and sometimes fatal, compressibility effects. By the mid-1930s, the aircraft industry was wading into uncharted water, and the NACA's high-speed research program became absolutely vital to the future progress of high-speed airplane design.

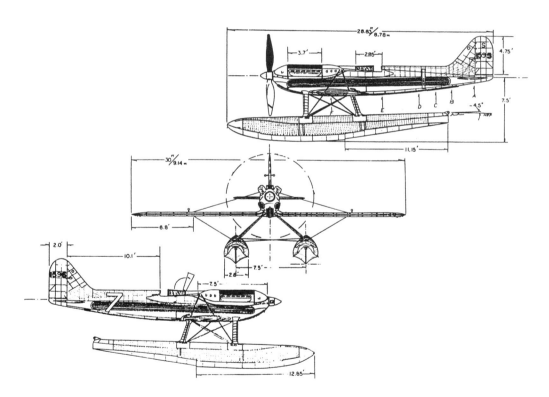

The Supermarine S.6B, the airplane used by the British to win the Schneider Trophy, 1931.

Stack became acutely aware of this new compressibility challenge. In 1933, he published in NACA TR 463 the first data to come from the newly modified, closed-throat High-Speed Tunnel. Although the airfoils were propeller sections, Stack wrote in the introduction, obviously referring to the Schneider Trophy racer:

> *A knowledge of the compressibility phenomenon is essential, however, because the tips speeds of propellers now in use are commonly in the neighborhood of the velocity of sound. Further, the speeds that have been attained by racing airplanes are as high as half the velocity of sound. Even at ordinary airplane speeds the effects of compressibility should not be disregarded if accurate measurements are desired.*[22]

For the most part, Stack's data in 1933 served to confirm the trends observed earlier. For example, Stack's measurements of the variation of drag coefficient with Mach number for a ten percent thick Clark Y airfoil are shown below; the large drag rise at high

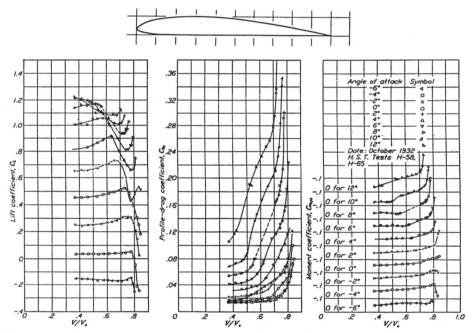

The first compressibility data published by John Stack. From NACA TR 463, 1933. The three graphs are, from left to right, the variations of lift coefficient, drag coefficient, and moment coefficient, respectively, versus the ratio of the freestream velocity to the speed of sound (the Mach number). The test model was a 3C1D airfoil, shown at the top of the figure. The adverse effects of compressibility are seen in the precipitous decease in lift coefficient and dramatic increase in drag coefficient as the Mach number is increased.

---

22. John Stack, "The N.A.C.A. High-Speed Wind Tunnel and Tests of Six Propeller Sections," NACA TR 463, 1933. At about the time of World War I, aerodynamicists were familiar with the fact that an airfoil stalled at high angle of attack because the flow separated from the top surface. The resulting drastic loss of lift was given the term "lift burble." Hence, after Briggs and Dryden had shown that the drastic loss of lift at high speeds, beyond the "critical speed," was also due to flow separation, it was natural to call this effect the "compressibility burble." This terminology, coined by the NASA in 1933, pervaded the high-speed aerodynamic literature throughout the 1930s.

speeds is clearly evident. He also confirmed that the onset of the adverse compressibility effects occur at lower Mach numbers as either or both the airfoil thickness and angle of attack increase. One of his conclusions reflected on the theoretical Prandtl-Glauert compressibility correction mentioned earlier. From his measurements, Stack concluded: "These results indicate that the limited theory available may be applied with sufficient accuracy for most practical purposes only for speeds below the compressibility burble." This conclusion presaged almost forty years of a theoretical void. The aerodynamic equations applicable to the transonic flight regime, Mach numbers between about 0.8 and 1.2 are non-linear partial differential equations that defied solution until the 1970s. And even then the solution was by brute force—numerical solutions using the power of the newly-developed discipline of computational fluid dynamics carried out on high-speed digital supercomputers.

By the way, the term "compressibility burble" was coined by Stack in the same NACA Technical Report. He wrote:

The lift coefficients increase as the speed is increased, slowly as the speed is increased over the lower portion of the range, then more rapidly as speeds above half the velocity of sound are exceeded, and finally at higher speeds, depending on the airfoil section and the angle of attack, the flow breaks down as shown by a drop in the lift coefficient. This breakdown of the flow, hereinafter called the compressibility burble, occurs at lower speeds as the lift is increased by changing the angle of attack of the model.[23]

Driven by the conviction and foresight of John Stack, the NACA now waved the red flag of compressibility problems to the whole world of aeronautical engineering. In January 1934, the first significant professional aeronautical society in the United States, the Institute of Aeronautical Sciences, published the first issue of its soon-to-be recognized premier journal, the *Journal of the Aeronautical Sciences*. It contained an article by Stack entitled "Effects of Compressibility on High Speed Flight."[24] In the first paragraph, Stack makes clear the theme that would be played out by the NACA for the next several decades:

The effects of compressibility have commonly been neglected because until the relatively recent development of the last Schneider trophy aircraft the speeds have been low as compared with the velocity of sound, and the consequent local pressures over the surfaces of high-speed airplanes have differed but slightly from atmospheric pressure. At the present time, however, the speeds associated with the fastest airplanes approach 60 percent of the velocity of sound, and their induced velocities over their exposed surfaces lead to local pressures that differ appreciably from the pressure of the atmosphere. When this condition exists, air can no longer be regarded as an incompressible medium. The effects of compressibility on the aerodynamic characteristics of airfoils have been under investigation by the N.A.C.A. in the high–speed wind tunnel, and it is the purpose of this paper to examine the possibility of further increases in speeds in the light of this relatively recent research.

By this time, it was clear that the NACA was the leading research institution in the world in the area of compressibility effects. Through its influence and sponsorship of the fledgling experiments in the 1920s by Caldwell and Fales at McCook Field, and by Briggs

---

23. *Ibid.*
24. John Stack, "Effects of Compressibility on High Speed Flight," *Journal of the Aeronautical Sciences* 1 (January 1934): 40–43.

and Dryden at the Bureau of Standards, and now by its own carefully conducted experiments at Langley, the NACA had been able to identify the first two aspects of the basic nature of compressibility effects, namely that (1) above a certain "critical speed," the lift decreased dramatically and the drag skyrocketed almost beyond comprehension, and (2) this behavior was caused by sudden and precipitous flow separation over the top surface of the wing or airfoil. There remained one question, the most important of all—*Why?*

John Stack and the NACA were responsible for the answer to this question—a breakthrough that occurred in 1934. By this time, Stack had a new instrument with which to work—a schlieren photographic system, an optical arrangement that made density gradients in the flow visible. One of nature's mechanisms for producing very strong density gradients is a shock wave; hence a shock wave ought to be visible in a schlieren photograph. Stack's boss, Eastman Jacobs, was familiar with such optical systems through his hobby of astronomy; it was in keeping with Jacob's innovative mind to suggest to Stack that the use of a schlieren system might make visible some of the unknown features of the compressible flow field over an airfoil, and might shed some light on the nature of the compressibility burble. It did just that, and more!

With the 11-inch tunnel running above the "critical speed" for an NACA 0012 symmetric airfoil mounted in the test section, and with the aid of the schlieren system, Stack and Jacobs observed for the first time in the history of aerodynamics a shock wave in the flow over the top surface of the airfoil. The shockwave was like that sketched in the figure below. It became immediately clear to these two experimentalists that the separated flow over the top surface of the airfoil, and the resulting compressibility burble with all its adverse consequences, was caused by the presence of a shock wave. The nature of this flow is sketched below, and it clearly shows that the shock wave interacts with the thin, friction–dominated boundary layer adjacent to the surface of the airfoil. This causes the boundary layer to separate in the region where the shock impinges on the surface. A massive region of separated flow trails downstream, greatly increasing the drag and decreasing the lift. One of the pioneering schlieren pictures of the flow over the NACA 0012 airfoil taken by Stack in 1934 is shown on the page 73.[25] The quality is poor by present-day standards, but it is certainly sufficient for identifying the phenomena. This is a historic photograph in the annals of the history of aerodynamics—one which led to the final understanding of the physical nature of the compressibility burble. This was a breakthrough of enormous intellectual and practical importance. And it was totally due to the work of two innovative and highly intelligent aerodynamicists at the NACA Langley Laboratory, John Stack and Eastman Jacobs, operating under the umbrella of an inspired creative atmosphere associated with the NACA in gen-

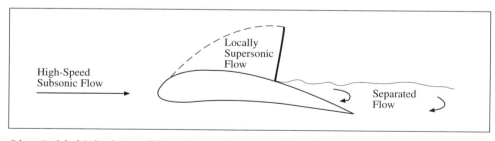

Schematic of shock-induced separated flow—the source of the compressibility burble.

---

25. Found by the author in the John Stack Files, NASA Langley Historical Archives.

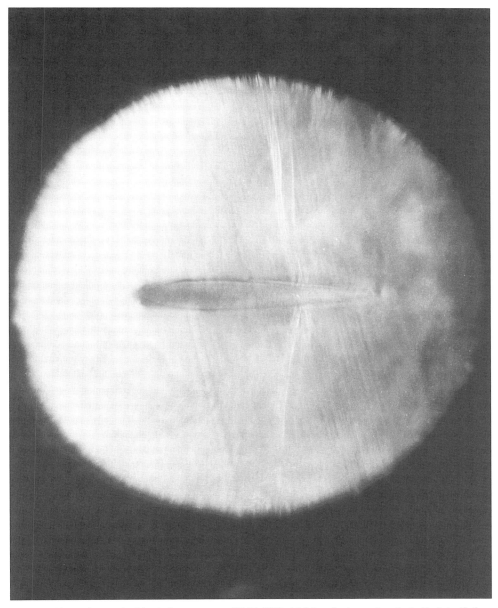

An early schlieren photograph of the shock pattern on an NACA 0012 airfoil in a freestream above the "critical speed". From the first group of schlieren photographs of the compressibility burble taken by John Stack, 1934. In this photograph the nature of the flow pattern causing the compressibility burble was seen for the first time. From the John Stack papers in the NASA Langley Archives. Courtesy of Richard Layman, Archivist.

eral, and the foresight of Joseph Ames and George Lewis at NACA Headquarters in Washington who placed priority on the NACA high–speed research program at a time when most airplanes of the day were lumbering along at 200 mph or slower.

Was this work by Stack and Jacobs engineering science? Absolutely yes! It provided the fundamental physical understanding of the root source of compressibility problems. This understanding was mainly qualitative at the time, but it allowed designers of high-speed airfoils to make more intelligent decisions about proper airfoil shapes—it helped to make the uncharted waters more navigable.

As with many new discoveries in science and technology, there are always those skeptical at first. One of those was Theodore Theodorsen, the best theoretical aerodynamicist in the NACA at the time, with a worldwide reputation for his pioneering papers on airfoil theory. John Becker, who joined the NACA in 1936 and who went on to become one of the most respected high-speed aerodynamicists at Langley, tells the following anecdote about Theodorsen's reaction to the schlieren photographs taken by Stack and Jacobs. It is repeated here because it reflects just how much a radical departure from the expected norm the results were.

> *The first tests were made on a circular cylinder about 1/2 inch in diameter, and the results were spectacular in spite of the poor quality of the optics. Shockwaves and attendant flow separations were seen for the first time starting at subsonic stream speeds of about 0.6 times the speed of sound. Visitors from all over the Laboratory, from Engineer–in–Charge H.J.E. Reid on down, came to view the phenomena. Langley's ranking theorist, Theodore Theodorsen, viewed the results skeptically, proclaiming that since the stream flow was subsonic, what appeared to be shockwaves was an 'optical illusion,' an error in judgement which he was never allowed to forget.*[26]

An interesting confluence of events occurred in 1935 that allowed the NACA in a timely fashion to inform the international research community of this intellectual breakthrough in understanding compressibility effects and the compressibility burble. One was the existence of the data itself—fresh, exciting, and revolutionary. The other was the scheduling of the fifth Volta conference in Italy.[27] Since 1931, the Royal Academy of Science in Rome had been conducting a series of important conferences sponsored by the Alessandro Volta Foundation. The first conference dealt with nuclear physics, and then rotated between the sciences and the humanities on alternate years. The second Volta conference had the title "Europe" and in 1933 the third conference was on the subject of immunology.

This was followed by the subject "The Dramatic Theater" in 1934. During this period, the influence of Italian aeronautics was gaining momentum, led by General Arturo Crocco, an aeronautical engineer who had become interested in ramjet engines in 1931, and therefore was well aware of the potential impact of compressible flow theory and experiment on future aviation. This led to the choice of the topic of the fifth Volta conference—"High Velocities in Aviation." Participation was by invitation only, and the select list included all the leading aerodynamicists at that time. Because of his reputation in the design and testing of the famous NACA four-digit airfoil series, and the fact that he was the Section Head of the NACA Variable Density Tunnel which had put the NACA on the international aerodynamic map in the 1920s, Eastman Jacobs received an invitation. He took the opportunity to present a paper on the new NACA compressibility research.

---

26. John V. Becker, *The High–Speed Frontier: Case Histories of Four NACA Programs, 1920–1950* (Washington, DC: NASA SP–445, 1980), p. 16.

27. Anderson, *Modern Compressible Flow*, pp. 282–84.

Hence, during the period between September 30 and October 6, 1935, the major figures in the development of high–speed aerodynamics of the 1930s (with the exception of John Stack) gathered inside an impressive Renaissance building in Rome that served as the city hall during the Holy Roman Empire, and discussed flight at high subsonic, supersonic, and even hypersonic speeds. The fifth Volta Conference was to become the springboard for new thought on the development of high–speed flight.

In the midst of all this discussion was Eastmann Jacobs representing the NACA. Jacobs' paper, entitled "Methods Employed in America for the Experimental Investigation of Aerodynamic Phenomena at High Speeds," was both tutorial and informative.[28] He took the opportunity to derive and present the basic equations for compressible flow assuming no friction and no thermal conduction. Then he described the NACA High–Speed Tunnel, the schlieren system, and the airfoil experiments carried out in the tunnel. Then came the blockbuster. He showed, for the first time in a technical meeting, some of the schlieren pictures taken at Langley. One of these was the photograph shown on page 73. Conscious of the NACA's penchant for perfection, especially in its publications, Jacobs apologized for the quality of the photographs, a very modest gesture considering their technical (and historical) importance: "Unfortunately the photographs were injured by the presence of bent celluloid windows forming the tunnel walls through which the light passed. The pictures nevertheless give fundamental information in regard to the nature of the flow associated with the compressibility burble."[29] With this, the NACA high–speed research program was not only on the map, it was leading the pack.

By this time, Stack had a newer, larger facility—the 24–inch High Speed Tunnel equipped with an improved schlieren system. The basic testing of compressibility effects on flows over airfoils continued in this facility. In 1938, Stack published the most definitive document yet on the nature of high–speed compressible flow over airfoils, including many detailed surface pressure measurements.[30] With this, the NACA continued to be the undisputed leader in the study of the effects of compressibility and the consequences of the compressibility burble.

The atmosphere at the Langley Laboratory during the 1930s allowed engineering science to flourish, although the laboratory never explicitly adopted this as a priority. It just happened when it needed to happen. The culture among its engineers was one of inquiry and free exchange of information; thoughts were readily shared on an interpersonal basis. Moreover, Langley had engineers who were adept at building new facilities, especially new wind tunnels. It was natural that a high–speed wind tunnel was built at Langely providing a unique facility for Langley engineers to unlock the secrets of high–speed aerodynamics. And the fact that the NACA had money, even during the depression years, allowed such wind tunnels to be first–class facilities. All this, in combination with first–class engineers and scientists, made Langley the leading research institution in high–speed compressibility effects during the 1930s.

Jacobs' paper at the fifth Volta conference represented in some sense a celebration of the second phase of the NACA research on high–speed flight. The first phase was the embryonic wind tunnel compressibility work of the 1920s, clearly oriented towards appli-

---

28. Eastman Jacobs, "Methods Employed in America for the Experimental Investigation of Aerodynamic Phenomena at High Speeds," NACA Misc. Paper No. 42, March 1936. A copy of this paper, which is the printed version of Jacobs' presentation at the fifth Volta conference, is available in the Technical Documents Section, Mathematics, Engineering and Physical Sciences Library, University of Maryland, College Park.
29. *Ibid*.
30. John Stack, W.F. Lindsey, and Robert E. Littell, "The Compressibility Burble and the Effect of Compressibility on Pressures and Forces Acting on an Airfoil," NACA TR 646, 1938.

cations to propellers. The second phase was the refocusing of this high-speed wind tunnel research on the airplane itself, complemented by a new initiative—the design and development of an actual research airplane.

The High-Speed Research Airplane: An NACA Idea

Hypothetical high-speed airplane conceived by John Stack, 1933.

The idea of a research airplane—an airplane designed and built strictly for the purposes of probing unknown flight regimes—can be traced to the thinking of John Stack in 1933. On his own initiative, Stack went through a very preliminary design analysis which, in his own words was "for a hypothetical airplane which, however, is not beyond the limits of possibility." The purpose of the airplane, as presented in his 1933 article in the *Journal of the Aeronautical Sciences*, was to fly very fast—well into the compressibility regime.[31] His design considered the airplane shown to the left; reproduced directly from his paper; here you see a highly streamlined airplane (for its time) with a straight, tapered wing having an NACA 0018 symmetric airfoil section at the center, and thinning to a 9 percent thick NACA 0009 airfoil at the tip. Stack even tested a model of this design (without tail surfaces) in the Langley Variable Density Tunnel. He estimated the drag coefficient for the airplane using the data he had measured in the eleven-inch High-Speed Tunnel. Assuming a fuselage large enough to hold a 2,300 horsepower Rolls-Royce engine, Stack calculated that the propeller-driven airplane would have a maximum velocity of 566 miles per hour—far beyond that of any airplane flying at the time, and well into the regime of compressibility. Stack's excitement about the possibilities for this airplane is reflected in the hand-drawn graph, reproduced on page 77. Drawn by Stack in 1933, this graph shows the horsepower required as a function of speed, comparing the results with and without the effects of compressibility. His hand sketch of the airplane is at the top of the graph (along with the aged rust marks of two paper clips). This graph was found by the author buried in the John Stack files in the Langley archives. The reason it is mentioned and reproduced here is that, barely distinguishable at the bottom of the reproduced graph, Stack had written "Sent to Committee Meeting, Oct. 1933." Stack was so convinced of the viability of his proposed research airplane that he had sent this quickly-prepared hand-drawn graph to the biannual meeting of the full committee of the NACA in Washington in October 1933. Ultimately the NACA did not act on helping Stack find a developer for the airplane, but in the words of Hansen, "the optimistic results of his paper study convinced many people at Langley that the potential for flying at speeds far in excess of 500 miles per hour was there."[32]

---

31. Stack, "Effects of Compressibility on High Speed Flight," pp. 40–43.
32. Hansen, *Engineer in Charge*, p. 256.

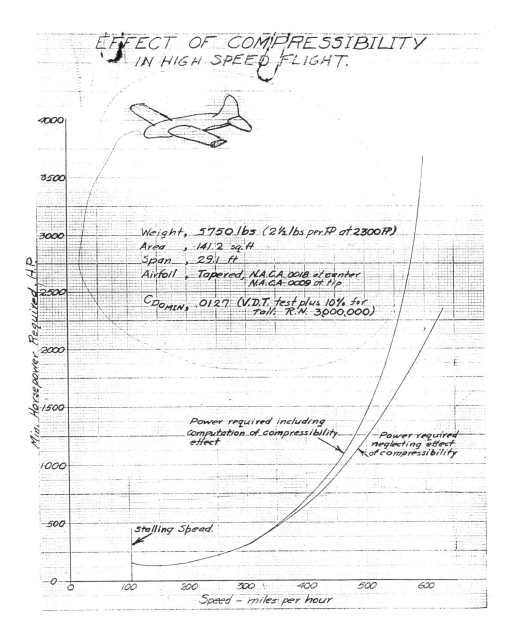

Graph and sketch hand-drawn by John Stack, 1933. The effect of compressibility on the power required for a hypothetical airplane. This sketch was subsequently sent to the October 1933 Committee Meeting of the NACA in Washington. From the John Stack papers at the NASA Langley Archives.

The High–Speed Research Airplane—For Real

The state of high–speed aerodynamics in 1939 can be illustrated by one graph, shown schematically in the figure below. Here, the variation of drag coefficient for an airplane is shown as a function of free stream Mach number. On the subsonic side, below Mach one, wind tunnel data indicated the familiar rapid increase in drag coefficient as Mach one is approached. On the supersonic side, ballisticians had known for years, supported by the results of linearized supersonic theory developed by Jakob Ackeret in Germany since 1928, how the drag coefficient behaved above Mach one.[33] Of course, all airplanes at that time were on the subsonic side of the curve shown in the figure below. John Stack nicely summarized the situation in 1938:

> *The development of the knowledge of compressible–flow phenomena, particularly as related to aeronautical applications, has been attended by considerable difficulty. The complicated nature of the phenomena has resulted in little theoretical progress, and, in general, recourse to experiment has been necessary. Until recently the most important experimental results have been obtained in connection with the science of ballistics, but this information has been of little value in aeronautical problems because the range of speeds for which most ballistic experiments have been made extends from the speed of sound upward; whereas the important region in aeronautics at the present time extends from the speed of sound downward.*[34]

In essence, the flight regime just below and just beyond the speed of sound was unknown—a transonic gap, as shown schematically below.

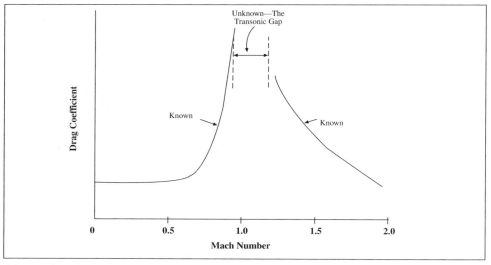

Schematic of the subsonic and supersonic variations of drag coefficient for an airfoil, illustrating the position of the transonic regime for which virtually no information was available in the 1930s and 1940s.

---

33. Anderson, *Modern Compressible Flow*, pp. 270–73.
34. Stack; Lindsey; and Littell, "Compressibility Burble and the Effect of Compressibility on Pressures and Forces Acting on an Airfoil."

The general aeronautics community was suddenly awakened to the realities of the unknown flight regime in November 1941, when Lockheed test pilot Ralph Virden could not pull the new, high-performance P-38 out of a high-speed dive, and crashed. Virden was the first human fatality due to adverse compressibility effects, and the P-38, shown below, was the first airplane to suffer from these effects. The P-38 exceeded its critical Mach number in an operational dive, and penetrated well into the regime of the compressibility burble at its terminal dive speed, as shown by the bar chart on page 80.[35] The problem encountered by Virden, and many other P-38 pilots at that time, was that beyond a certain speed in a dive, the elevator controls suddenly felt as if they were locked. And to make things worse, the tail suddenly produced more lift, pulling the P-38 into an even

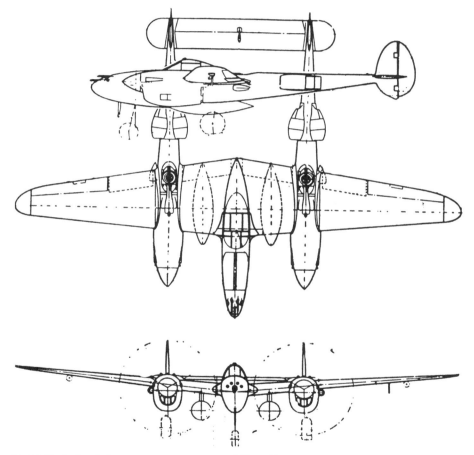

Lockheed P-38, the first airplane to encounter severe compressibility problems.

---

35. This chart is taken from the figure on page 78 of the article by R. L. Foss, "From Propellers to Jets in Fighter Aircraft Design," in Jay D. Pinson, ed., *Diamond Jubilee of Powered Flight: The Evolution of Aircrafeet Design* (New York, NY: American Institute of Aeronautics and Astronautics, 1978), pp. 51–64.

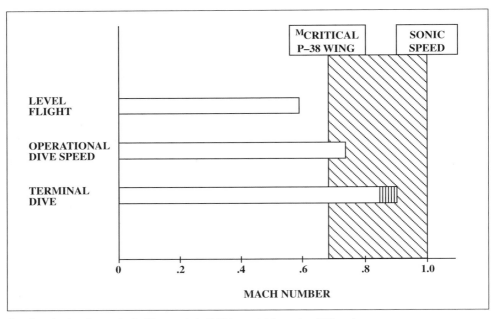

Bar chart showing the magnitude of how much the P-38 penetrated the compressibility regime.

steeper dive. This was called the "tuck-under" problem. It is important to note that the NACA soon solved this problem, using its expertise in compressibility effects. Although Lockheed consulted various aerodynamicists, including Theodore Von Kármán at Caltech, it turned out that John Stack at NACA Langley, with his accumulated experience in compressibility effects, was the only one to properly diagnose the problem. The wing of the P-38 lost lift when it encountered the compressibility burble. As a result, the downwash angle of the flow behind the wing was reduced. This in turn increased the effective angle of attack of the flow encountered by the horizontal tail, increasing the lift on the tail, and pitching the P-38 to a progressively steepening dive totally beyond the control of the pilot. Stack's solution was to place a special flap under the wing, to be employed only when these compressibility effects were encountered. The flap was not a conventional dive flap intended to reduce the speed. Rather, Stack's idea was to use the flap to maintain lift in the face of the compressibility burble, hence eliminating the change in the downwash angle, and therefore allowing the horizontal tail to function properly. This is a graphic example of how, in the early days of high-speed flight, the NACA compressibility research was found to be vital as real airplanes began to sneak up on Mach one.[36]

Indeed, it was time for *real* airplanes to be used to probe the mysteries of the unknown transonic gap. It was time for the high-speed research airplane to become a reality. The earliest concrete proposal along these lines was made by Ezra Kotcher, a senior instructor at the Army Air Corps Engineering School at Wright Field (a forerunner of today's Air Force Institute of Technology). Kotcher was a 1928 graduate of the University of California,

---

36. The "tuck–under" problem, and its technical solution, is described in John D. Anderson, Jr., *Introduction to Flight* (New York, NY: McGraw–Hill Book Co., 3rd ed., 1989), pp. 406–08.

Berkeley, with a B.S. degree in mechanical engineering. The same year that John Stack first walked through the gates of Langley as a junior aeronautical engineer, Kotcher first walked into the Air Corps Wright Field, also as a junior aeronautical engineer. These two engineers were contemporary with each other, and both had a driving interest in high-speed aerodynamics. The careers of these two people would come together for the development of the Bell X-1 in the 1940s. Kotcher's proposal, drafted during the period May–August 1939, was a response to Major General Henry H. "Hap" Arnold's request for an investigation of advanced military aircraft in the future. The proposal contained a plan for a high-speed *flight* research program. Kotcher pointed out the unknown aspects of the transonic gap, and the problems associated with the compressibility burble as elucidated by the NACA, and concluded that the next important step was a full-scale flight research program.[37] The Army Air Corps did not immediately respond to this proposal.

Meanwhile, back at Langley, the idea of a high–speed research airplane was gaining momentum. By the time the United States entered World War II in December 1941, John Stack had studied the behavior of the flow in wind tunnels when the flow in the test section was near or at Mach one. He found that when a model was mounted in the flow, the flow field in the test section essentially broke down, and any aerodynamic measurements were worthless. He concluded that the successful development of such transonic wind tunnels was a problem of Herculean proportions, and was far into the future. In order to learn about the aerodynamics of transonic flight, the only recourse appeared to be a real airplane that would fly in that regime. Therefore, during several visits by Dr. George Lewis, NACA's Director of Aeronautical Research, Stack seized the opportunity to mention the idea. Lewis, who liked Stack and appreciated the talent he brought to the NACA, was not immediately partial to the idea of a research airplane. But in early 1942, he left a crack in the door. In Hansen's words: "He left Stack with the idea, however, that some low-priority, back-of-the-envelope estimates to identify the most desirable design features of a transonic airplane could not hurt anyone, providing they did not distract from more pressing business."[38]

Given Stack's driving personality, this was all that was needed. With the blessing of the local management at Langley, Stack immediately formed a small group of engineers, and started to work on the preliminary design aspects of a transonic research airplane. By the summer of 1943, the group had produced such a design. Its principal features are listed below. This design established a mind-set for John Stack that guided NACA thinking on the transonic research airplane for the next five years—a mind-set that was to clash with the later ideas coming from Kotcher and the Army. The NACA design:

(1) was a small turbo-jet powered airplane,
(2) was to take off under its own power from the ground,
(3) was to have a maximum speed of Mach one, but the main feature was to be able to fly *safely* at high subsonic speeds,
(4) was to contain a large payload of scientific instruments for measuring the aerodynamic and flight dynamic behavior at near–sonic speeds, and
(5) was to start its test program at the low end of the compressibility regime, and progressively over time sneak up to Mach one in later flights.

---

37. Kotcher's role in the development of the high–speed research airplane is nicely presented by Hallion in *Supersonic Flight*, starting with p. 12, and continuing throughout the book. As stated in note 1 above, Hallion's book is still today the most definitive source on the circumstances leading to the Bell X–1.

38. Hansen, *Engineer in Charge*, p. 259.

The important goal was aerodynamic data at high subsonic speeds, not necessarily to fly into the supersonic regime. These features became [almost] a magna carta to Langley engineers, and to John Stack in particular.

The exigencies of wartime greatly accelerated research into high-speed aerodynamics; compressibility problems now had the attention not only of the NACA, but also of the Army and Navy as well. Stack, who had risen to be Eastman Jacob's Assistant Section Chief of the Variable Density Tunnel in 1935, and Head of the High-Speed Wind Tunnels in 1937, was made Chief of the newly formed Compressibility Research Division in 1943.[39] Stack now had his most influential position to date to push for the high-speed research airplane.

The Bell X–1: Point and Counterpoint

Although the NACA had the compressibility knowledge and technology, the Army and Navy had the money that would be necessary for the design and building of a research airplane. So it was appropriate that the Bell X-1 was conceived during a fateful visit by Robert J. Woods of Bell Aircraft to the office of Ezra Kotcher on 30 November 1944. Woods, who had NACA ties because he had worked at Langley during 1928–1929 in the Variable Density Tunnel, had joined with Lawrence D. Bell in 1935 to form the Bell Aircraft Corporation in Buffalo, New York. That day in November, Woods had dropped by Kotcher's office simply to chat. During the conversation, Kotcher relayed the information that the Army, with the help of the NACA, desired to build a special, non–military high–speed research airplane. After detailing the Army's specifications for the aircraft, Kotcher asked Woods if the Bell Corporation was interested in designing and building the airplane. Woods said yes. The die was cast.[40]

When Kotcher had been talking with Woods, he was operating with some authority. During 1944, Army and NACA engineers had been meeting to outline the nature of a joint research airplane program. Moreover, by mid-1944, Kotcher had received the Army's approval for the design and acquisition of such an airplane. However, the Army's concept of the high-speed research airplane was somewhat different than that of NASA. To understand this difference, we have to examine two situations in existence at the time.

The first situation was that of a common, public belief in the "sound barrier." The myth of the sound barrier had its beginning in 1935, when the British aerodynamicist W. F. Hilton was explaining to a newsman about some of the high-speed experiments he was conducting at the National Physical Laboratory. Pointing to a plot of airfoil drag, Hilton said: "See how the resistance of a wing shoots up like a barrier against higher speed as we approach the speed of sound." The next morning, the leading British newspapers were misrepresenting Hilton's comment by referring to "the sound barrier."[41] The idea of a physical barrier to flight—that airplanes could never fly faster than the speed of sound—became widespread among the public. Furthermore, even though most engineers knew differently, they still had uncertainty in just how much the drag would increase in the transonic regime, and given the low thrust levels of airplane powerplants at that time, the speed of sound certainly loomed as a tremendous mountain to climb.

---

    39.   Official NASA biographical and job description summary. From the John Stack files, Langley Historical Archives.
    40.   Hallion, *Supersonic Flight*, p. 34.
    41.   W.F. Hilton, "British Aeronautical Research Facilities," *Journal of the Royal Aeronautical Society* 70 (Centenary Issue, 1966): 103–104.

The second situation that colored the Army's thinking at that time was local, namely that Kotcher was convinced that the research airplane must be powered by a rocket engine rather than a turbojet. This stemmed from his experience in 1943 as project officer on the proposed Northrop XP–79 rocket–propelled flying wing interceptor, as well as the knowledge within the Army of Germany's new rocket–propelled interceptor, the ME–163.

Therefore, the Army viewed the high–speed research airplane as follows:

(1) *It should be rocket–powered.*
(2) *It should attempt, early in its flight schedule, to fly supersonically—to show everybody that the sound barrier could be broken.*
(3) *Later in the design process, it was determined that it should be air–launched rather than take off from the ground.*

All of these were in conflict with the NACA's more careful and scientific approach. However, the Army was paying for the X–1, and the Army's views prevailed.

Although John Stack and the NACA did not agree with the Army's specifications, they nevertheless provided as much technical data as possible throughout the design of the X–1. Lacking appropriate wind tunnel data and theoretical solutions for transonic aerodynamics, the NACA developed three stopgap methods for the acquisition of transonic aerodynamic data. In 1944, Langley carried out tests using the *drop–body* concept. Wings were mounted on bomb–like missiles which were dropped from a B–29 at an altitude of 30,000 feet. The terminal velocities of these models sometimes reached supersonic speeds. The data were limited, mainly consisting of estimates of the drag, but NACA engineers considered it reliable enough to estimate the power required for a transonic airplane. Also in 1944, Robert R. Gilruth, Chief of the Flight Research Section, developed the *wing–flow* method, wherein a model wing was mounted perpendicular at just the right location on the wing of a P–51D. In a dive, the P–51 would pick up enough speed, to about Mach 0.81, that locally supersonic flow would occur over its wing. The small wing model mounted perpendicular on the P–51 wing would be totally immersed in this supersonic flow region, providing a unique high–speed flow environment for the model. Ultimately, these wing–flow tests provided the NACA with the most systematic and continuous plots of transonic data yet assembled.[42] The third stopgap method was *rocket–model* testing. Here, wing models were mounted on rockets, which were fired from the NACA's facility at Wallops Island on the coast of Virginia's Eastern Shore. The data from all these methods, along with the existing core of compressibility data obtained by the NACA over the past 20 years as described in the earlier sections of this chapter, constituted the scientific and engineering base from which the Bell Aircraft Corp. designed the X–1.

Finally, we note that the NACA was responsible for the instrumentation that was housed *inside* the Bell X–1. This instrumentation and its location on the X–1 is illustrated on page 84. This is an example of one of those unseen aspects of technology upon which the acquisition of historic data depends. It is fitting that the NACA excelled in both aspects of the X–1 concept—the external configuration and the essential instruments mounted inside for the acquisition of quantitative knowledge.

---

42. Hansen, *Engineer in Charge*, p. 267.

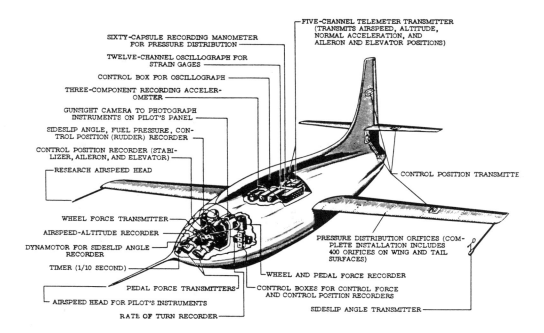

Schematic of the instrumentation mounted by the NACA in the Bell X-1.

Breaking the Sound Barrier

We began this chapter by transporting ourselves back to October 14, 1947, and riding with Chuck Yeager as he flew the Bell X-1 through the sound barrier, becoming the first human to fly faster than sound. The detailed events of 1946 and 1947 that finally resulted in this flight—the design, construction and early flight testing program by Bell, and the Army's intense preparations for the handling of the X-1 at Muroc—are nicely related by historians Richard P. Hallion and James O. Young.[43] Nothing is served by repeating them here. Rather, we return to the purpose of this chapter as stated in the introductory paragraphs. The first supersonic flight of the Bell X-1 represented the culmination of 260 years of research into the mysteries of high-speed aerodynamics. It was especially the fruition of 23 years of insightful research in high-speed aerodynamics by the NACA—research that represents one of the most important stories in the history of aeronautical engineering.

---

43. Hallion, *Supersonic Flight*; James O. Young, *Supersonic Symposium: The Men of Mach 1* (Edwards Air Force Base, CA: Air Force Flight Test Center History Office, September 1990), pp. 1–89.

On December 17, 1948, President Harry S. Truman presented the thirty-seven-year-old Collier Trophy jointly to three men for "the greatest aeronautical achievement since the original flight of the Wright Brothers' airplane."[44] The Trophy, officially the Collier Trophy for the year 1947, was the highest possible official recognition for the accomplishments embodied in the X-1. The announcement page from the December 25, 1948 issue of *Collier's* magazine is shown on page 86. Properly, John Stack was one of the three men, recognized as the *scientist*, along with Lawrence D. Bell, the manufacturer, and Captain Charles E. Yeager, the pilot. The citation to Stack read: "for pioneering research to determine the physical laws affecting supersonic flight and for his conception of transonic research airplanes." *A major purpose of this chapter was to bring meaning to this citation—* so much is hidden in these few words.[45] Unseen in this photograph, but present in spirit, is the team of NACA researchers who also worked towards determining the physical laws affecting supersonic flight, and to conceptualize the transonic research airplane. In this sense, the 1947 Collier Trophy was a "global" award to the entire NACA high–speed research program.

The 1947 Collier Trophy was also a recognition of the role of engineering science in the ultimate success of the Bell X-1. Note that in the award John Stack is explicitly recognized as a scientist (not an engineer). This is somewhat of a misnomer—Stack was performing as an *engineering scientist* in this activity, neither a pure scientist nor a pure engineer. The NACA had provided all the elements that allowed this engineering science contribution to occur.

At the time of this award, John Stack was Assistant Chief of Research at NACA Langley. In 1952, he was made Assistant Director of Langley. By that time he had been awarded his second Collier Trophy, the 1951 Trophy, for the development of the Slotted-Throat Wind Tunnel. In 1961, three years after the NACA was absorbed into the National Aeronautics and Space Administration, Stack became Director of Aeronautical Research at NASA Headquarters in Washington. Despairing of the de–emphasis of aeronautics in NASA, after thirty-four years of government service with the NACA and NASA, Stack retired in 1962 and became vice president for engineering for Republic Aircraft Corporation in Long Island. When Republic was absorbed by Fairchild Hiller in 1965, Stack was appointed a vice president of that company, retiring in 1971. On June 18, 1972, Stack fell from a horse on his farm in Yorktown, Virginia, and was injured fatally. He is buried in the churchyard cemetery of Grace Episcopal Church in Yorktown, only a few miles away from NASA's Langley Research Center. Today, F-15s from the nearby Langley Air Force Base fly over the churchyard—airplanes that can routinely fly at almost three times the speed of sound, thanks to the legacy of John Stack and the NACA high-speed research program.

---

44. *Collier's*, December 25, 1948.
45. John Stack files, NASA Langley Archives.

 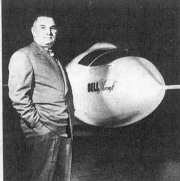 

SCIENTIST: John Stack, for the past 20 years a government research scientist with the National Advisory Committee for Aeronautics, is the first of the three men who share the award of the Collier Trophy for the achievement of human supersonic flight. It was because of Stack's awareness of the absolute necessity for ever superior aircraft, and his intensive study of problems of supersonic flight that a workable program for the construction of a research plane came into being.

MANUFACTURER: Lawrence D. Bell, president of Bell Aircraft Corporation, was awarded the contract by the Air Force to design and build the plane evolved from Stack's scientific presentation of supersonic flight. Bell has a reputation for taking on the unusual, the unconventional and what some called the impossible. The ship he designed and built was the Bell X-1 which, before delivery, was tested in 21 flights at a speed slightly less than that of sound.

PILOT: Captain Charles E. Yeager, USAF, was chosen from the nation's finest test-pilot talent as the man to fly the plane pioneered by Stack and built by Bell. Deemed "a natural airman, if there is such a thing," on October 14, 1947, Yeager became the first man to fly faster than the speed of sound. It is for the combined achievement of these three men in their successful penetration of the transonic barrier that the Collier Trophy for 1947 has been awarded.

The Collier Trophy
For Flight Beyond the Speed of Sound

By FREDERICK R. NEELY

For bringing about the achievement of human supersonic flight, John Stack, Lawrence D. Bell and Captain Charles E. Yeager, USAF, win America's highest aviation award

AMERICA'S highest aeronautical honor, the 37-year-old Collier Trophy, was presented by President Truman at the White House Friday, December 17th, to the three men adjudged most responsible for the attainment of human supersonic flight. The trophy is awarded annually by a committee selected by the National Aeronautic Association for "the greatest achievement in aviation in America, the value of which has been demonstrated by actual use during the preceding year." It will be shared equally for the ensuing year by:

John Stack, career government research scientist of the National Advisory Committee for Aeronautics "for pioneering research to determine the physical laws affecting supersonic flight and for his conception of transonic research airplanes."

Lawrence D. Bell, president of Bell Aircraft Corporation, "for the design and construction of the special research airplane X-1."

Captain Charles E. Yeager, U.S. Air Force, "who, with that airplane, on October 14, 1947, first achieved human flight faster than sound."

To those three men goes the honor of playing the major roles in an achievement which the Collier Trophy committee termed "the greatest since the first successful flight of the original Wright Brothers' airplane."

All three have been outstanding in their contributions to the vitally important science of supersonic flight—flight that is faster than sound, the speed of which at sea level, with a temperature of 59 degrees and in still air, is 761 miles an hour. However, at altitudes ranging between 40,000 and 100,000 feet, the speed of sound is reached at only 663 miles an hour. This is due to the fact that at such high altitudes the temperature is almost constantly 67 degrees below zero and sound travels more slowly in cold air. At just what altitude Capt. Yeager flew is as much of a secret as the actual supersonic speed he attained.

The problem that confronted Stack, Bell and Yeager was not so much that of flying faster than sound as it was successful flying at speeds between 600 and 900 miles an hour—the transonic range.

Aeronautical scientists were in grave doubt as to just what took place when conventional aircraft entered the transonic range in high-speed dives. They knew that both plane and pilot were kicked around unmercifully for seconds that seemed like centuries and that both were completely out of control. Badly and naturally frightened, the pilots were unable to bring back detailed scientific reports on the phenomenon, and they were usually unwilling to repeat their flights.

Wind tunnel tests with small-scale models revealed that the flow of air over a plane in the transonic range was partly subsonic and partly supersonic. Because of this, the conventional planes (usually fighter types) took on an extremely inconclusive and erratic behavior. But the tunnel findings were not conclusive and since supersonic tunnels large enough to mount a full-scale airplane are prohibitive in cost the scientists concluded they needed a special research airplane equipped with instruments capable of measuring and automatically recording all of the forces acting upon an airplane in transonic flight.

This was where John Stack came in. It was natural that he should have conducted the research phase for he had been working on the fundamental problems of high-speed flight in the wind tunnels and laboratories of the NACA at Langley Field, Virginia, since 1929, shortly after he had joined the government's great aeronautical research es-

Collier's for December 25, 1948

The first page of the Collier's *magazine announcement of the winners of the 1947 Collier Trophy, December 25, 1948.*

Chapter 4

The Transonic Wind Tunnel and the NACA Technical Culture

by Steven T. Corneliussen

When nuclear physics emerged as a compelling field of fundamental scientific inquiry during the 1920s, it needed new research tools, especially the invention of accelerators for probing nuclei with artificially energized subatomic particles.[1] Similarly, when the United States began expanding a national effort in applied aeronautical research during the 1920s, that too needed new research tools, especially improved wind tunnels for experiments using artificial airflows. Subsequent progress in both fields regularly resulted from research-tool advances—as subsequent Nobel Prizes regularly recognized, and as subsequent Collier Trophies did not.

By midcentury this contrast could be counted with the more obvious dissimilarities between the two fields. Though both nuclear physics and American aeronautics had continually required new empirical knowledge, their preeminent prizes since the 1920s had shown markedly differing esteem for advances in the means for generating it. In 1951, when particle-accelerator pioneers Sir John Cockroft and Ernest T. S. Walton won the Nobel Prize for physics, they joined previous laureates who had advanced nuclear science by inventing the cyclotron-type accelerator, the cloud chamber for making subatomic particle tracks visible, the magnetic resonance experiment method, further cloud chamber refinements, and a photographic technique for studying accelerator-generated nuclear processes. But until the Collier Trophy for that same year—save for the special case of 1947—the Collier's awarding committee had ignored research tools altogether, instead naming as the greatest advances in American aeronautics only aircraft equipment, air operations, heroic flights, and new airplanes. Yet aeronautical researchers with their continually improving research tools, especially the engineers and wind tunnels of the National Advisory Committee for Aeronautics, had contributed importantly to many of these advances. Thus the awarding committee for 1951 added importantly to the Collier's scope when it recognized the NACA's new transonic wind tunnels and the twenty NACA technical staff most closely associated with their advent.

---

1. Lord Ernest Rutherford, discoverer of the atom's nucleus, described in his 1927 "Anniversary Address as President of the Royal Society" a long-standing "ambition to have available for study a copious supply of atoms and electrons which have an individual energy far transcending that of the alpha and beta particles" available from naturally occurring radioactive sources in order to "open up an extraordinarily interesting field of investigation." Quoted in Mark Oliphant, *Rutherford: Recollections of the Cambridge Days* (Amsterdam, NY: Elsevier Publishing Company, 1972), p. 82. Daniel J. Kevles, *The Physicists: The History of a Scientific Community in Modern America* (New York, NY: Alfred A. Knopf, 1978), p. 227, cites Rutherford's desire during the 1920s for a "million volts in a soapbox." Concerning this and other topics, I am grateful for useful observations and information from historian of physics Catherine Westfall, whom I thank along with John V. Becker, Jay Benesch, Albert L. Braslow, H. Scott Butler, Francis J. Capone, Norman L. Crabill, James R. Hansen, J. D. Hunley, Peter Kloeppel, Richard T. Layman, Robert Riolo, Jim Spencer, Geoffrey Stapleton, and Walter G. Vincenti for reading this essay in manuscript form.

The Midcentury Need for Transonic Tunnels

In the 1947 special case, combat aviator Chuck Yeager flew manufacturer Lawrence Bell's new X-1 airplane faster than the speed of sound. Yeager thereby not only pierced the so-called sound barrier, but helped operate a transonic research tool conceived mainly by veteran NACA high-speed researcher and manager John Stack. The resulting Collier cited not only the heroic flyer and the airplane builder, but the NACA research-tool innovator as well.[2] Stack himself was not present in the California desert below the X-1 in its transonic research flights, but some of his NACA colleagues were. A detachment of engineers from Langley Memorial Aeronautical Laboratory in Virginia masterminded the experimental airplane's operation.[3] They instrumented it for data-gathering, planned and then observed each flight in detail, and assessed what was measured and recorded. They wanted new empirical knowledge of the bewilderingly complex, sometimes literally dangerous range of air speeds near the speed of sound, which varies with air temperature and can surpass 740 miles per hour.

Their NACA bosses at Langley Field and in Washington wanted transonic research advances too. Air speed had proven crucial in World War II, and jets were beginning to replace propeller-driven warplanes. In a high-profile 1946 assessment of the national defense program, Senator James M. Mead's special investigating committee had severely rebuked the NACA, charging past failures of "vision and imagination" concerning "revolutionary aeronautical developments" like Nazi Germany's missile technology and the jets that both Great Britain and Germany had developed in the 1930s, when the American aeronautical establishment still thought jets infeasible.[4] Accordingly, the NACA's 1946 annual report to Congress stated a resolve "to face the urgent necessity for renewed emphasis on fundamental research," as the NACA customarily called its practical-minded but scientifically grounded engineering studies. "Without certain essential design data," the report continued, "the development of very high-speed aircraft and guided missiles cannot proceed."[5] That word *urgent* recurs concerning transonics throughout NACA documents of the early postwar era, when air-war memories were fresh, Cold War worries were intensifying, and NACA bureaucratic-war strategies were beginning to target the Army Air Forces. Like the NACA, the AAF—soon to become the Air Force—

---

2. Richard P. Hallion, *Supersonic Flight: The Story of the Bell X-1 and Douglas D-558* (New York, NY: Macmillan Company, 1972), p. 176, notes that research airplanes like the X-1 were not "fabricated for setting records. Rather, they were designed as research tools. Though they set some spectacular records . . . their main function remained unchanged: the acquisition via flight instrumentation of data on a variety of areas." The NACA's 1954 annual report, p. 4, says the research airplanes' "prime justification was as tools to be used in developing necessary transonic information." (NACA annual reports are cited hereafter in the form *AR54.*) The 1947 Collier, following the frequent practice of the day, cited engineer Stack as a "scientist." But Stack's 1928 MIT degree was in aeronautical engineering, as reported in James R. Hansen, *Engineer in Charge: A History of the Langley Aeronautical Laboratory, 1917–1958* (Washington, DC: NASA SP-4305, 1987), appendix B. The influence of Hansen's engineering-centered interpretation of NACA research history pervades this essay.

3. To counter the notion of military control of "the research direction" of the X-1 program, Richard P. Hallion emphasizes the NACA's "virtual total control" in his review of Walter A. McDougall's *. . . the Heavens and the Earth: A Political History of the Space Age* (New York, NY: Basic Books, 1985); see *Technology and Culture* 28 (January 1987): 130–32.

4. Excerpt from Mead committee report, "Miscellaneous" folder, John Stack collection, Langley Historical Archive—hereafter called *LHA*—NASA Langley Research Center library. For LHA access and much else, I thank Langley historical program coordinator Richard T. Layman.

5. AR46, p. 2.

aspired to create and control expensive new national aeronautical research tools: large wind tunnels for experiments using artificial transonic and supersonic airflows.[6]

As a motivation for high-speed research, the urgency of international military competition—though not that of Washington political competition—shows in separate, representative pronouncements by the NACA and AAF research directors in 1947. "The urgency of aeronautical research results from the relation of air power to national security," reported Hugh L. Dryden to the NACA's main committee a few days after the X-1's famous October flight. "Aircraft having the highest speed dominate the air," he noted, adding—in a complete reversal of the NACA's cautious prewar belief—that it was "clear that there is no upper limit to the possible speed of aircraft." Dryden declared that "the nation that makes the best research effort to develop the new power plants and explore the problems of high-speed flight can lead the world in air power. That nation must be the United States. . . . It is the duty of the NACA to provide for the military services and the industry the basic data on aerodynamics and propulsion to make piloted supersonic flight not only possible, but safe and reliable."[7] In even more forceful terms, these themes had also appeared that April in a magazine editorial titled "We Must Furnish the Tools" by Maj. Gen. Curtis E. LeMay, the aggressive World War II strategic bombing leader who now headed the AAF Research and Development Agency. So emphatic was this two-page argument for new national high-speed wind tunnels that John Stack kept a photostat of it in personal papers now preserved in the NASA Langley Research Center historical archive. LeMay's editorial warned that for lack of proper research tools the United States risked losing the air-superiority race. In World War II, it said, the Nazis had been "at least five years ahead," though fortunately not in actually "applying the results of their technical superiority." In the postwar world, however, "even a one- or two-year lag" could probably "never be recovered."[8] Similar arms-race language concerning wind tunnels also appeared a few years later when the *Collier's* magazine article announcing the 1951 Collier Trophy headlined the awarding committee's assertion: "Now the U.S. has a two-year lead on the Communists in perfecting vital faster-than-sound planes."[9]

Harder to see in the late 1940s were the urgent political and bureaucratic motivations involved in the high-visibility push for new national aeronautical research facilities. Dryden and LeMay wrote only about the cooperation, not the rivalry, between the NACA and the Army Air Forces. But NACA historian Alex Roland has described a postwar NACA "at its nadir in reputation and influence" struggling "in deep and surreptitious competition" with the AAF.[10] Thus for Hugh Dryden in Washington and John Stack at Langley, the NACA's organizational self-interest must have accompanied the arms-race justification as a motivation to develop technology, and to seek construction funding, for new high-speed research tools.

---

6. Alex Roland, *Model Research: The National Advisory Committee for Aeronautics, 1915–1958*, 2 vols. (Washington, DC: NASA SP-4103, 1985), discusses the Mead committee and other postwar forces acting on the NACA; see chapters 9 and 10. The Stack collection, LHA, includes several folders of Stack's planning materials for postwar national wind tunnel facility construction. It must be noted that in an April 3, 1996, telephone interview, NACA and NASA high-speed research veteran John V. Becker recalled no particular urgency in the day-to-day postwar transonics work at Langley Field, whatever the outlook and motivations of the NACA itself might have been. I conducted all of the telephone interviews cited, retained electronic notes from each, sent a draft of this essay to every interviewee, and am indebted to all of them.

7. "Report of the Director of Aeronautical Research submitted to the National Advisory Committee for Aeronautics at its annual meeting, October 23, 1947," reprinted Roland, *Model Research*, 2:713–16; quotations from p. 714.

8. *Aero Digest*, April 1947, pp. 14–15; photostat in "Miscellaneous" folder, Stack collection, LHA.

9. *Collier's*, December 20, 1952, pp. 24–25.

10. Roland, *Model Research*, 1:259 and 1:214 .

In any case, wind tunnels were the desired tools. To most American aeronautical researchers it seemed clear that whatever the usefulness of research aircraft for transonics, truly comprehensive empirical knowledge in the long run would have to come mainly from these ground-test facilities with their convenient, versatile, relatively affordable, and safe laboratory conditions.[11] In the distinct NACA technical culture especially, airborne tests represented a component that could only complement, not replace, the wind-tunnel-test component. Although the airflow physics of a purely supersonic tunnel differs fundamentally from that of a subsonic tunnel, the NACA already had effective supersonic tunnels when the X-1 flew in 1947, and at Langley in the following month John V. Becker even began operating a small hypersonic tunnel that could reach speeds well beyond five times that of sound.[12] But in the airflow of high-subsonic, or near-sonic, wind tunnels—tools for the main transonic parts of the work that research directors Dryden and LeMay were emphasizing—complex troublesome effects arose, hampering tunnel operation and polluting or even ruining experimental data. No tunnel had yet been invented for overcoming these vexing transonic effects, despite NACA efforts dating back to the 1920s, despite efforts elsewhere, and despite a long-standing intuition that Stack and others shared about how to solve the problem.

So during the X-1's research flights in 1947, Stack—a high-speed wind tunnel innovator since 1928, and now a research manager—was not present in the California desert. Instead he was back at Langley, encouraging, smoothing the way for, and cajoling others who were trying to synthesize years of NACA experience to capitalize on that intuition and develop that solution. "Aeronautical experts swore it couldn't be done," the *Collier's* headline would trumpet once they had succeeded. But in reality engineers had long suspected that it could indeed be done, and that the answer would lie in somehow partly opening up a wind tunnel's walls. Just after the war Langley physicist Ray H. Wright, skilled in applied mathematics and widely knowledgeable concerning tunnel technology,[13] had used subsonic aerodynamic theory to calculate a solution: a tunnel with ventilation slots in the walls of its test section, the experiment chamber where the tunnel's artificial airflow moves across an instrumented test subject such as a scale-model segment of a wing. These test-section slots had to be precisely placed, paralleling the airflow direction, in the tunnel's interior surfaces above, below, and beside the test subject, which might either span a roughly cylindrical test section or be held in place by an apparatus behind it downstream. Wright and Stack and their colleagues hoped that these longitudinal openings could manipulate the complexities of air flowing at up to sonic speed, channeling the air around the test subject in just such a way as to yield valuable transonic research data.

In 1947 Langley was already trying out the slotted-wall idea in the test section of a small pilot tunnel, and had learned, apparently serendipitously, that the slots enabled smooth operation not just at very high subsonic speeds, but at low supersonic speeds too. By the time of Yeager's famous research flight that October, Stack had long since begun considering how to apply the slotted-wall results in two full-size high-speed tunnels—industrial-scale facilities with huge powerful fans and test-section diameters of eight feet and sixteen feet, sizable by any era's standards. With Ray Wright's specific design concept, Stack's vision and leadership, engineer Vernon G. Ward's technology-development contributions, and the NACA Langley technical staff's wind tunnel expertise and experience, the research-and-development effort relatively soon led to the conversion of these two

---

11. Hallion, *Supersonic Flight*, p. 45, reiterates in ch. 2 what he has made clear throughout ch. 1: "The principal reason" for transonic research aircraft "was the inability of existing wind tunnels to furnish satisfactory and reliable transonic aerodynamic data."
12. Hansen, *Engineer in Charge*, pp. 467, 471, and 344–47.
13. The end of this essay addresses conflicting interpretations of the breadth of Wright's technological awareness.

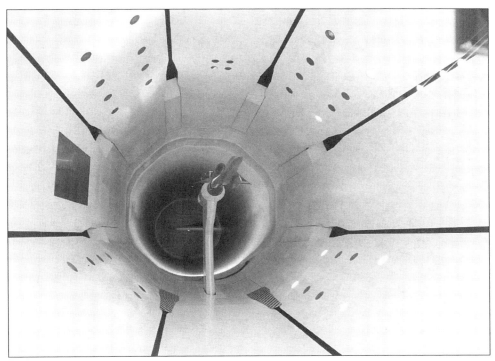

The present-day slotted-wall test section of the NASA Langley 16-Foot Transonic Tunnel, which began operating as the NACA Langley 16-Foot High-Speed Tunnel two days before Pearl Harbor. The tunnel's name derives in part from its test section's approximate diameter. The slotted-wall configuration shown here descends directly from the one in this tunnel that helped win the 1951 Collier Trophy. (NASA L-90-04029).

national research facilities: the now-retired 8-Foot High-Speed Tunnel, designated a national landmark in 1985, and the 16-Foot High-Speed Tunnel, later called the 16-Foot Transonic Tunnel and still operational with slotted walls in 1998. The resulting Collier Trophy for Stack and nineteen of his colleagues was the first ever awarded outright for a research tool, and the only Collier ever awarded for a ground-based one—even though, as with particle accelerators and detectors for nuclear science, wind tunnels have been crucially important for American aeronautics.[14]

---

14. "From the time of the Wright brothers, the wind tunnel . . . proved to be the essential piece of versatile experimental machinery on which much about the progressive evolution of aircraft depended," writes James R. Hansen in *Spaceflight Revolution: NASA Langley Research Center from Sputnik to Apollo* (Washington, DC: NASA SP-4308, 1995), p. 436, restating a main message of his earlier *Engineer in Charge*. "The wind tunnel dominates aeronautical research just as the microscope dominates biology, the telescope astronomy, and the particle accelerator nuclear physics," writes Roland in *Model Research*, 1:xiv. In this essay I do not address "tunnel vision"—Roland's name for a criticism of the NACA occasionally mentioned but seldom forthrightly leveled: that its engineers too often allowed research tools, especially wind tunnels, to dictate rather than merely serve research programs. In *Model Research* see especially 1:xiv–xv, but also 1:108, 220–21, and 309 and 2:507 and 520; see also Edward W. Constant, *Isis*, 73:4:269 (1982) 609–10.

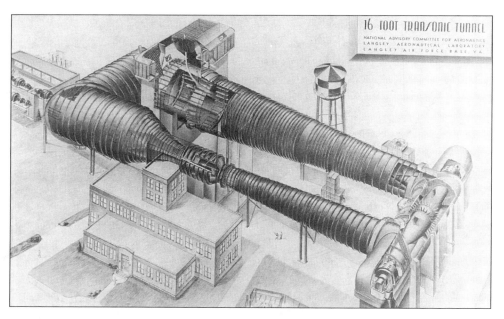

The 1951 Collier Trophy recognized a transonic-research-technology advance first applied in the two NACA Langley wind tunnels shown in these midcentury views.[15] Top: Air flows counterclockwise in the 16-Foot Transonic Tunnel, passing repeatedly through the test section linked to the topmost floors of the facility's brick office building. Bottom: In the 8-Foot High-Speed Tunnel's similar circuit, a concrete igloo enclosed the test section. (NASA photos L-90-3752 and NACA 12000.1).

15. The photographed artist's drawing is from the early 1950s, the actual photograph from earlier still. The modern 16-foot tunnel circuit has an air-removal system for enhanced low-supersonic operation.

In fact, given the wind tunnel component in the NACA's overall contributions, a Collier Trophy for an NACA tunnel seems fitting, as three low-subsonic examples from the 1920s and early 1930s illustrate. Each was the first of its kind in the world, and was soon copied elsewhere.[16] The Variable-Density Tunnel, or VDT, could, with fairly good success for the time, counteract *scale effects*—the skewing of test data inherent in testing scale models instead of full-scale aircraft or aircraft components. By the early 1930s, according to aeronautics historian Richard K. Smith, VDT-generated information published in formal NACA reports enabled aircraft designers to select a wing shape for a given application incisively, rationally, and conveniently.[17] That the VDT became an official national landmark in 1985 may help validate its historical significance. The Propeller Research Tunnel, or PRT, circumvented scale effects and other technical difficulties simply by being powerful enough, and large enough in its test-section diameter of twenty feet, to test at full scale a propeller and engine mounted on an actual fuselage or on a portion of a full-size wing. Several observers have noted that the NACA's first Collier Trophy, the one for the speed-enhancing engine cowling discussed in chapter 1, might well have recognized instead the PRT, the research tool that enabled the cowling's development.[18] The Full-Scale Tunnel, or FST, operational for nearly two-thirds of a century starting in the early 1930s, took the PRT's full-scale-testing principle one step further: in its thirty- by sixty-foot test section it could hold an entire small airplane. The FST was also designated a national landmark in 1985.

With a technical staff continually devising such tunnels and other research tools, the subsonic-era NACA became widely recognized for its applied aeronautical research. The organization became highly adaptable for fulfilling its statutory charge of finding practical solutions to the problems of flight—problems eventually defined as including the aerodynamics, and somewhat belatedly the aeropropulsion, of transonic and supersonic flight. In fact, during the 1920s and 1930s the NACA's earliest efforts in transonics began to grow out of its extensive subsonic efforts, and ultimately led to the transonic wind tunnel for which the 1951 Collier Trophy recognized "John Stack and associates at the Langley Aeronautical Laboratory, NACA." So besides celebrating the slotted-wall transonic tunnel's promise for jets, and beyond finally recognizing one representative NACA wind tunnel, the Collier Trophy for that year illuminates the effectiveness of the research-tool-centered NACA technical culture.

---

16. Useful sources on NACA wind tunnel history include Hansen's *Engineer in Charge* and Donald D. Baals and William R. Corliss, *Wind Tunnels of NASA* (Washington, DC: NASA SP-440, 1981). Roland, *Model Research*, 2:508–14, lucidly explains wind tunnels and tunnel technology.

17. "Better: The Quest for Excellence," in *Milestones of Aviation*, p. 241, ed. John T. Greenwood (New York, NY: Hugh Lauter Levin, 1989), 222–95.

18. Roland, *Model Research*, 1:117; Hansen, *Engineer in Charge*, p. 134; Hansen, chap. 1 in this volume; John V. Becker, *The High-Speed Frontier: Case Histories of Four NACA Programs, 1920–1950* (Washington, DC: NASA SP-445, 1980), p. 140. For the present essay and much else, Becker's book is centrally important as both a primary and a secondary source. In its introduction, Becker says he wrote it as a "participant-author" because the NACA's research solutions actually evolved as more than just "the inevitable result of wise management, inventive researchers, and unparalleled facilities," and because he believes that to "provide fundamental insights into the NACA's technical accomplishments the record should include the doubts and misconceptions that existed in the beginning of a project, the unproductive approaches that were tried and abandoned, the stimulating peer discussions that provided new insights, and the gradual evolution of the final solution. This kind of information is hard to find." Edward W. Constant, reviewing the book in *Isis*, 73:4:269 (1982) 609–10, calls it (p. 609) "an extraordinary glimpse into a whole category of technological knowledge not commonly covered either by the history of science or by the history of technology."

A Technological Organization's Group Achievement

Academic or Nobel Prize–like norms for assigning credit were only partly relevant in the Collier's recognition of the transonic tunnel achievement, for the cited triad of "conception, development, and practical application" of the slotted wall included effective work outside the purely intellectual realm. In fact, the Collier for 1951 required distinguishing among specific kinds of contributions as well as among contributors, including the technological organization itself—though the Collier committee at first adopted a simpler view. A look at how and where slotted-wall credit has been conferred, both by the Collier and by other means since, may show something about NACA-era views of the nature of technological achievement, and does show the central importance of a well-integrated technical culture in the NACA's work.

The slotted-wall achievement did have an important intellectual component, as Stack's technical peers have duly recognized in later citations and discussion in aeronautical publications. But Collier Trophy notwithstanding, they have not credited Stack. Although the Collier committee singled him out, and in fact originally intended the award for Stack alone, for over half a century Stack's professional peers have generally attributed the origin of slotted walls either by crediting the NACA generally or by citing the 1948 paper of Stack's Collier-winning "associates" Ray Wright and Vernon Ward, the engineer who spearheaded proof of the slotted-wall principle with the first small pilot tunnel.[19] Technical authors have left Stack not only uncited but unmentioned, even in passages that summarize historical background. It must be noted that Stack's rise within Langley management during the 1940s meant fewer papers from him and, when he did write, a broad-overview approach not conducive to academic citation.[20] And it must also be noted that Stack quite possibly intended not to take academic credit; *Wind Tunnels of NASA* author Donald D. Baals, one of Stack's Collier-sharing associates, said in 1996 that Stack might well have intended to send credit Wright's way.[21] Another associate, veteran NACA and NASA high-speed researcher John V. Becker, emphasizes the distinction between kinds of contributions. His book *The High-Speed Frontier: Case Histories of Four NACA Programs, 1920–1950* says unambiguously that the "first successful many-slotted transonic tunnel configuration was devised single-handedly by Ray H. Wright," that Wright was "the designer of the transonic tunnel," that "Wright's personal decision in 1945 to get down to cases" initiated the multiyear transonic tunnel effort, most of which "clearly bears the stamp of

---

19. NACA Research Memorandum L8J06, "NACA Transonic Wind-Tunnel Test Sections." The folder "Standardization of Wind Tunnels, October 13, 1948–Thru Feb. 1949" in the Research Authorization 70 file, LHA, contains this paper's approval and distribution paperwork as well as the October 6, 1948, final editorial copy. (Hansen, *Engineer in Charge*, pp. 572–74, explains the usefulness, and the use, of the LHA's research authorization files, hereafter cited in the form *RA70*. With two linear feet of documents, RA70 traces much of the evolution of wind tunnel technology from the early 1920s to the early 1950s.) The NACA republished the Wright-Ward paper in 1955 as Technical Report 1231, but changed it somewhat, mainly by deleting a paragraph near the end reporting lack of understanding of the low-supersonic capability and by slightly altering conclusions 4 and 6. The NASA Langley library holds the original 1948 RM version on microfiche. Key antecedents for the 1948 paper include Ray H. Wright, Physicist, and Vernon G. Ward, Aeronautical Engineer, to Compressibility Research Division Files, "Tunnel Wall Interference Effects in an Axially Slotted Test Section—Preliminary Tests," March 12, 1947 (Stack collection folder "New Types of Wind Tunnels, 1947," LHA) and Wright to Chief, Full-Scale Division, "Theoretical consideration of the use of axial slots to minimize wind-tunnel blockage," May 24, 1948 (Stack collection folder "Slotted-Throat Tests, 1946–48," LHA). The latter says the "theoretical investigation" it means "to record and preserve" may "later be combined and published with the results" of an experiment in progress, obviously the Wright-Ward pilot-tunnel experimentation—and indeed the eventual Wright-Ward paper reflects much from Wright's memo.
20. Becker, *High-Speed Frontier*, pp. 52, 53.
21. Telephone interview, April 7, 1996.

Wright's insights and personal integrity," but that it "is equally clear that without the enormous contributions of a quite different kind made by Stack, the achievement of the large slotted tunnels would not have happened" as soon as it did.[22]

The practice of excluding Stack from credit appears to have begun well before 1951, and it has continued for half a century. In October 1948, NACA research director Hugh Dryden began limited, high-priority circulation of the Wright-Ward paper. Within days, Clark Millikan of the Guggenheim Aeronautical Laboratory wrote to congratulate the NACA and to express hope for "following the lead given by Messrs. Wright and Ward." His letter does not mention Stack. Within weeks, Air Force wind tunnel expert Bernhard H. Goethert, formerly of the German aeronautical research establishment, visited Langley; Dryden had officially informed the military about the slotted wall's "revolutionary nature," and Goethert hoped to learn how to apply it. Wright, Ward, and Stack himself, together with engineer Eugene C. Draley, hosted Goethert's intensive visit and tour.[23] Yet Goethert wrote in his 1961 book *Transonic Wind Tunnel Testing* that the "first really successful transonic wind tunnel was investigated in the United States in 1947 in tests at the NACA." The passage footnotes Wright and Ward and leaves Stack unmentioned. Moreover, Stack's name barely appears at all in Goethert's book, an exhaustive survey of a research technology that the 1951 Collier Trophy credits Stack above all others with founding.[24] Similar attribution patterns appear in a 1955 NACA paper that in part reviews past NACA slotted-wall work, in a 1960 Air Force paper summarizing that service's wind-tunnel-development efforts, and in the 1965 textbook *High-Speed Wind Tunnel Testing*.[25] Stack's exclusion persisted in the mid-1990s at NASA Langley Research Center, where two papers addressed the slotted-wall issues that Wright and Ward first discussed in print. Both explicitly attribute the technology's origin to Wright and Ward. Neither mentions Stack, though upon inquiry, each principal author readily confirms clear awareness of him. One of these papers surveys the characteristics and technical history of what is now called the 16-Foot Transonic Tunnel, one of the two large Langley facilities where "practical application" of the slotted wall helped earn the 1951 Collier Trophy for Stack and his associates.[26]

---

22. Becker, *High-Speed Frontier*, pp. 99, 112, 115. In a July 15, 1988, letter to historian Hansen (copy in my files), Ward asserted a credit-claiming version of "the true facts in regard to the elimination of choking in wind tunnels and the developmental and design research of the NACA Transonic Wind Tunnel." Certainly his pilot-tunnel efforts did contribute importantly, and apparently he did personally discover the unexpected low-supersonic capability. However, his recollections conflict with the documentary record, discussed later in the present essay, concerning the clarity, and thus the priority, of Wright's 1946 expectations and intentions for the near-sonic significance of the theoretical work Wright began conducting before Ward became involved.

23. The RA70 folder "Standardization of Wind Tunnels, October 13, 1948–Thru Feb. 1949," LHA, contains a copy of Draley's December 21, 1948, memorandum reporting Goethert's visit in detail (clearer copy in folder "Special File, R.A. 70, April 1947 Thru Dec. 48") and a copy of Millikan to Dryden, October 19, 1948—of which a signed copy is in the "Research Authorization 70" folder, together with copies of Dryden's October 8, 1948, "revolutionary nature" letters to military research flag officers.

24. Goethert, *Transonic Wind Tunnel Testing* (New York, NY: Pergamon Press, 1961), p. 23, but see also p. 61. (Publication of the North Atlantic Treaty Organization's Advisory Group for Aeronautical Research and Development, edited by Wilbur C. Nelson, from a series by the NATO-AGARD Fluid Dynamics Panel, which under an earlier name had also published Goethert's important paper "Flow Establishment and Wall Interference in Transonic Wind Tunnels," AEDC-TR-54-44, pp. 247–292 in *AGARD Memorandum: Papers Presented at the Sixth Meeting of the Wind Tunnel and Model Testing Panel, Paris, France, 2–6 November 1954*, AG17/P7.)

25. B. H. Little, Jr., and James M. Cubbage, Jr., "The Development of an 8-Inch by 8-Inch Slotted Tunnel for Mach Numbers up to 1.28," NASA TN D-908, August 1961, originally published January 1955 as classified NACA RM L55B08; M. Pindzola and W. L. Chew, "A Summary of Perforated Wall Wind Tunnel Studies at the Arnold Engineering Development Center," AEDC TR-60-9, August 1960; and Alan Pope and Kenitth L. Goin, *High-Speed Wind Tunnel Testing* (New York, NY: John Wiley & Sons, Inc., 1965), pp. 103, 104.

26. Joel L. Everhart and Percy J. Bobbitt, "Experimental Studies of Transonic Flow Field Near a Longitudinally Slotted Wind Tunnel Wall," NASA Technical Paper 3392, April 1994, and Francis J. Capone, Linda S. Bangert, Scott C. Asbury, Charles T. L. Mills, and E. Ann Bare, "The NASA Langley 16-Foot Transonic Tunnel: Historical Overview, Facility Description, Calibration, Flow Characteristics, and Test Capabilities," NASA Technical Paper 3521, September 1995.

So why did the Collier committee members plan originally to cite Stack alone? Possibly they simply wanted a heroic interpretation like that in James Michener's 1982 novel *Space*, which attributes the transonic tunnel solely to "a genius named John Stack" who had a "brilliant idea" that led to "airplanes that could break through the sound barrier almost as undisturbed as a horse-drawn carriage heading for a country picnic in 1903." Possibly the committee's initial plan reflected a view like that of Orville Wright, who—no doubt remembering what actually led up to 1903—had complained in 1944 that Colliers were going too often to aviation organizations instead of innovative individuals. Possibly the intention reflected public relations aims of the NACA, whose executive secretary and chief propagandist John F. Victory chaired the Collier committee for 1951. The NACA apparently had a long-standing involvement in the award selection, and in at least one case—1947, when it seemed certain the NACA would be among those recognized—had calculated possible combinations of recipients to promote.[27]

If the committee members did intend the heroic interpretation, probably they wanted to lend a bit of romantic appeal to an award for an unromantic, ground-based research tool. Historian John William Ward has analyzed an analogous and much better known instance of credit-assigning in American aeronautics: the case of Charles Lindbergh. Concerning the adulation of Lindbergh, Ward observes that it is "strange that the long-distance flight of an airplane, the achievement of a highly advanced and organized technology, should be the occasion of hymns of praise to the solitary, unaided man." He describes a tension inherent in Americans' understanding of the new phenomenon of aviation: their identification with pioneering, self-reliant, free individuals versus their lack of interest in the collectivized, organized industrial society such individuals often actually represented. Possibly the Collier committee saw and sought to avoid such a tension in the choice between the pioneering Stack and the technological organization he represented. After all, this was already going to be the only Collier ever given for something so likely to be seen as inherently boring: not a heroic flight, not a new airplane, not a successful aviation program, not an improvement in airplane equipment. Just a wind tunnel, a noisy industrial plant for turning out research data. The NACA itself is the analog of the uninteresting and therefore uncredited collectivized industrial society in the Lindbergh achievement, but the analog of the lionized Lindbergh himself is John Stack, already identified by an earlier Collier as a pioneering individual for conceiving the plane that broke the sound barrier. A *Washington Post* article the week after that earlier award had said he didn't "look like a man of science" but was instead "a rather handsome fellow whom you'd take for a lawyer, a football coach, or even an actor."[28]

In any case, in public relations and other nontechnical realms the Stack-alone interpretation lived on even after the 1951 award actually did partly credit members of the technological organization that stood behind Stack. The 1954 NACA annual report tilts toward such a description, emphasizing Stack's primacy in the achievement. In a 1957 speech, NACA executive secretary Victory tilted all the way: he portrayed the accomplishment as an individual one, and flatly attributed it to Stack alone. At the 1962 ceremony

---

27. James A. Michener, *Space* (New York, NY: Random House, 1982), p. 175. *Model Research*, 1:351 n. 6 discusses Wright's complaint. In "George William Lewis," *Year Book of the American Philosophical Society*, 1948, pp. 269–78, NACA chairman Jerome C. Hunsaker notes longtime NACA research director Lewis's National Aeronautic Association life membership and says that Lewis had served on the Collier committee. *Model Research*, 1:383 n. 56 cites an "adamant" 1948 note from Hugh L. Dryden to Victory asserting the NACA's interest in promoting a three-way joint award of the then-impending Collier for 1947.

28. Ward, "Charles A. Lindbergh: His Flight and the American Ideal," in *Technology in America: A History of Individuals and Ideas*, 2d ed., ed. Carroll W. Pursell, Jr. (Cambridge, MA: MIT Press, 1990) pp. 211–26 (originally in *The American Quarterly*, spring 1958, as "The Meaning of Lindbergh's Flight"). "Intuition Brought Supersonic Flight," *Washington Post*, December 21, 1948.

awarding the Wright Brothers Memorial Trophy to Stack, the printed program declared that Stack had won two Colliers: one jointly for the X-1, and another "singly . . . for his development of the transonic wind tunnel." A 1993 history of the National Aeronautic Association, the organization that awards the Collier, mentions the associates and the teamwork, but names only Stack.[29]

But Stack himself knew better. When he learned of the Collier awarding committee's impending misassignment of credit, he took decisive steps to correct it. Recognition of the nineteen associates, a substantial partial cross section of the NACA technical culture, resulted from plain forthrightness in Stack, a product of that culture and in many ways an exemplar of its norms. *High-Speed Frontier* author John Becker, one of the nineteen himself, described Stack's reaction to word that he had won this second Collier to go with the one he had already shared with Yeager and Bell:

> *A few weeks before the second award was presented to him by President Harry S. Truman on December 17, 1952, Stack appeared unexpectedly in my office in a state of considerable agitation. He had just received notice of the award from J. F. Victory, chairman of the committee for the Collier Trophy. Stack said he was reluctant to accept the award as the sole recipient because so many others at Langley had contributed importantly. He wondered how the others would react. I believed they would feel as I did that he richly deserved this recognition. Without his aggressive leadership and promotional efforts there would have been no large transonic tunnels at Langley at that time. But Stack was insistent that the other principals should be included and we worked up a list of some 19 names.*

In the end Stack could not get his colleagues individually cited, but did manage to distribute some of the recognition by getting the words "and associates at Langley Aeronautical Laboratory, NACA" added to the formal citation. Before the award ceremony he issued a press release describing each person's participation and emphasizing the "teamwork, the pooling of scientific capacities in a research laboratory, that makes an idea successful." He also helped organize a dinner to recognize the nineteen. Even a decade later, Stack's official NASA biography sheet still made the point that in his 1952 acceptance of the Collier for 1951, he had "confirmed that NACA know-how and teamwork were largely responsible" for it.[30]

Like Stack in 1952, previous NACA individual Collier winners Lewis A. Rodert for 1946 and Stack himself for 1947 had also publicly declared NACA teamwork the real basis for their achievements. Rodert had said that his Collier was "awarded for the general work of all of us" and that he had been named "because only individuals [could] be so designated." Stack had emphasized a nearly identical sentiment.[31] This focus on the effective team rather than on any individual was entirely consistent with both the official outlook and the actual practice of the NACA. Aerospace historian James R. Hansen says that

---

29. AR54, p. 13; "Current Status of Aeronautical Research and Trends Towards Tomorrow," June 8, 1957, p. 6, Milton Ames collection folder "Victory, John F.," LHA; program for "Wright Memorial Dinner, Aero Club of Washington, December 17, 1962, Sheraton Park Hotel, Washington, D.C.," Stack collection folder "Awards and Biographical Information," LHA; Bill Robie, *For the Greatest Achievement: A History of the Aero Club of America and the National Aeronautic Association* (Washington, DC: Smithsonian Institution Press, 1993).

30. Becker, *High-Speed Frontier*, pp. 61, 62; Stack's press release was reprinted in the Langley *Air Scoop*, December 19, 1952, available in the LHA; Stack biography sheet, Stack collection folder "Awards and Biographical Information," LHA. Becker noted in an April 3, 1996, telephone interview that Stack was "very ill at ease" when he heard about the award, and that "it didn't cost him anything to add on" the associates, for he knew that in any case he would get most of the credit.

31. Langley *Air Scoop*, January 9, 1948, LHA; Hansen, *Engineer in Charge*, p. 304.

George Lewis, whose quarter-century tenure as the NACA's first research director lasted until after World War II, characteristically "emphasized teamwork over individual genius" and that Lewis believed in Thomas Edison's "nonheroic theory of invention and especially liked its emphasis on collective action." Lewis once asked that Langley frame and display a presidential tribute to Edison that he thought "aptly cover[ed] the aims and purposes" of the NACA. In the quotation, President Hoover—like Lewis, an engineer—had attributed "both scientific discovery and its practical application" to the "labor of a host of men" gradually "building up the structure of knowledge" in "great laboratories." Lewis's successor Hugh Dryden, coauthor in the 1920s of NACA reports on wind-tunnel-like experiments with transonic jets of compressed air, held similar views. His two-sentence letter transmitting the 1948 Wright-Ward report to Clark Millikan ends with a forthright attribution of slotted-wall "development"—a term in the eventual Collier citation's triad of "conception, development and practical application"—not to the unmentioned Stack, and not even to authors Wright and Ward, but to "the Committee's Langley Laboratory,"[32] where flourished what later came to be called the NACA technical culture.

As a management cliché, *teamwork* can obviously evoke skepticism or even cynicism, but NACA veterans have confirmed that this officially declared teamwork actually did flourish at the level of hands-on routine, and not just in managers' imaginations or public pronouncements. Stanford aeronautical engineering professor emeritus Walter G. Vincenti, for instance, who helped comprehensively define the transonic wind tunnel problem as an NACA engineer in the 1940s, and who writes on NACA history and the epistemology of engineering, has described the group dynamics of some important NACA flight research of about 1940 as exemplifying "the kind of fruitful melding of personal and group ambition and interest that can arise when talented technical people join in what they see as a demanding and worthwhile task. The whole was more than the sum of the parts." Becker, who joined the NACA Langley staff in 1936, says that a consequence of daily group discussions in the mid-1930s Langley lunchroom was that "often no one originator of an important new research undertaking could be identified. The idea had gradually taken form from many discussions and in truth it was a product of the group." He reiterated in 1996 that "seldom was there one clear, unequivocal route to a solution" to be found by one person alone; more often, he said, things really did happen by way of the group's interactions over time. Concerning the overall assignment of slotted-wall credit, Becker, who avoids expansive phrases and carefully distinguishes NACA public relations pronouncements from technical facts, tends to view the achievement as an important subset of all the late-1940s NACA transonics work—and he calls that overall program "one of the most effective team efforts in the annals of aeronautics."[33]

Four decades after the Collier Trophy for 1951, this teamwork-oriented, sometimes underappreciated NACA technical culture became a topic of some interest concerning NASA, especially in public-policy discussions of NASA's future. "NASA did not rise like a new creation from the sands of time when the space race began in 1957," declared

---

32. Hansen, "George W. Lewis and the Management of Aeronautical Research," in *Aviation's Golden Age: Portraits from the 1920s and 1930s*, ed. William M. Leary (Iowa City, IA: University of Iowa Press, 1989), 93–112, quotation from p. 106. Concerning the Hoover quotation, see Hansen, *Engineer in Charge*, pp. 132, 133, Roland, *Model Research*, 1:105, and Lewis's original request letter in the Milton Ames collection folder "George Lewis," LHA. In "Fact Finding for Tomorrow's Planes," *National Geographic*, December 1953, pp. 757–80, Dryden attributed "aeronautical progress [to] the growing store of human knowledge that underlies and makes possible the practical accomplishments," p. 758. Dryden to Millikan, October 8, 1948, "Research Authorization 70" folder in RA70 file, LHA.

33. Vincenti, *What Engineers Know and How They Know It: Analytical Studies from Aeronautical History* (Baltimore, MD: Johns Hopkins, 1990), p. 91; Becker, *High-Speed Frontier*, pp. 22, 23, 61; see also p. 74; telephone interview, April 3, 1996.

Howard E. McCurdy in his 1989 article "The Decay of NASA's Technical Culture." There and in a 1993 book, McCurdy describes the technical culture of the NACA as both an antecedent and a standard for that of NASA, which came into being in 1958 to combine, replace, and extend the NACA and other federal organizations. An underused means for adding to understanding of this technical cultural heritage is historical study, and one useful topic for such study is the NACA's handling of the transonic wind tunnel problem over the course of the three decades leading up to 1951.

To identify characteristics of the NACA and early NASA cultures, McCurdy's book relies primarily on observations and impressions of NASA staff, drawing secondarily on several historians of the NACA. The result, says sociologist Diane Vaughan in her book on the 1986 *Challenger* disaster, is an "unparalleled history of organizational culture" that shows NASA able during the 1960s "to maintain the strong technical culture that preexisted Apollo." Vaughan's own extensively researched study cites few directly NACA-related historical sources.[34] Other public discourse has also addressed NASA's NACA technical cultural heritage, sometimes with little reference to formal scholarship of any kind. In popular literature, Michener's *Space*, Tom Wolfe's *The Right Stuff*, and *Apollo: The Race to the Moon* by Charles Murray and Catherine Bly Cox presume the importance of the technical cultural link.[35] So do public-policy studies from Washington. A 1994 National Research Council report takes an explicitly historical approach involving the NACA to justify recommendations about NASA's building new national subsonic and transonic wind tunnels, but uses as its sole NACA source a self-serving, semiofficial historical summary ghostwritten in the 1950s for the NACA chairman by a public affairs officer. A 1994 Congressional Budget Office study of possible new NASA directions asserts that the agency's "organizational history is relevant to the criticism of its current conduct" and observes that among "NASA's institutional predecessors was the National Advisory Committee on [*sic*] Aeronautics. Its purpose was to develop useful aviation technology, a task that by most accounts it accomplished well." But beyond tying discussion of NASA's "original organizational culture" to McCurdy, the CBO study names no such accounts.[36] So there may well be room in the conventional wisdom, and a use in public-policy discussions, for an enlarged historical perspective concerning the NACA technical culture. Useful materials are available for it. Historian James R. Hansen's work, especially *Engineer in Charge: A History of the Langley Aeronautical Laboratory, 1917–1958* and *Spaceflight Revolution: NASA Langley Research Center from Sputnik to Apollo*, contributes substantially to elucidating the technical cultural link between the NACA and NASA. So does Alex Roland's *Model Research: The National Advisory Committee for Aeronautics, 1915–1958*.

Scholarly studies, both historical and sociological, occasionally attempt brief distillations concerning the NACA technical culture. Roland says that "the NACA by 1926 was committed to a research philosophy that valued process over prescience, the team over the individual, experiment over theory, engineering over science, incremental refinement

---

34. McCurdy, p. 304, "The Decay of NASA's Technical Culture," *Space Policy* (November 1989) 301–10; McCurdy, *Inside NASA: High Technology and Organizational Change in the U.S. Space Program* (Baltimore, MD: Johns Hopkins University Press, 1993); Vaughan, *The Challenger Launch Decision: Risky Technology, Culture, and Deviance at NASA* (Chicago, IL: University of Chicago Press), p. 499 n. 17 and p. 210. See also p. 502 n. 88.

35. Wolfe, *The Right Stuff* (New York, NY: Farrar Strauss Giroux, 1979); Murray and Cox, *Apollo: The Race to the Moon* (New York, NY: Simon and Schuster, 1989).

36. National Research Council, *Assessing the National Plan for Aeronautical Ground Test Facilities* (Washington, DC: National Academy Press, 1994). According to Roland (*Model Research*, 1:319), the historical summary document "Forty Years of Aeronautical Research," which NRC chap. 1 cites from *The Smithsonian Report for 1955*, was written by Walter T. Bonney for Jerome C. Hunsaker and "sings the Committee's [i.e., the NACA's] praises and ignores its problems and shortcomings." Bonney's summary also appeared in AR58. Congressional Budget Office, *Reinventing NASA*, March 1994, pp. 21 and 17.

of the existing paradigm over revolutionary creation of new paradigms." He then distills his own summary to six words: "the triumph of engineering over science," a variation of the thought that Hansen distills even further in his book title *Engineer in Charge*, a phrase that McCurdy in turn has appropriated to name the NACA cultural tradition. The McCurdy distillation of the original NASA technical culture that Vaughan selects to quote is consistent with Roland's, Hansen's, and others' historical scholarship: it "consisted of a commitment to research, testing, and verification; to in-house technical capability; to hands-on activity; to the acceptance of risk and failure; to open communications; to a belief that NASA was staffed with exceptional people; to attention to detail; and to a 'frontiers of flight' mentality."[37]

The history of the NACA's handling of the transonic wind tunnel problem may contribute to revising or refining such distillations. When NASA's antecedent technical culture began taking shape around 1920, a new research problem had arisen: on aircraft with increasingly powerful engines, longer propeller blades were traveling through larger arcs, their tips in some cases reaching sonic speed. Since a propeller is an airfoil, a complex, precise aerodynamic shape like a wing, this made transonic aerodynamics a practical aeronautical research issue, even though transonic flight—and the word *transonic*, for that matter—were still some distance in the future. So the NACA effort that eventually led to the slotted-wall transonic tunnel began. From the 1920s until the advent of NASA, this effort paralleled, reflected, and sometimes even partly constituted the development of the NACA itself. The effort's history suggests a few candidate modifications to distillations summarizing the NACA's technical culture: Its members conceived research, researcher, and research tool as organically interconnected. With an externally compelled applied-research focus, they sought what Stack came to call "physical understanding without mathematical weakness," but they kept in view the additional practical goal of fundamental scientific understanding. By continually enlarging their corporate technical and scientific memory and by continually developing craftsmanship in the arts of aeronautical research, they learned to exercise technical intuition deftly, and to adapt flexibly to new problems—though usually not until doing so accorded with the priorities of industry or the military.

Wind Tunnels, Transonics, and the NACA of the 1920s

The NACA's job was to supply American industry and the military with information for designing better airplanes. This information mainly took the form of more than 16,000 formal reports published and distributed during the research organization's forty-three years, an average of about one per day from 1915 to 1958.[38] Much of the NACA's information-generating research addressed the centrally important topic of aerodynamics, which means predicting the complex interactions between airplane and air, which in turn means understanding the nearly constantly changing *flow field*—the pressure, density, temperature, and relative velocity[39] at each point in the air affecting and affected by the airplane at each moment of flight. This predicting can be very hard. Flow fields differ for every contemplated aerodynamic configuration, and change with each airborne maneuver. Even at an airplane's slowest, its flow-field velocities match the wind speeds of a robust hurricane. The ideal form of flow-field understanding, using the mathematical language of the science of fluid dynamics, is a reliable theory—a comprehensive, systematized conceptual

---

37. Roland, *Model Research*, 1:98 and 99; McCurdy, *Inside NASA*, pp. 12, 134; Vaughan, *The Challenger Launch Decision*, p. 209 quotes p. 302 in McCurdy, "The Decay of NASA's Technical Culture."
38. *Model Research*, 1:xiii and 2:556.
39. John D. Anderson, Jr., calls these the "four fundamental quantities in the language of aerodynamics," p. 36, *Introduction to Flight: Its Engineering and History* (New York, NY: McGraw-Hill, 1978).

model applicable to the task of making correct performance predictions about possible airplane and airplane-component designs. Unfortunately, this level of understanding is hard to attain, especially for transonic speeds. It is possible, though, to get empirical information applicable to design problems by conducting wind tunnel tests or flight tests that replicate, or at least approximate, flow fields of interest. By 1920, the NACA had begun conducting both kinds.

A wind tunnel test replicates a flow field by moving air across a test subject instrumented for data-gathering, usually a scale model but sometimes an actual airplane or a full-scale component of one. The method is functionally equivalent to flight, for as Leonardo da Vinci pointed out, "what an object does against the motionless air, the same does the air moving against the object at rest."[40] Of course, da Vinci never tried establishing this functional equivalency in wind tunnel airflow near the speed of sound. A flight test, on the other hand, generates data by moving an instrumented test subject through the air. In 1919 the NACA began relatively low-speed flight experiments with ordinary biplanes. But to cite the more varied flight-testing examples from the NACA's 1940s-era efforts in transonics, a flight-test subject could be a piloted research airplane like the X-1, or it could be a scale-model airplane or wing shot skyward on a rocket, dropped from an airplane at high altitude, or fastened to the upper surface of, say, a P-51 Mustang's wing, where airflow could accelerate to sonic speed during steep subsonic power dives.

Both wind tunnel and flight tests generate useful information, but as the postwar NACA transonics effort illustrated, flight tests often require more time, effort, and resources, with each datum preciously won. A carefully crafted model dropped from altitude or launched on a rocket required an elaborate tracking system on the ground, had limited capacity to accommodate measuring and data-transmitting devices, and was expended in a single brief use. For wing-mounted models, the host airplane's own flow field often spoiled the smaller localized flow field under study. Transonic research airplanes, besides being expensive and requiring extensive support, also endangered their pilots: the NACA's Howard Lilly, third human to exceed the speed of sound, died in a May 1948 crash of the D-558-1 Skystreak, an aircraft comparable to the X-1. Although flight tests did contribute substantially in the midcentury attack on transonic aerodynamics, the postwar transonic-research-tool development goal was always to achieve the flexible, convenient, productive, and safe laboratory conditions of the wind tunnel. As the NACA had recognized even before 1920, in a tunnel's easily accessible test section, experiment setups are endlessly and comparatively cheaply reconfigurable, and results are comparatively easily observable and measurable. Of course, even if easily obtained, data from a tunnel's artificial conditions must still meet a verisimilitude criterion: they must correspond somehow with the actual flight conditions being replicated, either directly or by the application of reliable mathematical correction factors. Meeting this verisimilitude requirement was the central challenge of NACA wind tunnel history, and the NACA's best-known success in meeting it was the slotted-wall transonic tunnel.

Long before the 1951 Collier Trophy for that success, and long before there was an NACA, aeronautical researchers recognized the wind tunnel's advantages. The efforts of Orville and Wilbur Wright to engineer the first airplane included methodical studies of small aerodynamic shapes in artificial flow fields inside a six-foot-long wooden box with a fan at one end. By 1920, when the NACA began operating its first wind tunnel at Langley, several tunnels were in use in the United States, but the world standard was being set at Ludwig Prandtl's aeronautical laboratory in Göttingen, Germany, where the closed-circuit,

---

40. Quoted p. 91 in Hugh L. Dryden, "Aerodynamics—Theory, Experiment, Application," *Aeronautical Engineering Review* 12, No. 12 (December 1953): 88–95.

return-flow tunnel had been invented and refined. Its airstream cycled repeatedly, with power-saving efficiency, around its return circuit and through its test section. Years later at Langley, the precise placement of carefully calculated ventilation slots in the test-section walls of two high-speed versions of such tunnels made them the first capable of transonic testing. But in June 1920 at Langley, no world standard was set, or even met, by the NACA when its first wind tunnel started operating. Lacking a return circuit, it was "obsolete when it was built," according to *Wind Tunnels of NASA* author Baals.[41]

However, in that same year of 1920 the NACA, through its executive committee chairman Joseph S. Ames, did at least take steps to learn more about wind tunnels worldwide. In his capacity as NACA aerodynamics committee vice-chairman, Ames wrote to several prominent figures in American aeronautics to ask for help outlining "a program of tests to be made in the wind tunnels of this country and of Europe with a view to securing what one might call standardization, that is, information which would enable one to connect the data published, as obtained in these different wind tunnels." The immediate motivation was calibration. Analyzing the divergence of results from research tools carrying out identical experiments can improve interpretation of the results; by calibrating tunnels against each other, researchers could better extrapolate likely flow-field behavior aloft from artificial flow-field behavior on the ground. W. F. Durand of Stanford University, for decades a major figure in American aeronautics, answered Ames with strong support for the standardization tests. He offered several suggestions and specifically mentioned the need to include French and British results. The NACA did discuss the idea with Europeans; Prandtl sent five specific cross-comparison-test ideas, and the British and Dutch also sent suggestions.[42] In 1922, with Ames's cross-calibration testing program begun, the NACA's annual report included a section called "International Standardization of Wind-Tunnel Results." The young research agency's early-1920s efforts to correlate research results also included data from flight testing, which had started at Langley in 1919 at least partly for tunnel-comparison purposes.[43] Though the immediate motivation for all of this cross-comparison work was calibration to sharpen understanding of research results, the effort must also have calibrated and sharpened the NACA's understanding of its need for better research tools.

Already in 1920 that need had begun to extend to the transonic, as seen when Ames's letter elicited an expression of concern about wind tunnel results for the high-speed range that was not even yet called by that name. Elisha N. Fales of the Army's aeronautical laboratory at McCook Field near Dayton, Ohio—now Wright-Patterson Air Force Base—replied that standardization "has especial significance when, as in the McCook Field tunnel, speeds are attained which involve density changes."[44] Fales was bringing up the fundamental prob-

---

41. Baals, *Wind Tunnels of NASA*, pp. 9–14. This obsolescence assertion by wind tunnel expert Baals—himself notably loyal to the NACA for over half a century—contrasts with NACA publicist and chronicler George W. Gray's claim that Tunnel No. 1 "followed the best engineering practice of its day," *Frontiers of Flight: The Story of NACA Research* (New York, DC: Alfred A. Knopf, 1948), p. 34. Although open-circuit tunnels were indeed widely used circa 1920, within a few years the NACA replaced Tunnel No. 1. Gray is better valued for crystalline technical explanations than for impartiality. Roland, *Model Research*, 1:319 calls his book "as fine a summary of the NACA's claims for itself as is likely to be prepared."

42. In the folder "R.A.'s—Standardization of Wind Tunnels 1920–1926," RA70, LHA, are Ames to A. F. Zahm, L. J. Briggs, E. B. Wilson, W. F. Durand, E. N. Fales, J. G. Coffin, H. Bateman, and F. H. Norton, August 23, 1920; Durand to Ames, September 24, 1920; a copy of W. Knight's May 3, 1920, letter thanking Prandtl; and a two-page document with August 12, 1920, receipt notation titled "Suggested Aerodynamical Comparative Tests," which includes Prandtl's actual suggestions. Langley engineer Elton W. Miller, rebutting *Aero Digest* criticisms of the NACA in 1930, apparently did not know about Ames's letter; see *Model Research*, 2:658, item 7.

43. AR22, p. 36; "Fifty Years of Flight Research: A Chronology of the Langley Research Center, 1917–1966" (NASA TM-X-59314, apparently a republication of work by Michael D. Keller), p. 16.

44. Fales to Ames, August 31, 1920, in RA70 folder "R.A.'s—Standardization of Wind Tunnels 1920–1926," LHA.

lem of *compressibility*, a phenomenon already known in the field of fluid dynamics and beginning to require attention in the subfield of aerodynamics. Even at slow speeds, a flying object slightly compresses some of the air that it meets, raising that air's density and thus altering a key flow-field characteristic. At speeds approaching that of sound—that is, at transonic speeds much higher than those of the airplanes of 1920, but equal to those of some propeller tips of the day—this compressibility becomes significant and starts to degrade the performance of airfoils. For propellers, compressibility effects degrade the production of propulsive thrust. For airplanes themselves, compressibility effects can become disruptive and even dangerous, as indeed happened when airplanes began attaining much higher speeds in the late 1930s. Under NACA auspices, Fales in 1920 co-wrote "Wind Tunnel Studies in Aerodynamic Phenomena at High Speeds," a report on work that Stack later called "the earliest experimental investigation of airfoil characteristics as affected by compressibility," and that Becker says introduced two important compressibility terms: *critical speed* and *burble*.[45] At critical speed, some of the airflow accelerating across the airfoil surface reaches the speed of sound, creating a flow-field-disrupting compressibility burble, a discontinuity in the flow.

When Fales raised this high-speed research issue in answering Ames, little had yet been learned about how to study transonic phenomena. Becker notes, for instance, that Fales and report coauthor F. W. Caldwell did not even mention the centrally important ratio of flow-field speed to the local speed of sound—*Mach number*, as Swiss high-speed researcher Jakob Ackeret in 1929 proposed calling the ratio—even though the concept itself had been known to fluid dynamicists for decades.[46] Of course, much was still to be learned about how to study aeronautical questions in general. Research tools were often quite rudimentary and unsophisticated. The Propeller Research Tunnel at Langley, for example, originally had plain commercial platform scales for aerodynamic measurements. With air flowing around an engine-and-propeller configuration mounted on a framework atop the scales, researchers simply weighed the thrust and drag.[47]

Research tools were also rudimentary for the transonics studies the NACA at first contracted out during the 1920s, as future NACA research director Hugh Dryden learned firsthand. Becker says that with the high-speed work of Caldwell and Fales the "seeds of interest had been sown" in both the NACA and the National Bureau of Standards, another government agency with aeronautics interests. Accordingly, new high-speed studies began under NACA auspices. The work involved NBS aerodynamics section head Dryden, a 1919 Johns Hopkins Ph.D. in physics and mathematics whom Ames, in his capacity as a Johns Hopkins physics professor, had originally recommended to NBS. Ames once described Dryden as "the brightest young man . . . without exception" that he ever encountered. Like the transonic wind tunnel effort itself, Dryden was to contribute substantially over the years to defining the NACA technical culture. In 1947, after serving since 1931 on the NACA's aerodynamics committee, he joined the NACA staff to replace aging research director George Lewis just when slotted walls were being developed at Langley. Thereafter, in numerous articles in both the professional and popular press,

---

45. F. W. Caldwell and E. N. Fales, NACA Report 83, 1920; John Stack, NACA Report 463, "The N.A.C.A. High-Speed Wind Tunnel and Tests of Six Propeller Sections," 1933 (quotation on p. 416 in AR33); *High-Speed Frontier*, pp. 3–5.

46. Becker, *High-Speed Frontier*, p. 5. In Milton Ames collection folder "May 24, 1948, Transonic Wind Tunnels (Slotted Throats) Memo by Ray Wright," LHA, Hugh Dryden's June 25, 1948, lecture notes cite Ackeret's "'Air Resistance at Very High Speeds' in *Schweitzerische Bauzeitung* 94:179, 1929." Edward W. Constant, *The Origins of the Turbojet Revolution* (Baltimore, MD: Johns Hopkins, 1980), p. 288 n. 8, says Ackeret introduced the term in 1935 at the Volta high-speed conference in Italy.

47. Gray, *Frontiers of Flight*, pp. 54, 55. See also Roland, *Model Research*, 1:118 concerning the "primitiveness of early NACA research."

Dryden articulated the NACA's outlook on all aspects of aeronautical research. From 1958 until his death in 1965, he helped link the old NACA and the new NASA by serving as NASA's deputy administrator, bringing with him "the loyalty of the NACA's 8,000" employees, according to Richard K. Smith. It was Dryden's transonics experimentation of the 1920s that began these decades of contributions to the NACA and NASA technical cultural traditions. And that work involved rudimentary research tools, as Dryden recalled in an illustrative anecdote in a 1953 *National Geographic* article celebrating the research aspects of flight's first half-century:

> *As long ago as 1923 I was experimenting with propeller tip sections in a sonic-speed jet of air at General Electric's Lynn, Massachusetts, plant. Afterward when my colleagues and I walked out into the streets, we noticed that passers-by seemed unusually interested in our group. We later realized we had been unconsciously talking in very loud tones to compensate for the temporary deafness caused by working for several hours with our heads a few inches from a 12-inch sonic jet.*[48]

Dryden, Army Lt. Col. G. F. Hull, and Dryden's NBS colleague Lyman J. Briggs—a recipient of Ames's 1920 tunnel-standardization proposal letter, and years later the NACA's wartime vice-chairman—had gone to Lynn to use General Electric's huge centrifugal compressor, which, in Becker's words, "provided them in effect with a ready-made free-jet wind tunnel."[49] It could eject a jet of air at transonic speed from a circular nozzle just over a foot in diameter. The researchers took with them six three-inch-wide steel models, each representing the aerodynamic shape of a standard Army propeller blade, and each over seventeen inches long so as to completely span the high-speed jet of air, extending beyond its boundaries. So important was the precise construction of such models that Langley, developing its own aeronautical research craftsmanship, later bought the machining equipment that these particular models' Massachusetts maker also used for fashioning test subjects for the twenty-atmosphere pressure of Langley's Variable-Density Tunnel. The experimenters also took a specially constructed wind tunnel balance, an instrument with which they could hold a model airfoil in the airstream, incrementally change the airfoil's angle with respect to the airstream, and measure the resulting lift and drag forces. Their 1925 NACA paper "Aerodynamic Characteristics of Airfoils at High Speeds" reports that the investigation, carried out to obtain propeller-design information, showed that "the use of tip speeds approaching the speed of sound for propellers of customary design involves a serious loss in efficiency." Becker believes this work confirmed and extended that of Caldwell and Fales, offered the first useful attempt at explaining compressibility phenomena, and provided "the first statement of the relation between the critical speed and the known low-speed velocity distribution about the airfoil"—a piece of fundamental understanding "resurrected and exploited" a decade later in Langley's efforts to improve high-speed airfoils by designing them to have higher critical speed and thus a delayed compressibility burble.[50]

---

48. Becker, *High-Speed Frontier*, p. 7. Richard K. Smith, *The Hugh L. Dryden Papers 1898–1965* (Baltimore, MD: Milton S. Eisenhower Library, Johns Hopkins University, 1974), pp. 20–28. Elizabeth A. Muenger has also noted Dryden's public advocacy of NACA research; see p. 64, *Searching the Horizon: A History of Ames Research Center, 1940–1976* (Washington, DC: NASA SP-4304, 1985). Dryden, p. 762, "Fact Finding for Tomorrow's Planes," *National Geographic Magazine*, December 1953, pp. 757–80.

49. Becker, *High-Speed Frontier*, p. 8.

50. Report 207 appears pp. 465–79 in AR25; p. 466 discusses W. H. Nichols, the purchase of whose equipment Hansen reports on p. 83, *Engineer in Charge*. Becker, *High-Speed Frontier*, pp. 8, 9, 20.

Still, the methods and tools were rudimentary. For example, the experimenters made some unquantified, purely qualitative observations based on airflow patterns that appeared in oil they had placed on the model airfoils to keep them from rusting in bad weather—an apparently serendipitous ad hoc technique in the wind tunnel art of flow visualization. More significantly, expert observers later noted several limitations in the open jet of air,[51] some of which the experimenters themselves addressed in a section of their report called "Precision of Results":

> *The large power consumption of the compressor (5000 horsepower at high speeds) and the high cost of operation have made it impossible to repeat observations at will. In the interest of economy, many of the measurements were made while the [compressor equipment was] being put through shop tests. During such tests, the speed of the air stream was not under our control, and would often vary before a complete set of observations could be made. The noise of the air stream was so great that it was difficult for observers to communicate with each other while the compressor was running, so that modification of the program to meet changing conditions was difficult.*

Besides these bothersome impediments to proper scientific procedure, the jet of air also imposed an important fundamental limitation—a version, in fact, of the problem that Wright, Stack, and their associates overcame years later at Langley: *jet boundary effects*, or, more simply, wall interference. An enclosed test section's walls can distort the artificial flow field and thereby also the test results, particularly at transonic speeds. Similarly, even though an open jet has no solid walls to degrade flow-field verisimilitude, distortions comparable to those in a closed test section nonetheless arise because of the de facto boundary between the open jet and the surrounding air it hurtles through. An open jet does not constrict its artificial flow field within actual walls, but it still introduces measurement-distorting boundary effects.

So complex are boundary effects in the transonic range, wrote Bernhard Goethert in *Transonic Wind Tunnel Testing* in 1961, that the late-1940s effort to invent slotted walls could not have succeeded based on experimentation alone, but required an "orientation of theoretical calculations." This notion too—like Fales's introduction of the compressibility issue—arose concerning tunnels in general in Ames's 1920 tunnel-standardization discussion. American wind tunnel pioneer Albert F. Zahm, replying to Ames's letter, suggested beginning the cross-calibration project by having "the ablest theoretical aerodynamicists," such as Prandtl, "discuss the mathematical theory of the flow in a wind tunnel." Without "adequate theory, furnished before hand," wrote Zahm, "it seems improbable that all the observations and precautions would be taken that are necessary to make wind tunnel data strictly comparable."[52] In contrasting the gathering of empirical information with the larger issue of erecting a comprehensive theoretical framework into which it can fit, Zahm raised a question that engaged members of the NACA technical culture throughout the forty-three years the agency existed. The question has also engaged observers, critics, and historians both during and after those years—especially Hansen, not only in the essay that opens this volume, but in other works including his NACA Langley history *Engineer in Charge*. Usually

---

51. Becker, *High-Speed Frontier*, pp. 8, 11; Eastman Jacobs, p. 341, "Experimental Methods—Wind Tunnels: Part 2," in William F. Durand, ed., *Aerodynamic Theory*, Vol. 3 (New York, NY: Dover, 1963; republication of 1935 version), pp. 319–348.

52. Goethert, *Transonic Wind Tunnel Testing*, p. 236; Zahm to Ames, September 17, 1920, in the RA70 folder "R.A.'s—Standardization of Wind Tunnels 1920–1926," LHA. It must be noted that Becker says that Ray Wright "agree[d]" in a 1978 interview that systematic experiments might also have worked; see *High-Speed Frontier*, p. 100.

the question is seen in terms of the science and engineering of aircraft themselves, but as Zahm's letter shows, and as the NACA's transonic wind tunnel achievement highlights, it also applies to the science and engineering of the primary research tools of aeronautics. Fluid dynamics is as fundamental for wind tunnels as it is for airplanes. Thus it was that Ray Wright, a physicist and applied mathematician among NACA engineers, eventually used what Zahm in 1920 called "the mathematical theory of the flow in a wind tunnel" to provide for the accurate replication of transonic flow fields.

A tension between empiricism and theory existed from the start in the NACA. The agency's first annual report in 1915 lamented a general "distrust of mathematical formulae" and "a natural tendency on the part of designers and constructors to assume that mathematical theories are of use only to those who are mathematically inclined."[53] Such distrust seems to have been more common in American aeronautics than in European. Theodore von Kármán, a longtime leader in American aeronautics trained by Prandtl at Göttingen, reminisced in the 1960s about the contrast of "the practical inventor vs. the theoretical mathematician" he had found "characteristic of American scientific life in the twenties," and about the need, as he had long seen it, "to draw mathematics and engineering closer together" in this country.[54] The NACA's Max M. Munk, the former Prandtl student who proposed the Variable-Density Tunnel in the early 1920s, worried that those desiring efficient mathematical condensing of empirical experience would encounter not only a distrust of mathematical formulae but an even deeper antipathy to theoretical approaches and understanding in general. In an influential 1922 paper on airfoil design theory, Munk revealed acute defensiveness concerning the place of theory in aeronautics: "Is it really necessary to plead for the usefulness of theoretical work? This is nothing but systematical thinking and is not useless as sometimes supposed, but the difficulty of theoretical investigation makes many people dislike it." Ironically, the new theoretical ideas in Munk's paper led in the 1930s at Langley to Theodore Theodorsen's further theoretical work, and then to the theory-based, wind-tunnel-refined wing-design successes of Eastman Jacobs and others, including Stack—work that produced low-drag NACA laminar-flow airfoils, contributed to NACA advances in shaping airfoils for delaying to higher speed the onset of compressibility effects, and illustrated the utilitarian NACA's ever-present practical interest in enlarging fundamental understanding. Walter Vincenti has observed that complexity precluded experiment-based success in this wing-design work, just as Goethert has observed it did in the invention of slotted walls: both efforts required that orientation of theoretical calculations.[55] By the late 1930s, the NACA commonly incorporated such an

---

53. AR 1915, p. 14 and p. 13.
54. Theodore von Kármán with Lee Edson, *The Wind and Beyond: Theodore von Kármán, Pioneer in Aviation and Pathfinder in Space* (Boston, MA and Toronto, Canada: Little, Brown and Co., 1967), p. 124. Longtime von Kármán student, friend, and colleague William R. Sears, p. 36, "Von Kármán: Fluid Dynamics and Other Things," *Physics Today*, January 1986, pp. 34–39, wrote: "I think von Kármán believed that any problem in engineering (and perhaps in a much broader category) could profitably be attacked mathematically."
55. Report 142, "General Theory of Thin Wing Sections"; Walter G. Vincenti, "The Davis Wing and the Problem of Airfoil Design: Uncertainty and Growth in Engineering Knowledge," *Technology and Culture* 27 (October 1986): 717–58, especially pp. 740–44, 749, and 750; Hansen, *Engineer in Charge*, pp. 81 and 111–18. In an April 3, 1996, telephone interview, Becker confirmed that there was cross-pollination at Langley concerning theoretical understanding of laminar-flow airfoils and closely related high-speed airfoils. Vincenti calls for more scholarship on airfoil design on pp. 738 and 739 in his article, which also appears as a chapter in *What Engineers Know and How They Know It*.

orientation in much of its research.[56] Like other NACA work, NACA transonics efforts came to rely on empirical approaches mainly, but as Zahm had recommended for subsonic tunnels back in 1920, not exclusively.

Nonetheless, forceful criticisms of the NACA's general focus on applied research rather than on deeper scientific questions have appeared from time to time, and bear on the history of NACA transonics. For the early NACA, perhaps the best-known general statement of the charge came in 1930, when *Aero Digest* accused the agency of being far too narrowly and myopically empirical, never seeking to apply test results "to any logical system, to digest them, and to interpret their general significance in the sum of general knowledge."[57] Among historians, perhaps the best-known leveling of this charge comes from Edward W. Constant in *The Origins of the Turbojet Revolution*, a 1980 analysis of the pre–World War II convergence of technological developments, comprehensive scientific understanding, and combined scientific and technological imagination that resulted in the first jet aircraft—in Britain and Germany, but notably not in the United States. Constant says that before World War II the U.S. aeronautical research establishment, including the NACA, "had no interest in fundamental aerodynamic science," as shown in part by the "unimaginative" George Lewis's lack of interest in Theodore von Kármán's recommendation that a large supersonic tunnel be built. Constant's overall formulation of the charge, however, specifies more than the mere malfeasance of dwelling on the production of engineering data for near-term application, and more than the mere nonfeasance of failing to seek comprehensive theoretical understanding. Beyond these sins of commission and omission, Constant believes, was a more fundamental failure, a utilitarianism so narrowly focused on existing technology and so unimaginative as to constitute a sort of tragic flaw in the character of American—and therefore NACA—science and technology. Unlike the British and the Germans, the fundamentally flawed prewar American aeronautical research establishment could not even see, and therefore could not act upon, the synthesis possibilities that had gradually become implicit for aeropropulsion in the areas of turbomachinery, aerodynamics, and aircraft streamlining and structures. Like von Kármán, Constant sees differing "national patterns in the pursuit and utilization of aerodynamic science," and he observes that they "may reflect fundamentally differentiated cultural traditions. No later than 1900 Germany certainly had an unequalled tradition of mathematical and theoretical excellence in science and also had developed a deliberately close relationship between science and industry. Britain shared a similar if more empirical and less mathematically rigorous tradition in science. In contrast, the United States still was possessed of a scientific tradition extreme in its empiricism and utilitarianism."[58]

Whatever the validity of such criticisms, the early NACA did not employ its empiricism and utilitarianism unaware. In 1915, future NACA chairman (1941–1956) Jerome C. Hunsaker noted that experiments designed to answer current practical questions could also, over time, supply answers to deeper scientific questions, much as George Lewis believed. In *Model Research*, Roland says this principle became de facto NACA research policy by the late 1920s. In *Engineer in Charge*, Hansen shows how the principle applied in the matter of the cowling: the NACA first provided a quick practical solution and won the Collier Trophy, but in the longer term also worked for and achieved a genuine depth of theoretical understanding. In 1923, Joseph Ames used a courtroom simile to describe the principle: when the NACA conducted its practical tests, said Ames, it was "also doing fun-

---

56. Hartley A. Soulé, "Synopsis of the History of Langley Research Center, 1915–1939," p. 37 (item CN-141,573, Langley technical library); Becker, *High-Speed Frontier*, p. 23.
57. Frank A. Tichenor, "Why the N.A.C.A.?" *Aero Digest*, December 1930, pp. 47ff.; reprinted in *Model Research*, 2:652–57; quotation on p. 657.
58. Constant, *Turbojet Revolution*; quotations from pp. 159 and 176.

damental scientific work continuously, exactly as a justice of a high court expresses his deepest thoughts as *obiter dicta*."[59]

Certainly Ames's obiter dicta principle applied in the evolution of the NACA's understanding of the fluid dynamics of wind tunnels—the scientific component that supplemented engineering experience and technical craftsmanship in the overall wind tunnel expertise that began to grow in the NACA from about the time of Ames's 1920 initiative. The epistemological task of isolating and identifying this scientific component belongs to followers of Walter Vincenti, who has engaged similar questions about American aeronautical history. That such a component was indeed present, however, can be seen in Goethert's firsthand observation that the slotted-wall invention required an orientation of theoretical calculations. Possibly the scientific component was still small in 1922, when the NACA's annual report listed five technical papers on wind tunnels, one of them a Prandtl translation. Possibly it was small in 1925, when Joseph Ames told the NACA executive committee that Munk had developed a theory of tunnel wall interference. Possibly it was still small in 1930, when the available body of formal wind tunnel knowledge had grown large enough that an NACA report about correcting test data for subsonic open-jet boundary effects could cite four NACA and three European works on wind tunnel technology, along with one American and four European works on related aerodynamics topics—with only one source predating the 1920s. And certainly the scientific component was overrated in the NACA's 1934 annual report, which claimed that with the appearance of an NACA subsonic study called "Experimental Verification of the Theory of Wind-Tunnel Boundary Interference," the problem of fundamentally understanding wall interference could "for all practical purposes be considered solved." The problem had been solved "for all types of wind tunnels," the annual report said, even though the technical report in question carefully noted that only "conventional" and "ordinary" tunnels had been involved[60]—as well it should have noted, given that in that same year of 1934 Langley built its second small high-speed tunnel in part to investigate the far-from-conventional, far-from-ordinary transonic boundary effects that had been revealed in its first one, built in 1928.

That first high-speed tunnel had indeed raised lots of questions. The NACA built it to begin conducting "in-house" the kinds of studies Dryden and others had been conducting under NACA auspices elsewhere. It resembled a pipelike metal chimney, as for an open circular fireplace, with an eleven-inch-diameter test section about where such a chimney would have a flue damper. Compressed air powered it, tapped from an ideal reservoir at twenty atmospheres of pressure: the much larger Variable-Density Tunnel, which had to be depressurized occasionally anyway. The small vertical tunnel used the induction-jet principle, suggested by George Lewis based on a cursory contemporary Langley study of thrust augmentation, an antecedent of jet propulsion. In a rush lasting just long enough to yield some test data, piped-in air entered the tunnel just above the test section from an opening that ringed the pipe's circumference. This motion entrained a more massive flow of air

---

59. Hunsaker as quoted by Hugh L. Dryden, p. 93 in "Aerodynamics—Theory, Experiment, Application," *Aeronautical Engineering Review* 12, No. 12 (December 1953): 88–95; Roland, *Model Research*, 1:103; Hansen, *Engineer in Charge*, ch. 5, especially p. 137; Ames as quoted in Model Research, 1:349 n. 20. At least two NACA research engineers of the pre–World War II era have disagreed with the charge that engineering unduly triumphed over science in the NACA: Roland, reporting disagreement with his own interpretation in *Model Research*, quotes Ira Abbott, 1:345, n. 60, and John Becker, 1:346, n. 65. Langley engineer Elton W. Miller, rebutting *Aero Digest*, restated the obiter dicta principle, though without Ames's figure of speech, in a memorandum to the engineer in charge, December 19, 1930, reprinted in *Model Research*, 2:657–59.

60. AR22, pp. viii and 45; "Fifty Years of Flight Research: A Chronology of the Langley Research Center, 1917–1966" (NASA TM-X-59314, apparently a republication of work by Michael D. Keller), p. 27; Montgomery Knight and Thomas A. Harris, Report 361, "Experimental Determination of Jet Boundary Corrections for Airfoil Tests in Four Open Wind Tunnel Jets of Different Shapes"; AR34, p. 16; Theodore Theodorsen and Abe Silverstein, NACA Report 478.

upward from the room, generating a high-speed flow field around a small model facing downward in the test section. Both closed and open test sections were tried, giving Langley engineers a sense of the contrast between a walled-in jet of high-speed air and an open one. Despite some open-jet advantages, an enclosed test section was chosen for permanent use. This comparatively modest research tool, called the 11-Inch High-Speed Tunnel, began operation in mid-1928, about when John Stack completed his aeronautical engineering degree at Massachusetts Institute of Technology and, in Becker's words, arrived in Virginia "to dominate Langley high-speed aerodynamics for the next 30 years."[61]

A Measured Pace in the 1930s

"It is gratifying," the NACA modestly proposed in opening its 1933 annual report to Congress, "to report that the past year was notable as witnessing the greatest advance in airplane performance and efficiency accomplished in any single year since the Great War. This is largely the cumulative result of years of organized scientific research conducted by this Committee and of the practical application of the results by the Army, the Navy, and the aircraft industry." Apparently this expansive claim had substantial legitimacy. Richard K. Smith has written that between 1928 and 1938 "no other institution in the world contributed more to the definition of the modern airplane" than the NACA. Smith's aero-

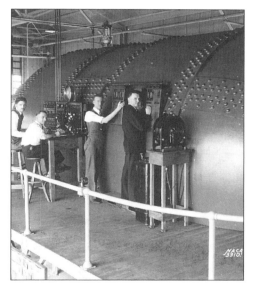 

Both figuratively and literally, the low-speed, pressure-tank-enclosed Variable-Density Tunnel breathed life into early NACA high-speed research. With the leadership of Eastman Jacobs, far left, the VDT helped establish the NACA's wind tunnel credentials and its confidence for further innovation. It also provided blasts of high-pressure air to power the NACA's original 11-Inch High-Speed Tunnel and later the 24-Inch High-Speed Tunnel shown here. The later vertical tunnel, with its twenty-four-inch-diameter test section, worked in the same way as the eleven-inch tunnel, but accepted larger models and had better data-gathering instruments. (NASA photos NACA 3310 and NACA 11443).

---

61. Stack, NACA Report 463, "The N.A.C.A. High-Speed Wind Tunnel and Tests of Six Propeller Sections," 1933; Becker, *High-Speed Frontier*, p. 13.

John Stack, NACA research craftsman and research leader. Left: Research craftsman Stack in the 1930s, reaching to help W. F. Lindsey adjust instrumentation inside Langley's briefly disassembled 24-Inch High-Speed Tunnel. When closed, the pipelike vertical apparatus could channel a many-hundred-mile-per-hour flow of ascending air through its twenty-four-inch-diameter test section and across a tiny downward-facing aerodynamic model linked to measuring and recording devices. (Photo courtesy John V. Becker) Right: Research leader Stack after World War II, when—in colleague John Becker's words—he was widely "recognized not only as the NACA's leading expert in aerodynamics, but also as an unusually colorful character" with "tough assertive characteristics" who "was at his best in the midst of conflict, crusading passionately for some cause such as a new wind tunnel." For three decades Stack helped define the NACA technical culture, but unlike NACA director Hugh Dryden, he found himself excluded from helping to do the same in NASA after the NACA years ended in 1958.[62] (NASA photo 48,989).

nautical history colleagues Hallion, Hansen, and Roland, as well as physics historian Daniel J. Kevles, have made similar assessments. Even Constant, in *Turbojet Revolution*, mildly praises the interwar NACA for its subsonic work. Continuing the annual report's self-congratulation, however, the NACA entered a realm where gaining later endorsements for its work in the 1930s has been hard, but incurring criticism has been easy: speed. Calling speed "the most important single factor" for improving airplanes, the report proclaimed that "primarily as a direct result of the Committee's researches there have been great increases in speed and efficiency during the past year, which have opened a new era in the development of both military and commercial aircraft."[63]

Of course, with no serious thought yet given in American aeronautics to jets, the NACA merely meant that propeller-driven airplane speed would continue to be developed

---

62. *High-Speed Frontier*, pp. 34, 14, and 13. On p. 176 in *The Birth of NASA: The Diary of T. Keith Glennan* (Washington, DC: NASA SP-4105, 1993), NASA's first administrator, writing about a July 1960 visit from Stack, called him "an interesting character—almost ready for retirement, outspoken and somewhat lacking in common sense," adding that although he was "one of the very best men in the aeronautical field" it was "obvious" he should not become associate director of Langley.

63. AR33, p. 1; Smith, "Better: The Quest for Excellence," p. 240; Hallion, *Test Pilots: The Frontiersmen of Flight* (Garden City, NY: Doubleday & Company, Inc., 1981), p. 50; Hansen, "George W. Lewis and the Management of Aeronautical Research," p. 94; Roland, *Model Research*, 1:xiii; Kevles, *The Physicists: The History of a Scientific Community in Modern America* (New York, NY: Alfred A. Knopf, 1978), pp. 292–93; Constant, *Turbojet Revolution*, ch. 6, especially pp. 156, 159, and 175. A mildly positive assessment appears on p. 75 of Walter A. McDougall, . . . *The Heavens and the Earth: A Political History of the Space Age* (New York, NY: Basic Books, Inc., 1985).

in this "new era." So a better term for the NACA's 1930s now appears to be *plateau*, as used by NACA and NASA aeronautical engineer Laurence K. Loftin, Jr., in *Quest for Performance: The Evolution of Modern Aircraft*. Airplane development, he wrote in 1985, "has been characterized by a series of technological levels, or plateaus, that extend over a period of years. Each level has been exemplified by an aircraft configuration type that is gradually improved by a series of relatively small refinements, without any major conceptual change." The mid-1930s forerunner of the P-47 Thunderbolt fighter, for instance—with stressed-skin metal construction, low cantilever wing with trailing-edge landing flaps, fully cowled radial engine with controllable-pitch propeller and geared single-speed supercharger, enclosed cockpit, and retractable landing gear with wheel brakes—represented, along with the DC-3 and the B-17, "the definitive and final configuration of the propeller-driven aircraft concept." Room remained, of course, for additional smaller refinements, like improvements in propeller-blade design. The NACA contributed substantially to reaching this plateau, but a new era, Loftin wrote, would actually require a "revolutionary breakthrough or new concept."[64]

For American aeronautical researchers as opposed to certain imaginative technologists in Europe, then, the idea of a "new era" in aviation speed in the 1930s suggested differing sets of research questions: those for propeller planes and those for jets. And since the research question generally dictates the need for the research tool, this difference was reflected in the NACA's high-speed wind tunnel development during the 1930s. Marching in time with conventional technology, and a few but not too many steps ahead, it proceeded at a conservative, measured pace.

Long before 1933, in fact, some European technologists had begun considering the possibilities for breakthroughs leading to very high-speed aircraft, and the possibilities for corresponding high-speed wind tunnels as well. Constant, alert to instances of foresight concerning radical technology change, closes *Turbojet Revolution* by alluding to a 1922 discussion among French and English engineers concerning the possibility of flying "with incredible speed in the stratosphere."[65] In 1924 in France, E. Huguenard's paper on high-speed wind tunnels predicted airplane speeds beyond 500 miles per hour, and conjectured that although speeds up to almost 750 miles per hour had formerly seemed "fabulous . . . as in Jules Verne," they now appeared "realizable, not in a remote future, but immediately." This nearly quarter-century-early conjecture of almost sonic flight speed may suggest why Becker calls Huguenard "overly sanguine." Whatever the excesses of Huguenard's enthusiasm, though, it is plain that in 1924 he squarely addressed a future that actually started arriving in the late 1930s—and that by 1925 his paper and its ideas were noted in the United States. The NACA published a translation that year, well before the agency used versions of two preexisting tunnel-technology ideas that Huguenard discussed: a compressed-air reservoir for driving a high-speed tunnel, and, for observing high-speed phenomena, an optical technique based on the way light behaves in air of changing density. Also in 1925, *Scientific American* favorably summarized Huguenard, reporting his prediction of 500-mile-per-hour speeds, his speculation about the need for

---

64. Laurence K. Loftin, Jr., *Quest for Performance: The Evolution of Modern Aircraft* (Washington, DC: NASA SP-468, 1985), pp. ix, x, 95, and 96.

65. Constant, *Turbojet Revolution*, p. 246.

some form of reaction propulsion, and his emphasis on the coming importance of wind tunnels for high-speed flight.[66]

For any NACA high-speed researchers inclined to consider the possibilities Huguenard had proposed, however, the late 1920s and early 1930s, with their official focus on propeller-tip studies, would have presented a certain tension. A 1929 report of an NACA-sponsored study of tiny airfoil models in an open, two-inch-wide transonic jet of air provides a typical example of the focus: "If a propeller is mounted directly on the shaft of a modern high-speed airplane engine," wrote Lyman Briggs and Hugh Dryden, explaining the practical engineering design question motivating their study, "the outer airfoil sections of the propeller travel at speeds approaching the speed of sound. It is possible by the use of gearing and a somewhat larger propeller to reduce the speed of the propeller sections, but only at the expense of additional weight and some frictional loss of power. In order to determine whether gearing is desirable, it is necessary to know the loss of efficiency due to high tip speeds and to compare this loss with that due to gearing." In other words, in their tests at speeds involving compressibility, they merely sought airfoil performance data to use in determining the optimum tradeoff, or balance, between competing design choices. The report mentions nothing about applications of the work to wings for very high-speed flight.[67]

Even in the mid-1930s, in fact, a forward-looking NACA engineer would have been aware that the NACA officially believed the trend to higher flight speeds would level off not too far above 500 miles per hour. Among Huguenard's enthusiasms, on the other hand, had been a willingness to project continuation of the upward trend. Observing that aircraft speeds had regularly doubled nearly twice per decade, Huguenard criticized those who always found "formulas" to show that "each new performance" in this trend would be the last. He even gave these doubters a name that fit the official NACA: *pessimistic calculators*. For the NACA, research director George Lewis seems to have exemplified this restrained outlook, at least in his public statements. In 1932 he predicted that the impressive upward trend in flight speeds would end for "airplanes as they are now constructed" at about 500 miles per hour. "At that speed," Lewis added, "the resistance of the air against the plane becomes so great that it would be physically impossible to obtain an engine giving enough added horsepower to pull the plane through the air at a greater speed." Although Lewis did note, by way of qualification, that "no one knows what the airplane of the future will resemble," his 1932 emphasis corresponded entirely with Loftin's 1985 concept of the plateau. John Becker arrived at Langley in 1936; when asked in 1996 if Lewis and Stack in those days might have harbored some hidden belief in a sonic future, he responded with confidence that he believed they had not.[68]

---

66. NACA Technical Memorandum 318, June 1925, translation of E. Huguenard, "High-Velocity Wind Tunnels: Their Application to Ballistics, Aerodynamics and Aeronautics," from *La Technique Aéronautique*, November 15 and December 15, 1924; quotations on p. 28. Huguenard refers (p. 15) to "a report by the American Lieutenant Sewall, to the United States War Department (S. Sewall, 'Report on high-velocity wind tunnels,' November 12, 1918)." Becker, *High-Speed Frontier*, p. 12. Both William F. Durand and Hugh L. Dryden later cited Huguenard: Durand in a reference list recommended on p. 252 and appearing on p. 349 of Volume III of *Aerodynamic Theory: A General Review of Progress*, ed. Durand (Dover Publications Inc.: New York, 1935, 1963) and Dryden, pp. 2 and 3, lecture notes, "Sixty Years of Experimental Supersonic Research" in Milton Ames collection folder "May 24, 1948, Transonic Wind Tunnels (Slotted Throats) Memo by Ray H. Wright," LHA. "High-Speed Wind Tunnels," *Scientific American* 133 (October 1925): 275–77. It is interesting to note *Scientific American*'s claim, in its 150th anniversary issue, September 1995 (p. 58), that "technology and the future have always been the province of this magazine." The issue boasts (p. 14) that the magazine covered the Wrights almost two years before Kitty Hawk, and quoted Robert Goddard saying in 1920—six years before the first liquid-fueled rocket flight, the Kitty Hawk of rocketry—that "a rocket capable of reaching the moon could be built."

67. NACA Report 319, "Aerodynamic Characteristics of Twenty-Four Airfoils at High Speeds," 1929.

68. Huguenard, TM 318, p. 28. "How Fast Can We Fly?" *The Sunday Star*, September 11, 1932, in Milton Ames collection folder "George Lewis," LHA. Telephone interview, July 18, 1996.

In any case it would be difficult to establish that in the 1930s the NACA could have pushed high-speed research or high-speed tunnel technology much faster than it did, even if it had wanted to. Industry and military energies compelled its focus on the technology of today and tomorrow but not the day after. Becker states flatly that even as late as 1940, the research-agenda-setting aircraft industry considered Mach 0.8—roughly 600 miles per hour—"a rather optimistic upper limit for the future." He also says that most "NACA veterans believe that it would have been quite impossible in the prewar period to have obtained any major support from the military, industry, or Congress for research and development aimed at such radical concepts as the turbojet, the rocket engine, or transonic and supersonic aircraft." One such veteran, who helped build Langley's 24-Inch High-Speed Tunnel in the mid-1930s, believed it "certain that if the NACA had had the foresight to do research on the turbine engine in the decade before World War II, the agency would have met with such technical ridicule and criticism about wasting the taxpayers' money that it would either have had to drop it or have been eliminated." And indeed the prewar NACA did face political perils difficult enough to negotiate without the agency's also seeking to venture too boldly beyond or above the technology plateau of the day.[69]

It is worth noting, moreover, that the prewar NACA in many ways did plan for the future, within the limits of a political reality in which lend-lease had eventually to be concocted to help the British halt the Nazi onslaught. In the mid-1930s, for instance, the NACA—advancing at the steady, measured pace of the times in American aeronautics—built not only the twenty-four-inch tunnel but the 500-mile-per-hour 8-Foot High-Speed Tunnel, later repowered for still more speed. This strategic resource was to become in 1950 the first large facility to operate with slotted walls. In the late 1930s the NACA began planning Langley's 16-Foot High-Speed Tunnel, the other large facility later converted. Alarmed years in advance about war's likelihood—in part thanks to George Lewis's visits to Europe—the NACA also sought to build new research laboratories, and indeed had managed to get funding to start a pair by the time of Pearl Harbor. For two years in the late 1930s "after learning of the frantic pace of aeronautical research in Europe, especially in Germany," wrote Alex Roland, "the NACA was unable to convince the Congress or the Bureau of the Budget that a crisis was in the making, a crisis requiring a crash program in aeronautical research." Yet the postwar Mead committee charged that the prewar NACA knew "of the need for increased personnel and facilities to carry on its research work" but "did not request sufficient funds from Congress." However, more than three years before Pearl Harbor the NACA did include in its annual report to Congress a frank plea for expansion—a plea highlighted, analyzed, and endorsed by a January 1939 editorial in the *New York Times*.[70] Thus for the prewar NACA and the country, an apt analogy might be that of the so-called next-quarter syndrome, in which a corporation's stockholders compel a shortsightedness that its critics contrast with the foreign competition's supposed longer view. It is true that the prewar American aeronautical establishment failed to invent jets and guided missiles. But it is also probable that the failure originated at a cultural level deeper than that of the scientific and technological choices actually available to American aeronautical researchers and their managers—as even one of their main critics, Constant himself, all but proposed in conjecturing about those "fundamentally differentiated cultural traditions" of Europe and America.

---

69. Becker, *High-Speed Frontier*, pp. 162 and 31 (see also p. 147); Hansen, *Engineer in Charge*, p. 184. Concerning the NACA's prewar travails, see Roland, *Model Research*, ch. 6 and 7.

70. Lewis's "Report on Trip to Germany and Russia, September–October, 1936" is in the Milton Ames collection folder "George Lewis," LHA. Roland, *Model Research*, 1:147. Excerpt from Mead committee report, Stack collection folder "Miscellaneous," LHA. On January 10, 1939, the *Times* quoted and editorialized (p. 18) about statements in AR38, and on p. 8, along with articles about the "nation's rearming," featured an article headlined "Our Air Supremacy Is Held Endangered; National Advisory Committee Says Intensified Research Is Necessary to Retain It."

Therefore it is also worth noting, concerning researcher Stack and manager Lewis, that in the 1940s Stack sometimes implied or even claimed that Lewis and the NACA had actually shown substantial foresight early on concerning flight at very high speeds. A 1948 newspaper story quoted Stack claiming that the "NACA's supersonic flight project really [went] back 20 years" to when Lewis, "with his long nose for the future, put in the first high-speed wind tunnel."[71] But if a supersonic-flight motivation for building the 11-Inch High-Speed Tunnel really did exist in 1928, it was apparently completely hidden. In 1945, Stack claimed in his formal paper for the prestigious Wright Brothers Lecture of 1944 that in the 1920s, when "a few foresighted aeronautical scientists" had planned ahead for very high-speed flight, Lewis had shown "great foresight" in sponsoring Langley's brief, cursory jet propulsion study.[72] Maybe such claims only represent what Constant has called the NACA's "habitual but mythic retrospective attribution of foresight to itself." Certainly Stack understood the NACA public relations juggernaut and could often be part of it; Roland says that by the 1950s he became too much a part of it. In any case, a draft of the Wright Brothers paper shows that Stack also considered claiming that "probably the first practical application of jet propulsion in aeronautical work" was Langley's, and long-nosed Lewis's, adaptation of the cursory jet study's induction-jet principle for the eleven-inch tunnel. By permanently deleting that claim, Stack avoided its justifying any "long nose" descriptions of himself—not for prescience, but for exaggeration.[73]

By the time of the NACA's 1933 boasting about speed, Stack himself was calculating at least somewhat optimistically about future propeller-driven high-speed flight, but there is evidence he felt constrained from pressing even that topic too far. The tension shows in a pair of historically significant papers he wrote, early contributions in his substantial compressibility research output during the years before he rose high in management. One, published as an article in the January 1934 *Journal of the Aeronautical Sciences*, reflected mainly his own outlook. The other, published officially as NACA Report No. 463, reflected mainly the organization's outlook. The journal article described a possible high-speed airplane and addressed its high-speed-flight potential. The NACA report described the 11-Inch High-Speed Tunnel and emphasized its usefulness in propeller-tip studies.

The journal article, "Effects of Compressibility on High-Speed Flight," presented performance predictions Stack had computed for a highly aerodynamically refined propeller-driven airplane that he called "hypothetical" but "not beyond the limits of possibility." Stack's computations showed that speeds much higher "than those so far attained" were "possible and likely," in part by using wings of a compressibility-effects-delaying shape derived from experiments in the eleven-inch tunnel. Some of this new design information, Stack wrote, was "already available to designers." With compressibility ignored, Stack's computations predicted a top speed of 566 miles per hour for the hypothetical airplane.

---

71. "Intuition Brought Supersonic Flight," *Washington Post*, December 21, 1948.
72. Stack, p. 128, "Compressible Flows in Aeronautics: The Eighth Wright Brothers Lecture," *Journal of the Aeronautical Sciences* 12, No. 2 (April 1945): 127–48. See also pp. 2 and 3 of the hand-annotated 36-page double-spaced typescript, apparently by Stack, titled "Report of the NACA Executive Committee: Supersonic Center Project," Stack collection folder "Revised Unitary Program, 1946–48," LHA. It claims that "sonic and supersonic tunnels" operated at Langley in the late 1920s.
73. Constant, reviewing *High-Speed Frontier* in *Isis*, 73:4:269 (1982): 609–10, says (p. 610) that Becker generally "debunks" the NACA's retrospective claims of foresight. Roland's strongest criticism of Stack's exaggerations appears on pp. 261–63 and in the accompanying n. 6 on p. 384, *Model Research*. Stack's crossed-out 1944 exaggeration is on p. 9 of the typed draft (corresponding to p. 128 in the article version), Stack collection folder "Wright Bros. 1944 Lecture," LHA. Study of the relation between transonic wind tunnel development and NACA public relations practices is mainly beyond the scope of the present work as it evolved, but is also the principal desideratum it generated. One example: the episode of the Annular Transonic Tunnel, which the newly security-minded postwar NACA advertised in such a way as to confuse outsiders concerning NACA progress in transonics.

With compressibility considered, that would fall to 524, but the new wing shape, Stack computed, could raise the top speed to 544 miles per hour "due to the delayed compressibility burble." At one point the journal article enthused about long-standing NACA foresight and leadership in high-speed-flight studies, but cited as evidence only the 1925 report of Briggs, Hull, and Dryden—which solely addressed propellers, though its analysis could be transferred and applied also to wings and thus to Stack's optimistic subject. And he calculated even more optimistically in an early handwritten draft, where a line he ultimately did not publish went so far as to say—as Huguenard did say back in 1924, and as the NACA did not say until the postwar world was upon it—that it was "dangerous to predict a maximum speed beyond which increases may be impossible."[74]

In NACA Report No. 463, "The N.A.C.A. High-Speed Wind Tunnel and Tests of Six Propeller Sections," Stack addressed high-speed technology with an entirely different slant, conservatively emphasizing propeller tips and not the airplane itself. "Speeds common to most aircraft" were low compared to the speed of sound, the introduction admitted, but knowledge of compressibility was nonetheless "essential" because propeller tip speeds commonly did reach the "neighborhood" of sonic speed. The introduction mentioned that racing airplanes had been attaining speeds "as high as half the speed of sound," and that "even at ordinary airplane speeds, the effects of compressibility should not be disregarded if accurate measurements are desired." But the report did not squarely address compressibility's overall future implications for the entire airplane—the very subject of Stack's roughly concurrent journal article—until near its end, where the statement appears that compressibility "is of considerable importance in the structural design of fast-diving airplanes," affecting distribution of loads. One of the report's conclusions also made the qualitative prediction that "errors may be expected in the estimated design loads for airplanes which attain speeds such as those attained by diving bombers when in a dive if the effects of compressibility on the wing moment coefficient are neglected." Nothing in the report's title, its lengthy opening summary, or its introduction suggested the presence of this kind of information. Yet that kind of information was to become very important at about the time of Pearl Harbor, when Stack and others at Langley helped solve serious, sometimes fatal, structural problems compressibility was causing in warplanes.

But the sharp contrast between this pessimistically calculating official report and Stack's optimistically calculating journal article, though it illustrates the NACA's conservative early-1930s research priorities, shows only one of the ways in which the report is significant in the multidecade evolution of the transonic wind tunnel. There are others. In focusing far more on the research tool than on the data obtained with it, the report introduced to the aeronautical world the NACA's first high-speed wind tunnel, including the early test-section-development work that Becker says strongly influenced slotted-wall development years later. The report called for a larger wind tunnel, and then served throughout the 1930s as the standard reference to cite for describing how experiments were performed not only in the eleven-inch tunnel, but in the larger twenty-four-inch apparatus that indeed did ensue and that was operated in the same way. And the report correlated high-speed wind tunnel data with results from full-scale propellers operated at high tip speeds in the low-speed airflow of the Propeller Research Tunnel—a notable instance of technical cross-pollination between the NACA's subsonic and transonic research efforts.

In yet three more ways, three particularly important ones, Stack's 1933 report illuminates transonic wind tunnel evolution and its NACA technical cultural implications. First, it defined the engineering science of NACA transonics—"physical understanding without

---

74. *Journal of the Aeronautical Sciences* 1 (January 1934): 40–43. The October 1933 manuscript and preceding draft materials are in an untitled folder, Stack collection, LHA; the paper's title is written on the front of the folder, not its tab.

mathematical weakness," to borrow a distillation Stack would use in 1942—by addressing the difficulties of attaining theoretical understanding of compressible flow, by claiming comprehensive accommodation of what little theoretical understanding was already available, and by showing Stack's acute determination to respect and use theory but not to let "mathematical complications" impede attainment of the physical kind of understanding an engineer often wants to visualize. Second, the report generally identified the vexing transonic tunnel issues that Langley later won the Collier Trophy for solving: "the effect of the tunnel walls," the test-data-skewing "constriction effect at the test section due to the presence of the model," the relation between model size and test-section size, and the question of a mathematical "constriction correction" to make transonic test results for artificial, ground-bound flow fields correspond with physical reality aloft. Third, as an approach for confronting these issues, it introduced as potentially useful what Goethert, looking back in 1961, called the indispensable "orientation of theoretical calculations." The report suggested conducting "a theoretical analysis of the flow in the tunnel with a view to determining the constriction correction," and added that the "analysis should include an examination of the effects of compressibility"—an important stipulation, the report said, but one that "because of the mathematical difficulty involved" seemed "improbable" in 1933. In 1944, however—when transonics had become a top research priority, thus making theoretical study of transonic tunnel flow a priority too—NACA research engineers H. Julian Allen and Walter Vincenti conducted just such a theoretical analysis at the new Ames Laboratory in California. Their report's title echoed Stack's 1933 language: "Wall Interference in a Two-Dimensional Flow Wind Tunnel, with Consideration of the Effect of Compressibility."[75] But what they showed was that in fact there could be no correcting of test data for the worst conditions of transonic flow within solid-boundary tunnels. Although their report apparently did not directly influence Ray Wright's search for a way to circumvent any need for corrections, Stack's idea of addressing the transonic wind tunnel problem via theory obviously did. If the awarders of the 1951 Collier were right in their original intention to credit Stack alone, if they perhaps really just meant to take the longest view of Stack's overall contributions to transonic wind tunnel development, the justification might well start with "The N.A.C.A. High-Speed Wind Tunnel and Tests of Six Propeller Sections" of 1933.

In 1933 Langley Field's runways were not yet paved. The term *sound barrier* was not yet sensationalized; that happened in 1935 following a casual remark to a journalist from British high-speed researcher W. F. Hilton. In 1933, the NACA's newly updated compilation of standard aeronautical nomenclature still included lots of biplane terms, but not *compressibility*, *Mach*, or any word with the suffix *sonic*.[76] Nonetheless, the NACA's high-speed research program, at its measured pace, continued advancing in sophistication during the mid-1930s, led by Eastman Jacobs and Stack.

Stack's papers trace the progress. In 1934, he and Albert E. von Doenhoff published an NACA report on airfoil research in the eleven-inch tunnel. The stated focus was still propellers, but wings and high-speed flight were now slightly more visible within the official field of view. According to Hilton's 1951 book *High-Speed Aerodynamics*, this was "Stack's classic paper, which exerted great influence by virtue of its early publication." But Stack and von Doenhoff had relied on experimental parameter variation, the systematic empirical method that James Hansen emphasizes as centrally important in the NACA's

---

75. Report 782.
76. Langley's runways were not paved until 1937, according to Robert I. Curtis, John Mitchell, and Martin Copp, *Langley Field: The Early Years, 1916–1946* (Langley Air Force Base, Virginia, 1977), p. 101. Hansen, *Engineer in Charge*, p. 253, discusses the sensationalized remark. AR33 contains one of the periodic updates of the NACA's report on standard nomenclature.

engineering science. The compressibility burble itself remained mysterious. Stack's 1935 report "The Compressibility Burble" declared that although the eleven-inch tests had "yielded much valuable information for design problems," they had also shown the necessity of a "more fundamental investigation."[77]

The 1935 report itself described early stages of such an investigation, conducted "to determine the physical nature of the compressibility burble." The experiments took place in the new twenty-four-inch tunnel, where improved instruments could simultaneously gather for correlation two kinds of data about transonic air interacting with a test model's surfaces: pressures and photographic images of the accompanying compressibility shock patterns. A schlieren optical system generated the photographable images by exploiting the behavior of light passing through air that is changing abruptly and radically in density. The report overlapped substantially with the paper famously presented by Jacobs that year at the international Volta conference on high-speed aeronautics in Italy. Later, in 1938, Stack, W. F. Lindsey, and Robert E. Littell published a refined and extended version of Stack's 1935 report: "The Compressibility Burble and the Effect of Compressibility on Pressures and Forces Acting on an Airfoil." Becker says that together with Jacobs's Volta paper, "these publications proclaimed the first major contribution of NACA in-house high-speed research—the fundamental understanding of the burble phenomena derived in large part from the revelations of the schlieren photographs."[78]

The 1938 report's research focus expressly included "future high-speed aircraft," and by this point in the prewar decade the research-methods focus had also widened: though still primarily empirical, it now included substantial overlap with airfoil theory, as Stack's 1939 "Tests of Airfoils Designed to Delay the Compressibility Burble" shows.[79] The 1939, report's antecedents included work by Langley theorist Theodore Theodorsen, which itself built in part on Max Munk's 1922 airfoil theory paper—the one in which Munk lamented the general distaste for theory he perceived in others. The overlapped work notably included Jacobs's new computational method for designing drag-reducing laminar-flow airfoils, for the physics involved in sustaining laminar flow is similar to that involved in delaying the compressibility burble: both require shaping the airfoil to control the way pressure changes in air flowing across its surface. To devise his computational design method, Jacobs had inverted Theodorsen's theoretical approach. The work Stack reported in his 1939 paper incorporated closely related analysis.

But even with its sophistication in high-speed research methods, Stack's 1939 report maintained the NACA's long-standing conservative outlook on high-speed research purposes. Its introduction, after noting that "high-speed aircraft" themselves needed "serious consideration," added that it was "important to realize, however, that the propeller will continue to offer the most serious compressibility problems." Of course, with the world war starting, this technology prediction had genuine merit within the context of continued refinements crucial for the conventional warplanes that would soon swarm from American factories. But overseas, jets were also in development. The NACA's high-speed research—and its high-speed research tools and methods—had advanced during the 1930s at a measured pace. Within a few years, Hugh Dryden and Curtis LeMay would be calling for an urgent one.

---

77. "Tests of 16 Related Airfoils at High Speeds," Report 492; W. F. Hilton, *High-Speed Aerodynamics* (New York, NY; London, England; Toronto, Canada: Longmans, Green and Co., 1951), p. 81; "The Compressibility Burble," Technical Note 543, October 1935.

78. Report 646; *High-Speed Frontier*, p. 19. The never-capitalized word *schlieren* has long vexed authors and editors. A. C. Kermode, *Mechanics of Flight* (London: Pitman Publishing, 1972), p. 317, noted that "it is not the name of some German or Austrian scientist, but simply the German word for streaking or striation, which is descriptive of the method."

79. Technical Note 976, December 1944, reprint of ACR of June 1939.

The Pace Hastens

Before Pearl Harbor some warplanes could already dive fast enough to encounter dangerous compressibility effects—as roughly predicted by that brief, unemphasized conclusion in Stack's official NACA report of 1933, when propeller tips constituted the NACA's only official compressibility research focus.[80] In transonic airflow, as Hugh Dryden explained in a 1948 *Physics Today* article, "disturbances known as shock waves" arise. These "abrupt changes in pressure and temperature" can lead to "a violently fluctuating motion shaking or buffeting the wing, and if the wake of the wing strikes the tail, the tail structure may be subjected to loads varying with violent irregularity sufficient to damage it."[81] Vulnerable airplanes during the war included the Bell P-39 Airacobra, the Curtiss P-40 Warhawk, and the Republic P-47 Thunderbolt.[82] At about the time of Pearl Harbor, when the problem had just arisen, Stack and others used Langley's twenty-four-inch and eight-foot high-speed tunnels in a rush effort to learn how to counteract the disturbances and stop the Army's new P-38 Lightning from occasionally breaking up and crashing. They quickly showed that a special under-the-wing flap could be developed to do the job.[83] The P-38, of which over 10,000 were ultimately built, went on to shoot down more Japanese aircraft than any other fighter.[84] The NACA went on to encounter still more complex transonic research problems during the 1940s, and to invent research tools—including the slotted-wall tunnel—for solving them.

By October 1948, when Dryden explained transonic research problems to a broad audience with the *Physics Today* article and advertised a new transonic research tool to a tiny audience with the Wright-Ward paper, the NACA's compressibility research focus had long since expanded. Maybe Stack had been prudent in 1933 to delete from his high-speed-airplane journal article the claim that it was "dangerous to predict a maximum speed beyond which increases may be impossible," but now the NACA itself officially gloried in seeing no "definite limit to the speed that may be attainable."[85] The late-1930s goal of refining airfoil shapes to delay the onset of compressibility had been replaced: "Regardless of how high the critical Mach number may be raised," asserted Stack in his 1944 Wright Brothers Lecture, "flight at supercritical speeds must eventually be solved."[86] Devising airfoils suitable not just for delaying the burble but for negotiating the entire transonic range would only be part of the solution. Effective transonic aircraft would also have to stay stable and controllable in an aerodynamically complex environment.[87] Moreover, researchers since the 1930s had been aware that separate high-speed tests of individual components—a cowling and a wing both meant for the same fuselage, for instance—could not always predict the components' performance in use together. Therefore solving supercritical flight required seeing the "integrated whole," as NACA main committee member Edward P. Warner called the principle of conceiving transonic

---

80. Conclusion 6, "The N.A.C.A. High-Speed Wind Tunnel and Tests of Six Propeller Sections."
81. Dryden, "Faster Than Sound," *Physics Today* 1, No. 6 (October 1948): 6–10 (see p. 8).
82. Hallion, *Test Pilots*, p. 187.
83. John Stack, "Compressible Flows in Aeronautics: The Eighth Wright Brothers Lecture," *Journal of the Aeronautical Sciences* 12, No. 2 (April 1945): 127–48 (see pp. 141–42).
84. John D. Anderson, Jr., "Faster and Higher: The Quest for Speed and Power," *Milestones of Aviation*, ed. John T. Greenwood (New York, NY: Hugh Lauter Levin, 1989), 78–147 (see p. 127).
85. AR46, p. 2.
86. "Compressible Flows in Aeronautics," p. 140. Becker, *High-Speed Frontier*, p. 35, sees this December 17, 1944, Washington lecture as marking expansion of the NACA's research focus in transonics.
87. Dryden, "Faster Than Sound," p. 8.

aircraft in an organic rather than a modular way.[88] For example, jet engines needed to be integrated into airframes specifically designed for the task. Much later the area rule, the transonic design principle described in chapter 5, grew out of NACA research engineer Richard Whitcomb's integrated view of the whole aircraft, and was nurtured by his experiments in one of the original slotted-wall transonic wind tunnels—a tunnel he helped to commission and refine.

In 1948, however, NACA transonic researchers' tools were mainly research aircraft, rocket- and airplane-borne models, and a few partly effective, sometimes even makeshift adaptations of high-speed tunnels. Some progress had been made in designing jet-propelled warplanes. Loftin says that the P-80 (later F-80) Shooting Star climbed and flew faster than the first U.S. jet, the P-59, thanks to "a careful synthesis of weight, size, and thrust parameters, as well as close attention to aerodynamic refinement." In April 1948, a swept-wing F-86 reached supersonic speed in a dive. Jet-propelled bombers were being developed.[89] But judging by the summer 1948 responses of thirteen aircraft manufacturers, the Air Force, and the Navy to an urgent NACA survey, these efforts only helped stimulate more desire for transonic data—as well as interest in the research tools with which the data would be obtained.

The NACA aerodynamics committee's survey asked the agency's industrial and military clients how the NACA could best use its research tools to aid transonic aircraft design. The answer: numerous practical-minded requests for empirical data on wing planforms, airfoils, controls, and complete three-dimensional—that is, integrated whole—models, with secondary interest in air inlets, buffeting effects, pilot escape, bomb bays, and aircraft stability. The Air Force and eleven of the thirteen companies also addressed research tools and methods. One consensus recommendation called for increasing rocket-borne model tests by a factor of three. Another pleaded that "the NACA continue under as high a priority as possible the study, development, and procurement of test facilities for obtaining [transonic data] in a manner equivalent to that followed in the best available low-speed wind tunnel testing"—that is, in convenient, versatile, relatively cheap, and completely safe laboratory conditions. Of the fifteen respondents, only three even mentioned theory; one of these few, Benedict Cohn of Boeing, urged that the NACA "obtain very fundamental data on the aerodynamics of transonic flow rather than attempt solutions of small specific items." Although the survey-sponsoring NACA aerodynamics committee formally agreed with the respondents' decidedly empirical majority view, it pointedly emphasized as well that the "NACA should also continue to give careful consideration to results of theoretical work."[90]

Indeed the aircraft industry and the military in the pressure of 1948 may generally have had little interest in theory. Dryden apparently gauged the military that way concerning wind tunnel theory, in any case. Even though Wright and Ward had translated theoretical ideas into a useful research tool, Dryden's October 1948 letters transmitting their paper to military research authorities carefully cautioned against letting the "considerable amount of background theoretical material . . . obscure the practical significance of the

---

88. W. S. Farren, "Research for Aeronautics—Its Planning and Application," *Journal of the Aeronautical Sciences* 11, No. 2 (April 1944): 95–109, addresses the "integrated whole" idea; the phrase itself appears in Edward P. Warner's appended remarks, p. 108. See also Loftin, *Quest for Performance*, p. 248. Concerning the preliminary sense of the idea in the 1930s, see the last paragraph of Stack's "Effects of Compressibility on High-Speed Flight," p. 12 in AR39, and Becker, *High-Speed Frontier*, p. 26.

89. Loftin, *Quest for Performance*, pp. 288, 295, 357.

90. "Summary of Recommendations on Research Problems of Transonic Aircraft Design, Compiled by Aerodynamic Research Branch, NACA Headquarters, for the Special Subcommittee on Research Problems of Transonic Aircraft Design," July 1948, Stack collection, LHA—where the thicker copy's extra appendix contains all the individual letter responses, including Cohn to NACA, July 26, 1948. Both copies contain the NACA aerodynamics committee's formal answer to the survey responses and recommendations.

work."[91] Nonetheless the postwar NACA itself, insofar as it could, sought to stay mindful of the benefits that improved research tools would represent for aerodynamic theory in general—the obiter dicta benefits, in Joseph Ames's 1923 courtroom simile—and in turn of theory's benefits for aeronautical engineering. Stack elucidated the NACA's wind-tunnel-centered version of this awareness the following June. Studies of transonic flows with models sent skyward on rockets or dropped from high altitude, he wrote, "have defined fundamental problems of fluid mechanics. Experimentation with standardized equipment, nonexpendable models, under closely controlled conditions permitting detailed measurements"—that is, in wind tunnels—"still appears to be a most important key to progress toward the attainment of the ultimate goal, that is, successful complete calculation of such flows."[92]

Not that Stack himself had never exhibited a decidedly empirical outlook. In 1942, soon after leading the somewhat dramatic applied-research solution of the P-38 problem, he taught a University of Virginia night school postgraduate course called "Compressibility Effects in Aeronautical Engineering," held for Langley staff only. Without the usual NACA public relations constraints, his opening lecture proclaimed that he would "exclude, insofar as possible, the mathematical exercises which though elegant are frequently so meaningless to the engineer," and that he would try instead "to adhere more closely to the discussion of physical concepts, introducing mathematical methods only as necessary to aid in understanding the physical concepts. . . . I think that it is well if we realize in the beginning that in this field the engineer is leading the mathematical scientist. The present state is such that the engineer is projecting himself perhaps to some extent blindly into difficulties, and by physical reasoning without mathematical weakness"—the phrase that distills Stack's approach to transonics—"arriving at the expedient solution of his difficulties." But the course syllabus somewhat belied this energetic introductory emphasis on empiricism, citing the objective of covering "the fundamentals of compressible flows, the status of present knowledge on the subject, and its application to engineering problems," and naming "summary of significant theories" the subject of six of thirty-two scheduled hours—three hours each for the "subcritical range" and the "supercritical range." And indeed the opening lecture, once past the introductory remarks, did immediately invoke in some detail compressibility's fundamental fluid dynamics context.[93] Two years later in Washington, Stack's 1944 Wright Brothers Lecture on compressibility mainly addressed experimentation, but it too rested distinctly within a scientific, theoretical context. Here is what we have done, that lecture said, in circumstances where little prospect has existed for advancing theoretically. This leading NACA aeronautical research engineer primarily sought near-term physical understanding, but secondarily, and for practical ends in the longer term, he wanted to see it attained by "physical reasoning without mathematical weakness" within the formal scientific realm of fundamental understanding.

In 1951, W. F. Hilton reemphasized the long-standing common belief that whatever theory's long-term potential, it held little near-term prospect for advancing transonics.[94] Hugh Dryden, however, maintained a formally scientific outlook about transonics in the

---

91. October 8, 1948; copies in RA70 folder "Research Authorization 70," LHA.
92. Typescript of "Methods for Investigation of Flows at Transonic Speeds: Paper for Presentation at the Naval Ordnance Laboratory Aero-Ballistics Research Facilities Dedication Symposia, June 27–July 1, 1949," pp. 13 and 14, in Stack collection folder of the same name, LHA. This passage reappears verbatim on p. 573 of Stack's 1951 revision of the 1949 paper, "Experimental Methods for Transonic Research," pp. 563–592j, Third Anglo-American Conference 1951, in a Stack collection folder labeled with the revised paper's title. The 1951 version was obviously published; the citation for reference 54 on p. 172 in Becker, *High-Speed Frontier*, says the 1951 conference took place September 7–11 in London.
93. See materials in the Stack collection folder "Defense School 1942," LHA. Quotations from pp. 1 and 2 of typescript "Introduction, Orientation, and Summary."
94. *High-Speed Aerodynamics*. See especially the preface and p. 9.

late 1940s anyway. As an employee of the National Bureau of Standards, he had served on and helped lead the NACA aerodynamics committee since 1931. During the war he managed a large guided-missile research and development project for the military.[95] After the war he was deeply involved in NACA transonics as high-speed aerodynamics committee chairman. In September 1947 he joined the NACA staff and took over from George Lewis as director of aeronautical research, a title shortened in 1950 to *director*.[96] His scientific outlook on transonics was an extension of his general view that the "discovery of how to make better aircraft results from the discovery of rational theories firmly supported by experimental evidence."[97] At Langley in early 1947, he chaired a conference on high-speed, aerodynamic theory attended by luminaries including Theodore von Kármán, who briefly summarized the state of compressible flow theory, and Tsien Hsue-shen, known for later famously leaving the United States and leading China's development of missile technology. After the theory conference Dryden reported, apparently with some disappointment, that despite worthwhile exchanges between theorists and experimentalists, "the hoped-for result of a rather concrete definition of the direction which future theoretical research in the field should take was not achieved."[98] But this veteran of early experiments with open transonic jets of air could also adopt the outlook of Stack and other practical-solutions-seeking NACA experimentalists and say that "progress in those aspects of aeronautics for which a rational theory has not yet been developed proceeds by the recognition of the common features of complex flow patterns."[99]

Thus it was that in engineer Stack at Langley and in physicist Dryden in Washington, the early postwar NACA had leadership well suited for fostering conception, development, and practical application of the slotted-wall transonic wind tunnel. Each had extensive personal experience in practical-solutions-oriented transonic experimentation, but each also understood and genuinely valued the formal fluid dynamics context. From long membership, each knew and had confidence in the NACA technical culture with its accumulated technological and scientific understanding and its highly developed tradition of aeronautical research craftsmanship. In such a setting Stack could follow his intuition concerning physicist and applied mathematician Ray Wright's theoretical ideas, and Wright, in the words of historian Hansen, could benefit "from the collective knowledge and experience of the engineers working around him" and from his own "good intuitions."[100]

Intuition was important. Hilton in 1951 called transonic aircraft design "more a product of trained intuition than the result of applying exact scientific principles."[101] Instances of similar intuition pervade NACA research history, according to historians of the three laboratories existing or begun by the time of Pearl Harbor. Such instances also pervade NACA transonic research history. Hansen makes intuitive technological artistry the theme of "The Slotted Tunnel and Area Rule," chapter 11 in his Langley history.[102] Becker

---

95. Richard K. Smith, *The Hugh L. Dryden Papers: A Preliminary Catalogue of the Basic Collection* (Baltimore, MD: Milton S. Eisenhower Library, Johns Hopkins University, 1974), p. 23.
96. Roland, *Model Research*, 1:247, 2:713, and 2:490.
97. See Dryden's addendum to W. S. Farren, "Research for Aeronautics—Its Planning and Application," *Journal of the Aeronautical Sciences* 11, No. 2 (April 1944): 95–109 (see p. 106).
98. Minutes, "Informal Conference on High-Speed Aerodynamic Theory," February 3, 1947, and p. 2, "Minutes of Meeting of Subcommittee on High-Speed Aerodynamics," April 29, 1947, in Stack collection folders "Conf. on High-Speed Aerodynamic Theory" and "Subcommittee on High-Speed Aerodynamics 1946–1948," LHA.
99. Remarks appended to Stack, "Compressible Flows in Aeronautics" (see p. 145).
100. *Engineer in Charge*, p. 318.
101. *High-Speed Aerodynamics*, p. 9.
102. Hansen, *Engineer in Charge*. See also Elizabeth A. Muenger, *Searching the Horizon: A History of Ames Research Center, 1940–1976* (Washington, DC: NASA SP-4304, 1985), p. 39, and Virginia P. Dawson, *Engines and Innovation: Lewis Laboratory and American Propulsion Technology* (Washington, DC: NASA SP-4306, 1991), p. 74.

believes that what motivated "initiation of in-house NACA research in high-speed aerodynamics" in the first place was intuition, not the "great foresight" Stack mentioned in his 1944 Wright Brothers Lecture. In that lecture Stack said that he and his colleagues devised the first NACA schlieren flow-visualization apparatus in early 1933 when they "had in the airfoil experiments temporarily exhausted [their] intuition as regards methods for improving aerodynamic shapes."[103] Stack's 1952 Collier press release says that in large part his and his colleagues' "faith in the probability of a solution" in 1946 had rested in Wright's "subsonic high-speed theoretical studies,"[104] a statement about acting on scientifically framed intuitive faith that calls to mind Stack's 1942 classroom remark about an engineer's "projecting himself perhaps to some extent blindly into difficulties, and by physical reasoning without mathematical weakness arriving at the expedient solution of his difficulties." In December 1948, following the award of the X-1 Collier to Stack, Bell, and Yeager, the opening lines of a *Washington Post* article set forth a version of Stack's philosophy on the relation between intuition and technological success:

> *"Intuitive research" brought about successful supersonic flight many months ahead of schedule, states the man most responsible. John Stack, designer of the first plane to fly faster than sound, says that "believing what you couldn't prove and trusting it" paid off by speeding up normal scientific processes. He puts it this way: "You say to yourself, if these things are true, then this must be true. You haven't an exact answer but you do have an intuitive answer. So if you want to make a big step forward, you take a chance of falling flat on your face and trust your intuition."*[105]

Surely at that moment in late 1948—two weeks after Bernhard Goethert's formal visit concerning Langley's modest initial proof of the slotted-wall principle—Stack must have had at least partly in mind the chance of falling flat on his face not with the X-1, already proven in the sky, but with the slotted wall, as yet proven only in miniature. He later said there had been "no turning back" once a construction contract had been signed earlier in 1948 for installing a slotted wall in Langley's huge 16-Foot High-Speed Tunnel.[106] And surely a reason for trusting his intuition was the NACA itself, a technical culture with broad general experience building wind tunnels and long-standing specific experience replicating high-speed flow fields in some of them.

Precisely Defining the Transonic Tunnel Problem

Although the NACA had been accumulating understanding of the difficulties of replicating transonic flow fields since the 1920s, the overall problem was apparently not comprehensively defined anywhere until the mid-1940s. Even in 1947, the textbook *Wind-Tunnel Testing* could only note somewhat vaguely that the "proper procedure for testing and correcting the results of high-speed tests has not been completely established" and that it "appears that the accentuated blocking and the shock-wave reflection off the tunnel walls contribute to the uncertainty."[107] NACA translations of European papers partially addressing the difficulties had been available since the mid-1930s to augment Langley's own growing understanding. In 1935, for instance, Swiss supersonics

---

103. Becker, *High-Speed Frontier*, p. 13; Stack, "Compressible Flows in Aeronautics," pp. 128 and 130.
104. Reprinted in the Langley *Air Scoop*, December 19, 1952, available LHA.
105. "Intuition Brought Supersonic Flight," *Washington Post*, December 21, 1948.
106. Langley *Air Scoop*, December 19, 1952, available LHA.
107. Alan Pope, *Wind-Tunnel Testing* (New York, NY: John Wiley & Sons, Inc., 1947), p. 207.

expert Jakob Ackeret discussed the blocking effects of test models on tunnel capabilities near Mach 1, contrasted the near-sonic-speed performance of open jets of air with that of airstreams enclosed within solid tunnel walls, and noted shock wave reflection in tests at low supersonic speed—problems addressed also in a 1938 paper by Italian aerodynamicist Antonio Ferri, whose own accumulated understanding about solving them was put to good use when the NACA managed to import him at the end of the war.[108] In November 1943 the Army formally requested that the NACA define the overall problem. A preliminary report of special work in the 24-Inch High-Speed Tunnel ensued in short order, for apparently the work had begun in advance; Stack had even discussed its main conclusions at an October meeting of the NACA aerodynamics committee in Washington. The Army asked for copies of the preliminary report to send to aircraft manufacturers including Douglas Aircraft Corporation, Curtiss-Wright Corporation, General Motors, and Northrop. Langley's Robert W. Byrne completed a full technical report during 1944, "Experimental Constriction Effects in High-Speed Wind Tunnels,"[109] one of several NACA studies to define the problems of replicating transonic flow fields in a tunnel, and the one Stack customarily cited retrospectively in later years.

One such study was that 1944 theoretical one at Ames Laboratory by Allen and Vincenti: "Wall Interference in a Two-Dimensional Flow Wind Tunnel, with Consideration of the Effect of Compressibility," the kind of analysis Stack first called for in his 1933 NACA report about the eleven-inch tunnel. Allen and Vincenti may not have directly influenced Ray Wright, but fifty years later their report remained useful for technical study.[110] For its comprehensive explanation of the fundamental problems of closed-wall wind tunnel operation at transonic speeds, it also remained useful for historical study.

The paper begins—as Stack's 1933 report had begun—by alluding to two numerical indicators aerodynamicists use, among other purposes, to score the similarity of a wind tunnel's flow field to an actual flow field aloft: "The need for reliable wind tunnel data for the design of high-performance aircraft has led in recent years to attempts to make the conditions of tunnel tests conform more closely with the conditions prevailing in flight, especially with regard to the Reynolds and Mach numbers." *Reynolds number* combines measures of an aerodynamic object's size and of its flow field's density, speed, and viscosity into a simple ratio expressed as a whole number. Ideally in a test with a scale model, this score should be high enough to conform with that of the simulated full-size airplane or component in its actual flight conditions. But most wind tunnel tests mismatch the full-size value of the Reynolds number by using a model of considerably reduced scale. The

---

108. NACA Technical Memorandum 808, "High-Speed Wind Tunnels," November 1936 translation of a paper Ackeret presented at the Fifth Convention of the Volta Congress, Italy, September 30 to October 6, 1935; see Part B, "Wind Tunnels for Subsonic Velocities." NACA Technical Memorandum 901, "Investigations and Experiments in the Guidonia Supersonic Wind Tunnel," July 1939 translation of a paper Ferri presented in Berlin in October 1938. Stack's early-1960s professional biography ("Awards and Biographical Information" folder, Stack Collection, LHA), p. 4, asserts that Stack "initiated action" that led to the postwar importation of both Ferri and Adolf Busemann, the German credited with suggesting in the 1930s the benefits of swept wings; see also Hansen, *Engineer in Charge*, pp. 318–20.

109. "Minutes of Meeting of Committee on Aerodynamics, October 12, 1943," pp. 21 and 22, in Stack collection folder "Committee for Aerodynamics Minutes 1943," LHA. In the two folders constituting the RA1204 file, LHA, are "Preliminary Data for Army Air Forces, Materiel Command: Constriction Effects in High-Speed Tunnels," January 31, 1944, letters requesting copies of the preliminary data report for manufacturers, and the official January 8, 1944, research authorization document, which cites a November 16, 1943, Army request. Stack's October synopsis and the January preliminary report directly echo Byrne's Advance Confidential Report L4L07a, December 1944.

110. Joel L. Everhart and Percy J. Bobbitt, "Experimental Studies of Transonic Flow Field Near a Longitudinally Slotted Wind Tunnel Wall," NASA Technical Paper 3392, April 1994, cites in its introduction H. Julian Allen and Walter G. Vincenti, NACA Report 782.

low-speed Variable-Density Tunnel counteracted this mismatch by using pressurized air, which of course meant higher air density in the flow field, and therefore also a higher density term in the ratio—which in turn meant improved verisimilitude as indicated in the higher score. Another way to raise a test's Reynolds number is simply to diminish the mismatch by using a larger-scale model. In fact, in that way the old Propeller Research Tunnel and the Full-Scale Tunnel simply canceled the mismatch: they were large enough for tests not at reduced scale, but at full size. The second verisimilitude indicator, *Mach number*—a shorthand term not yet used in 1933 by Stack, who still called it *compressibility factor*—compares flow speed with the speed of sound for the given conditions. It leads to a simple ratio too: for example, Mach 0.8 for a speed eight-tenths that of sound. In subsonic tunnels, the fundamental physics of Englishman Osborne Reynolds had long framed the problem of achieving flow similarity; for tunnel airflows involving compressibility, the physics of Austrian Ernst Mach now required attention as well.[111]

However, because of "practical limitations in size and power," Allen and Vincenti continued, "most existing wind tunnels, whether high speed or low speed, are not capable of providing full-scale Reynolds numbers for all flight conditions." Their readers would not need reminding that to enlarge a tunnel's airflow channel size for larger models, and thus for higher Reynolds numbers, or to increase its airflow speed for higher Mach numbers, leads with exponential quickness to a prohibitively expensive power bill—assuming enough power is available at all. An obvious partial answer, the authors said, was to use as large a model as possible in a given tunnel. But in the case of a high-subsonic-speed tunnel, the larger the model, the more magnified the problems of testing it. As Mach number rises, there is a "tendency of the [compressible] flow pattern . . . , if unrestrained, to expand." But since the tunnel walls indeed do restrain expansion of these streamlines of flowing air, the resulting test data need correcting—that is, need artificial adjustment by some formula or mathematical procedure—"if they are to be applied with confidence to the prediction of free-flight characteristics." This analysis led Allen and Vincenti to the centrally important issue of correcting results from solid-wall tunnel tests at the still higher subsonic Mach numbers where the complication known as *choking* arises—the problem that *Aviation Week* later reported "had effectively bottlenecked" transonic tunnels until NACA researchers "licked" it by inventing slotted walls.[112]

Concerning choking, Allen and Vincenti's readers would recall a fundamental airflow-physics principle: subsonic air moves faster when its channel constricts, but a supersonic airstream must expand to go faster. A test model, by constricting the channel, creates in effect the *nozzle* of a supersonic tunnel: a convergence of the flowing air followed by a divergence. The result is that "sonic velocity is reached at all points across a section of the tunnel at the position of the model, and the flow in the diverging region downstream of this section becomes supersonic. When this occurs, increased power input to the tunnel has no effect upon the velocity of the stream ahead of the model, the additional power serving merely to increase the extent of the supersonic region in the vicinity of the model. At this point the tunnel is said to be 'choked' and no further increase in the test Mach number can be obtained." That is, choking cannot be overcome by brute force, and for a

---

111. I am grateful to veteran NACA and NASA aeronautical engineer Albert L. Braslow for suggestions about this passage and much else in the essay. Concerning flow similarity, in 1934 in NACA Report 492, "Tests of 16 Related Airfoils at High Speeds"—a "classic paper which exerted great influence" according to W. F. Hilton (p. 81, *High-Speed Aerodynamics*)—John Stack and Albert E. von Doenhoff wrote: "It has been shown that the speed of flow expressed in terms of the speed of wave propagation, or the speed of sound, in the fluid is an index of the extent to which the flow is affected by compressibility. Thus, the ratio of the flow velocity to the velocity of sound, V/V_c, is a parameter indicative of the pattern similarity in relation to compressibility effects just as the Reynolds number is an index of the effects of viscosity."

112. "NACA Tunnels Bare Secrets of Transonic," *Aviation Week*, May 28, 1951, p. 13.

given model and solid-wall tunnel, the choking Mach number is the speed limit. Moreover, the authors' theoretical analysis confirmed what had been long suspected,[113] that "at the choking Mach number, the flow at the airfoil in the tunnel cannot correspond to any flow in free air. It follows that, at choking, the influence of the tunnel walls cannot be corrected for. Further, in the range of Mach numbers close to choking where the flow is influenced to any extent by the incipient choking restriction, any correction for wall interference may be of doubtful validity." In other words, once very near or at choking speed in a solid-wall tunnel, there is no translating the test data into usefulness, for these results do not correlate with actual flight conditions, not even in some hidden way.

In the end, the point was that only very small models—with very low Reynolds numbers, and thus with little verisimilitude—could be tested at near-sonic speeds in enclosed, solid-wall tunnels. Bernhard Goethert once cited an illustrative case involving a complete three-dimensional model rather than the tunnel-spanning "two-dimensional" case Allen and Vincenti addressed. For test speeds up to Mach 0.95, the model could be large enough to block head-on only one-fifth of one percent of the tunnel's airflow. This meant it could have "a maximum diameter of no more than 5.5 inches in a 10-foot-diameter wind tunnel"—a relative size like that of a softball inside a transport airplane fuselage. "It is apparent," Goethert concluded, "that transonic testing in a closed wind tunnel is very impractical."[114]

Ray Wright, Principal Agent of a Collective Solution

A 1994 NASA Langley technical paper identifies the ultimate source of the slotted-wall solution that NACA Langley devised for the transonic tunnel problem in the late 1940s: "The first 30 years of wind tunnel wall-interference research yielded an important fact for modern wind tunnels; that is, theoretically and experimentally, solid-wall corrections are opposite in sign from those of open-jet test sections. Thus, if a wall is partially open, an adjustment to the geometric openness should be possible to obtain a near-zero wall-interference correction and thereby allow a more realistic simulation of free-air conditions."[115] In even plainer terms, ventilation openings placed in just the right way in a tunnel's walls can cause the complex data-polluting effects of open-wall and closed-wall interference to cancel each other. The statement echoes similar ones by Becker, Stack, and Wright and Ward. It also echoes Goethert, who had served in the Nazi-era German aeronautical research establishment, and whose 1961 book asserted that Germany, Italy, and Japan "produced theoretical correction-free slot arrangements" but failed actually to build slotted tunnels for high-speed compressible flows only "because of the circumstances connected with and following World War II."[116] In different circumstances possibly the NACA could have found the solution earlier itself, though certainly there was no prewar call for it from industry or the military. In any case, Becker says that early experience with open jets in the eleven-inch tunnel "more than any other single factor encouraged Stack and his cohorts 15 years later to embark on the further developments which produced the transonic slotted tunnels," and that "Stack often referred to this early work as the genesis of transonic facility development."[117]

---

113. Becker, *High-Speed Frontier*, p. 66 says that the NACA by 1938 had begun to see that "there was no hope of 'correcting' data taken in the choked condition."
114. Goethert, *Transonic Wind Tunnel Testing*, p. 49.
115. Everhart and Bobbitt, "Experimental Studies of Transonic Flow Field Near a Longitudinally Slotted Wind Tunnel Wall," p. 1.
116. Becker, High-Speed Frontier, pp. 38, 98, 100, 114; Stack, "Experimental Methods for Transonic Research," p. 592a; Wright and Ward, pp. 1, 2; Goethert, *Transonic Wind Tunnel Testing*, pp. 21, 22.
117. Becker, *High-Speed Frontier*, p. 65.

At war's end two tunnel-technology studies in particular helped motivate Langley's translation of this open-closed idea into a specific proposal for longitudinally slotted walls: Antonio Ferri's high-speed tests in an Italian semi-open tunnel, presented in a report he wrote upon arriving at Langley, and Coleman duP. Donaldson's comparisons of open and closed high-subsonic-speed airflows, presented in a report Ray Wright wrote after Donaldson left for military service. Ferri investigated the performance of a rectangular test section of about sixteen inches by twenty-one inches, with solid side walls but no top or bottom to restrain the airstream. Becker calls the work "the first real demonstration that partly open arrangements could be used successfully" near Mach 1, and says it helped motivate the Donaldson study. Donaldson tested a postage-stamp-sized airfoil in both open and closed three-inch-wide jets of compressed air, much as Dryden and others had done with small open jets in the 1920s—only this time under genuine laboratory conditions, with good instrumentation for taking data. Donaldson's tests were intended generally "to show the nature of the jet-boundary interference" in both the open configuration up to Mach 1 and the closed configuration up to choking at just under Mach 0.8. Donaldson concluded that open jets "should be advantageous for tests at high Mach numbers." Becker later wrote that this study helped spur Langley's conversion of a small high-speed tunnel to the semi-open configuration. Stack later wrote that it served "to show, in principle, the possible difference in choking limitations for open- and closed-throat tunnels." Thus it was that Ray H. Wright, the man who committed Donaldson's study to paper, entered the year 1946 fully mindful of this crucial difference for the laboratory replication of compressible flow fields up to Mach 1.[118]

The question of what Ray Wright was mindful of in 1946 is important for two reasons. The less important one has to do with proportioning credit for the slotted-wall transonic tunnel. The more important one has to do with assessing the effectiveness of the NACA technical culture.

Both Baals, in *Wind Tunnels of NASA*, and Hansen have portrayed Wright as having a solely subsonic and somewhat technically naive outlook in proposing the longitudinally slotted wall that year. "Strictly speaking, Wright's analysis was applicable only to low-speed flows," Baals wrote, "but Langley aerodynamicists, led by John Stack, immediately recognized in this simple proposal the possibility of solving the serious problems they had been having with wind tunnel testing near Mach 1." This interpretation conflicts not only with the story as Becker tells it, but with the record of Wright's activities up to 1946. Becker portrays Wright exercising both technological initiative and scientific imagination in an effort purposefully targeting the wind tunnel replication of transonic flows. That Wright's theoretical work happened to be subsonic, Becker says, simply derived from the constraints of the available mathematical techniques.[119]

But it is Wright's formative activities at Langley during the decade leading up to 1946 that really matter, for they show that Wright, like Stack, was a genuine product of the

---

118. See Becker, *High-Speed Frontier*, pp. 39, 79, and 99, and Stack, "Experimental Methods for Transonic Research," p. 580, concerning Antonio Ferri, "Completed Tabulation in the United States of Tests of 24 Airfoils at High Mach Numbers (Derived from Interrupted Work at Guidonia, Italy, in the 1.31- by 1.74-Foot High-Speed Tunnel)," ACR L5E21, June 1945 (also called WR L-143) and concerning Ray H. Wright and Coleman duP. Donaldson, NACA Technical Note 1055, "Comparison of Two-Dimensional Air Flows About an NACA 0012 Airfoil of 1-Inch Chord at Zero Lift in Open and Closed 3-Inch Jets and Corrections for Jet-Boundary Interference," May 1946 (but actually, and significantly for Ray Wright's education, completed in early January, according to p. 34). Becker, *High-Speed Frontier*, pp. 79 and 99, treats Donaldson as the latter's main author, even though Wright's name appeared first in the heading. Donaldson described his and Wright's contributions in an April 11, 1996, telephone interview.

119. Baals, *Wind Tunnels of NASA*, p. 61; Hansen, *Engineer in Charge*, pp. 316, 317; Becker, *High-Speed Frontier*, chap. III, especially pp. 99–104; see p. 100 concerning the mathematics.

NACA technical culture, and in the case of the transonic wind tunnel, its important agent. Less importantly, these activities also demonstrate his entirely sophisticated awareness of his slotted-wall proposal's implications. In the late 1930s, he worked alongside 8-Foot High-Speed Tunnel designer and veteran high-speed research engineer Russell G. Robinson on the airfoil design problem of delaying the compressibility burble, building on work going back to the 1920s. By the end of the war he was working on wall interference in the eight-foot tunnel, which was being repowered for sonic speed, and he helped establish a new, minimally flow-field-disrupting method for holding its test models in place—a system first used in 1946 tests of research airplane models including the X-1. By early 1946 he had written up Donaldson's comparative investigation of transonic open and closed boundary effects, which linked directly to what he was about to propose. Thus the pre-1946 activities of physicist and applied mathematician Ray H. Wright constituted something like an apprenticeship in the engineering art and science, such as they then stood, of transonic wind tunnel testing. But by far the most revealing formative activity of this soon-to-be agent of accumulated NACA understanding took place in August 1946, when he wrote a memorandum.[120]

Wright's lengthy, detailed memorandum to Langley's compressibility research chief advocated synthesizing what the NACA already knew about high-subsonic and near-sonic wind tunnel research. "As a result of work on wind-tunnel interference and of other experiences gained over the past several years," it began, "ideas and information have been accumulated for a number of useful report projects that could be carried out with a minimum of time and effort." The point was to assess the organization's corporate store of technical and scientific knowledge about transonics up to Mach 1, and to determine how to exploit it—at minimal expense, and with the practical goal of improved research capabilities. For each of sixteen possible report project topics, Wright wrote a paragraph-length synopsis drawing on his overall awareness of existing NACA work. The topics included "general consideration of the effect of compressibility on wind tunnel interference," "wind tunnel interference at Mach numbers greater than the critical," and "flow conditions and tunnel-wall interference near choking." To be based on these three in particular, together with another closely related three, he projected among the sixteen prospective projects a "general report on wind tunnel interference at high speeds," for which a "considerable amount of material [was] already in existence" that he said "should be compared, sifted, and collated."

In this transonics-focused memorandum Wright also suggested precisely the famous project in which he himself was apparently already engaged: "wind tunnels with zero or negligible interference." For this one the accompanying synopsis is the memorandum's lengthiest, amounting to a prospectus for the theoretical and experimental work that would lead to slotted-wall tunnels. Thus it also amounted to a plan for finally realizing Langley engineers' long-held intuitions about an open-closed solution to the transonic tunnel problem. To circumvent the difficulties of high-speed wall interference, it said, "as well as to prevent choking, the wind tunnel may be so designed as to minimize the interference. If the interference can be entirely prevented, the obtaining of model data can be simplified by abolishing the necessity for making tunnel-wall corrections." The tunnel would use "an automatically compensating method" of "multiple-sided open-closed test sections." Mathematical techniques, Wright wrote, were "available for investigating this problem," and if a "mathematical investigation indicated a probability of success," small-scale, principle-proving model wind tunnels "incorporating the automatically compensating features should be designed and tested. The possible usefulness of such an investigation,"

---

120. Becker, *High-Speed Frontier*, pp. 27, 74, 75, 99; "Memorandum for Chief of Compressibility Research Division: Possible Report Projects that Could be Completed with a Minimum of Time and Effort," August 27, 1946, signed "Ray H. Wright, Physicist," in Stack collection folder "Research Problems & Questions (Reid's trip to Europe) 44–46," LHA.

added the technologically sophisticated, NACA-engineer-trained physicist, "suggests that it should be carried out as soon as personnel can be spared. Only a bare start has been made on the calculations."

Stack's 1952 press release crediting the slotted-wall contributions of his nineteen associates begins by describing his "old written notes" from 1946 showing that "for some time" he and others had had a "faith in the probability" that a transonic tunnel solution was in hand, that "a good part" of this faith "rested in the subsonic high-speed theoretical studies of Ray H. Wright," and that "in the late summer of 1946" the arrangements began for the small proof-of-principle pilot project that Vernon G. Ward spearheaded.[121] How or even whether these notes relate to Wright's August 1946 memorandum is not clear, but it is clear that Wright comprehensively understood the problems of replicating transonic flows up to Mach 1, and that much of his ability to contribute importantly to their solution derived directly from a formative decade of immersion in the technical culture around him.

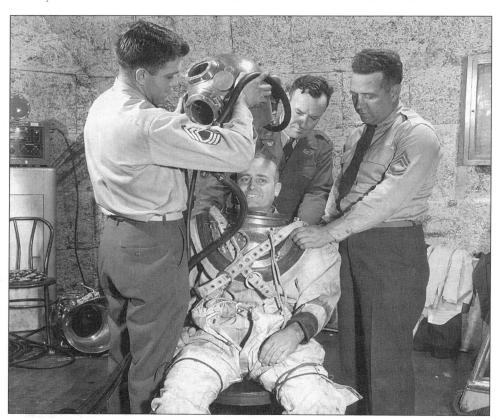

Physicist Ray Wright's decade-long immersion in the practical-solutions-seeking NACA engineering culture prepared him to propose a workable theoretical solution for the transonic wind tunnel problem. Later, his participation in the solution's hands-on realization in the 8-Foot High-Speed Tunnel required immersion in the harsh conditions that slots caused in the chamber beside the formerly entirely enclosed test section. To withstand these conditions, Wright wore a diving suit. (NASA photo L-64110).

---

121. Langley *Air Scoop*, December 19, 1952, available LHA.

Relatively soon, NACA Langley developed slotted walls well enough to apply them in two national wind tunnel facilities, all under the general guidance and technopolitical shepherding of Stack, who according to Becker "was adamant regarding schedules, at times ruthless in dealing with any interference, and always able to inspire, to make quick decisions, and to give effective orders." The newly converted tunnels were valuable; Loftin says they "provided a new dimension in transonic testing."[122] But like other useful research tools, they were imperfect too. NACA advertising notwithstanding, difficulties persisted between Mach 0.98 and Mach 1.05, part of the range from Mach 0.95 to Mach 1.2 that the NACA's 1948 survey participants had unanimously agreed was where "the real fundamental lack of information occurs." The difficulties remained in slotted-wall transonic tunnels even a half-century later. In the eight-foot tunnel in 1950, Langley engineers spent months making improvements to the initial slotted-wall installation.[123] For example, Richard Whitcomb remembers coordinating directly with Langley woodworkers to devise an apparatus at the downstream end of the test section to reintroduce the air that had gone through the slots—an efficient, focused, red-tapeless way of working that he says became "totally verboten" before he retired. By 1953, Langley high-speed researchers had commissioned a new eight-foot tunnel, this time with slotted walls planned from the outset, and with other improvements including pressurization at two atmospheres for higher Reynolds numbers, a test section designed for easier data-gathering, and modifiable slot shapes.[124] History, or public relations, might momentarily have highlighted the original two slotted-wall tunnels, but transonic research questions continued to arise, and NACA researchers like experimentalist Whitcomb continued devising research tools for answering them.

Over the years, though, NACA researchers tended not to advertise their research tools, possibly contributing to the Collier Trophy's tardiness in recognizing a wind tunnel. Roland says that even though research tools were among the NACA's chief accomplishments from the time of the Variable-Density Tunnel, NACA research director George Lewis feared sharing information about them with the NACA's competitors. Possibly this secrecy has exacted a cost in the understanding not just of the research tools, but of the technical culture from which they derived. To explain the secrecy, Lewis once compared NACA research tools to Stradivarius violins. "Antonio Stradivari," he wrote, "made a success by making the world's finest violins, and not by writing articles on how others could construct such instruments."[125] But Stradivari could only have learned to make such fine instruments where he did learn: among the Cremonese masters, a technical culture whose corporate technical memory, scientific understanding, and shared traditions of craftsmanship[126] enabled its members to build devices that move air in just such a way as to produce beautiful music. Much the same can be said for the technical culture of the NACA, where engineers—and engineering-minded physicists—learned to build devices that move air in just such a way as to produce useful knowledge.

---

122. Becker, *High-Speed Frontier*, p. 109; Loftin, *Quest for Performance*, p. 252.
123. Becker, *High-Speed Frontier*, p. 113; "Summary of Recommendations on Research Problems of Transonic Aircraft Design," p. 8 in "Report of Special Subcommittee" section; p. 18, Francis J. Capone, Linda S. Bangert, Scott C. Asbury, Charles T. L. Mills, and E. Ann Bare, "The NASA Langley 16-Foot Transonic Tunnel: Historical Overview, Facility Description, Calibration, Flow Characteristics, and Test Capabilities," NASA Technical Paper 3521, September 1995; Hansen, *Engineer in Charge*, p. 110.
124. Richard Whitcomb, telephone interviews, April 1 and 19, 1996.
125. Roland, *Model Research*, 1:246; Dawson, *Engines and Innovation*, p. 32.
126. Thomas Levenson, "How Not to Make a Stradivarius," *The American Scholar* 63, No. 3 (Summer 1994): 351–78, describes Stradivari as "essentially a craftsman of science, one with considerable, demonstrable knowledge of mathematics and acoustical physics," who attained his skills in an instrument-making culture of "old masters" with a science-based "accumulation of craft technique" in Cremona, Italy.

Chapter 5

The Whitcomb Area Rule: NACA Aerodynamics Research and Innovation

by Lane E. Wallace

As the 1940s came to a close, military aircraft manufacturers in the United States faced a disturbing problem. The Bell X-1 had broken the so-called "sound barrier," and both the Air Force and the Navy were looking for next generation aircraft that could operate at supersonic speeds. But preliminary tests of models indicated that even the best designs put forth by industry engineers were not going to be able to achieve that goal. A sharp increase in drag at speeds approaching Mach One was proving too much for the limited-power jet engines of the day to overcome.

The solution to this frustrating impasse was found by Richard T. Whitcomb, a young aerodynamicist at the National Advisory Committee for Aeronautics (NACA) Langley Research Center in Hampton, Virginia. His development of the "area rule" revolutionized how engineers looked at high-speed drag and impacted the design of virtually every transonic and supersonic aircraft ever built. In recognition of its far-reaching impact, Whitcomb's area rule was awarded the 1954 Collier Trophy.

Yet it is not just the significance of the concept that makes the discovery and application of the area rule interesting. The story of its development provides insights on how innovations are "discovered" and how, even at a time when research projects were growing bigger and more complex in scope, a single, creative individual could still play a critical role in the development of new technology. In addition, while the area rule concept was applied almost universally to supersonic aircraft designs, that "success" also illustrates some of the factors that influence whether industry applies a given technology, regardless of its inherent worth.

The Transonic Drag Problem and the Area Rule

Researchers in the Langley Research Center's wind tunnels had begun working with transonic airflows and the problem of transonic drag (at speeds approaching and surpassing the speed of sound) even before the end of World War II. In 1943, John Stack, head of Langley's Eight-Foot High-Speed Tunnel branch, obtained approval to increase the power in the tunnel from 8,000 horsepower to 16,000 horsepower. The upgrade, completed in the spring of 1945, allowed researchers to produce reliable airflow data in the tunnel for speeds up to Mach .95.[1]

One of the researchers working with Stack in the Eight-Foot High-Speed Tunnel was a young engineer named Richard Whitcomb. Whitcomb had been fascinated with airplanes and aerodynamics since he was a young boy, building and testing airplane models

---

1. James R. Hansen, *Engineer in Charge: A History of the Langley Aeronautical Laboratory, 1917–1958* (Washington, DC: NASA SP-4305, 1987), pp. 313–14.

NACA/NASA Langley engineer Richard T. Whitcomb was awarded the 1954 Collier Trophy for his development of the "area rule," an innovation that revolutionized the design of virtually every transonic and supersonic aircraft ever built. Here Whitcomb inspects a research model in the 8-Foot Transonic Tunnel at Langley. (NASA photo no. LAL 89118).

made out of balsa wood. He was hired by the Langley Research Center in 1943, after receiving an engineering degree from the Worcester Polytechnic Institute. The Langley managers initially wanted him to work in the Flight Instrument Division, but Whitcomb stubbornly insisted that he wanted to work in aerodynamics. Fortunately, he was granted his preference and was assigned to Stack in the 8-foot wind tunnel.

Initially, Whitcomb was assigned the task of performing test monitoring for other researchers. But for an eager young engineer, the key to advancement was to "run the tests and keep your eyes open, your ears open," Whitcomb recalled. "I kept coming to Gene (Draley, Stack's replacement as head of the 8-foot tunnel) and saying maybe it ought to be done this way. Let's try this. And somewhere along the way, Gene says 'OK, go try it,' and that's where I got started."[2]

By July 1948, Whitcomb had developed a reputation as "someone who had ideas"[3] and was starting to pursue his own research experiments. He proposed a series of wind tunnel tests in the repowered 8-Foot High-Speed Tunnel for a variety of swept wing and fuselage combinations. He hoped the tests would uncover a configuration with significantly lower

---

2. Richard T. Whitcomb, interview with Walter Bonney, March 27, 1973.
3. Richard T. Whitcomb, telephone interview with author, May 2, 1995.

drag at transonic speeds. The tests were run in late 1949 and 1950, but the results were both perplexing and discouraging. None of the combinations had much effect on reducing the drag of the models as they approached Mach One.[4] Clearly, the researchers needed to know more about the behavior of airflow in the transonic region in order to figure out what was causing such a stubborn drag problem. Unfortunately, this data was difficult to obtain. Even the upgraded eight-foot wind tunnel at Langley could only reach speeds of .95 Mach.

Because of the limitations of the available wind tunnels, researchers in the mid-1940s had resorted to several "stopgap" methods to try to learn more about transonic airflow. One series of experiments involved dropping instrumented test missiles from a B-29 Superfortress. Test airfoils were also mounted on the wing of a P-51 Mustang fighter plane that was then put into a high-speed dive. With this configuration, the airplane's speed remained subsonic but the airflow over the portion of the wing holding the test airfoil surpassed the speed of sound. A third approach used rocket models launched from Wallops Island, a remote beach location across the bay from the Langley Research Center.

All three methods had their drawbacks, however. The falling-body and wing-flow techniques offered less precise data than that obtained in a wind tunnel. The rocket tests produced more precise data, but they were "100 times as expensive as a wind tunnel test" and could only explore a single parameter at a time. Furthermore, the Schlieren photographs that illustrated the shock wave patterns of high-speed airflow could only be obtained in a wind tunnel.[5]

Consequently, it was not until Stack and his team of engineers, which included Whitcomb, developed a "slotted-throat" modification for the 8-foot wind tunnel in 1950 that transonic flows could be thoroughly explored.[6] The slotted-throat modification prevented the choking that had limited the speeds in the test section of the tunnel and allowed the air to go through the speed of sound. For the first time, researchers had a tool to investigate precisely what airflow did in that speed range and what might be causing the puzzling drag they had observed.

Actually, the slotted throat wind tunnel was only one of the tools Whitcomb and his associates used to investigate transonic airflows. But once that was in place, they could then employ other existing research tools to look at what the airflow was doing. In late 1951, Whitcomb tested a swept-back wing-fuselage combination in the now-transonic Eight-Foot High-Speed Tunnel.[7] Tuft surveys, which used small pieces of yarn taped onto airfoil and fuselage sections, were conducted to look at airflow disturbances. Coverings with pressure-sensitive openings were put on model sections to determine the velocity of the air over particular areas, and Schlieren photographs were used to look at the shock wave characteristics of the model at transonic speeds.[8]

---

4. Richard T. Whitcomb, "A Proposal for a Swept Wing Fuselage Combination with Small Shock Losses at Transonic Speeds," Langley Central Files, AH 321–1, July 1948; Hansen, *Engineer in Charge*, pp. 332–33.

5. Richard T. Whitcomb, telephone interview, May 2, 1995; Hansen, *Engineer in Charge*, pp. 261–70.

6. The development of the slotted–throat transonic wind tunnel at the Langley Research Center proved important enough to merit its own Collier Trophy, awarded to Stack and his associates in 1951.

7. The time delay between each of Whitcomb's initial ideas and the actual wind tunnel tests of them was a result of Langley's typical but long process of designing and building wind tunnel models. It was not at all unusual for that process to take fifteen–eighteen months. Nevertheless, the time delay was frustrating and Whitcomb sometimes worked directly with wind tunnel technicians to incorporate modifications in the tunnel to avoid the delay of going through normal channels.

8. Richard T. Whitcomb and Thomas C. Kelly, "A Study of the Flow over a 45–degree Sweptback Wing–Fuselage Combination at Transonic Mach Numbers," NACA RM L52DO1, June 25, 1952; Dr. Richard T. Whitcomb, "Research on Methods for Reducing the Aerodynamic Drag at Transonic Speeds," address presented at the ICASE/LaRC Inaugural Eastman Jacobs Lecture, Hampton, VA, November 14, 1994, pp. 1–2; Hansen, *Engineer in Charge*, pp. 332–33.

The results, especially those revealed by the Schlieren photographs, showed that the shock waves created as the airflow approached the speed of sound were different and bigger than anticipated. Undoubtedly, it was the losses from these unexpected shock patterns that was causing the sharp increase in drag at transonic speeds. But the question of what was causing the shockwaves still had to be answered before researchers could try to find a way to combat the phenomenon.

Several weeks later, a world renowned German aerodynamicist named Dr. Adolf Busemann, who had come to work at Langley after World War II, gave a technical symposium on transonic airflows. In a vivid analogy, Busemann described the stream tubes of air flowing over an aircraft at transonic speeds as pipes, meaning that their diameter remained constant. At subsonic speeds, by comparison, the stream tubes of air flowing over a surface would change shape, become narrower as their speed increased. This phenomenon was the converse, in a sense, of a well-known aerodynamic principle called Bernoulli's theorem, which stated that as the area of an airflow was made narrower, the speed of the air would increase. This principle was behind the design of venturis,[9] as well as the configuration of Langley's wind tunnels, which were "necked down" in the test sections to generate higher speeds.[10]

But at the speed of sound, Busemann explained, Bernoulli's theorem did not apply. The size of the stream tubes remained constant. In working with this kind of flow, therefore, the Langley engineers had to look at themselves as "pipefitters." Busemann's pipefitting metaphor caught the attention of Whitcomb, who was in the symposium audience. Soon after that Whitcomb was, quite literally, sitting with his feet up on his desk one day, contemplating the unusual shock waves he had encountered in the transonic wind tunnel. He thought of Busemann's analogy of pipes flowing over a wing-body shape and suddenly, as he described it later, a light went on.

The shock waves were larger than anticipated, he realized, because the stream tubes did not get narrower or change shape, meaning that any local increase in area or drag would affect the entire configuration in all directions, and for a greater distance. More importantly, that meant that in trying to reduce the drag, he could not look at the wing and fuselage as separate entities. He had to look at the entire cross-sectional area of the design and try to keep it as smooth a curve as possible as it increased and decreased around the fuselage, wing and tail. In an instant of clarity and inspiration, he had discovered the area rule.

In practical terms, the area rule concept meant that something had to be done in order to compensate for the dramatic increase in cross-sectional area where the wing joined the fuselage. The simplest solution was to indent the fuselage in that area, creating what engineers of the time described as a "Coke bottle" or "Marilyn Monroe" shaped design. The indentation would need to be greatest at the point where the wing was the thickest, and could be gradually reduced as the wing became thinner toward its trailing edge. If narrowing the fuselage was impossible, as was the case in several designs that applied the area rule concept, the fuselage behind or in front of the wing needed to be expanded to make the change in cross-sectional area from the nose of the aircraft to its tail less dramatic.[11]

---

9. A venturi, named after the 19th century Italian physicist G.B. Venturi, is one method used to generate the suction or vacuum power necessary to drive aircraft instruments. A venturi is mounted on the outside of an aircraft, paralleling the fuselage. As the speed of airflow through the cinched neck portion of the venturi increases, it is accompanied by a decrease in air pressure, creating suction that runs the instruments connected to the system inside the plane.

10. Whitcomb, interview, March 27, 1973.

11. Richard T. Whitcomb, "A Study of the Aero–Lift Drag–Rise Characteristics of Wing–Body Combinations Near the Speed of Sound," NACA Report 1273, Langley Aeronautical Laboratory, Langley Field, Virginia, 1956, pp. 1, 20–21; Whitcomb, interview, March 27, 1973; Whitcomb, "Research on Methods for Reducing the Aerodynamic Drag at Transonic Speeds," p. 3.

The Pieces of the Puzzle: Creative Innovation

Although the pieces may have come together in a flash of insight, there were actually several important elements and processes that contributed to Whitcomb's discovery. Whitcomb had developed a reputation as something of a "Wunderkind" at Langley because of his unique combination of knowledge and intuition about airflows; a combination that undoubtedly contributed to his discovery of the area rule.[12] The intuition may have been a gift, but his knowledge of airflow behavior was certainly enhanced by his seven years of experience working with Langley's 8-foot wind tunnel.

The discovery of the area rule concept was also dependent on the previous invention of the slotted-throat tunnel design. Without that piece of technology, Whitcomb could not have gathered the information necessary to understand the causes of transonic drag. In fact, the very existence of the wind tunnels at Langley was a critical factor in allowing a new approach in design to surface and be tested. If the information had to be obtained through an elaborate, expensive flight test program, fewer ideas could have been investigated, and Whitcomb might not have had the opportunity to test his innovative theory.

At Wallops Station, in tidewater Maryland, in 1953, Langley's Pilotless Aircraft Research Division (PARD) tested rocket-powered models of the delta-winged Convair F-102 before (left) and after (right) modification to take advantage of Whitcomb's "area rule." (NASA photo).

---

12. Eugene S. Ferguson, *Engineering and the Mind's Eye* (Cambridge, MA: MIT Press, 1992), p. 54; Hansen, *Engineer in Charge*, p. 332.

In addition, the projects conducted at Langley were still fairly small, individual research efforts that allowed for experimentation. This kind of atmosphere, while not entirely unique among government-funded facilities in the early 1950s, was becoming more unusual. At one time, individual or small-group research efforts had characterized many research laboratories. But the exponential growth of technology and complex technological research during World War II began to change that. The Manhattan Project, responsible for the development of the atom bomb, symbolized for many a significant shift in technological research from small, independent projects conducted by single laboratories to large, complex research programs involving many people, broad resources and funding, and multiple disciplines.[13]

In a bigger and more complex research environment, with approvals and decisions dependent on higher-level program managers, Whitcomb might not have had the latitude or opportunity to develop and test the area rule concept. But the NACA Langley environment offered a middle ground between a small, independent laboratory and a large research program. Whitcomb had expensive technological tools at his disposal, such as the slotted-throat wind tunnel, but he still had the independence and flexibility to develop and test a radical new concept on his own.[14]

Whitcomb was also assisted by the informal management environment and the orientation toward experimental research at the Langley Research Center, both of which were conducive to individual innovation. As John Becker explained in his case histories of four NACA programs,

> *Management (at Langley) assumed that research ideas would emerge from an alert staff at all levels. . . . On a problem of major proportions such as transonic facilities, any scheme for research that survived peer discussions and gained section and division approvals was likely to be implemented . . . and very little (paperwork) was required in the simple NACA system. Occasional chats with his division chief or department head, or a brief verbal report at the monthly department meeting were about all that was required of the NACA project engineer.*[15]

This kind of environment was particularly well-suited to an introspective thinker like Whitcomb. Managers knew he was a talented aerodynamicist, and they were wise enough to keep his paperwork to a minimum and give him the space and freedom to think, experiment, and explore.[16]

Langley's orientation toward hands-on, experimental research was a significant factor in Whitcomb's discovery, as well. As opposed to research centers that focused more on theoretical research, Langley encouraged exploratory experiments such as the wind tunnel tests Whitcomb devised to investigate wing-body combinations and airflow at transonic speeds. The breakthrough on the transonic wind tunnel itself, in fact, was a result of a researcher asking himself, "I wonder what would happen if I turned up the power?" That simple question—"I wonder what would happen if . . ." instigated numerous experiments at Langley that, in turn, led to significant discoveries.[17]

---

13. James H. Capshew and Karen A. Rader, "Big Science: Price to the Present," *OSRIS*, 2nd series 7 (1992): 19; Thomas P. Hughes, *American Genesis: A Century of Invention and Technological Enthusiasm* (New York, NY: Penguin Books, 1989), pp. 440–42.
14. John V. Becker, *The High–Speed Frontier: Case Histories of Four NACA Programs. 1920-1950* (Washington, DC: NASA SP-445, 1980), pp. 117–18.
15. Ibid.
16. Hansen, *Engineer in Charge*, p. 341.
17. Whitcomb, interview, May 2, 1995; information on transonic wind tunnel development also in Hansen, *Engineer in Charge*, p. 322; and in Ch. 1 of this book.

This curiosity-driven, experimental approach was especially significant in discovering the area rule, because there was no available theory to explain the unusual drag encountered at transonic speeds. Researchers had to come up with a creative way of reaching beyond the known, and the exploratory experiments conducted by Whitcomb and others yielded the data that allowed him to understand the cause of the transonic drag and shock wave phenomena. Conducting hands-on experiments with an aircraft model in a wind tunnel also helped Whitcomb "see" the airflow behavior in a way mathematical formulas would not have.

Still, these factors only provided the tools and environment that made Whitcomb's discovery possible. The breakthrough still required the insight of a creative mind; a mind able to "see" the problem and able to step back from accepted rules of design to contemplate a solution based on an entirely new approach. The process by which Whitcomb was able to do that offers insight itself as to how scientific or technological innovation occurs.

Science and technology are often viewed as fields completely divorced from any of the arts. Common phrases that distinguish something as "a science, not an art" and describe "the scientific method" as a way to discern an unassailable truth indicate our collective view of science as a rational, logical, linear, mathematical and precise process. Yet since almost the beginning of time, artistic vision has played a critical role in the advancement of technology and science. Undoubtedly, even the first cave dweller to invent the wheel first had a picture in his or her mind of what the device would look like.

Albert Colquhoun, a British architect, asserted that even scientific laws are "constructs of the human mind," valid only as long as events do not prove them wrong, and applied to a solution of a design problem only after a designer develops a vision of the solution in his head.[18] This artistic vision becomes even more important when a scientist or engineer needs to go beyond the leading edge of knowledge, where existing theories cease to explain events. At this point, a designer's imagination is critical in envisioning potential new solutions. As one analyst of technological development said, "The inventor needs the intuition of the metaphor maker, some of the insight of Newton, the imagination of the poet, and perhaps a touch of the irrational obsession of the schizophrenic."[19]

Whitcomb was not the only person to look at the problem of transonic drag. As early as 1944, German aerodynamicist Dietrich Kuchemann had designed a tapered fuselage fighter plane that was dubbed the "Kuchemann Coke Bottle" by American intelligence personnel. Kuchemann's design was not aimed at smoothing the curve of the cross sectional area to displace the air less violently, however. He had simply observed the direction of air flow over a swept-wing design and was trying to design a fuselage that would follow the contours of that flow.[20]

Whitcomb's area rule was also, in retrospect, said to be implicit in a doctoral thesis on supersonic flow by Wallace D. Hayes, published in 1947. But the mathematical formulas employed by Hayes, as well as several other researchers working on the general problem of transonic and supersonic air flows, did not lead their creators to the necessary flash of inspiration that crystallized the area rule for Whitcomb. Why didn't they see what Whitcomb did? The answer, in part, may lie in the precise fact that they were working with mathematical formulas, instead of visual images. The answer may have been imbedded in the numbers in front of them, but they couldn't see it.

---

18. Ferguson, *Engineering and the Mind's Eye*, p. 172.
19. Hughes, *American Genesis*, p. 76; Hansen, *Engineer in Charge*, p. 311; Ferguson, *Engineering and the Mind's Eye*, pp. 172–73.
20. David A. Anderson, "NACA Formula Eases Supersonic Flight," *Aviation Week & Space Technology* 63 (September 12, 1955): 13.

What led to Whitcomb's insight was his talent to see and work with visual metaphor—a skill described by Aristotle as a "sign of genius" and an important tool for seeing things from a fresh perspective, or discovering new truths about existing objects or ideas.[21] In his history of American technological progress, Thomas Hughes also stressed the importance of visual metaphors in developing innovative ideas, noting that "although they are articulated verbally, the metaphors of inventors have often been visual or spatial. Inventors, like many scientists, including Albert Einstein, Niels Bohr, and Werner Heisenberg, show themselves adept at manipulating visual, or nonverbal, images."[22]

When Adolf Busemann used his "pipefitting" metaphor to describe the behavior of transonic air flow, Whitcomb painted a vivid picture in his mind of air "pipes" flowing over an aircraft. He then incorporated into that image the other information he had obtained through his experiments with transonic air flow. Suddenly, he "saw" what was causing the unusual shock waves and what could be done to combat the problem.

In order to see a solution that went beyond existing theory, however, Whitcomb also had to be willing to break free from accepted rules, or paradigms, of aerodynamics.[23] In the late nineteenth century, Ernst Mach had shown that a bullet-shaped body produced less drag in flight than any other design. This accepted "paradigm" of aircraft design led to the basic fuselage shape employed by transports, World War II fighter planes, and even the Bell X-1 rocket plane. It was also still the accepted rule of thumb as engineers began to design the first turbojet-powered supersonic aircraft. The assumption that a bullet-shaped fuselage was the most efficient aerodynamic shape, however, led researchers to look elsewhere for elements that could be modified to reduce the drag of aircraft at transonic speeds. To see the solution that Whitcomb envisioned—indenting the fuselage in the area of the wing to reduce the dramatic changes in the aircraft's overall cross-sectional area from nose to tail—required going against a "truth" that had worked and had been accepted for over fifty years.

The same paradigm that had helped advance aircraft design for half a century became, ironically, one of the barriers that kept researchers from advancing aircraft design beyond subsonic flight. Why was Whitcomb able to step back and consider an approach that broke this accepted rule? For one thing, the circumstances required it. Kuhn noted that "the failure of existing rules is the prelude to a search for new ones."[24] Certainly, the stubborn problem of transonic drag presented Whitcomb with a situation where existing theories and rules were not working.

Secondly, Kuhn observed that "almost always, the men who achieve...fundamental inventions of a new paradigm have been either very young or very new to the field whose paradigm they change."[25] When he came up with the area rule concept, Whitcomb was only 30 years old. Possibly, the fact that he had not spent twenty years designing bullet-shaped fuselages contributed to Whitcomb's ability to conceive of a different design. He was also something of an introspective thinker and individualistic researcher, which may have made him more able to contemplate a "fringe" idea that broke from his peer group's assumptions. In any event, Whitcomb was willing to step back from accepted truths and

---

21. Aristotle, *Poetics*, translated by Ingram Bywater, in *The Rhetoric and the Poetics of Aristotle* (New York: Random House, 1954), p. 255.

22. Hughes, *American Genesis*, p. 82.

23. Thomas Kuhn described paradigms as "familiar notions," or "examples that provide models from which spring particular coherent traditions of scientific research." On the one hand, these accepted notions can help lead to more detailed further research in a particular area. But Kuhn cautioned that paradigms could also insulate the research community against seeing new solutions. From: Thomas S. Kuhn, *The Structure of Scientific Revolutions*, 2nd ed., Foundations of the Unity of Science Series: Vol. II, Number 2 (Chicago: University of Chicago Press, 1970), pp. 10–11, 24, 37.

24. *Ibid.*, p. 68.

25. *Ibid.*, p. 90.

simply look at what his data was showing him; paint a visual picture of it in his mind and see not what he expected to see, but what was really there.

While this may seem a simple and obvious solution to outsiders with forty years of hindsight, Whitcomb's ability to break free of the design doctrines that dominated aeronautics in his day was, in fact, a unique and remarkable ability that truly set him apart from many others in his field. Once someone comes up with an answer, it often seems obvious. But the researchers struggling with transonic drag were not aware they were caught in a paradigm that did not work. They were focused on trying to cut a workable path through a dense forest they knew as real and immutable. Whitcomb's genius was his ability to see that the problem was not the path, but the forest itself.

From Idea to Application

When Whitcomb presented his concept of the area rule to some of his colleagues at Langley, he encountered skepticism. After all, it was a radical approach to aircraft design. But division chief John Stack still allowed Whitcomb to present the idea at the next technical seminar. And listening to Whitcomb's presentation, this time, was Adolf Busemann, whose stature in the aerodynamics community was such that his opinion carried a great deal of weight. Busemann, whose visual pipefitting metaphor had provided the catalyst to Whitcomb's discovery, understood what Whitcomb had seen. He told the others present that Whitcomb's idea was "brilliant." The skepticism among some of the others, including Stack, remained. But the support from Busemann was enough to get Whitcomb the go-ahead to test his theory.[26]

Throughout the first quarter of 1952, Whitcomb conducted a series of experiments using various area-rule based wing-body configurations in Langley's 8-Foot High-Speed Tunnel. As he expected, indenting the fuselage in the area of the wing did, indeed, significantly reduce the amount of drag at transonic speeds. In fact, Whitcomb found that "indenting the body reduced the drag-rise increments associated with the unswept and delta wings by approximately 60 percent near the speed of sound," virtually eliminating the drag rise created by having to put wings on a smooth, cylindrical shaped body.[27]

In a simple world, this validation of Whitcomb's theory would have been sufficient for the principle to be applied to all new industry designs. All that would have been necessary would have been to notify the aircraft manufacturers that a better design approach had been developed. The world is not that simple, however, and the inherent worth of an innovation is rarely enough for it to be incorporated into commercial products. As Louis B.C. Fong, director of the Office of Technology Utilization at NASA (National Aeronautics and Space Administration) commented in 1963, "In this age of automation, there is nothing automatic about the transfer of knowledge or the application of an idea or invention to practical use...there is resistance to new ideas and new technologies; part psychological, part practical...and often economic."[28]

---

26. Whitcomb, interview, May 2, 1995; Hansen, *Engineer in Charge*, p. 336.
27. Whitcomb, "A Study of the Aero-Lift Drag-Rise Characteristics of Wing-Body Combinations Near the Speed of Sound," pp. 20–21.
28. Louis B.C. Fong, Dir., NASA Office of Technology Utilization, "The NASA Program of Industrial Applications," address at the Third National Conference on the Peaceful Uses of Space, Chicago, IL, May 8, 1963, NASA Historical Reference Collection, NASA History Office, NASA Headquarters, Washington, DC.

NACA or NASA engineers tend to measure the success of a new idea or technology strictly in terms of technical objectives met. Industry, on the other hand, measures innovative success in terms of profit dollars generated within a specified payback period.[29] Consequently, a new approach or technology, even if it is technically "better," may be rejected by industry if its use involves extra costs for the manufacturer. These costs can be in retooling for a new design, replacing machinery, or even in retraining employees or changing the traditional ideas and approaches of its engineers. All of these factors can produce resistance to a new idea or technology within a company, and overcoming that resistance can be a difficult process.[30]

There are a couple of situations in which new technology may be rapidly assimilated into commercial products, however. One is if it can be incorporated with minimal extra cost, and a second is if it solves a problem that a manufacturer needs to solve.[31] When Whitcomb developed his area rule, there was a manufacturer in each of these situations, and that fact played a significant role in the speed with which his innovation began to impact the design of new aircraft.

While Whitcomb was conceiving and testing his area rule concept, the Convair Division of General Dynamics was developing what it hoped would be the company's first supersonic aircraft. The Convair F-102 "Delta Dagger" was designed to be a long-range interceptor, with delta wings and the most powerful turbojet engine available at that time, the Pratt & Whitney J-57. Early test results of an F-102 model in Langley's 8-Foot High-Speed Tunnel, however, seemed to indicate that the design's transonic drag might be too high for the aircraft to surpass Mach One.

The NACA had immediately classified any information pertaining to the area rule, as it had the research on the slotted throat wind tunnel that allowed the area rule to be developed. In 1952, the United States was engaged in heated and high-stakes competition for military superiority with the Soviet Union, and NACA realized the importance of transonic research in developing superior military aircraft. Although the classification was necessary, it made dissemination of information about the area rule more difficult. Fortunately, NACA's history of successful technology transfer efforts had been less a product of published writings than the various levels of informal NACA-industry cooperation and researcher-to-engineer discussions.[32] The area rule would prove no exception.

In mid-August 1952, a group of Convair engineers were at Langley to observe the performance of the F-102 model in the Eight-Foot High-Speed Tunnel. Shown the disappointing test results, the engineers asked the Langley engineers if they had any suggestions. Whitcomb's first research memorandum on the area rule would not be published for another month, but he had completed his tests on the various wing-body combinations using indented fuselage shapes. He explained his findings and the area rule concept to the Convair team.

Intrigued, the Convair engineers worked with Whitcomb over the next few months to experiment with modifying the F-102 design and building a model that incorporated the area rule concept. At the same time, however, the company continued work on the original F-102 prototype. The engineers may have been open to exploring a possible new

---

29. Denver Research Institute, "NASA Partnership with Industry: Enhancing Technology Transfer," NASA CR-180-163, July 1983, pp. xx, Appendix D-3; William D. Mace and William E. Howell, "Integrated Controls for a New Aircraft Generation," *Astronautics & Aeronautics* 16 (March 1978): 48–53.

30. Denver Research Institute, "NASA Partnership with Industry," pp. xx, Appendix D-3; R. P. Schmitt, et al., "Technology Transfer Primer," Wisconsin University–Milwaukee, Center for Urban Transportation Studies, FHWA/TS–84/226, July 1985, pp. x, 1–5.

31. Schmitt, et al, "Technology Transfer Primer," p. 5.

32. Denver Research Institute, "NASA Partnership with Industry," p. xiv.

option, given the uncertainty produced by the wind tunnel tests of the original F-102 model, but the company had already made a commitment to the Air Force to build two prototypes of the original F-102. In addition to any mental and institutional resistance Convair might have had to changing a design which it had touted so highly and had already made a commitment to build, the company's commitment also created an issue of cost.

By mid-1952, when Convair tested the F-102 model at Langley, the company had already begun setting up a production line at its San Diego, California, facility for manufacturing the aircraft. To change the design would mean not only delays and additional engineering costs, but revamping the production line, as well. Consequently, far from being receptive to a new design approach, Convair had a significant stake in proving that its new aircraft could perform just fine without it.[33]

Nevertheless, the company could not totally ignore the doubtful test results of its original design, so its engineers began working on a "Plan B" with Whitcomb while production of the prototype F-102s continued. Starting in May 1953, the Convair engineers and Whitcomb began testing models of a modified, area rule-based, F-102 design in Langley's wind tunnel. By October 1953, they had developed a model that could meet the Air Force performance specifications. Convair noted the results but continued working on the original F-102 prototype, which flew for the first time on October 24, 1953.[34] The first prototype was severely damaged on its maiden flight, so test flights had to be postponed until January 11, 1954, when the second prototype flew for the first time. The results of the flight tests, however, proved to be largely the same as those predicted by the wind tunnel tests of the F-102 model in 1952. The aircraft performed below expectations and could not attain supersonic speeds in level flight.[35]

Even at that point, Convair might have continued to press for production of the design as it was, given that the tooling and production line in its San Diego plant was already set, except for one crucial factor. The Air Force officials working on the F-102 design were aware of Whitcomb's area rule and the fact that a modified F-102 model, based on that concept, had achieved supersonic speeds in wind tunnel tests. Consequently, the Air Force realized that the F-102 was not the best that Convair could do. Whitcomb's experiments had proven that a supersonic airplane was possible, and the Air Force decided to settle for no less. The F-102 program manager at Wright Field in Ohio informed Convair that if the company did not modify the F-102 to achieve supersonic flight, the contract for the fighter/interceptor would be cancelled.[36]

Incorporating Whitcomb's innovative design approach involved extra expense, but nothing compared to the cost of losing the entire F-102 contract. Convair immediately halted the F-102 production line and began working on the modified design Whitcomb and the company engineers had developed and tested. In only 117 working days, the company had built a new, area rule-based prototype, designated the F-102A. The F-102A flew for the first time on December 24, 1954, and surpassed the speed of sound not only in level flight, but while it was still in its initial climb. The area rule had improved the speed of the F-102 design by an estimated twenty-five percent.[37]

---

33. Donald D. Baals and William R. Corliss, *Wind Tunnels of NASA* (Washington, DC: NASA SP-440, 1981), p. 62; Hansen, *Engineer in Charge*, p. 337; Whitcomb, interview, May 2, 1995.
34. Bill Gunston, ed., *The Illustrated History of Fighters* (New York, NY: Simon and Schuster, 1984), p. 194.
35. Baals and Corliss, *Wind Tunnels of NASA*, p. 63.
36. Whitcomb, interview, May 2, 1995; Whitcomb, "Research on Methods for Reducing the Aerodynamic Drag at Transonic Speeds," November 14, 1994; Hansen, *Engineer in Charge*, pp. 337–39.
37. Baals and Corliss, *Wind Tunnels of NASA*, p. 63; Hansen, *Engineer in Charge*, p. 338; Whitcomb, interview, May 2, 1995.

While Convair was struggling with its F-102 design, the Grumman Aircraft Engineering Corporation was also working to develop its first supersonic carrier-based fighter, the F9F/F-11F Tiger.[38] Although the area rule research was classified, the NACA released a confidential Research Memorandum on the subject to appropriately cleared aircraft manufacturers in September 1952. Just two weeks after receiving that memorandum, Grumman sent a group of its engineers to Langley to learn more about it. The information they brought back to Bethpage, New York, was immediately incorporated into the design, and in February 1953, Whitcomb was flown in to review the final design plans before construction on the prototype was begun. On April 27, 1953, the Navy signed a letter of intent with Grumman for the fighter, based on the Whitcomb-approved design. On August 16, 1954, the Grumman F9F-9 Tiger "breezed" through the sound barrier in level flight without the use of the afterburner on its Wright J-65 turbojet engine.[39]

The enthusiastic incorporation of Whitcomb's innovation by Grumman stands in stark contrast to the qualified experimentation and resistance that characterized Convair's response. But the two companies were in different situations. Convair had already completed a design for the F-102 and had begun construction of two prototypes and a production line. Grumman, on the other hand, was still working to design the F11F Tiger when Langley published its confidential report on Whitcomb's area rule breakthrough. It was the perfect time to incorporate a better design idea, and involved few extra costs to the company. At the same time, the Navy had not yet contracted for the fighter, and Grumman may well have recognized that its chances of winning the contract would be improved by incorporating any available new technology into its design; especially something that might improve its speed.

In any event, Whitcomb's innovative idea was incorporated into two production military aircraft only twenty-four months after he completed his initial wind tunnel tests on the concept. This incredibly "successful" example of technology transfer was a result of two important factors. First and foremost, there was a "problem looking for a solution"[40] that the area rule was able to solve. Transonic drag was a real and seemingly unsurmountable obstacle to supersonic flight. Whitcomb's area rule was not one of a number of potential solutions; it was the *only* approach anyone had developed that had proven itself capable of overcoming that barrier. It also had the backing of a very powerful customer: the United States military. When the Air Force decided to hold firm on its demand that Convair's aircraft fly supersonically in level flight, Convair could not simply sell its F-102s to another customer. The Air Force was its only client, just as the Navy was for Grumman.

But another important element, especially with regard to Convair, was the cooperation and individual relationships that existed between the Langley researchers, including Whitcomb, and the industry engineers. The modified F-102A model that proved to the Air Force that a fighter could achieve supersonic flight was a cooperative effort between Whitcomb and Convair engineers. Without that cooperation, or the informal discussions at Langley that launched that work, the fate of the F-102 might have been different.

---

38. The prototype was designated first as the F9F-8, and then as the F9F-9, although the original Grumman F9F-2 design was the straight-wing Panther jet, and the F9F-6 was the swept-wing Cougar. The Tiger was really an unrelated design, but the prototypes were still labeled as variants of the F9F design. The production model Tigers, however, were called F11Fs.

39. Michael J.H. Taylor, ed., *Jane's Encyclopedia of Aviation* (New York, NY: Portland House, 1989), pp. 447–48; Gunston, *Illustrated History of Fighters*, p. 192; Hansen, *Engineer in Charge*, pp. 339–40.

40. Numerous NASA and industry engineers, including Whitcomb himself (Whitcomb, interview, March 27, 1973), have used this phrase to describe the kind of situation that tends to lead to quick acceptance of a new technology.

The area rule undoubtedly would have been incorporated into aircraft designs eventually, regardless of the individuals involved. But that timeframe could have been different, which could have had an impact on the kind of air defenses the United States had at its disposal in the early days of the Cold War.

As it was, the success of the area rule-based F-102 and F11F was followed by the incorporation of the area rule in virtually every supersonic aircraft built after that point. The Vought F8U "Crusader" fighter and the Convair B-58 "Hustler" bomber, both of which were on the drawing board at the time the area rule was developed, were redesigned using Whitcomb's approach. The F-106, which was Convair's follow-on design to the F-102A, adhered even more to the area rule. It was able to incorporate a much deeper indentation in the fuselage than its predecessor, because it was an entirely new aircraft, unencumbered by existing design elements.

The fuselage of the Republic F-105 "Thunderchief" fighter/bomber, which flew for the first time in 1955, incorporated the area rule in a slightly different manner. It could not be indented because of its complex engine inlets, so a bulge was added to the aft region of the fuselage to reduce the severity of the change in the cross sectional area at the trailing edge of the wing. The Rockwell B-1 bomber and the Boeing 747 commercial airliner also used the addition of a cross-sectional area to reduce their drag at transonic speeds. Both the B-1 and the 747 have a vertical "bump" in the forward section of the fuselage ahead of the wing. It is perhaps more visible in the 747, where it houses the airliner's characteristic second story, but both airframe modifications were added to smooth the curve of the design's cross-sectional area.[41]

The Collier Trophy

Whitcomb's Area Rule research was classified until September 1955, so he did not receive any immediate accolades or press on his discovery. But two months after his work was made public, Whitcomb received the National Aeronautic Association's Robert J. Collier Trophy in recognition of his achievement the previous year, when the Grumman F9F-9 Tiger and the Convair F-102A prototypes demonstrated just how significant the area rule was. The Collier Trophy citation read, "For discovery and experimental verification of the area rule, a contribution to base knowledge yielding significantly higher airplane speed and greater range with the same power."[42]

Conclusion

Although an engineering design approach using formulas or algorithms does not lend itself to the kind of notoriety that a project like the X-1 generated, the development of the area rule was no less significant. The X-1 proved the sound barrier could be broken. The area rule made that discovery practical by enabling production aircraft to operate at that speed.

The fact that the area rule was discovered by an engineer sitting with his feet up on his desk, contemplating a vision in his mind, also shows the importance of creativity and the individual in advancing technology. Postwar science and research projects may have been growing in complexity and size, but Whitcomb's discovery was a reminder that the

---

41. Whitcomb, interview, May 2, 1995; Whitcomb, "Research on Methods for Reducing the Aerodynamic Drag at Transonic Speeds," November 14, 1994, p. 3.

42. Bill Robie, *For the Greatest Achievement: A History of the Aero Club of America and the National Aeronautic Association*, (Washington, DC: Smithsonian Institution Press, 1993), p. 232; Richard T. Whitcomb, telephone interview with author, May 15, 1995.

individual researcher was more than a cog in a scientific, process-driven wheel. Experimentation and the visions in the mind of an individual able to put available information together in a new way have led to many innovative "breakthroughs" in technology and knowledge.

The history of the area rule research also illustrates that even a "breakthrough" discovery does not always win immediate acceptance by those who might implement it. As opposed to projects that were wholly funded, developed and implemented by the NACA and its successor, the National Aeronautics and Space Administration (NASA), or other government agencies, Whitcomb's breakthrough was just an idea. It may have been developed at a NACA laboratory, but it was not up to NACA to apply it. In order for the innovation to have any impact at all, industry had to agree to use it, which is not always as simple a process as it might seem. Whitcomb's area rule was the answer to a tremendous problem that industry needed to solve, but the enthusiasm with which it was received differed greatly between Convair and Grumman. The advantages offered by the innovation were the same; the costs of implementing it differed.

But even in the application of the area rule concept, individuals played an important role. An Air Force demand was the primary reason Convair incorporated the area rule into the F-102, despite the added cost. But the Air Force might not have had the confidence to make that demand if it had not been for the model work performed by a small number of individuals at Langley and Convair. As scientific and engineering research and projects became more expensive, complex, and systems-oriented, it was easy to lose sight of the individuals that made those systems work. Richard T. Whitcomb, in developing and helping to win acceptance for a concept that revolutionized high-performance aircraft design, was a reminder that the individual still mattered.

Chapter 6

The X-15 Hypersonic Flight Research Program: Politics and Permutations at NASA

by W. D. Kay

Despite the fact that it is one of the most celebrated experimental aircraft ever flown, most historical writings have always had a rather peculiar blind spot regarding the X-15 program.[1] The citation for the 1961 Collier Trophy, for example, noted that the vehicle had made "invaluable technological contributions to the advancement of flight." It also commends "the great skill and courage" of its test pilots.[2] In his letter nominating the program for the award earlier that same year, NASA Deputy Administrator Hugh L. Dryden struck the same general themes, albeit in greater detail:

> *To the X-15 Research Airplane Team, the scientists, engineers, technicians, and pilots of the National Aeronautics and Space Administration; the Department of Defense; and North American Aviation, Incorporated for the conception, design, development, construction, and flight operation of the X-15 research airplane, which contributed valuable research information in the supersonic and hypersonic speed regime up to the fringes of space, and who have thereby made an outstanding contribution to American leadership in aerospace science and technology and in the operation of manned space flight.*[3]

These two features—an outstanding piece of machinery, flown by exceptionally brave and proficient pilots—still stand as the primary legacy of the X-15.

Certainly, all of this fame is well-deserved. Considering its technical achievements, as well as its contribution to knowledge about the upper atmosphere, hypersonics, high-altitude piloted flight, and so on, the X-15 clearly stands as one of the most successful research programs in the history of aviation. Similarly, the men who flew the craft into the fringes of space at six times the speed of sound proved themselves time and again to be extraordinary individuals. These elements of the program have been recognized repeatedly, with the X-15 and its members receiving sixteen awards in addition to the Collier Trophy.

---

1. Because it was designed to penetrate into the lower fringes of what is commonly agreed to be where "space" begins (about 100 kilometers), some accounts refer to the X-15 as a "spacecraft" or "spaceplane" (or even "America's first spaceship"). See Milton O. Thompson, *At the Edge of Space: The X-15 Flight Program* (Washington, DC: Smithsonian Institution Press, 1992); Jonathan McDowell, "The X-15 Spaceplane," *Quest: The Magazine of Spaceflight History* 3 (Spring 1994): 4–12. Since most of its flight activity occurred within the Earth's atmosphere, this essay usually will use the term "aircraft,"

2. Bill Robie, *For the Greatest Achievement: A History of the Aero Club of America and the National Aeronautic Association* (Washington, DC: Smithsonian Institution Press, 1993), pp. 192, 233; "NAA's Collier Award: A Rose Garden Affair," *National Aeronautics*, September 1962, pp. 12–13. See also Robert C. Seamans, Jr., "Objectives and Achievement of the X-15 Program," remarks at X-15 Awards Ceremony, July 18, 1962, in NASA Historical Reference Collection, NASA History Office, NASA Headquarters, Washington, DC. The award was officially presented to four pilots representing the program's major participants: Robert M. White of the Air Force, Joseph A. Walker of NASA, A. Scott Crossfield of North American Aviation, and Forrest N. Petersen of the Navy.

3. Hugh L. Dryden, NASA Deputy Administrator, to Martin M. Decker, President, National Aeronautics Association, May 2, 1962, Deputy Administrator Files, NASA Historical Reference Collection.

In a White House ceremony, July 18, 1961, President John F. Kennedy presented the Collier Trophy to X-15 pilot Major Robert M. White (shown standing next to the Trophy). Also receiving the award were Commander Forrest S. Petersen, and Dr. Joseph A. Walker (not pictured). (NASA photo no. 62-X-15-19).

The problem with the prevailing view of the X-15 is not so much that it is wrong, but rather that it is incomplete. For more than three decades, the vehicle's technical design, its scientific accomplishments, contributions to aerospace engineering, its flight records, and even the personal stories of its pilots have been extolled repeatedly in books, articles, monographs, and lectures.[4] Very little, however, has been written about how the program *was actually run*, and virtually nothing has ever been recorded about its overall management.[5] Most historical accounts begin with the National Advisory Committee for Aeronautics' (NACA) decision in the early 1950s to pursue development of a high-altitude research plane, describe the technical aspects behind the selection of the contractors, and then skip over to the October 1958 rollout of the first vehicle.[6]

---

4. See, Myron B. Gubitz, *Rocketship X-15: A Bold New Step in Aviation* (New York, NY: Julian Messner, 1960); Joseph A. Walker, "I Fly the X-15," *National Geographic*, September 1962, pp. 428–50; John V. Becker, "The X-15 Project," *Astronautics & Aeronautics*, February 1964, pp. 52–61; Wendell H. Stillwell, *X-15 Research Results* (Washington, DC: NASA SP-60, 1965); Irving Stone, "The Quiet Records of the X-15," *Air Force/Space Digest*, June 1968, pp. 62–66, 71; "The X-Series," *Aerophile*, March/April 1977, pp. 72–93; Curtis Peebles, "X-15: First Wings into Space," *Spaceflight*, June 1977, pp. 228–32; Thompson, *At the Edge of Space*; McDowell, "X-15 Spaceplane."

5. The major exceptions here are U.S. Air Force, Air Force Systems Command, *The Rocket Research Program, 1946–1962* (Edwards AFB, CA: AFSC Historical Publications Series, 1962), pp. 62–110; and Robert S. Houston, Richard P. Hallion, and Ronald G. Boston, "Transiting from Air to Space: The North American X-15," in Richard P. Hallion, ed., *The Hypersonic Revolution: Eight Case Studies in the History of Hypersonic Technology*, 2 Vols. (Wright–Patterson AFB, OH: Special Staff Office, Aeronautical Systems Division, 1987), 1:1–183, neither of which has ever been published (both are available at the NASA Historical Reference Collection). There is also a brief discussion of some aspects of the program's management in Richard P. Hallion, *On the Frontier: Flight Research at Dryden, 1946–1981* (Washington, DC: NASA SP-4303, 1984), pp. 106–29.

6. Not surprisingly, this is especially true of U.S. government publications. See "Brief History of the X-15 Project," NASA news release, April 13, 1962, NASA Historical Reference Collection; Stillwell, *X-15 Research Results*; "X-15 to Enter Smithsonian," NASA news release, April 27, 1969, NASA Historical Reference Collection. Many discussions, however, will briefly mention the problems with the vehicle's main engine.

Not only is this view largely incomplete, but it also tends to give the impression that the X-15 experience was completely smooth and trouble free. Even the program's most serious technical problems are seldom described in any detail, and some difficulties, such as the fact that the project ran significantly over its budget, have never really been discussed at all.[7]

To take one example, which will be explored further below, the development of the vehicle's main XLR-99 rocket engine fell considerably behind schedule, at one point posing a significant threat to the entire program. Ultimately, after much wrangling with the engine contractor, Air Force and NACA officials opted to conduct initial flight tests with two smaller XLR-11 engines. Most X-15 histories, however, dispose of this affair in a couple of sentences, almost suggesting that it was nothing more than a brief annoyance. Indeed, in remarks made at the Collier Trophy Award ceremony in July 1962, Robert C. Seamans, Jr., portrays it as a routine decision, virtually planned in advance, rather than forced by necessity: "In January, 1958, the project management decided to continue the development of the 57,000-pound thrust engine, but to use a small engine as the power plant for initial X-15 flights."[8]

This account of the X-15 is unfortunate for a number of reasons. To begin with, the historical literature—laudatory as it has been—actually understates the magnitude of the program's accomplishments. Technical malfunctions, delays, and cost overruns are a normal part of any "cutting edge" research and development (R&D) program, and those in charge of the vehicle's development and operation deserve even more credit than they have received for working around such difficulties. Their efforts are especially impressive in view of the fact that the X-15 represented the NACA's (and later NASA's) first efforts at managing a large-scale project.[9]

Secondly, because most discussions of the X-15 have been so idealized, current United States space policy, and particularly NASA itself, have sometimes suffered by comparison. For years, observers have contrasted the cost, reliability, and performance of the X-15 with the ongoing problems of the space shuttle fleet.[10] Since the history of the shuttle's development has been explored rather thoroughly, the extent to which such comparisons are warranted can only be determined by examining the full history of the earlier program in greater detail.[11]

Finally, a full understanding of the X-15's administrative and managerial history can provide some important insights into the problems of the present United States space program. Given that practically all that the vehicle is known for today is its superb design, it is hardly surprising that pilots and engineers who speak of the "lessons learned" from the X-15 experience confine themselves exclusively to technical questions.[12]

---

7. Once again, Houston, et al., "Transiting from Air to Space" is an exception, although this matter is also touched upon in Dennis R. Jenkins, *The History of Developing the National Space Transportation System: The Beginning through STS-50* (Melbourne Beach, FL: Bradfield Publishing, 1992), pp. 5–9

8. Seamans, "Objectives and Achievement of the X-15 Program," pp. 2–3. See also, "Brief History of the X-15 Project," p. 3.

9. Roger E. Bilstein, *Orders of Magnitude: A History of the NACA and NASA, 1915–1990* (Washington, DC: NASA SP-4406, 1989), p. 51

10. See, for example, an April 16, 1973, memorandum to the Deputy Associate Administrator (Programs), Office of Aeronautics and Space Technology on "Comparing the X-15 and Space Shuttle Programs," See also Gregg Easterbrook, "NASA's Space Station Zero," *Newsweek*, April 11, 1994, pp. 30–33.

11. John M. Logsdon, "The Space Shuttle Program: A Policy Failure?" *Science 232* (May 30, 1986): 1099–1105; Thomas H. Johnson, "The Natural History of the Space Shuttle," *Technology in Society* 10 (1988): 417–24; W. D. Kay, "Democracy and Super Technologies: The Politics of the Space Shuttle and Space Station Freedom," *Science, Technology, and Human Values*, April 1994, pp. 131–51.

12. William H. Dana, "The X–15: Lessons Learned," Presentation to the Society of Experimental Test Pilots Symposium, September 1987, notes in NASA Historical Reference Collection. See also "Lessons from X-15s to Assist X-30," *Antelope Valley Press*, June 9, 1989, p. 8.

The X-15 rocket airplane, designed to fly at speeds near 4,000 miles per hour and to altitudes above 50 miles, shown in Rogers Dry Lake at the NASA Flight Research Center, Edwards, California, where the research vehicle underwent an extensive flight test program. (NASA photo no. 60-X-31).

As this chapter will show, the program still has a great deal to teach about the administration, and especially the politics, of large-scale and complex R&D programs. After a brief overview of the facts about the X-15 that are already generally known, it will examine some of the less celebrated aspects of the project, and show what administrative and especially political factors played a role in its great success.

Overview

The original mission of the X-15 was to explore the phenomena associated with hypersonic flight. Three of the rocketplanes were built by the North American Aviation Corporation. Each was constructed out of a newly-developed nickel alloy known as Inconel X, and measured fifteen meters long, with a wingspan of nearly seven meters. Missions took place within the specially constructed High Test Range, an aerodynamic corridor that stretched 780 kilometers (by 80 kilometers) from Utah across the Nevada and California deserts to Edwards Air Force Base, complete with radar tracking stations and emergency landing sites. During a typical mission, the X-15 vehicle was carried to an altitude of 14 kilometers by a modified B-52 (of which two were built) and released. The single pilot would ignite the XLR-99 engine, which would burn for approximately ninety seconds, accelerating to an average speed of Mach 5. After flying a parabolic trajectory into the upper atmosphere, the pilot would bring the craft in for a glide landing on the Rogers dry lake bed at Edwards.

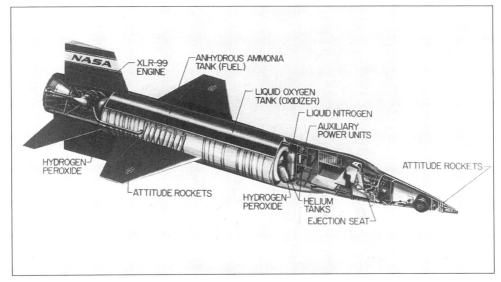

The X-15 rocket airplane, showing its major components. (NASA photo no. 62-X152-22).

Serious planning for the X-15 began in the early 1950s, when the NACA began to consider the problems that were likely to be encountered in piloted space flight.[13] By early 1954, the agency had identified four technical areas of concern: the materials and structures needed to resist the high temperatures of reentry, a better understanding of the aerodynamics operating at hypersonic speeds, systems to maintain vehicle stability and control, and the ability of pilots to work effectively in the space environment.

The NACA's Langley Aeronautical Laboratory, Ames Aeronautical Laboratory, and the High-Speed Flight Station began studying the feasibility of developing a research airplane capable of exploring these critical issues. By the middle of the year, NACA engineers had settled upon the basic design configurations for a craft capable of speeds up to 6,600 feet per second (Mach 6) and an altitude in excess of 250,000 feet.

The agency quickly realized that developing such a plane would be too large and expensive an undertaking for the NACA alone. Accordingly, in July 1954 officials met with representatives of the Air Force and the Navy, both of which were considering developing similar vehicles and saw the NACA proposal as a reasonable compromise.

Thus, in December 1954, representatives from the NACA, the Air Force, and the Navy signed a Memorandum of Understanding (MOU) for the development and testing of a winged hypersonic vehicle. The MOU called for the NACA to have technical control over the project, and for the Air Force and Navy to fund the design and construction phases, under Air Force supervision. After contractor testing was completed, the vehicle would be turned over to the NACA, which would conduct the actual flight tests.[14] The Air Force

---

13. The basic history of the X-15 can be found in the sources listed in notes 4, 5, and 6. For a discussion of the "prehistory" of the program (i.e., the period before 1954), see U.S. National Aeronautics and Space Administration, Langley Research Center, *Conception and Background of the X-15 Project*, June 1962 in NASA Historical Reference Collection; U.S. Air Force, *The Rocket Research Program, 1946–1962*; and Hallion, *On the Frontier*, pp. 106–108.

14. By the time the first X-15 was ready for flight, the agency had become the National Aeronautics and Space Administration (NASA).

Air Launch of X-15 #1 from Boeing B-52 Stratofortress. (NASA photo).

would also oversee (and pay for) construction of the High Test Range. The Navy was in charge of the simulation and training portions of the program.[15] An interagency body, the Research Airplane Committee (known by participants as the "X-15 Committee"), consisting of one representative from each of the sponsoring organizations, was formally in charge of supervising the project, although it appears to have played a largely symbolic role.[16] On January 17, 1955, the plane was officially designated the X-15.

The Air Force sent out invitation-to-bid letters to twelve prospective contractors on December 30, 1954, and a bidder's conference was held at Wright-Patterson Air Force Base on January 18, 1955. Proposals were received from four companies on May 9. By August, the Air Force's Wright Air Development Center and the NACA had concluded that North American Aviation's proposal had the greatest merit. Negotiations with North American were stalled, however, by the company's concern over the proposed timeframe (it was at that time also building the F-107A and F-108 aircraft). Project managers agreed to extend the program from thirty to thirty-eight months, and in November (following price negotiations), the Air Materiel Command Director of Procurement and Production issued the formal contract letter to North American for the development and construction of three X-15 aircraft.[17]

---

15. Memorandum of Understanding, "Principles for the Conduct by the NACA, Navy, and Air Force of a Joint Project for a New High-Speed Research Airplane," December 23, 1954, NASA Historical Reference Collection.
16. See, Hallion, *On the Frontier*, p. 109.
17. A thorough discussion of all contract negotiations associated with the X-15 can be found in Houston, et. al., "Transiting from Air to Space," especially Ch. 1.

Separate invitations-to-bid were issued to four potential engine contractors on February 4, 1955, and the final contract for the X-15 engine, the XLR-99, was issued to Reaction Motors on September 7, 1956. By mid-1958, when it became clear that the XLR-99 would not be ready in time for the first round of test flights, Air Force project managers directed that two smaller XLR-11 engines (also built by Reaction Motors) be used for the initial tests.

Construction on the first X-15 began in September 1957. It was delivered (without the XLR-99 engine) to the Flight Test Center at Edwards on October 17, 1958.[18] Scott Crossfield, an engineering test pilot for North American (who had earlier been a Navy pilot and NACA research engineer) flew the contractor demonstration flights, including the first captive flight on March 10, 1959, the first glide flight on June 8, and the first powered flight (with the XLR-11 engines) on September 17. The first government mission, with NASA pilot Joseph A. Walker, took place on March 25, 1960. Crossfield made the first flight with the XLR-99 engine on November 15, 1960.

By the end of 1961, the X-15 had achieved its design goal of Mach 6 and had achieved altitudes in excess of 200,000 feet. On August 22, 1963, Walker achieved an altitude record for piloted aircraft, taking the X-15 to 354,000 feet (more than 67 miles). On October 3, 1967, Captain William J. "Pete" Knight set a world speed record of 4,520 miles per hour (Mach 6.7), which would stand until the first mission of the space shuttle *Columbia* in 1981.[19]

In March 1962, the X-15 Committee approved an "X-15 Follow-on Program," a series of flights in which the vehicle was converted into a testbed for use in a variety of scientific observations and technological development projects. These flights produced a wealth of scientific information in such areas as space science, solar spectrum measurements, micrometeorite research, ultraviolet stellar photography, atmospheric density measurements, high-altitude mapping. The final flight of the X-15 program, the 199th, took place on October 24, 1968.[20]

Most of those involved with the project had expected that work with the X-15 would lead directly to an even more ambitious craft, the X-20, or Dyna-Soar (short for "Dynamic Soaring" vehicle), which would actually fly to and from Earth orbit. That project, however, was canceled in the 1960s.[21] It would not be until the Space Shuttle program that NASA would turn to the use of winged vehicles for piloted space flight.

Even an abbreviated listing of the X-15's accomplishments is truly impressive.[22] As noted above, the program achieved, and in some cases surpassed, all of its initial objectives. Its top speed of Mach 6.7 exceeded the original goal of Mach 6.0. Similarly, its record altitude flight was far above the intended 250,000 feet.

In the area of technology development, the X-15 saw the first use of a "man-rated," "throttleable" rocket engine, the XLR-99 (once again, the performance of this engine would only be surpassed by those of the shuttle). It was the first vehicle to employ a reaction control sys-

---

18. The second vehicle arrived in California April 1959. X-15 number 3 was almost completely destroyed in June 1960 during a ground test of the troubled XLR-99. After being rebuilt, it was delivered to NASA in June 1961.

19. Stone, "The Quiet Records of the X-15"; Jenkins, *The History of Developing the National Space Transportation System*, pp.7–8. For a complete listing of X-15 flights, see "X-15 to Enter Smithsonian," NASA News Release, April 27, 1969, pp. 14–21. For a list that includes aborted missions, see McDowell, "The X-15 Spaceplane," pp. 8–12.

20. Several efforts were made to complete mission number 200 before the program ended. The final attempt, on December 20, 1968, was canceled due to snow at Edwards.

21. See Jenkins, *The History of Developing the National Space Transportation System*; Clarence J. Geiger, "Strangled Infant: The Boeing X–20A Dyna–Soar," in Hallion, *Hypersonic Revolution*, 1:185–370.

22. For a thorough listing, see John V. Becker, "Principal Technology Contributions of X-15 Program," NASA Langley Research Center, October 8, 1968 (in NASA History Office); and the somewhat dated Stillwell, *X-15 Research Results*.

tem for attitude control in space, a device that would be used by all the spacecraft that followed. The program saw the development of advanced bioastronautics instrumentation (including, for the first time, the ability to gather "real time" biomedical data) and an improved full-pressure suit. Finally, the X-15 provided an essential testing ground for advances in areas such as thermal protection, guidance, and navigation. All of these new technologies were to be used later in development of the Gemini, Apollo, and shuttle programs.[23]

With regard to human factors, the project demonstrated that a pilot could function at hypersonic velocities, high altitudes, and during periods of weightlessness. In particular, it showed that it was possible for a pilot to fly a reentry path, that is, to cross the region between relatively airless space and the thicker lower atmosphere. The Navy's portion of the program—pilot training—marked the first extensive use of motion simulators, such as its human centrifuge at the Naval Air Development Center in Johnsville, Pennsylvania.

Given the magnitude of its objectives, as well as the vehicle's sheer complexity, the total development time of five years from project approval to first powered flight (and two years from construction start) is quite impressive. The estimated costs of the program appear similarly modest, particularly when compared to the space-related projects that followed. The program's total cost, including development and eight years of operations are usually estimated at $300 million in 1969 dollars. Each flight is estimated to have cost $600,000.[24]

By the time it became fully operational, the X-15 could be turned around in less than thirty days. Using all three craft, NASA was able to fly an average of four missions per month. More important, the program had an exceptionally low casualty rate. In November 1962, the landing gear on craft number two collapsed, flipping the vehicle over on its back and injuring pilot Jack McKay (who recovered and was to fly the X-15 again). On November 15, 1967, pilot Mike Adams was killed in a crash that destroyed craft number 3. These tragedies notwithstanding, for nearly 200 missions in a high-performance aircraft operating at the fastest speeds ever attained in a region of the upper atmosphere about which little was known, the X-15's record for safety and reliability was really quite extraordinary. Indeed, the most common reason for mission delays and aborts was weather (which had to be clear along the entire High Test corridor).[25]

Finally, the program captured the popular imagination at a time when many Americans, and much of the world, believed that the United States had fallen behind in the space race with the Soviet Union. Public interest (and media coverage) of the initial flights was quite high, although it dissipated quickly after the beginning of Project Mercury. Nevertheless, the success of the X-15 provided the first tangible evidence to the country after *Sputnik* and Vanguard that American science and technology were on a par with that of the Soviet Union.

Administrative Achievements; Technical Problems

Even under ideal conditions, a successful R&D program of the scope of the X-15 represents an extraordinary managerial challenge. In addition to the sheer complexity of the technology, project officials had to overcome a number of unique administrative difficulties:

As already noted, this was NASA's first foray into full-scale project management. As a program, the X-15 involved far more than the development and flying of the aircraft itself.

---

23. "Brief History of the X-15 Project."
24. See "Comparing the X-15 and Space Shuttle Programs." It is important to keep in mind, however, that although these figures appear nominal by the standards of the current space program, they were far in excess of the program's original estimates. The issue of X-15 cost overruns will be discussed further below.
25. Hallion, *On the Frontier*, p. 117.

Managers also oversaw the preparation of the two B-52 bombers, the construction of an 800 kilometer-long test range, and the design of the advanced full-pressure suit and the other new biomedical equipment. A completely new pilot training regime was developed and implemented. Indeed, in many respects the range of activities associated with the program (including dealing with intense media coverage) seem to foreshadow the practices and procedures the agency (as NASA) would employ in the Mercury, Gemini, Apollo, and shuttle programs.

The X-15 is also notable for being a *successful joint program*, bringing together the efforts of the NACA, NASA, the Air Force, and the Navy. The fact that this collaboration worked as well as it did is remarkable for a number of reasons. To begin with, the later half of the 1950s generally was characterized by a high degree of interservice and interagency rivalry, particularly on matters related to space flight.[26] Indeed, it is difficult to reconcile the military's solicitousness in building and testing a multimillion dollar experimental aircraft (*and* a test range on which to fly it) only to hand it over to (what by then had become) NASA, while it was at the same time fighting with President Eisenhower over the transfer of most of its space facilities to the same agency.[27] Certainly, the whole arrangement seems unimaginable today.

Joint program experiences of NASA and the Department of Defense (DOD) generally have proved disappointing. In fact, the project to which the X-15 is most often compared—the Space Shuttle—is one of the more recent cases where NASA and DOD collaboration was less than successful. Critics of the program have charged that modifying the shuttle orbiter to carry out military missions was one factor in that craft's largely unsatisfactory performance.[28]

Conventional wisdom holds that a joint project ought to have each participant's roles clearly articulated. One of the more striking features of the X-15 MOU, however, is that the division of responsibility for the craft's design—e.g., that the NACA had "technical control" under the Air Force's "supervision"—does not seem to be all that well spelled out. Such ambiguity is almost always a potential source of trouble for any joint project, particularly in view of the fact that the Air Force was providing the bulk of the program's funding.

As was noted earlier, the interagency X-15 committee was formally in charge of the project, but it does not appear that this body had much involvement in day-to-day decision-making, or in settling disputes among the participants. One observer has described its role as that of offering high-level sanction to lower-level decisions.[29] There were exceptions: on one occasion, when the Air Force had started to protest over building the High Test Range only to hand it over to the NACA (like the X-15 craft itself), the committee's endorsement of the original agreement served to end the dispute.[30] For most other areas of potential conflict, however, there is no evidence that the X-15 committee ever played any substantive role.

---

26. See John M. Logsdon, *The Decision to go to the Moon: Project Apollo and the National Interest* (Cambridge, MA MIT Press, 1970).

27. Robert L. Rosholt, *An Administrative History of NASA, 1958–1963* (Washington, DC: NASA SP-4101, 1966); Bilstein, *Orders of Magnitude*. Historical discussions of the X-15 program can sometimes become confusing due to the fact that one of the principal participants changes its identity. Thus, it was the National Advisory Committee for Aeronautics that signed the MOU, but the National Aeronautics and Space Administration that accepted the final delivery and conducted the test flights and later experimental missions. It will be the practice throughout this chapter to refer to the two organizations contemporaneously, that is, to use "NACA" when referring to events prior to 1958, and "NASA" thereafter.

28. Logsdon, "The Space Shuttle"; Kay, "Democracy and Super Technologies."

29. Hallion, *On the Frontier*, p. 109.

30. Houston, et. al. "Transiting from Air to Space," pp. 117–18.

Dr. Joseph A. Walker stands beside the 1961 Collier Trophy, awarded to him and the other X-15 pilots by President John F. Kennedy. (NASA photo no. 62-X-20).

Commander Forrest S. Petersen, USN, standing beside the 1961 Collier Trophy presented by President John F. Kennedy. (NASA photo no. 62-X-15-21).

The situation was further complicated by the fact that responsibility for the development and manufacture of the X-15's systems was spread across an exceedingly large number of contractors and sub-contractors. These included not only North American Aviation and Reaction Motors, but also General Electric (which was responsible for the Auxiliary Power Units), David Clark Co. (developer of the pressurization suit), the International Nickel Company (creator of the Inconel X nickel alloy for the fuselage), Bell Aircraft (supplier of the ballistic control rockets), Sperry Gyroscope (developer of the in-flight electronic indicator systems), and many, many others. In all, more than 300 private firms participated in the project.[31]

Fortunately—and surprisingly—the internal conflicts that did occur were minor, and appear to have had no impact on the program overall. Early in the design process, for example, the NACA's request for a modification to allow for testing different types of "leading edges" was rejected by the Air Force.[32] In late 1955, during the negotiations with Reaction Motors, the Navy's Bureau of Aeronautics made a bid to take over responsibility for the development of the XLR-99. The Navy based this claim on the fact that it had been working with Reaction Motors for the past three years developing the XLR-30 rocket engine, the design of which was to serve as the basis of the X-15 power plant. The Air Force rejected this argument, citing (somewhat ironically) the need to keep management responsibility within a single agency.[33] Finally, as already noted, in 1955 the Air Force sought to retain control over the High Test Range.

One area of conflict, once again between the Air Force and the NACA, did prove to be rather serious, but in some respects may actually have been somewhat beneficial. The problem involved the development of the XLR-99, which proved be the most serious technical (and administrative) obstacle in the entire program.[34] The NACA had already complained to the Air Force in late 1955 that the procurement process for the engine was taking too long, prompting the latter to write a letter of reassurance. Then, in April 1956, a representative of the Lewis Laboratory who had visited the Reaction Motors facility reported the company's efforts on the engine to be "inadequate" on several fronts. He felt that the development program was already behind schedule and that some of its time estimates were too optimistic by as much as a year.

Although it is not clear what immediate impact this report had on the Air Force project managers, subsequent events were to bear out the NACA's concerns. In August 1956, an Air Force representative noted in a letter to Reaction Motors that a test of the engine's thrust chamber, which had been scheduled for April, had not yet taken place. By early 1957, North American had begun to complain about the pace of the engine development. The prime contractor found that not only was the program four months behind schedule, but that the weight of the engine was increasing while its projected performance appeared to be declining.

The difficulties arising from divided authority can be illustrated by the responses to North American's criticisms. In February 1957, two sets of meetings were held between Reaction Motors personnel and representatives of the Air Force (February 12 and 18) and the NACA and North American (February 19). For its part, the Air Force appeared to come out its meetings assured that "every effort [would be] expended to prevent further engine schedule slippages."

---

31. See "X-15 History," North American Aviation Press Release, n.d. pp. 7–8, in NASA Historical Reference Collection.
32. Houston, et. al. "Transiting from Air to Space," pp. 51–52.
33. *Ibid.*, pp. 65–67.
34. By far, the most in-depth account of this affair is in *ibid*, Ch. 3, from which the following discussion is taken.

As was the case the previous April, however, the NACA was far more pessimistic. Its report of the February 19 meeting expressed doubt that the new schedule could be met (although the agency agreed to accept a delay of four months in delivery and a weight increase from 588 to 618 pounds). More significantly, this report for the first time mentioned the possibility of using an interim engine in order to maintain the X-15's flight test schedule.

Once again, the NACA's gloomy assessment proved to be correct. In July 1957, Reaction Motors advised the Air Force that it would need a nine-month extension (it also reported another weight gain, from 618 to 836 pounds). The following December, it reported another six-month slippage. Needless to say, there were substantial cost increases as well: by January 1958, costs for the engine's development were almost double the amount estimated just six months earlier. At this point, Air Force project managers seriously considered canceling the Reaction Motors contract and bringing in a new firm, which would have delayed full-power flights until at least 1961 (and might even have resulted in the outright cancellation of the program). By February 1958, however, the decision was made to continue with the current contractor, but to procure two smaller XLR-11 engines for the initial test flights.

The timetable of the main engine development seems to have been the only area of disagreement among the project's participants involving a major subsystem on the X-15, and even this was only a matter of timing, since all parties ultimately reached the same conclusion.[35] It is also worth noting here that the NACA's and the Air Force's primary concerns were with the engine's performance and completion date. Staying within the original budget does not appear to have been a major consideration in the government's dealings with Reaction Motors, even though this phase of the program already was incurring massive cost overruns.

All in all, each of the principal organizations worked very well together. Rather than fall into competitive wrangling (a common danger of joint programs, particularly when problems arise), each of the partners provided a measure of much-needed redundancy and in-depth checking.

In considering difficulties like those surrounding the XLR-99, it is important to remember that it was the most sophisticated rocket engine built up to that time, in some respects even more complex than the Saturn V.[36] For there to be significant delays and technical problems with such a system is only to be expected. In fact, the project team's eventual response to the XLR-99 issue demonstrates yet another of its impressive management features, namely that it was able to absorb a number of delays and still maintain something approaching an orderly test schedule.

As it turned out, the main engine was not ready for flight until November 1960, more than two years after delivery of the first vehicle. The decision to substitute the two smaller engines, rather than wait on the XLR-99, allowed at least part of the initial flight tests to go forward; other aircraft systems could be checked out and the pilots could gain some familiarity with the vehicle.

This robustness, the ability of the program to adapt to inevitable technical failures, was seen time and again throughout the life of the X-15. No doubt much of this was due to the exceptional technical skills of North American and NASA engineers. During the first glide flight of craft number one on June 8, 1959, pilot Scott Crossfield experienced wild pitching motions just prior to landing; the ground team quickly (and successfully; it never occurred again) corrected the problem, and Crossfield was able to make the first

---

35. The only other incident involved a brief period of confusion between the Air Force and North American was over which was responsible for procuring the pressurization suit (i.e., whether it was to be a government or a contractor procurement). See *ibid.*, pp. 93–101, and American Institute of Aeronautics and Astronautics, "History of Rocket Research Airplanes" program, July 28, 1965, 2:21–29, transcripts available in NASA Historical Reference Collection.

36. Jenkins, *History of Developing the Space Transportation System*, p. 7.

powered flight (in craft number two) less than three months later. On November 5, an engine fire broke out on X-15 craft number two, forcing Crossfield to make an emergency landing, which, in turn, literally broke the craft's back; that particular vehicle was grounded for only 98 days.

One of the more serious incidents of the demonstration phase occurred during the first ground tests of the XLR-99 engine in June 1960. A stuck pressure regulator caused X-15 craft number three to explode. The airplane essentially disintegrated aft of its wing. Despite the fact that it needed to be rebuilt completely, craft number three was returned to NASA and made its first successful flight eighteen months later. The first use of the XLR-99 in flight occurred on November 15, 1960.

The X-15 experienced technical difficulties and malfunctions of varying degrees of severity for much of the remainder of the program, but these seldom affected its overall flight schedule. Problems with different components and subsystems were repaired or even completely replaced whenever necessary, and the vehicle returned to duty relatively quickly. As noted earlier, the engineering prowess of the flight team deserves a great deal of credit, but it would also appear that the X-15 operations crew benefitted from the same lack of economic constraints enjoyed by Reaction Motors during the development of the main engine. NASA engineers at the Flight Research Center were routinely rejecting twenty-four to thirty percent of manufactured space parts as unusable.[37] As was the case with the XLR-99, the primary emphasis was on reliability and performance, rather than staying within a budget.

Discussion

This last point suggests that, the extraordinary performance of the X-15 project team (managers as well as engineers) notwithstanding, the program benefitted from a number of external factors that were not necessarily under any of the participants' direct control.

To begin with, it appears that the X-15 succeeded as a joint undertaking primarily because of the consensus on its specific objectives among all of the parties involved, a fortunate circumstance that clearly could not have been dictated by any one member. Whenever an interagency project fails to meet its intended goals, it is usually because each organization has brought to it a different (and sometimes even contradictory) set of priorities.

This is essentially what occurred in the Space Shuttle program. In attempting to design the shuttle in a way that satisfied both its own objectives and those of the Department of Defense (as well as meeting the cost requirements imposed by the Office of Management and Budget and the Congress), NASA engineers were forced to make too many compromises in the spacecraft's design, with severe consequences for the long-run success of the program. Similar sorts of problems have plagued the space station as well.[38]

The reference to OMB and Congress suggests another important difference between the X-15 and the shuttle (or, for that matter, the space station). The history of the earlier program shows virtually no involvement in the project (especially its design) on the part of outside political or budgetary agencies. Indeed, one major advantage that the X-15 program had over many later U.S. space projects (and one which is seldom mentioned in any X-15 histories) was the highly favorable political, economic, and social environment that surrounded most of the period of its development and the early phases of its flight operation.

---

37. Hallion, *On the Frontier*, p. 117.
38. See W. D. Kay, "Is NASA to Blame for Confusion in Space Effort?" *Forum for Applied Research and Public Policy* 7 (Winter 1992): 36–43.

The X-15 was never forced through in-depth hearings before congressional committees or protracted negotiations with the Bureau of the Budget (as it was then known), let alone subjected to outside scrutiny each year of its existence. Although responsibility for the project was spread across a number of government agencies and private firms, these actors—the military, the NACA, the NASA, and the aerospace contractors—represented a fairly uniform set of concerns: all wished to build a high-altitude, hypersonic experimental aircraft, and there was substantial agreement on what specific design and performance criteria the vehicle was to meet. This ensured that the major design decisions on the project would be made primarily according to technical, rather than political or economic considerations.

This is most clearly evident with regard to the question of the program's original cost estimates and time frame. It is seldom acknowledged in the historical literature, but the X-15 program was a victim of what has become a fairly common occurrence in the U.S. space program, namely substantial delays and overruns. Three hundred million dollars does seem small in comparison to the cost of, say, Apollo or the shuttle, but it is still more than *seven times* the original estimate of $42 million.[39] The final development costs of the engine alone were more than $68 million (plus a $6 million fee to Reaction Motors), a *tenfold* increase over what was expected when the project began.[40] In addition, the complete vehicle, including the large engine, was ready for flight more than two years behind schedule. Despite all of this, development during the 1955–1957 period was never held up by a lack of funds, although in some years needed funding did not come through until the last minute.

After the launch of *Sputnik 1* in 1957, interest in the project on the part of the military, political leaders, and the public at large grew rapidly. As already noted, media coverage of the first flights was the most intense ever seen at Edwards, and even led to some public relations mix-ups between NASA and the Air Force.[41]

Once the first Mercury flights got underway, public attention shifted to the events at Cape Canaveral. This might, however, have ultimately worked to the program's benefit. A major contributor to the X-15's success over the long run was its emphasis on incremental development and its use in highly specialized scientific and technical research.[42] As experience with many later space projects (including Apollo after *Apollo11*, the shuttle, etc.) has shown, the general public tends to lose interest in such "routine" undertakings rather quickly. In short, it appears as though the X-15 got a needed boost of public fanfare at precisely the right point in its history—the later development and early flight test stage—and then became regarded as a low-key effort worthy of only occasional interest just as it was entering its less "flashy" research phase. These shifts in external perception probably could not have been planned any better.

The lack of external (i.e., outside the aerospace community) scrutiny very likely contributed to one more important effect. As seen repeatedly in the case of the XLR-99, as well as in actual flight operations, project officials from both the Air Force and NASA were never hesitant to point out—and more important, work to correct—potential (or actual) technical flaws, even when this resulted in increased costs. Recently, critics of the shuttle program have accused NASA of ignoring—or even covering up—such problems for fear of the political ramifications.[43]

---

39. Houston, et. al., "Transiting from Air to Space," pp. 13–15.
40. *Ibid.*; see also Jenkins, *History of Developing the National Space Transportation System*, p. 7.
41. Houston, "Transiting from Air to Space," pp. 118–20.
42. Dana, "The X–15: Lessons Learned."
43. See Gregg Easterbrook, "The Case Against NASA," *New Republic*, July 8, 1991, pp. 18–24; and "NASA's Space Station Zero." See also Joseph J. Trento, *Prescription for Disaster: From the Glory of Apollo to the Betrayal of the Shuttle* (New York, NY: Crown Publishers, 1987).

To the extent that this claim has any validity, the larger question it raises is whether NASA officials are simply more timid now than they were forty years ago, or whether the prevailing political and economic climate creates conditions more conducive to error detection and recovery. This is a particularly important point since, the claims of some critics of current U.S. space policy notwithstanding, one of the most interesting aspects of the X-15 program is that, far from being substantially different from later NASA enterprises, it is in many respects a familiar story: rampant cost increases, serious delays, technical failures, and even loss of life.

To be sure, the management of the X-15 was superb, particularly given the difficulty of its mission. There was some degree of infighting, which usually was settled quickly. As expected on a project of this nature, technical difficulties arose, necessitating design compromises, additional costs, and schedule slippages. Because the program was surrounded by a supportive political and economic environment, however, NASA officials and their counterparts in the Air Force were able to face these problems squarely, and develop solutions, some of them quite innovative.

Nevertheless, given all of the controversy besetting the present U.S. space program, it is today a cause for wonder that an undertaking that had as many serious problems as the X-15 was not only tolerated at the time, but is now touted as one of the great aerospace success stories. In this context, perhaps even more now than then, the X-15 deserved the Collier Trophy as the program for the most outstanding aerospace achievement of its time.

Chapter 7

The Collier as Commemoration: The Project Mercury Astronauts and the Collier Trophy

by Jannelle Warren-Findley

On October 10, 1963, the seven astronauts of the National Aeronautics and Space Administration's (NASA) Project Mercury gathered in the Rose Garden of the White House in Washington, DC, to receive the Collier Trophy for 1962. In the brightness of that autumn morning, President John F. Kennedy relished the opportunity to award what the newspapers referred to as "one of the nation's highest space honors"[1] to Lt. Commander M. Scott Carpenter, United States Navy (USN); Capt. Leroy Gordon Cooper, Jr., United States Air Force (USAF); Lt. Col. John H. Glenn, Jr., United States Marine Corps (USMC); Capt. Virgil I. Grissom, USAF; Lt. Commander Walter M. Schirra, Jr., USN; Commander Alan B. Shepard, Jr., USN; and Capt. Donald K. Slayton, USAF. In addition to the astronauts' wives, "picture-pretty" according to one newspaper account,[2] 150 guests, including Vice President Lyndon B. Johnson, cabinet officers, and representatives from the aerospace industry heard Kennedy urge Americans to a "greater appreciation of the space program and its potential benefits to the United States and mankind."[3]

Kennedy's remarks put Project Mercury in the context of spacefaring plans in 1963. The excitement of launch and recovery, the tickertape parades and media coverage were behind them. The President said he was particularly glad to be awarding the Collier Trophy to the Mercury astronauts, because "I hope this award, which in effect closes out a particular phase of the program, will be a stimulus to them and to the other astronauts who will carry our flag to the moon and perhaps, some day, beyond."[4]

The Collier Trophy was awarded to the pilots of Project Mercury "for pioneering manned space flight in the United States."[5] The 1962 award differed by definition from earlier Collier honors in several ways. For one, Project Mercury was the first American space mission to receive such kudos; this was the first time that the Collier Trophy could in fact be considered, as the newspapers claimed, one of the nation's highest *space* honors. It was, moreover, the first of several awards to NASA during the forthcoming decades.[6] The award honored solo performance in space of the sort demonstrated by airplane test pilots; after Project Mercury, space forays always utilized teams of astronauts.[7]

---

1. *New York Times*, October 11, 1963, clipping, NASA Historical Reference Collection, NASA History Office, NASA Headquarters, Washington, DC.
2. *Ibid.*
3. *Chicago Tribune*, October 11, 1963, clipping, NASA Historical Reference Collection.
4. New York Times, October 11, 1963, *ibid.*
5. Bill Robie, *For the Greatest Achievement: A History of the Aero Club of America and the National Aeronautic Association* (Washington, DC: Smithsonian Institution Press, 1993), Appendix A, p. 233.
6. See Appendix A in Robie, *For the Greatest Achievement.*
7. By August 1959, the New Projects Panel of the Space Task Group was recommending the development of a three-person spacecraft for transport to and from a space laboratory and circumlunar flights. See Alan J. Levine, *The Missile and Space Race* (Westport, CT: Praeger, 1994), pp. 113–14.

In 1962 astronauts Lr. Cdr. M. Scott Carpenter, U.S. Navy; Maj. L. Gordon Coope, U.S. Air Force; Lt. Col. John H. Glenn, Jr., U.S. Marine Corps; Maj. Virgil I. Grissom, U.S. Air Force; Cdr. Walter M. Schirra, Jr., U.S. Navy; Cdr. Alan B. Shepard, Jr., U.S. Navy; and Maj. Donald (Deke) K. Slayton, U.S. Air Force, received the award for pioneering human space flight in the United States. (NASA photo no. M-278, ASTRO 17).

In addition, unlike earlier practices, the Collier Trophy for 1962 honored the men rather than the machines. Collier awards in earlier years usually went to designers, engineers and inventors of innovative aviation hardware rather than to those who flew the new machines.[8] The 1962 award to the Project Mercury astronauts could have been presented for achievements in "big technology" if not in "big science," as more recent analysts have characterized twentieth century developments in large-scale technological undertakings.[9] Project Mercury counts as "big technology" because of the vast numbers of designers, engineers, managers, test pilots, and workers, both military and civilian, from government and industry who fabricated and flew the hardware. Project Mercury was in fact born of and flown by President Eisenhower's worrisome military-industrial complex.[10] Its story is a case study of the development, in the face of enormous time, political, and collective psychological pressures, of a specifically military-*aerospace* complex. Project Mercury was, from this big technology perspective, the opening shot into a new world: and the award ceremony marked "the end of the beginning" of the space age.

    8.   See Appendix A in Robie, *For the Greatest Achievement.*
    9.   See the discussion in James H. Capshew and Karen A. Rader, "Big Science: Price to the Present," *Osiris*, 2nd Series, 1992, 7:3–25 for background to the concept of "Big Science." I am arguing that Project Mercury played at best a transitional role in the development of big science.
    10.   Eisenhower warned in his farewell address of the dangers of economic and political concentration in a military-industrial complex. See reference in T. Keith Glennan, *The Birth of NASA, The Diary of T. Keith Glennan*, J.D. Hunley, ed. (Washington, DC: NASA SP-4105, 1993), p. 308.

But, in the face of a compelling story of creative engineering and important technological and administrative developments, the 1962 Collier Trophy was in fact awarded to the test pilots who rode the Mercury capsules. If there is a real anomaly in the award of the Collier Trophy to the Mercury astronauts, it is that from the perspective within NASA, the Project Mercury astronauts appear to have had relatively little to do with the development of the program and of the agency whose patch they wore. The collective space-age versions of "Lucky Lindy" and their wives and children were certainly the most public part of Project Mercury. Within an agency, and a federal government, in which competing visions of strategies for carrying out space undertakings surfaced regularly, politics and public relations dictated that the heroes serve as a focus for the program and be celebrated accordingly.

The Collier Trophy had been awarded annually since 1911 "for the greatest achievement in aviation in America, the value of which has been thoroughly demonstrated by use during the preceding year"[11] By 1963, the trophy had been presented nearly fifty times (the war years 1917–1920 had been skipped). The contrast between the early winners and the group of seven astronauts lined up behind President Kennedy in the Rose Garden in October 1963 symbolized profound changes in the ways that the United States thought of aeronautics and awards for achievements in aviation. To understand the way the Mercury program developed is to map (as of the early 1960s) a series of changes in public administration and management; undertakings in science, engineering, and technology; developments in economic organizations; and changes in popular culture that, spurred by World War II and the turmoil and tensions of the Cold War period, transformed many elements of life in the United States.

Much has been written about Project Mercury, from newspaper coverage to *Life* magazine and other popular magazines during the period, to works like the monumental in-house history of the undertaking, *This New Ocean: A History of Project Mercury* (1966).[12] To examine Project Mercury as a test case of the emergence of big technology in the space field in the mid-twentieth century, a large number of materials from a variety of disciplines and perspectives were examined. Three particular sets of questions emerged from this literary exploration, and they shape the following essay.

The first questions relate to the time in which Project Mercury took place. How did the political, economic, cultural, and diplomatic competition of the Cold War affect American technological developments, particularly in the realm of military defense? In contrast, how did that play out in the fledgling civilian space agency and Project Mercury itself? The second set of questions concerns the transitions from the predecessor agencies and traditional practices in design, testing, production methods, and management techniques: how were very different ways of organizing and interpreting data of various sorts melded into an agency and a program that relied for its success on a common approach and focus? The third set of questions focus on the Project Mercury astronauts. The lack of analyses of these early adventurers from broad cultural, gendered, or social perspectives leave the reader with the impression of a group of mostly one-dimensional pilots with a collective death wish. Yet they played a series of roles in American life and culture, and the award of the Collier Trophy to them certainly reflects some element of that public presence. What were those roles and how did the public persona of each astronaut play out during Project Mercury?

---

11. Robie, *For the Greatest Achievement*, pp. ix–x.
12. Loyd S. Swenson, Jr., James M. Grimwood, and Charles C. Alexander, *This New Ocean: A History of Project Mercury* (Washington, DC: NASA SP-4201, 1966).

The Context for Project Mercury: The Cold War

Project Mercury lasted four and a half years from initial announcement to the twenty-two-orbit flight of Gordon Cooper in *Faith 7*. The overall cost was estimated to be $384,131,000, "of which thirty-seven percent went for the spacecraft, thirty-three percent for the tracking network, and twenty-four percent for launch vehicle procurement. Flight operations and 'R and D' costs made up the remainder."[13] Those funds paid "a dozen prime contractors, some seventy-five major subcontractors, and about 7,200 third-tier sub-subcontractors and vendors, all of whom together employed at most about two million persons who at one time or another had a direct hand in the project."[14] With NASA employees, military and civilian employees of the Department of Defense, and employees of other civilian institutions, including educational institutions, the number of people employed on the program probably peaked at just over two million.[15]

America's first human space flight program was announced by the new National Aeronautics and Space Administration administrator, T. Keith Glennan, on Wright Brothers Day, December 17, 1958. It was fifty-five years after the events at Kitty Hawk.[16] A unit transferred to NASA from its predecessor agency, the National Advisory Committee for Aeronautics (NACA), headed by Robert R. Gilruth and physically located at Langley Memorial Aeronautical Laboratory in Virginia, had done preliminary planning for a human space flight project. The project involved suborbital flights using a Redstone rocket; longer suborbital flights using a Jupiter missile, which were later eliminated from the program; and orbital human flights using an Atlas booster.[17] The program, reduced to a slide show in the period of planning from September 1958 to January 1959, could be described in shorthand:

Objectives
1. Orbital flight and recovery
2. Man's capabilities in environment

Basic Principles
1. Simplest and most reliable approach
2. Minimum of new developments
3. Progressive build-up of tests

Method
1. Drag vehicle
2. ICBM booster
3. Retrorocket
4. Parachute descent
5. Escape system[18]

The plan was refined as time passed, and changed or developed as necessary. But the basic strategy of building incrementally and using techniques and technologies already

---

13. Swenson, Grimwood, and Alexander, *This New Ocean*, p. 508.
14. *Ibid.*
15. *Ibid.*
16. Swenson, et al., *This New Ocean*, p. 132.
17. Roger D. Launius, *NASA: A History of the U.S. Civil Space Program* (Malabar, FL: Krieger Publishing Company, 1994), p. 39.
18. Swenson, Grimwood, and Aexander, *This New Ocean*, p. 134.

available whenever possible continued for the life of Project Mercury. That approach clearly differentiated Project Mercury from the Manhattan Project-type approach of the later Apollo program.

The space program in the United States developed as it did because of the Cold War. World War II brought significant change to the NACA, to its mission, and to the spheres of aeronautics, science, and technology. New developments like atomic energy, radar, large rockets, jet engines, radio telemetry and the computer all had the potential to reshape American life in the latter half of the twentieth century.[19]

But the aftermath of the second world war did not bring peace and measured development of these new technologies. Rather, the rise of the Soviet Union and its spheres of influence and the fall of what Winston Churchill called "the Iron Curtain" across central Europe led to the international political and technological competition known as the Cold War. The American military played an active role and influenced directly or indirectly many Cold War developments. Indeed, as one study observes:

> *The essential feature of the mid-century military-political landscape was the Cold War—a type of strife radically unlike any other in history. Weapons for the first time were designed not to be used; they were sought for their preemptive value. Each combatant had to continually improve its arsenal, so as to deter the other from using its arms. Fewer and fewer units of each successive weapon were made, but each was much more technically sophisticated than the last. A process of institutionalized innovation was set in motion. The new form of warfare, atmospheric rather than ground or sea, radically altered both the conduct of war-making and the production complex that fashioned the weapons and support equipment.*[20]

Even where the military was only indirectly involved—and the NASA program, and Mercury as the first human space flight program, were self-consciously non-military undertakings—this new form of warfare shaped the whole notion of a "space race." And the "process of institutionalized innovation" shaped the space program. The development of that process began with the announcement of Project Mercury.

World War II is recognized as the catalyst for organized, national rocket development because the war effort demanded new weapons and Russia, the United States, and Germany began to develop missiles as weapons.[21] The captured German rocketeers continued their work with captured V-2s and parts which they had brought out of the Reich, first at White Sands Proving Ground in New Mexico, and after 1950, at the U.S. Army's Redstone Arsenal in Alabama.[22]

During the same period of development that would lead to Project Mercury, the Naval Research Laboratory began to work with sounding rockets, launching *Viking I*, built by the Glenn L. Martin Company, from White Sands on May 3, 1949. The Army's Project Bumper joined a Jet Propulsion Laboratory—produced WAC Corporal missile to a V-2. The one

---

19. Legacy Resource Management Program, *Coming in From the Cold the Department of Defense Cold War Project*, draft manuscript history of the Cold War, 1992, in possession of the author.

20. Ann Markusen, et al, *The Rise of the Gunbelt: The Military Remapping of Industrial America* (New York, NY: Oxford University Press, 1991), p. 30.

21. The history of the German rocket program prior to and during World War II is fully described in Michael J. Neufeld, *The Rocket and The Reich: Peenemünde and the Coming of the Ballistic Missile Age* (New York, NY: The Free Press, 1995); Levine, *Missile and Space Race*, studies American developments during and after World War II.

22. Launius, *NASA*, pp. 33–34.

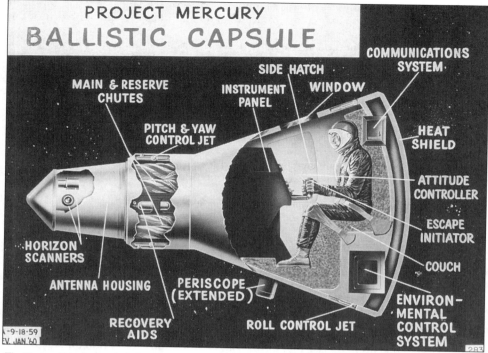

The Mercury Capsule, America's first true spacecraft, shown in a cutaway drawing made in January 1960. (NASA photo no. M-278, ASTRO 17).

fully successful launch took place on February 24, 1949.[23] During the 1950s, the Germans working for the U.S. Army and building on the V-2, developed the Redstone missile, first launched on August 20, 1953, from Cape Canaveral, Florida. Redstone carried enormous importance, for Project Mercury, because it was utilized to launch the first Mercury astronauts.[24] It was in fact the ICBM booster listed on the briefing slide in the original planning presentation mentioned above.

After the Korean War, the development of an intercontinental ballistic missile, or ICBM, took priority among the military services. The Army's Redstone was the nation's first operational ballistic missile, although its military importance proved to be nil. "Along with the H-bomb," notes Alan J. Levine, "the ICBM was the critical weapons development of the Cold War era. And . . . the principal launch vehicles of the space program in the 1960s and later (and even today) were products or byproducts of the ICBM effort."[25] Development was speeded up during the Eisenhower administration partly due to the recommendations of the Strategic Missiles Evaluation Group, or the Teapot or von Neumann Committee. The committee warned that the Russians might be ahead in the development of such missiles and urged a crash program to give the United States operational weapons

---

23. Launius, *NASA*, pp. 12–14; Levine, *Missile and Space Race*, pp. 2–8, 11–16.
24. Launius, *NASA*, p. 15; Levine, *Missile and the Space Race*, pp. 27–28.
25. Levine, *Missile and Space Race*, p. 28.

in six to eight years.²⁶ In January 1951, Convair had received a research contract to work on what became the Atlas ICBM. By 1954, the Atlas was on the developmental fast-track, with 2,000 companies and 40,000 workers involved in its design and production.²⁷

The military's push for ICBMs helped pave the way for civilian exploration of space, as did the development of reconnaissance satellites in the mid-1950s. In contrast to military space weapons efforts, what would be the Eisenhower administration's *civilian* space program began in 1952 with the establishment, by the International Council of Scientific Unions, of the International Geophysical Year (IGY) from July 1, 1957, to December 31, 1958. The use of rockets with instrument packages to help study the scientific issues of the IGY was recommended in 1952; by 1954, the organization called for the orbiting of artificial satellites to help map the Earth's surface.²⁸ The National Security Council voted on May 26, 1955, to approve a plan to orbit a scientific satellite as part of the IGY activities. President Eisenhower announced in July that small, pilotless satellites would be launched for those purposes. The Naval Research Laboratory and the Army's Redstone Arsenal went into immediate competition to develop the capacity to launch the satellite. The Navy's Project Vanguard was chosen in September 1955 and launched the first Vanguard mission on December 8, 1956. But Vanguard was slow to develop and starved for funds.²⁹

As Roger Launius points out, the United States in the mid-1950s thus had two separate space efforts underway. The high-priority military program, to build ICBMs and to work on reconnaissance satellites, was kept under wraps as much as possible. The IGY program, on the other hand, was public and focused on the need to encourage the free access to space of all spacefaring nations. The Vanguard program was struggling toward orbit with limited financial support. The Eisenhower administration, vitally concerned with achieving the goal of free access to space, was willing to push the Vanguard program in order to accomplish the launch of Earth's first artificial satellite even at the expense of the military plans.³⁰

Sputnik 1, however, changed everything on October 4, 1957. The Earth's first artificial moon weighed 183 pounds and orbited the Earth every hour and a half in an elliptical orbit. The Eisenhower administration's reaction to this historic event was restrained; although the Russians were congratulated for their historic and scientific achievement, officials downplayed the strategic meaning of the launch and successful orbit.³¹

The American people, on the other hand, were shocked, horrified and frightened by the news. In contrast to the administration's facade of calmness and lack of concern, wrote Walter McDougall. "The public outcry after Sputnik was ear-splitting. No event since Pearl Harbor set off such repercussions in public life."³² The Russians confirmed their ability to launch large objects and to carry biological passengers, in that case a dog, when they orbited *Sputnik 2* on November 3.

---

26. Levine, *Missile and Space Race*, pp. 30–31.
27. Ibid.
28. Launius, *NASA*, pp. 21–22.
29. *Ibid.*, pp. 21–25.
30. Launius, *Ibid.*, pp. 23–24.
31. *Ibid.*, p. 24.
32. Walter A. McDougall, . . . *the Heavens and the Earth: A Political History of the Space Age* (New York, NY: Basic Books, 1985), p. 142.

Project Mercury and the Transition to Big Programs

The reactions to *Sputnik 1* and *2* fell into a number of categories. Some critics called for an immediate improvement in American scientific education despite the fact that the Sputniks were predominantly engineering feats.[33] Senator Lyndon B. Johnson opened hearings in a subcommittee of the Senate Armed Services Committee in November, and the investigation found too little being spent on space-related activities, and considerable diffusion of effort among the military services. As Johnson's aide, George Reedy pointed out, in a graphic description of how the perception of world power had changed, "The simple fact is that we can no longer consider the Russians to be behind us in technology. It took them four years to catch up to our atomic bomb and nine months to catch up to our hydrogen bomb. Now we are trying to catch up to their satellite."[34]

In a now more urgent response to Russia's gauntlet, the administration scheduled a test launch of a Project Vanguard booster on December 6, 1957. That test, televised nationally, was disastrous and embarrassing when the rocket rose briefly and fell back to the pad, disintegrating in flames. "Flopnik," the press called it; "Kaputnik." The second Vanguard launch, in February, was no more successful; the rocket got off the launch pad but came apart at an altitude of four miles.

Despite the earlier decision to allow the Navy to put the initial U.S. satellite into orbit, the Administration turned to the Army program and the Germans in Huntsville to prepare a backup launch. The Jet Propulsion Laboratory repackaged instruments from Vanguard, including a cosmic-ray experiment designed by James Van Allen of the University of Iowa. On January 31, 1958, the Jupiter-C launched *Explorer 1* and soon after made arguably the first important scientific discovery of the Space Age by locating the Van Allen radiation belts. In addition, the bringing of the Army's team of German rocket experts to the center of America's space efforts represented an important developmental shift for human space flight programs like Project Mercury.[35]

President Eisenhower continued to exhibit calm in the face of the Soviet triumphs and American failures, and he continued to try to hold Federal spending down. Space policy, in both the military and civilian spheres, changed during 1957–58, however, and it seems clear in retrospect that the development of both programs were shaped by the Sputniks. Military developments included the firming up of plans for nine squadrons of Atlas missiles; the approval of plans for the Air Force's Minuteman, a missile which could be kept in a hardened missile silo and fired when necessary; the acceleration of the reconnaissance satellite program, pointedly perhaps, named Sentry. The go-ahead was given for work on the Army's Pershing and the Nike-Zeus. The Strategic Air Command was further strengthened and work on Distant Early Warning (DEW) line construction sped up. Development of the Navy's submarine-launched Polaris was advanced three years, so that the missile would be ready for operations in 1960. In management developments, the Advanced Research Projects Agency (ARPA) was established in the Pentagon in February 1958. ARPA's role was to act as a clearinghouse and evaluation center for ideas and efforts from all the services. In time, the agency was assumed to be slated to take over military space undertakings, presumably including American piloted space efforts.[36]

Changes in response to Sputnik which were more obvious to the general public occurred, as did the establishment of ARPA, in the organizational area. Senator Lyndon B. Johnson (D-Texas) convened hearings on the United States' space program, or lack

---

33. McDougall, *Heavens and the Earth*, p. 143.
34. George Reedy, quoted in Launius, *NASA*, p. 26.
35. Levine, *Missile and Space Race*, pp. 68–69.
36. *Ibid.*, pp. 69–70.

thereof, in November 1957. Johnson's goal was to push the administration to support the technological developments necessary to the new Space Age and to acknowledge the need for international power and recognition that a space program would entail. Johnson's congressional hearings found the current American space program seriously wanting, and long-term planning virtually useless. As a result, on February 6, 1958, the Senate voted to establish the Special Committee on Space and Aeronautics. Its task specifically would be to craft legislation to create a new national space agency. The House of Representatives soon followed suit.[37]

While the Legislative branch deliberated, the administration also took steps to address the space crisis. In November 1957, Eisenhower established the President's Science Advisory Committee (PSAC) and named James R. Killian his Science Advisor. In February 1958, Eisenhower asked the PSAC to create a plan for a new civilian space agency. The next month, Killian and his committee proposed that all nonmilitary space activities be merged into an expanded National Advisory Committee on Aeronautics (NACA), the Federal agency which had been responsible for basic research into aeronautical problems since 1915. On the basis of that advice, the administration drafted legislation establishing the National Aeronautics and Space Administration.[38] President Eisenhower signed it on 29 July, and NASA began to function on October 1, 1958.[39]

Howard McCurdy argues that the cultures which various precursor agencies brought to NASA helped to shape the organizational culture within which Project Mercury developed.[40] As the agency took shape during the early phase, when Project Mercury was the prime human space flight mission, it seems clear that the melding of diverse groups of engineers, scientists, and managers into one organization changed the way that all did business.

When the shift from the pre-World War II military arsenal system to the Air Force's contracting system is also factored into the developmental period of the space program, the changes not only within NASA's constituent groups but outside are wide-ranging. The military arsenal system was established early in U.S. history. Both the Army and the Navy developed, designed, and fabricated the weapons used by their troops in government facilities. The Redstone Arsenal in Alabama, the site where the German rocket team was installed in 1950, was one among many such facilities, a fair number of which were established in the late 1930s or early 1940s for weapons production during World War II. The Navy Yard in Washington, DC, was one of the nation's oldest arsenals for Navy work.

That system served until the Air Force undertook in 1953 to develop, on an emergency basis, an ICBM capability for the United States. Using the Manhattan Project as a model of all-out development (a significant and perhaps flawed model which turns up repeatedly in the early days of space activity), the Air Force "adopted a system of parallel contracting, whereby hundreds of privately owned companies simultaneously designed and fabricated program components. The Air Force even relied upon contractors to help coordinate other contractors."[41]

The assumption was that a "national" effort of this sort demanded different or differently adapted strategies for organizing, planning, building, launching, and evaluating activities. T. Keith Glennan, NASA's first administrator, decided to build a program similar

---

37. Launius, *NASA*, p. 29.
38. *Ibid.*, p. 30; See "National Aeronautics and Space Act of 1958," Public Law 85–568, 72 Stat., 426, in John M. Logsdon, gen. ed. *Exploring the Unknown: Selected Documents in the History of the U.S. Civil Space Program, Volume I: Organizing for Exploration* (Washington, DC: NASA SP–4407, 1995), pp. 334–45.
39. McDougall, *Heavens and the Earth*, pp. 172–76; Launius, *NASA*, pp. 31–32.
40. Howard E.McCurdy, *Inside NASA: High Technology and Organizational Change in the U.S. Space Program* (Baltimore, MD: Johns Hopkins University Press, 1993).
41. *Ibid.*, pp. 15, 36–50.

to the Air Force's contracting program. This decision came partly from his support of the restrained Federal spending advocated by the Eisenhower administration. But Glennan, the president of Case Institute of Technology (later Case Western Reserve University) in Cleveland, came to the job of chief of the new organization understanding that more Federal spending and larger government staffs would be expected by old NACA hands. As he wrote:

> *Having the conviction that our government operations were growing too large, I determined to avoid excessive additions to the federal payroll. Since our organizational structure was to be erected on the NACA staff, and their operation had been conducted almost wholly 'in-house,' I knew I would face demands on the part of our technical staff to add to in-house capacity . . . but I was convinced that the major portion of our funds must be spent with industry, education and other institutions.*[42]

When James E. Webb succeeded Glennan as administrator of NASA for John Kennedy's administration in January 1961, his intent was much the same, though his focus was not the size of government bureaucracy. Rather, Webb had a grand vision of using NASA and its work to build science and technological education in the United States. He wanted, as his biographer pointed out, "to use NASA as a vehicle to move the whole nation to a 'new frontier' of enhanced technology-based educational and economic development." Space policy was to be integrated with economic and industrial policy. Webb's vision came too late for the Project Mercury program, however.[43]

In the beginning of the Project Mercury period, in fact, the system was barely developed and transitions of organizational culture unfinished; much of what was done relied on the earlier work and established organizational cultures of the various units transferred into NASA. The most important initial group transferred into NASA was the organization of the National Advisory Committee on Aeronautics. At least one historian of the period asserts that Hugh Dryden, then NACA director, actively campaigned for the role. The scholar adds,

> *NACA was not an inevitable choice. A small applied-research agency oriented mainly to work on aircraft, it had no experience in developing hardware or managing big programs. . . . But it was already at least on the fringes of space with the X-15 research craft, and its Pilotless Aircraft Division and the Lewis Flight Laboratory were doing significant research on space (the latter campaigned actively for space activities).*[44]

This new program was shaped against a complicated backdrop of technological developments, cultural change, and political imperatives that had come about in an exceedingly short period of time.

The Pilotless Aircraft Research Division (PARD) at the Langley Research Center was renamed NASA's Space Task Group, and in 1962 relocated as the Manned Spacecraft Center in Houston. These engineers were charged with the responsibility of Project Mercury. Langley itself became a NASA field center and early activities, including the early training of the seven Mercury astronauts, took place in Virginia. Administrator Glennan described the beginnings of the undertaking later by pointing out that "the philosophy of the project was to use known technologies, extending the state of the art as little as neces-

---

42. Glennan, *Birth of NASA*, p. 5.
43. W. Henry Lambright, *Powering Apollo: James E. Webb of NASA* (Baltimore, MD: Johns Hopkins University Press, 1995), pp. 99–101.
44. Levine, *Missile and Space Race*, p. 103.

sary, and relying on the unproven Atlas. As one looks back, it is clear that we did not know much about what we were doing. Yet the Mercury program was one of the best organized and managed of any I have been associated with."[45]

As the nucleus of NASA, the NACA employees brought with them elements of their former institutional culture. As Howard McCurdy describes it,

> *[NACA] employees believed thoroughly in the importance of research and testing. They insisted on seeking technical solutions to space flight problems, with a minimum of outside interference. Associated as they were with the test pilots of the astronaut corps, they adopted the ethic of taking risks to push performance frontiers. In only one respect did [the Space Task Group] depart significantly from the Langley research culture: it relied significantly upon contractors for spacecraft fabrication and technical assistance.*[46]

The NACA had never been part of the military's arsenal system, although work that it had done for the military services before World War II may ultimately have ended up in military fabrication shops. The NACA operated to some extent like an arsenal, however; it worked with its own machine shops and "hands-on" engineering work was a trademark of employment there. Thus this change had important implications for the ways that work was done in the new organization.

"Hands-on" work was a tradition which NACA employees tried to transfer into their jobs as NASA engineers. Years of research had prepared the Langley engineers to design space capsules that could safely carry their human cargo out of and back into Earth's atmosphere. Dr. Maxime A. Faget of the Langley team designed the Mercury spacecraft, and a contract to build it was awarded in late 1959. But the contract did not end the work of the Langley engineers with the spacecraft. McCurdy points out that they performed airdrop studies and tested escape rockets. They made blunt-body wind tunnel studies. They examined landing techniques. "Tests like these," McCurdy concluded, "kept NASA employees directly involved in the mechanics of space flight."[47] Much of this testing could have been carried out and required of the contractor, but NASA engineers were determined to retain control of as much of the engineering process as possible.

The Army's German missile team at the Redstone arsenal, transferred to NASA in 1960 after much effort by NASA officials,[48] had developed the launcher. The seeds of considerable conflict within NASA were sown by attempting to link together the NACA engineers and the German rocket team. As Howard McCurdy pointed out, "although the two groups shared many cultural norms, such as their belief in research and testing, they derived those norms in different ways." Former NACA employees hailed from a proud tradition of American aircraft design and testing. The Germans reflected a similar deep pride in German approaches and techniques. Both groups worked hard to retain control of their work, although the Germans proved to have little faith in American aerospace companies.[49] The issue of control of the process and of the product shaped many conflicts between the forces within the youthful space agency, and between them and the outside world.

The technical culture which both predecessor agencies brought into NASA worked well as long as the projects undertaken in common were limited in scope. When Project Mercury began, however, the multinational engineers of the new NASA faced a new universe of problems. "We now had to build something," Howard McCurdy quoted one of

---

45. Glennan, *Birth of NASA*, p. 13; Launius, *NASA*, pp. 38–39.
46. McCurdy, *Inside NASA*, p. 18.
47. *Ibid.*, p. 44.
48. Glennan, *Birth of NASA*, pp. 9–12.
49. McCurdy, *Inside NASA*, pp. 18–19.

them as saying. "We now had to fly something that we built. We now had to interface with the contractors to get that done. We had to build an organization. We had to make things happen that we had not ever been associated with before."[50]

Project Mercury paled in comparison to placing Americans on the Moon and returning them safely to Earth, but for the period 1958–1961 at least, the complexity offered by Project Mercury would stretch NASA's staff to its limits. NASA workers had to figure out what needed to be done, and instead of walking across the street to the machine shop with a sketch in hand, they now wrote detailed specifications for its manufacture. Contractors bid on work using the specifications, and queries from outsiders had to be considered and answered. The entire contracting section of NASA had to be developed from scratch because its predecessor research agency had never needed such an organization. NASA staff had to work with the contractors, overseeing work as well as testing it when it came off the assembly line.

A final challenge to the old ways of research and development came from the need, once Project Mercury got underway, to work directly with a network of aerospace contractors and other government agencies to make the whole space flight process work. For Project Mercury, the capsule (designed in-house) came from McDonnell Aircraft Corporation and the Redstone rocket from the Army's German missile team in Huntsville who had them fabricated by Chrysler Corporation. Later, Atlas launchers were used, courtesy of the Air Force but made by Convair Corporation. Project Mercury flights were launched from the Air Force's facilities at Cape Canaveral, Florida. The Navy picked up the astronauts. Thus, as Project Mercury developed, the sharing of responsibilities with competing organizations took considerable effort to organize. But the long-term effort led to the development of a technocratic organization capable of carrying off the Moon landing a decade later.[51]

The research and development phase of Project Mercury lasted roughly from October of 1958 to April 1961. In that time, the space capsule, designed by Max Faget and built primarily by the McDonnell Corporation, was readied. The spacecraft, an example of which is on display at the Smithsonian's National Air and Space Museum in Washington, DC, was designed to carry a lone astronaut for an orbital journey of about a day's duration. Integration of boosters and capsules began in 1960. Construction of a complex worldwide communications system, tracking systems, and a vastly expanded launch complex at Cape Canaveral, Florida, accompanied the fabrication of the launch vehicle.

Project Mercury and the Human Dimension of Space Flight

In addition to developing in a context of Cold War urgency and facing the enormous task of integrating varied and formative work cultures from civilian and military engineering organizations, Project Mercury propelled NASA into human space flight operations. The first seven American astronauts were introduced to the press on April 9, 1959.[52] The Mercury astronauts, recipients of the 1962 Collier Trophy, presumably stood, in the mind of the American public, for the agency, the engineers, and the contractors. Thus, the part they took in the Mercury program and in American culture of the late 1950s and 1960s deserves a close look.

The role that the astronauts played in the early history of NASA, of American space flight, and in American culture depended to some extent on the meaning assigned to the

---

50. Quoted in *ibid.*, p. 91.
51. *Ibid.*, p. 134.
52. Swenson, Grimwood, and Alexander, *This New Ocean*, pp. 109–338; Launius, *NASA*, pp. 38–41.

enterprise. If space exploration is seen entirely as a mechanical exercise (which was the perception of many engineers and scientists involved with Project Mercury as well as the Eisenhower administration) then the astronaut's role is relatively minor: he goes along for the ride and to make minor adjustments to the equipment. Testing human reactions is simply part of the technological testing process.

But a second understanding of the meaning of Project Mercury was also possible. Beyond the sheer technological basics, a more romantic notion, of individual challenge and courage in exploration of the universe or defense of the homeland can be seen as the reason for making machines that will carry explorers. In the second case, the focus is decidedly different. NASA, with its staff of machine-makers from the old NACA and Peenemünde and administrators in a conservative Republican administration may have begun the project with the first, straightforwardly technical vision. NASA and the Kennedy administration, for political reasons in Congress and among the American people, helped to shift the focus to the second. A memo written by James E. Webb, the second NASA administrator, to President John F. Kennedy, described the uneasy alliance:

> *The extent to which we are leaders in space science and technology will in large measure determine the extent to which we, as a nation, pioneering on a new frontier, will be in a position to develop the emerging world forces and make it the basis for new concepts and applications in education, communications and transportation, looking toward viable political, social, and economic systems for nations willing to work with us in the years ahead.*[53]

All of the rhetoric of the early years of space flight emphasized the technological nature of the competition in space between the Soviet Union and the United States. "The launching of *Sputnik 1* had a 'Pearl Harbor' effect on American public opinion . . . ," wrote NASA's chief historian. "The event created an illusion of a technological gap."[54] As the Soviet Union achieved more space spectaculars, the fear grew in the United States that Russia's perceived technological prowess had made the United States a second-class nation. Worries arose that the nations of the non-aligned world would choose to follow the Soviets because of their technological superiority. Thus the American space program strove to reestablish the preeminence of American science and technology in a world changed by Soviet scientific and engineering challenges since World War II and the early days of the atomic age.[55]

In that sense, as Mark E. Byrnes pointed out, the compelling images for the early days of space related to nationalism and the need for the United States to reassert itself as the most important world power. As he said,

> *In its most general form, nationalism has emphasized that America must be active in space in order to protect its national interest, however defined. NASA has named the space program's broadest and most important objective as "the establishment and maintenance of a strong national capability to operate in space and to use space fully in the national interest." Such a capability would give the nation "freedom of choice to carry out whatever missions the national interest may require—be they for national prestige, military requirements, scientific knowledge, or other purposes." Proficiency in space would also "prevent any other power from denying us the utilization of space in our interests."*[56]

---

53. Webb to Kennedy, quoted in William D. Atwill, *Fire and Power: The American Space Program as Postmodern Narrative* (Athens: University of Georgia Press, 1994), pp. 7–8.
54. Launius, *NASA*, p. 25.
55. Mark E. Byrnes, *Politics and Space: Image Making by NASA* (Westport, CT: Praeger, 1994), pp. 25–38.
56. *Ibid.*, p. 7.

In this context, big science and big technology will unite to support and strengthen American power on the international scene.

If, on the other hand, the viewpoint through which the U.S. space program is studied starts with the notion that a new set of pioneers will ride NASA's technological wonders to the stars, a different kind of narrative comes into play. Added to the imperative of nationalism are various ways of seeing the potential of the space program. The first is the space frontier as metaphor, the view expressed by scholars like William Goetzmann and Stephen Pyne.[57] In Pyne's clearest formulation of the role of the space frontier in world cultural development,[58] he argued that the International Geophysical Year (IGY) in 1957 "announced a new epoch of exploration, a Third Great Age of Discovery. Like its predecessors, the Third Age would claim special realms of geography, interact with distinctive syndromes of thought, pose immense new problems of assimilation for politics, economics and scholarship, and demand a new moral drama to give it legitimacy."[59]

NASA itself early began to use the imagery of exploration and of the wonders and possibilities of a new western frontier for the United States. A number of authors, including Patricia Nelson Limerick,[60] examined the frontier imagery used by NASA. Limerick pointed out the perils of using historical analogies badly.[61] Where Stephen Pyne finds solace and indeed appeal in the interior-exterior journey produced from the melding of modernism and exploration, Limerick finds cautionary tales. Running away from home, she notes, is an "inefficient way of leaving one's individual and collective problems behind." Settlements dependent on one form of transportation—settlers and railroads, or space station astronauts and shuttles—are likely to find themselves economically depressed and victimized by a sole source that has the ability to control their economic and personal agendas. Equal distribution of the fruits of colonization and settlement are rare; Limerick noted that,

> *far more often, the frontier comparisons shows, one person's benefit means another person's loss. Anglo-Americans acquired property, while Indians and Hispanics lost it; nineteenth century mineowners got the profits, while local miners got limited wages, considerable physical danger, frequent layoffs and little insurance or other protection. Just as clearly, the interests of various resource-users competed.*[62]

As Stephen Pyne argues that the new age of discovery will bring different challenges, Limerick makes the case that learning from the past may involve understanding a different set of stories.

Others have examined the methods and mythologies by which the space program was marketed by NASA and by its supporters. Michael L. Smith, in "Selling the Moon: The U.S. Manned Space Program and the Triumph of Commodity Scientism"[63] examined the

---

57. See, for example, William H. Goetzmann, *New Lands, New Men: America and the Second Great Age of Discovery* (New York: W.W. Norton and Co., 1986) and Stephen J. Pyne, *The Ice: A Journey to Antarctica* (Iowa City, IA: University of Iowa Press, 1986).
58. Stephen J. Pyne, "Space: a Third Great Age of Discovery," *Space Policy* 4 (August 1988): 187–99.
59. *Ibid.*, p. 187.
60. Patricia Nelson Limerick, "Frontier Life," revised version of speech to Second Annual Colorado Space Policy Workshop, Boulder, Colorado, September 8, 1989, p. 2. In addition, see works like Beverly J. Stoeltje, "Making the Frontier Myth: Folklore Process in a Modern Nation," *Western Folklore* 46 (October 1987): 235.
61. Patricia Nelson Limerick, Chair, "Session 3: What is the Cultural Value of Space Exploration?" in What is the Value of Space Exploration? A Symposium sponsored by the Mission from Planet Earth Study Office, Office of Space Science, NASA Headquarters and the University of Maryland at College Park, July 18–19, 1994, p. 13.
62. Limerick, "Frontier Life," p. 15.
63. Michael L. Smith, "Selling the Moon: The U.S. Manned Space Program and the Triumph of Commodity Scientism," in Richard Wightman Fox and T. J. Jackson Lears, eds., *The Culture of Consumption: Critical Essays in American History, 1880–1980* (New York, NY: Pantheon Books, 1983), pp. 177–209.

rhetoric of program supporters and the packaging of the astronauts. Starting at more or less the same point as Pyne and Limerick—with the explorer-scientist who mapped the West and filled museums with artifacts and images—Smith argued that by the 1950s national advertising agencies in the United States had created three particularly significant patterns of technological display: dramatic unveiling of products; the transferring of the special attributes of the product to the customer ("transitivity") usually by using actors in ads; and establishing through those actors a character type with which to identify. In Smith's view, a major 1950s image was "the helmsman, whose mastery over his environment through the products of technology provides a model for consumer aspiration." The helmsman and his machine—like Lindberg and "The Spirit of Saint Louis"—made a pair, the attributes of each enhancing the abilities of the other. As Smith noted,

> *Each of the helmsman's display qualities conveyed value to the product, which in turn appeared to reinforce precisely those qualities in its owner. Foremost among them was his masculinity. In a male-dominated society in which mechanization has been perceived alternately as a source of power and a threat to independence, advertisers forged an alliance between technological and gender display. . . . Technological sophistication and socially admired masculine traits were conveyed each through stylized variations of the other.*[64]

When *Sputnik* was launched in 1957, the ground was prepared to combine the advertising images of helmsmanship and technology in a Cold War race for space.

The first group of astronauts, the helmsmen of Smith's advertising world, was introduced to the press and the American people on April 9, 1959. While the debate about their role or lack thereof in Project Mercury continued among the engineers working to design a capsule and man-rate the rocket, the astronauts *became* the space program for most of their fellow citizens. As the official history of Project Mercury notes,

> *These personable pilots were introduced in civilian dress; many people in their audience forgot that they were volunteer test subjects and military officers. Their public comments did not class them with an elite intelligentsia. Rather they were a contingent of mature Americans, average in build and visage, family men all, college-educated as engineers, possessing excellent health, and professionally committed to flying advanced aircraft.*[65]

Others saw them somewhat differently. They had "the right stuff," observed writer Tom Wolfe, including the political sense in the case of one successful candidate, to recruit his estranged wife from their separation to the cause of his successful career as an astronaut.[66] Alan J. Levine argued that the Project Mercury astronauts were in fact presented in a way designed to make space travel as mechanical and ordinary, as risk-free as possible. "NASA's publicity machine and the *Time-Life* empire, which gained the rights to the astronauts' stories," Levine commented, "contrived to show them, and to some extent, the Mercury project as a whole, in a misleading way."[67] NASA, in this view, wanted to minimize risk; Henry Luce, of *Time-Life*, on the other hand, worked to show them as "typical middle-class white Protestants."[68] Michael Smith's process of presentation fits the introduction of the astronauts to the press.

---

64. *Ibid.*, p. 184.
65. Swenson, Grimwood, and Alexander, *This New Ocean*, p. 160.
66. The astronaut who was separated during the selection process was Gordon Cooper, according to Tom Wolfe, *The Right Stuff* (New York, NY: Farrar, Strauss & Giroux, 1979), pp. 146–47.
67. Levine, *Missile and Space Race*, p. 110.
68. *Ibid.*

Time-Life played a particular role in the presentation of the Project Mercury astronauts because the astronauts signed an exclusive contract with the company on August 5, 1959, for their "personal stories." The deal apparently originated inside NASA. As the Project Mercury history comments,

> *partly because of . . . natural public interest and partly because the civilian space agency had a statutory mandate to conduct educational publicity, NASA Headquarters, after investigation and decision, encouraged the astronauts to stay together and to accept the fringe benefits of a single private enterprise publishing offer arranged in outline even before their selection. This precluded eventual competitive bidding for individual story rights.*[69]

The astronauts were to receive $500,000, to be divided equally, without regard to who was to be the first American—and, it was hoped, the first human—in space. The stories, to be written by *Life* staff, were to be presented under first-person bylines, and the astronauts and their wives had final approval over the contents. *Life's* intention was to make the astronauts and their families look good. The astronaut's wives were full partners in the deal and in the stories that were told. The arrangement was immediately and continuously controversial; as *This New Ocean* observes, "Few other peripheral policy decisions regarding Project Mercury were to become so controversial in the long run."[70] The contract, unthinkable in later generations, guaranteed a continuous flow of information about the new space pioneers and their families throughout the 1960s. *Life* could not send its photographers into space aboard *Friendship* 7 or *Faith* 7 so that its ability to tell the story as *Life* saw it depended on the exclusive sharing of the stories and experiences of others.

The *Life* contract and the unexpected worldwide interest in the astronauts played into the cultural crisis set off by Russian space spectaculars. The Eisenhower administration took little public notice of *Sputnik's* orbits overhead but Members of Congress and the public reacted. *Sputnik* represented unexpected prowess on the part of the U.S.S.R. and a measure of military might; but it also revealed, in the view of many, Americans as soft, flabby, adrift in a sea of material goods. In the period during which the Space Act was passed, NASA was organized, and the 1960 election was held, the nation embarked on a search for national goals—a Presidential commission, a Special Studies Project funded by Rockefeller money, and *Life* magazine itself all devoted attention to national renewal and discovered some version of a loss of a sense of purpose or of mission. The timing fit perfectly and, in Michael Smith's view, "from the outset, then, the architects of the space program viewed it as a new source of national iconography."[71] The iconography of science and exploration masked the political (at home) and diplomatic (abroad) importance of national prestige as the national need impelling plans for a space program forward.[72]

Helmsmen were needed for the great adventure, and the fighter pilot astronaut emerged as the figure most worthy of carrying America's banner to the stars. People were necessary to the program in order to achieve "projection of the national imagination into space," to stand in for others who could share the dream vicariously. "Machines alone will not suffice if men are able to follow," observed a journalist writing in *The Nation*. "The difference is [of] that between admiring a woman's photograph and marrying her."[73]

---

69. Swenson, Grimwood, Alexander, *This New Ocean*, pp. 237–38.
70. *Ibid.*, p. 238.
71. Smith, "Selling the Moon," pp. 192–93.
72. *Ibid.*
73. Quoted in *ibid.*, p. 199.

The launch of John Glenn's "Friendship 7" spacecraft on February 20, 1962, atop the Atlas launch vehicle. The first American orbital flight, Glenn made three orbits of Earth. (NASA photo no. 62-MA6-112).

In the early days of space launchings, as they were then called, the Mercury astronauts appeared to be superfluous; their "functional role in the flight was not unlike that of a rather elaborate hood ornament."[74] Yet Alan Shepard, Gus Grissom, and John Glenn personified the rural farm youth or small-town white male daredevil image necessary to popular myth. Later astronauts in programs that followed Project Mercury, sharing their capsules with one or two companions took on a different look; *U.S. News and World Report* observed that "A new breed of cosmic explorer has emerged. Gone is the earlier image of the rocket-riding daredevil, the superman of the 'wild blue yonder.' The astronaut now is seen as a dedicated

---

74. *Ibid.*, p. 201.

scientist concerned more with discovery than with setting orbiting records."[75] The fighter-pilot was now transformed into an explorer-scientist ready to convert the void into an American landscape (ironically, only one scientist, geologist Harrison Schmitt, set foot on the moon or worked in the early programs of the space age). But for Project Mercury, the helmsman as quintessential American hero was the job description, and NASA and the media worked to make the candidates fit the profile.[76]

There was, in fact, no formal job description for the first astronauts because nobody working with research and development expected there to be a job for the men. As a consequence, the roles they played in Project Mercury developed as the program developed. One of the key points which mark the transition from aviation to space flight resides in the role of the human being on board. Joachim P. Kuettner, one of the German rocket technicians brought to the United States after World War II, described the difference. "While it is admittedly an oversimplification," he wrote,

> *the difference between the two technologies may be stated in the following general terms. From an aviation standpoint, man is not only the subject of transportation, and as such in need of protection as a passenger; but he is also a most important integral part of the machine over which he truly has control. . . .*
>
> *In contrast, rocket technology has been for twenty years a missile technology governed by the requirements of target accuracy and maximum range. As such, it had to develop automatic controls. Unlike a human payload, a warhead has no use except on the target. Once a missile fails, it may as well destroy itself during flight.*
>
> *The development of manned space flight is not just a matter of replacing a warhead by a manned cabin. Suddenly a switch is thrown between two parallel tracks, those of missile technology and those of aviation technology, and an attempt is made to move the precious human payload from one track to the other. As in all last-minute switchings, one has to be careful to assure that no derailment takes place.*[77]

Although the naming of astronauts assured that the space race would involve human space flight, the role of that human being vis-a-vis the role of automatic controls of all aspects of spacecraft and launcher operations became a subject of some controversy. Many, including the astronauts' fellow test pilots in the X-series tests at Edwards Air Force Base in California, considered the Mercury astronaut to be "Spam-in-a-can," a passive passenger in the space flight. Many of the engineers working on Project Mercury preferred that option, believing that automatic controls could protect human cargo more effectively than the human cargo could control the mission.

The astronauts, in contrast, had strong views about what they thought they ought to be doing. Astronaut Deke Slayton spoke to the issue before the Society of Experimental Pilots, when he observed that "Objections to the pilot range from the engineer, who semi-seriously notes that all problems of Mercury would be tremendously simplified if we didn't have to worry about the bloody astronaut, to the military man who wonders whether a college-trained chimpanzee or the village idiot might not do as well in space as an experienced test pilot. . . ." Slayton argued that the human role was vital: the astronaut should be "not only a pilot, but a highly trained experimental test pilot is desirable . . . as in any

---

75. Quoted in *ibid.*, p. 203.
76. See Peter Biskind, *Seeing is Believing: How Hollywood Taught Us to Stop Worrying and Love the Fifties* (New York, NY: Pantheon Books, 1983), pp. 101–59.
77. Quoted in Swenson, Grimwood, and Alexander, *This New Ocean*, pp. 171–72.

scientific endeavor the individual who can collect maximum valid data in minimum time under adverse circumstances is highly desirable."[78]

After Project Mercury had ended, Christopher C. Kraft, Jr., chief flight director for the Space Task Group, described the shift in thinking that gave Project Mercury astronauts a larger role in spacecraft control, and later astronauts a larger role still. "The real knowledge of Mercury," Kraft remembered, "lies in the change of the basic philosophy of the program. At the beginning, the capabilities of Man were not known, so the systems had to be designed to function automatically. But with the addition of Man to the loop, this philosophy changed 180 degrees since primary success of the mission depended on man backing up automatic equipment that could fail."[79] This shift in perspective was fundamentally an engineering decision, and did not at all mean that the astronauts were being given a green light to drag race in space. Yet, as the official history of Project Mercury notes, it had implications for the way the space race and spacefaring activities undertaken by the United States were understood by the general public. "The field managers of Mercury had ruefully discovered," reports the history, "that people, or at least reporters, were more interested in people than machines, so they allowed 'Shorty' Powers to skew publicity toward machine-rating the men rather than man-rating the machines."[80]

Like deep-sea divers, spacefarers had to take their environment with them. In the case of the Mercury capsule, the environment was two-tiered: the suit, which was a mini-environment within itself; and the capsule, sealed against all the stresses and extremes of launch, orbit and reentry. The B. F. Goodrich Company was awarded the contract to design the spacesuit on July 22, 1959. Suit design went through numerous changes and modifications during 1959 and 1960, until the model finally met the approval of astronauts and program managers in May 1960. The model for the suit—coveralls, helmet and gloves—came from outfits already fabricated for those piloting high-flying aircraft. But even with formal approval of specifications for the model space suit, the design continued to evolve. It was, in fact, one of the elements of Project Mercury that changed most often during the life of the undertaking.[81]

As designers prepared to envelop the pilots in their protective garb, other engineers developed plans for what role the astronauts would actually play in flight. At this point in project organization, the astronaut began to resemble less the passive "hood ornament" of earlier concepts, and more a physiologically conditioned integral element of the space flight system. This was by no means a move to a concept of "piloted flight"; but the amount of control that the pilots could assume of the craft, particularly in an emergency, expanded somewhat. A list of activities showed that the Mercury astronaut would be expected to "communicate with ground stations, make scientific observations, monitor onboard equipment, control capsule attitude, navigate and fire retrorockets, initiate emergency procedures, activate the escape system if necessary, and deploy the landing parachute if required."[82]

Other undertakings, which had origins in the test pilot programs of the military services, included test pilot and astronaut inspections of the equipment which would carry them into Earth orbit, and intensive training for and simulation of in-flight experiences. Training was particularly complex because of previously unfamiliar conditions of space flight, such as weightlessness. Although initial concepts of astronaut training included an extensive academic course, most of the activities ended up as hands on—or human being

---

78. Slayton, quoted in *ibid.*, p. 177. The US was not alone in this debate. When Yuri Gagarin made his historic one–orbit in *Vostok 1*, the controls were locked, in case he went crazy. He had the directions for unlocking them in a sealed envelope. See Levine, *Missile and Space Race*, p. 119.
79. Quoted in Swenson, Grimwood, and Alexander, *This New Ocean*, p. 177.
80. *Ibid.*, p. 221.
81. *Ibid.*, pp. 225–31.
82. *Ibid.*, p. 234.

in—work in mechanical aids. Simulation of weightlessness, disorientation, exposure to loud noises, acceleration patterns conditioned the astronauts to a range of experiences.[83] That the training was successful seemed to be verified when each astronaut went aloft and confirmed that every planned-for sensation felt familiar. As Tom Wolfe observed of Al Shepard's sub-orbital flight, "he was introducing the era of precreated experience. His launching was an utterly novel event in American history, and yet he could feel none of its novelty . . . he could only compare it to the hundreds of rides he had taken on the centrifuge at Johnsville. . . . "[84]

As the astronauts trained in the summer and fall of 1960, however, Project Mercury's existence became increasingly doubtful because the boosters necessary to insert a capsule into orbit keep failing. And NASA's rocket failures were very public, particularly in a time of high political interest, as the 1960 campaign for the Presidency of the United States between John F. Kennedy and Richard M. Nixon intensified. The first Mercury-Atlas flight on July 29, 1960, took off as scheduled but, above a thick bank of clouds over the Cape Canaveral launch pad, it apparently disintegrated. The effect of this failure only intensified the next month, when the Russians launched a satellite with a biological cargo: "muttniks" Streika and Belka; rats; mice; flies; plants; fungi; and seeds. After 18 orbits, the Russians recovered the dogs and their travelling partners. The next month, Soviet Premier Nikita Khrushchev attended meetings at the United Nations and told the press that the Russians were ready to orbit a human.[85]

The next spring, they did. The Soviet Union orbited Major Yuri Alekseyevich Gagarin in *Vostok 1* on April 12, 1961. Gagarin, thus, became the first human in space by making one full revolution of the Earth.[86] *Life* magazine sent reporters to cover the victorious welcome of Gagarin back to Moscow. The capsule in which Gagarin rode weighed 10,417 pounds, almost three times bigger than the Mercury capsule being readied for American astronauts. Flight apogee was 203 miles and perigee, 112 miles. Gagarin was weightless for 89 minutes of his 108 minutes of flight.

The pictures from Moscow taken by *Life* magazine reporters differed profoundly from the photos taken around American space launchings. There were no views of rockets on launchers or hardware tracking the flight, none of observers searching the sky. Rather, these photos were clearly after the fact; Gagarin and most of his fellow Soviet officials were dressed in military uniforms or Politburo winter wear and Gagarin was pictured walking down a vast red carpet to receive the congratulations of his country's ruling group. Nothing in the photographs of the celebration indicated that a flight into outer space had occurred, that human space flight was now a reality, or that human history had been profoundly altered.[87]

Project Mercury had nonetheless been eclipsed by the Russian achievement. Still, as Astronaut John Glenn commented, "I am, naturally, disappointed that we did not make the first flight to open this new era. The important goals of Project Mercury, however, remain the same—ours is peaceful exploration of space. These first flights, whether Russian or American, will go a long way in determining the direction of future endeavors. There is certainly work for all to solve the tremendous problems involved."[88]

---

83. *Ibid.*, pp. 238–48.
84. Wolfe, *The Right Stuff*, p. 256.
85. Swenson, Grimwood, and Alexander, *This New Ocean*, pp. 272–81.
86. Peter Bond, *Heroes in Space: from Gagarin to Challenger* (London: Basil Blackwell, 1987), pp. 12–17.
87. *Life*, April 21, 1961, cover and interior photos.
88. Quoted in Swenson, Grimwood, and Alexander, *This New Ocean*, p. 335.

Project Mercury Operations and the Astronauts

April 1961 brought the end of the pilotless test phase of the Project Mercury program, and fortunately, also the last major flight failure in Mercury.[89] Delays plagued the program and the decision to fire one last automated test in March 1961, rather than begin the piloted tests, gave the Russians the opportunity to launch Gagarin first.

The test of human space flight brought to fruition all the various processes and perspectives that had characterized the astronauts and their programs since their initial naming on April 9, 1958. Whether they were to prove to be "Spam-in-a-can" as detractors saw the space pilot role, or active pilots of new kinds of craft would become clear in practice. Whether the agency view of their relative unimportance to the program or the press' and public's view of their centrality would become the historical account of Project Mercury would now be tested as thoroughly as the hardware that they rode into space. Whether they would prove to be paper heroes or celebrated as the nation's finest would depend partly on how their test flights turned out.[90]

As a consequence, the first American piloted flight carried a good deal of symbolic weight, even though it was not the first human flight into space. The first American to be lobbed sub-orbitally was Navy Commander Alan Shepard. Chosen from two other finalists—Gus Grissom and John Glenn—Shepard's historic flight was postponed three times before its final launch on May 5. At 9:34 a.m., the Mercury-Redstone combination left the launch pad while about forty-five million Americans watched on television and many more tuned in on the radio and held their collective breath.[91]

Shepard's flight in the capsule named *Freedom 7* took fifteen minutes and twenty-two seconds. Its altitude was 116.5 miles, its maximum speed was 5,180 mph, and it travelled 302 miles from Cape Canaveral. *Freedom 7* was the last version of the capsule designed before the astronauts began to shape the process of capsule design. It had only portholes instead of a window, and Shepard made his Earth observations through a periscope. He clearly distinguished between cloud masses and land masses and recognized various landmarks including Lake Okeechobee in Florida, and the islands of the Bahamas. Shepard took control of the spacecraft twice for brief periods; his flight plan called for him to manually position the capsule for retrofire, and he corrected a slight pitch problem on the positioning of the craft in one instance. In the second, he took control for a brief period during reentry of the capsule's attitude. Shepard withstood space flight conditions well, including five minutes of weightlessness and everything about the flight, from ignition to recovery and debriefing, went without a hitch.

Machine-rating the humans had begun. This initial foray proved to American engineers and technical designers that humans could function in space, even while weightless, thus bolstering the argument that the astronauts should be part of the working systems of the spacecraft rather than passive passengers. Moreover, as an open news event, covered by the media throughout, the flight earned Cold War propaganda points by illustrating the openness of the American space program, in contrast to that of the Soviets. Although in strictly comparative terms, the first American in space did not come close to matching the feats of his Soviet counterpart, Shepard nonetheless became an immediate, full fledged American hero through his competent, laconic performance. President Kennedy awarded him the Distinguished Service Medal on May 8.[92]

---

89. Several of the first test flights of Mercury spacecraft without human passengers failed on both the Redstone and the Atlas boosters.
90. See Launius, *NASA*, p. 40, as an example of these kinds of pressures.
91. Swenson, Grimwood, and Alexander, *This New Ocean*, p. 341.
92. *Ibid.*, pp. 352–61; Bond, *Heroes in Space*, pp. 27–33.

The contrast between the accomplishments of *Vostok 1* and *Freedom 7*, however, demonstrated that the Soviets had what looked like an overwhelming technological advantage; Shepard's flight had been in fact little more than a man shot out of a cannon on a very large scale. To address the fears and concerns of the American public who perceived a real danger in American second-place status in the space race, Kennedy had ordered Vice President Lyndon B. Johnson to make an overall survey of the possibilities for American space triumphs after Gagarin's initial flight in mid-April. After surveying the viewpoints of NASA staff, Congress, aerospace contractors, and experts in science and technology, Johnson produced recommendations for Kennedy on April 28. Johnson argued that world leadership increasingly depended on "dramatic accomplishments in space"[93] and that continual Soviet dominance in that realm would ultimately lead to their presumed dominance in other international arenas. Human exploration of the Moon, Johnson thought, would be an effort worth a great deal, and it was possible that the United States could get there first.

Following this memo, Johnson submitted another report to Kennedy, with NASA director James Webb and Secretary of Defense Robert McNamara's concurrence.[94] That led, in turn, to a shift in the American government's space policy as Kennedy committed the nation to going to the Moon and back in the decade of the 1960s. Kennedy announced his decision, and the new policy, in a major speech to a joint session of Congress on May 25, 1961, by saying, "I believe that this nation should commit itself to achieving the goal, before this decade is out, of landing a man on the moon and returning him safely to Earth."[95]

For Project Mercury, the policy change coming on the heels of its first successful human flight added urgency to immediate accomplishments and at the same time, diverted the attention of NASA and its supporters and suppliers to larger goals and more complicated projects. The Eisenhower-era basis on which Project Mercury had been designed, to use the simplest and most reliable approach with a minimum of new developments and incremental steps was replaced by Kennedy Cold War "urgent national needs" and an explosion of congressional budgetary support. In moving to the Manhattan Project approach to space efforts, Project Mercury became part of a program larger than itself. Just as Alan Shepard's flight was eclipsed by Yuri Gagarin's, so was Project Mercury, in some senses, eclipsed by the opening of the race to the Moon.

America's original seven astronauts may have been the only part of the NASA hierarchy that remained focussed almost solely on Project Mercury, but the American public remained focussed on the astronauts. The second of seven planned suborbital tests carried Virgil L. "Gus" Grissom aloft in *Liberty Bell 7* on July 21, 1961. Grissom profited from lessons learned on Shepard's flight and other astronaut comments early in the capsule design process. *Liberty Bell 7* had a central window instead of portholes. An improved attitude control system allowed for more astronaut piloting. A new hatch, armed with explosive bolts, was another improvement. Grissom flew for 15 minutes and 37 seconds, at speeds of as much as 5,300 miles an hour to an apogee of 118 miles. The flight was virtually flawless until the recovery phase when, in a process that has never been fully explained, the hatch bolts suddenly blew. Grissom found himself in the water, with the oxygen inlet valve in his space suit open. He nearly drowned before the rescue helicopter picked him up, and the capsule *Liberty Bell 7* was lost.[96]

---

93. Lyndon B. Johnson, "Memorandum for the President," April 28, 1961, quoted in Logsdon, gen. ed., *Exploring the Unknown*, p. 427–29.

94. See James E. Webb, NASA Administrator, and Robert S. McNamara, Secretary of Defense, to the Vice President, May 8, 1961, with attached: "Recommendations for Our National Space Program: Changes, Policies, Goals," in Logsdon, gen. ed., *Exploring the Unknown*, pp. 439–52.

95. John F. Kennedy, "Urgent National Needs" speech to a Joint Session of Congress, May 25, 1961, in *ibid.*, p. 453.

96. Bond, *Heroes in Space*, pp. 33–35.

Five more suborbital flights were planned, but the Russians increased the pressure on NASA to launch an orbital flight when they sent Gherman Titov around the world 17 times in 24 hours in August. Three of those orbits were over the United States. As a result of this pressure and the more general pressure to work towards the more complex Apollo program, only one more suborbital launch took place, to test the Mercury-Atlas combination, with a chimp named Enos aboard. After that success, on August 18 NASA announced that the Mercury-Redstone sub-orbital program had achieved its objectives and was, thus, ended.[97]

As a result of that cancellation, John Glenn, originally scheduled to be only the third sub-orbital astronaut, instead became the first American to orbit the Earth. That fact was somewhat ironic, since Glenn had made no secret of his fury at being passed over for the initial suborbital space mission.[98] "First," he said, "is first." Glenn had worked hard to build a public persona and become the best-known of the astronauts, but his ambition and straitlaced personal life contributed to a lack of popularity among the other seven space pioneers. Apparently, this lack of support played some role in his being chosen as backup pilot and his quasi-public complaints caused NASA higher-ups to suggest that he show some restraint lest he not fly at all.[99]

John Glenn's three orbits renewed the faith of Americans in their culture's scientific and technological know-how. But when his turn finally came, the assignment demanded a good deal of patience. The first piloted orbital flight was postponed repeatedly in a period starting in December 1961 and only finally resulting in launch at 9:47 a.m. on February 20, 1962. Glenn's five-hour space flight saw three sunsets and three sunrises; as a *Life* magazine researcher estimated, he spent four Tuesdays and three Wednesdays in orbit.[100] Dust storms and clouds obscured much of Africa. When Glenn passed over Australia the first time, it was night and the citizens of Perth turned their house lights on. To Glenn, Perth looked from space like a small town seen from an aircraft. As the sun rose for the first time, Glenn noticed thousands of particles swarming around the capsule. "It is," he said, "as if I were walking backward through a field of fireflies."[101]

Although the flight began routinely, a number of problems developed during the mission of *Friendship 7*. An unexpectedly rough ride into orbit, caused partially by nearly empty Atlas fuel tanks, caused Glenn to comment, "They really boot you off" as he entered orbit, free of the Atlas superstructure at last.[102] More seriously, an attitude control problem developed in the first orbit, in which first one and then the second of two yaw-controlling autopilots stuck. Glenn, proving the efficacy of having humans on board to compensate for automatic system malfunction, took over the attitude control manually and controlled it for the remainder of the flight. Finally, ground controllers had received what turned out to be an erroneous signal that the landing bag on *Friendship 7* had deployed in orbit. Such deployment would mean that the capsule's heat shield, crucial equipment to keep the capsule from burning up during reentry, could rip off during the reentry process. Ground controllers, in consultation with Maxime Faget, the capsule's designer, instructed Glenn not to follow the regular procedures and jettison the retrorocket package that held the landing bag in place after retrorocket firing. These instructions turned out not to be necessary, since the signal received was erroneous, but Astronaut Glenn was not fully informed of the situation until just before he began the reentry process. He took great

---

97. *Ibid.*, p. 36; Swenson, Grimwood, and Alexander, *This New Ocean*, pp. 370–78.
98. Bond, *Heroes in Space*, p. 37; Loudon Wainwright, *The Great American Magazine: An Inside History of Life* (New York, NY: Alfred A. Knopf, 1986), pp. 272–73.
99. Wainwright, *Great American Magazine*, p. 273.
100. Dora Jane Hamblin, *That was the Life* (New York, NY: W.W. Norton, 1977), p. 81.
101. Quoted in Bond, *Heroes in Space*, p. 42.
102. Quoted in *Ibid.*, p. 39.

"Launch of Mercury-Atlas," watercolor by John McCoy, NASA art program, Space Art-1. (NASA photo 67-HC-617).

exception to being treated as a passenger rather than as a pilot, and ground controllers from that flight on were much more open with the astronauts about the status of their capsules and flights while they were aloft.[103]

John Glenn's return to Earth was cause for enormous celebrations in the United States. President Kennedy telephoned his congratulations to the astronaut aboard the recovery ship *Noa*. Kennedy also made a statement to the Nation, in which he said

> *I know that I express the great happiness and thanksgiving of all of us that Colonel Glenn has completed his trip, and I know that this is particularly felt by Mrs. Glenn and his two children.*
>
> *I also want to say a word for all of those who participated with Colonel Glenn at Canaveral. They faced many disappointments and delays — the burdens upon them were great—but they kept their heads and they made a judgment, and I think their judgment has been vindicated.*
>
> *We have a long way to go in this space race. But this is the new ocean, and I believe the United States must sail on it and be in a position second to none.*[104]

Although Glenn got the bulk of the praise from the media, those of the technical teams came in for some note; one periodical praised the "leaders of this technical team who did their work on civil service pay and sold no serial rights to national magazines."[105]

Glenn and the other astronauts paraded past an estimated 250,000 people in Washington, on their way to a twenty-minute speech before a joint session of Congress. New York City held a tickertape parade and proclaimed March 1 "John Glenn Day" in the Big Apple. The headquarters of the United Nations held a reception in his honor. Glenn was greeted by 75,000 people who turned out in New Concord, Ohio, his home town.[106] And, as the NASA historian noted, "NASA discovered in the process of this hoopla a powerful public relations tool that it has employed ever since."[107]

Three more Mercury launches took place between February 1962 and May 1963. Malcolm Scott Carpenter was launched on May 24, 1962, in *Aurora 7* for a three-orbit mission. The launch went perfectly, but numerous problems developed during the flight because Carpenter's flight plan was too full and he was too interested in observing the Earth to calculate carefully the amount of fuel he was using. Still, two three-orbit missions completed successfully indicated to NASA administrators that the Project Mercury mission might be lengthened for the next flight.[108]

Walter Schirra was next, but in the meantime, the Russians orbited *Vostoks 3* and *4* for six days in space, with a combined total of 112 orbits. Schirra's flight in *Sigma 7*, in comparison to the freewheeling orbits of Scott Carpenter, was to be all test piloting with few additional scientific experiments. October 3, 1962, was the launch date for Schirra's flight, six orbits with a splashdown in the Pacific. Schirra's mission produced very low fuel consumption and clear proof that a pilot could fly the Mercury spacecraft in an efficient, very accurate manner.[109]

The final flight in the Project Mercury series was *Faith 7*, piloted by L. Gordon Cooper, Jr., who was launched on May 15, 1963. The last launch of an Atlas with a human aboard,

---

103. *Ibid.*, pp. 42–47; Swenson, Grimwood, and Alexander, *This New Ocean*, pp. 429–30; Launius, *NASA*, p. 80.
104. President John F. Kennedy, quoted in Swenson, Grimwood, and Alexander, *This New Ocean*, p. 434.
105. *Ibid.*, p. 435.
106. *Ibid.*
107. Launius, *NASA*, p. 81.
108. Swenson, Grimwood, and Alexander, *This New Ocean*, pp. 446–60.
109. *Ibid.*, pp. 461–86; see also Captain Walter M. Schirra, Jr. with Richard N. Billings, *Schirra's Space* (Boston, MA: Quinlan Press, 1988), pp. 79–95.

this mission lasted for 22 orbits. Cooper's ground observations and the surprising level of detail he reported being able to see had implications for security undertakings in space. The major difficulty in this last flight involved the loss of the automatic control system in the twenty-first orbit. Cooper had to position the capsule manually for reentry and fire the retrorockets manually. He did so with great accuracy and showed once again that the human presence could save a mission with serious mechanical failures.[110]

Cooper's Project Mercury flight was truly "the end of the beginning"[111] of piloted space undertakings. Some discussion had already occurred within NASA as to the form that commemoration of these early days of space exploration might take.[112] Almost a year before Cooper's flight, in September 1962, Webb decided to commission artists to capture each stage of the launch-and-return process. As Webb commented at the time, "important events can be interpreted by artists to give a unique insight into significant aspects of our history-making advance into space."[113]

With more hyperbole, H. Lester Cooke, curator of paintings at the National Gallery of Art who served as the first head of the program, said in 1963 that "not since the lungfish slithered out of the oolitic ocean have living creatures sought to change their basic environment, and it was felt that this epic step must be recorded in every way possible . . . what if Queen Isabella had sent along a top-flight artist with Columbus? or artists had been at Kitty Hawk? or at the White Sands Proving Grounds? And what a stroke of genius to send Winslow Homer to the Civil War front!" And he added, "Perhaps this project will help to prove to future generations that the United States in the sixties produced not only engineers and scientists capable of shaping the destiny of our age but also artists worthy to keep them company."[114]

The first artists arrived at Cape Canaveral in May 1963 to cover Gordon Cooper's Mercury flight. Seven artists worked at the Cape itself; one flew out to the Pacific to cover the splashdown. Peter Hurd's visual sense manifested itself from the airplane:

> *Activity on the Cape is continuous throughout the day and night and my impression from the air was of a vast and deceptively festive display. Whether by design, by chance, or from technical need, the score or more of enormous gantry cranes, which seemed to stride in a great marching procession along the shore, were painted an intense and subtly beautiful shade of red. The cranes are of open steel work, an interlacing maze of girders and tubing, lavishly lighted from inside and out, giving an unbelievable realistic effect of incandescent filigree.*

The routine devised for the artists began with a tour of the Cape. The first night was the last of relative quiet. At the Mercury Control Center, "moonlight ruled the stage, making pools of deep shadow from which emerged a long narrow scaffold of crisscrossed girders. This in turn was surmounted by a profusion of television antennae like fragile spangles of silver gleaming in lost-and-found pattern against the night sky. . . ." A substantial portion of information about NASA was imparted during the tour including a description of safety procedures and a recounting of the "shake test" in the White Room, where the Mercury capsule being assembled was shaken to reveal loose bolts, screws, filings and hairs.[115]

---

110. Swenson, Grimwood, and Alexander, *This New Ocean*, pp. 488–501.
111. *Ibid.*, p. 492.
112. Memo of James E. Webb to Dr. Hiden Cox, March 16, 1962, Impact-Art, NASA Historical Reference Collection.
113. Quoted in Robert Schulman, "Space Art," *Space Times*, November–December 1989, p. 5.
114. H. Lester Cooke, "Countdown at Canaveral," *Art in America*, No. 5, 1963, p. 60.
115. *Ibid.*, p. 62.

On May 15, the "Big Morning" as Hurd called it, the artists were driven to a roadside viewing area where they were surrounded by rescue craft and news crews. At T minus 60 seconds, Hurd stopped being able to write; as he explained later, "I pick up from memory the suspense of those last seconds—terrifying for us, for we were each of us in that capsule whether conscious of it or not." In trying to explain his feelings, he added, "Perhaps it was in witnessing a supreme gathering of forces, the sight of so many individuals engaged in a wide range of techniques, all addressed to achieving one objective: the successful completion of another orbital flight. The thought kept occurring to me that a similar mass effort built the great cathedrals; the same desire of man to attain to his ultimate capacity."[116]

The artist captured the awe that most civilian observers felt for the seven daring test pilots of Project Mercury and the structures and scientific and engineering feats that sent them aloft into that new ocean. James Webb was right to capture that in art as well as in photographs, and in what might be called big culture as well as in big science and technology. The award of the Collier Trophy to the original astronauts confirmed their importance to the engineering fraternity in NASA and to American culture. Webb commented about the award of the Collier Trophy that "The recognition of these outstanding Americans by the representatives of the National Aeronautics Association is indeed a high honor, and I feel that this honor will be one of the highlights of their careers and a highlight in the growth and development of the National Aeronautics and Space Administration."[117]

The award was undoubtedly a highlight for the National Aeronautic Association as well, for the organization honored the pioneer pilots rather than the engineering hardware and joined the general celebration of human effort in Project Mercury. Not quite hood ornaments, not passive passengers, but not test pilots free to follow their own instincts either, the Mercury Seven were the last group of astronauts to solo in Earth orbit. The transition group between airplane test pilots and astronauts of the later programs, they were honored for being the first—and in a real sense, for being the last.[118]

---

116. *Ibid.*, p. 64.
117. James E. Webb to Martin M. Decker, President, National Aeronautic Association, September 10, 1963, NASA Historical Reference Collection.
118. Bond, *Heroes in Space*, p. 63.

Chapter 8

Managing America to the Moon: A Coalition Analysis

by W. Henry Lambright

Without question, the lunar landing program—Project Apollo—was one of the greatest examples of technological achievement in history. It was also a great managerial feat. Finally, the fact that it sustained political support long enough to implement the Kennedy goal was also remarkable. Today it is difficult to maintain momentum for any governmental program, it seems, beyond a single presidential election.

The Collier trophies have understandably gone in NASA's direction a number of times for extraordinary Apollo–related accomplishments. This essay explores the story behind the following awards:

- *1965, award to James E. Webb, NASA Administrator, and Hugh Dryden, Deputy Administrator, as representatives of NASA working on a project that significantly advanced the human experience in space flight;*
- *1967, award to Lawrence A. Hyland "representing the Surveyor Program Team at Hughes Aircraft Company, the Jet Propulsion Laboratory, and associated organizations that put the eyes and hands of the United States on the Moon";*
- *1968, award to Col. Frank Borman, USAF; Capt. James A. Lovell, Jr., USN; and Lt. Col. William A. Anders, USAF, representing the entire space flight team of Apollo 8, for the successful execution of the first manned lunar orbit mission in history;*
- *1969, award to Neil A. Armstrong; Col. Edwin E. Aldrin, Jr., USAF; Col. Michael Collins, USAF, "For the epic flight of Apollo 11 and the first landing of man on the surface of the moon, July 20, 1969"; and,*
- *1971, award to Col. David R. Scott, USAF; Col. James B. Irwin, USAF; and Lt. Col. Alfred M. Worden, USAF, "For demonstrating superb skill and courage and to Robert Gilruth as representative of the engineering genius of the manned space flight team culminating in Apollo 15—man's most prolonged and scientifically productive lunar mission."*

The awards went for discrete accomplishments along the way, but they were part and parcel of a huge program that began in 1961 with the Kennedy goal "before this decade is out, of landing a man on the moon and returning him safely to the Earth." This program, which extended to 1975, when the final Apollo flight took place (*Apollo–Soyuz*) could have garnered any number of other awards for any number of other achievements.

It is somewhat arbitrary to single out any particular event, since what is at issue is a program in which one decision built upon another. The Collier awards really identify particular feats that are representative of the many. While they mention managers, or astronauts, or a government–industry team, what they are really about is a large group of individuals and organizations that combined for a relatively brief period of time to accomplish what in retrospect seems extraordinary. What made the lunar landing program possible? Who did what? How?

Approach

This chapter approaches the subject of Apollo as an achievement in program management. There is a great deal of academic and practical interest in this subject.[1] In her introduction to this volume, Pamela Mack raises the issue of "big science."[2] "Big science" stands as a symbol for billion-dollar research and development projects. Today, large-scale programs include the Space Station, the Genome Project, the Hubble Space Telescope, and the Earth Observation System. The Manhattan Project and Apollo are historical examples of "big science." For the most part, big science is a misnomer. What is entailed is technology, huge machines such as the Hubble Space Telescope. This is especially true for Apollo, with its $24 billion cost. Apollo was a large-scale technological program whose rationales were pride, prestige, and Cold War competition, not specifically science.

There are many ways such enterprises can be analyzed. Many sociologists have favored a "social constructionist" approach, in which societal forces shape science and technology. John Law and Michel Callon, in contrast to other social constructionists, see society and technology as affecting one another. In studying particular big science programs, they have isolated certain "actors" who are protagonists behind programs. In constructing programs, such actors build "local" and "global" networks.[3]

Thomas Hughes, an historian, has written of society and technology as "a seamless web."[4] But he also finds that certain actors, pursuing order and control, become "system builders" providing technological trajectories and momentum. These concepts have much in common with my own approach, rooted in political science, which focuses on the dynamics of large-scale technology as a political process. That is, the progenitors build coalitions of support behind these programs, coalitions of internal ("local") and external ("global") actors. Coalitions can grow, change, be strengthened, weakened, or unravel. The shape, scale, and direction of the program depends on the coalition—its size, cohesion, and leadership. Leadership in particular matters greatly in coalition building, for leaders are the coalition builders. Their strategies make the big science program go.[5]

Large-scale programs take time to be implemented, often a decade or more. Along the way politicians in the White House and Congress change, the economy goes through cycles, and international and domestic crises alter national priorities. What is possible for administrative leaders at one time may be impossible at another.

This chapter looks at Apollo as a long-term, large-scale program that had a beginning, middle, and end. The Collier awards reflect NASA achievements along the way as well as

---

1. There are many books on Apollo. They include: Courtney G. Brooks, James M. Grimwood, and Loyd S. Swenson, Jr., *Chariots for Apollo: A History of Manned Lunar Spacecraft* (Washington, DC: NASA SP-4205, 1979); Arnold S. Levine, *Managing NASA in the Apollo Era* (Washington, DC: NASA SP-4102, 1982); John Noble Wilford, *We Reached the Moon* (New York, NY: Bantam, 1969); Charles Murray and Catherine Bly Cox, *Apollo: The Race to the Moon* (New York, NY: Simon and Schuster, 1989); Alan Shepard and Deke Slayton, *Moon Shot: The Inside Story of America's Race to the Moon* (Atlanta, GA: Turner Publishing, 1994); and Andrew Chaikin, *A Man on the Moon: The Voyages of Apollo* (New York, NY: Viking, 1994).
2. James H. Capshew and Karen A. Rader, "Big Science: Price to the Present," *OSIRIS*, 2nd series, 7 (1992): 3–25. See also Peter Galison and Bruce Hevly, eds., *Big Science: The Growth of Large-Scale Research* (Stanford, CA: Stanford University Press, 1992).
3. John Law and Michel Callon, "Engineering and Sociology in a Military Aircraft Project: A Network Analysis of Technological Change," *Social Problems*, Vol. 35, No. 3, 1988, pp. 284–97. See also Wiebe Bijker and John Law, eds. *Shaping Technology/Building Society: Studies in Sociotechnical Change* (Cambridge, MA: MIT, 1992).
4. Thomas Hughes, "The Seamless Web: Technology, Science, Etcetera, Etcetera," *Social Studies of Science*, Vol. 16, 1986, pp. 281–92.
5. Jameson W. Doig and Erwin C. Hargrove, eds. *Leadership and Innovation*, (Baltimore, MD: Johns Hopkins University Press, 1987). See also Robert Smith, *The Space Telescope: A Study of NASA, Science, Technology and Politics*, (New York, NY: Cambridge University Press, 1989).

NASA Administrator James E. Webb and Deputy Administrator Dr. Hugh L. Dryden received the 1965 Collier Trophy for effective management of a large-scale research institution. Shown here are, left to right, Administrator Webb, Vice President Hubert H. Humphrey, and Mrs. Hugh Dryden, representing her husband and accepting the award for all of the Gemini teams who significantly advanced human experience in space flight. (NASA photo).

the total achievement. This process had an internal dimension (getting the work done) and an external dimension (getting the resources and political support). Behind the rise and completion of this program was a coalition building process. Behind this process were NASA leaders.

Seen historically, Apollo marked the culmination of many trends. It entailed the NACA tradition, set long before World War II, which linked government with huge in-house facilities and laboratories. It expanded upon trends set during and after the war by which government accomplished big science through contracts with industry and universities. It added the visual drama of real-time reporting of man-in-space through satellite-based television. More than any R&D program before, Apollo merited the title of "national" endeavor, for the nation truly was involved and engaged, at least in the 1960s. There has been nothing similar since Apollo, for the conditions that made it possible have not repeated. The space program has continued, and built on some of Apollo's legacies, but it has never had the national priority it enjoyed then. What is significant is that NASA took advantage of that priority, making the most of the historic confluence of political and technological circumstances.

The Apollo Decision and Its Impetus

The Apollo decision of May 1961 was a reaction to the large political forces then at play. The Soviet Union was clearly ahead of the United States in space, and Congress, media, and the public were deeply concerned. American pride and prestige were bent. President Kennedy, who won an election by promising to get the country moving again, was depressed. The space flight of Gagarin and the Bay of Pigs fiasco symbolized Kennedy's own frustration. He—and the country—needed something dramatic as an assertion of national will. NASA had conceptualized Apollo in its plans since the Eisenhower years. But there had been no match between what NASA wished to propose and what Eisenhower wished to receive. There was now such a match in the case of Apollo and Kennedy. Apollo solved the immediate need the President and the nation had for a bold action.[6]

James Webb, NASA administrator, understood that he had to use the impetus of the decision to maximize NASA's administrative and political advantages for the long haul. Prior to Kennedy's announcement, he had told Vice President Lyndon Johnson that reaching the moon would require political support throughout the decade. Webb knew that this was the primary factor in success or failure. Some people spoke of a national "commitment," but not Webb. He regarded the decision as a beginning, one that gave him a brief honeymoon period he could use to get Apollo off to a fast start.[7]

Before 1961 was over, NASA had let many of its most important contracts: the three Saturn rocket stages that would sequentially boost Apollo beyond Earth's gravity went respectively to Boeing, North American, and Douglas; the Apollo spacecraft to North American; and the Apollo guidance system to MIT. In addition, the decision was made to create a new Manned Spacecraft Center (MSC) and locate it in the area of Houston, Texas.

These decisions and others were all justifiable on technical grounds. However, they also had the effect of building a coalition of support for NASA in various regions of the country and among those legislators who represented them. Most notable was the decision to locate MSC in the Houston area, home of the Chairman of the House appropriations subcommittee responsible for NASA's budget, Representative Albert Thomas.

NASA leadership was indeed trying to accomplish multiple objectives, with coalition building being one of them. For example, NASA needed to work with universities to accomplish its objectives. Webb established a special program, Sustaining University Program (SUP), with broad goals. NASA had been accused by critics of taking scientists and engineers from other nationally-important endeavors. Webb reasoned that NASA would replenish the coffers through SUP, a program with a robust Ph.D. fellowship component. In addition, this program helped link potential critics in the nation's scientific community more closely to the space program. Winning support and neutralizing opposition were key to managing big science. NASA was a relatively weak and insecure agency before the President's Apollo decision. It had to be technically and administratively bolstered internally and externally afterward for the lunar mission. To be strong technically and administratively, it had to be strong politically.[8]

---

6. John Logsdon, *The Decision To Go To The Moon: Project Apollo and the National Interest* (Cambridge, MA: MIT Press, 1970).

7. For the Webb role in Apollo, see W. Henry Lambright, *Powering Apollo: James E. Webb of NASA* (Baltimore, MD: Johns Hopkins University Press, 1995).

8. For Webb's perspective on "Big Science," see his *Space Age Management: The Large-Scale Approach* (New York, NY: McGraw Hill, 1969).

NASA's Managerial Coalition

The coalition concept was applicable to NASA's management, used at the very top of the agency and extending outward. At the apex of NASA was a management "triad" of three men. First was Webb, whose background was law and administration. He had been Budget Director and Under Secretary of State in the Truman Administration. At NASA, Webb had overall responsibility, but particularly concentrated on external political relations: President and Congress. The Deputy Administrator, Hugh Dryden, was a long–time civil servant and leader of NASA's predecessor organization, the National Advisory Committee for Aeronautics (NACA). A physicist, Dryden was "Mr. Science" in the triad, and took special interest in NASA's international science activities. The third member was Robert Seamans, an engineer with both university and industry experience. He was Associate Administrator and "General Manager." There was thus a blending of skills at the top of NASA; the three men complemented one another and got along well. Webb was definitely the dominant personality, but he brought the other two into the most important decisions and used the triad to help his own credibility within the technical organization that was NASA. The three men presented a united leadership on decisions, a stance particularly strengthening the position of Seamans.[9]

Below the triad were the program managers. For Apollo, the key actor was the Director of the Office of Manned Space Flight (OMSF) an office created by the 1961 reorganization when NASA was changed to better match the Apollo priority. The OMSF Director had various subordinates. As OMSF developed, there were managers for the major manned projects: Mercury, Gemini, Apollo, and Apollo Applications (the planned interim follow-on to Apollo).

Then there came the centers—the huge operations that served as in–house laboratories and technical managers for most of the contracts. The three OMSF centers were Manned Spacecraft Center, Marshall Space Flight Center, and what came to be called Kennedy Space Center. The Manned Spacecraft Center focused on spacecraft development and astronaut training, Marshall on rocket development, with Kennedy Space Center being responsible for the actual launches.

This was the formal NASA organizational arrangement. Informally, NASA was composed of different cultures: the National Advisory Committee for Aeronautics culture, the German rocket culture led by Wernher von Braun, Director of Marshall, and the systems engineer culture which largely ran OMSF at headquarters. What united the aeronautics and rocket cultures were their affinity for "hands–on" technical work. They liked to perform research and/or build hardware, and had to be prodded to become contract managers. The systems engineers liked to pull men and machines together on a large scale. They were accustomed to contracting out and managing. NASA's systems engineers were mainly drawn from industry and the Air Force, specifically to meet the administrative demands of Apollo.[10]

Under the first NASA Administrator, T. Keith Glennan, the decision had been made to contract out most of NASA's work to industry and universities. Webb continued this pattern, and often pointed out that 90–95 percent of NASA's Apollo work was spent outside government.[11]

---

9. On NASA's leadership, see Lambright, *Powering Apollo*. See also Robert Seamans, *Aiming at Targets* (Beverly, MA: Memoirs Unlimited, 1994).

10. Howard McCurdy, *Inside NASA: High Technology and Organizational Change in the American Space Program* (Baltimore, MD: Johns Hopkins University Press, 1993).

11. Levine, *Managing NASA in the Apollo Era*.

Hence, the NASA organizational coalition included headquarters, centers, and contractors. It was diverse and competitive. What these elements had in common was a goal—to get to the Moon. They also had a NASA administrator, Webb, who believed that a management system had to have hierarchy, but also checks and balances. Thus, OMSF would have its own support contractors to give it technical strength to cope with centers and the major hardware companies. The centers would be kept strong institutionally as the technical core of NASA so they could deal confidently with industry. Worried that an imbalance could exist between NASA in–house expertise and industrial contractors in the electronics field, Webb established a new Electronics Research Center in Cambridge, Massachusetts.[12]

Webb supplemented this vast system with personal consultants to himself, men he called "scouts," who would rove around the NASA-industry-university system and give him early feedback on problems. The informal supplemented the formal chain of command. At its height, the NASA system included 400,000 governmental and non-governmental personnel, held together by mutual dependencies and the lunar vision.[13]

Technological Choice and the President as Coalition Member

The President was the most important member of the NASA external coalition. Kennedy's Apollo decision made him as dependent on NASA as NASA was on him. NASA's success or failure redounded to his own prestige. The debate over how to get to the Moon illustrates the relationship. It also shows how technological, political, and administrative factors converged in specific decisions regarding Apollo.

There had been enough internal studies for NASA to know that the lunar landing was within the realm of scientific and engineering feasibility. Indeed, Apollo was much more an engineering than scientific enterprise, as scientific critics would continually complain. However, the engineering of Apollo was technological development in the most daunting sense. There had to be substantial advances in rocketry, heat–resistant materials, and computers if NASA was to succeed. The unmanned spacecraft NASA would send to photograph lunar landing sites and ultimately land on the lunar surface also pushed the state of technical art. This was technology at the frontier of innumerable fields. Technical success was possible, but by no means assured, and the addition of man to the equation added a host of novel technical requirements and immense risk. Webb felt he had to shield his technical organization and contractors from political interference and financial instability to give them a fighting chance to succeed. In order to accomplish this, he needed the President on his side.

When Kennedy announced the Apollo decision, NASA did not know precisely what approach it would use to get to the Moon. There were three options. One was called direct ascent, via a gigantic new rocket to be developed, named Nova. A second, called Earth orbit rendezvous (EOR), entailed assembling equipment in the Earth's orbit to go to the Moon. The third, lunar orbit rendezvous (LOR), also involved assembly, but in lunar orbit. Direct ascent was soon rejected because such a rocket would take too long to develop. The contest was between EOR and LOR. Webb allowed the debate to rage within his agency, feeling that this was the most critical technical decision in Apollo and his agency

---

12. *Ibid.* See also Robert Rosholt, *An Administrative History of NASA, 1958–1963* (Washington, DC: NASA SP-4104, 1966).

13. Lambright, *Powering Apollo*; Leonard Sayles and Margaret Chandler, *Managing Large Systems: Organizations for the Future*, (New York, NY: Harper and Row, 1971).

had to be united behind it. Eventually, the agency went with LOR because it promised the most savings in weight over the total mission without adding significant costs.

This decision was challenged by the President's Science Advisor, Jerome Wiesner, largely on risk grounds. It became an early test of who was in charge of Apollo. Webb's view was that NASA had to prevail on such an important technical decision. The conflict went to the President who backed Webb, and thus NASA.[14]

In 1962-63, there were other disputes within NASA and between NASA and outside forces. Perhaps the most important internal dispute was between Webb and D. Brainerd Holmes, the head of the Office of Manned Space Flight, an executive recruited to run Apollo. In 1962, Holmes felt Apollo was falling behind schedule and needed a substantial infusion of funds. He asked Webb to go to Congress for a supplement to the money already provided. Webb was anxious to show Congress that when NASA presented a budget request, it was a credible number. Congress did not know much about the details of space; it had to trust that NASA was well-managed. Credible budgets were critical to the management image Webb wished to convey. Webb said "no." Holmes then asked Webb to take the money from less important parts of the NASA program. Again, Webb declined. Holmes took his case to the media. The dispute escalated, reaching the President, who again backed his administrator. Holmes soon departed NASA.[15]

In gauging the management factors critical to the success of Apollo, there is no underestimating the important role Kennedy played as a supportive member of NASA's implementation constituency. While Webb was not a member of Kennedy's inner circle (as was Robert McNamara, Secretary of Defense), he was a man Kennedy regarded highly for his accomplishments. Kennedy was anxious for NASA to succeed, and believed he had a good administrator in Webb. The NASA Administrator told Kennedy that if they worked together, NASA would succeed, but if they did not, he could not guarantee that would be the case. Kennedy chose to stick with Webb. However, Kennedy also hurt Apollo and Webb at one point in 1963 when he announced that instead of competing, the U.S. and U.S.S.R. might cooperate in space. This brought a negative reaction from NASA critics in Congress who were anxious to cut the agency's budget. Webb worked intensely and closely with Rep. Thomas to turn back this assault. NASA's budget went up over what it had been the year before, but the requested raise was reduced substantially. Meanwhile, the Kennedy U.S.S.R. initiative did not go anywhere, and the race to the Moon continued.

NASA and the Defense Department

Getting to the Moon required the cooperation of the Department of Defense (DOD), and for the most part, NASA received the cooperation it sought. However, the Air Force was a rival of NASA for space programs and Defense Secretary Robert McNamara, while not a "space buff," saw space as a place where DOD had an appropriate role, at least in regard to using near-Earth orbit for manned reconnaissance.

McNamara wanted control of the Gemini program, which could give DOD the capability he sought. Gemini had been formulated after the lunar decision to fill a technological gap between Mercury and Apollo. Mercury ended in 1963. Apollo flights were scheduled to commence in 1967. Gemini (which carried two men) would be more complex than Mercury (which carried one man) and lead the way to Apollo (three men). With the LOR decision, Gemini was critical to NASA for learning how to operate in space and developing docking

---

14. See accounts of this dispute in Murray and Cox, *Apollo: The Race to the Moon*, and Seamans, *Aiming at Targets*.

15. See note above.

and rendezvous techniques. Finally, Gemini was important in keeping the media and general public part of the NASA coalition. It would show activity in the critical middle years of the lunar program, between Mercury and Apollo flights, and keep NASA before the public eye.

Webb and McNamara met, with Webb determined to hold the line on control. McNamara, who had to fight various battles on other bureaucratic fronts he accorded higher priority, backed off. Webb, for his part, compromised by permitting DOD experiments to be carried by Gemini. This agreement symbolized the basic relationship on space where NASA and DOD were concerned: NASA was the senior partner, even though DOD was the more powerful agency. DOD was subtly enlisted in the NASA coalition, largely on Webb's terms, and gave NASA important logistic support. It also supplied a number of key managers to Apollo after Holmes left in 1963, including Air Force General Sam Phillips. Phillips was appointed director of Apollo, reporting to George Mueller, head of the Office of Manned Space Flight.

President Johnson as Coalition Member

After Kennedy was assassinated and Lyndon Johnson became President in November 1963, NASA continued to have an ally in the White House. As Vice President, Johnson had been a strong advocate of NASA's going to the moon. This support continued through his presidency.

However, Johnson also wanted to build a Great Society and Apollo seemed far afield from this new national priority. Not so, said Webb, who argued that the space program was fully part of the Great Society, indeed embodying its deeper meaning. Webb's rhetoric soared as he described the Sustaining University Program as showing how investments in space could pay off in enhanced technological spinoff on Earth, how regional economic development and educational advancements could transform domestic America into a "Space Age America."[16] Johnson was elated and had other agencies look to NASA as a model for linking technology, education, and economic development through university-based science centers throughout the country.[17] Going to the Moon and creating a Great Society were linked rhetorically and strategically by Webb for LBJ.

Johnson also wanted to use the space program to project his image as a man of peace. He sent astronauts to foreign countries as ambassadors of good will. Johnson increasingly looked for ways to improve his peacemaking image. Vietnam, a relatively modest confrontation when Johnson came into office, was escalating steadily by 1965. He hoped his association with civilian space would counter some of the negative publicity Vietnam brought him.

Catching Up to the Soviet Union

What really helped NASA with Johnson, Congress, and the American people was the step–by–step, highly visible success of Apollo. The first Collier award for the lunar landing program came in 1965 when Gemini (firmly identified as a NASA program) was achieving one spectacular flight after another. Television was now capable of enhancing the space program through images transmitted from space to each American's living room. The

---

16. Walter A. McDougall, . . . *the Heavens and the Earth: A Political History of the Space Age* (New York, NY: Basic Books, 1985).

17. W. Henry Lambright, *Presidential Management of Science and Technology: The Johnson Presidency* (Austin, TX: University of Texas Press, 1985).

American people participated as space technology advanced and men and machines seemingly worked to perfection. The coalition behind the lunar landing program extended to the media and general public.

Gemini was intended to be a technological bridge between Mercury and Apollo, and that it was. Major lessons were learned and transmitted by Gemini about rendezvous and docking in space, about human beings operating outside of their craft ("extra-vehicular activity"). NASA learned that men could live up to two weeks in space and how astronauts could work with operators "on the ground, in the control room, around the tracking network, and in industry."[18] Gemini advanced technology in propulsion, "fuel cells, environmental control systems, space navigation, spacesuits, and other equipment. In the development stage of Apollo, the bank of knowledge from Gemini paid off in hundreds of subtle ways. The bridge had been built."[19]

Gemini, of course, was not without mishaps, but NASA seemed capable of turning problems into opportunities.[20] All this contributed to an image of computer-guided managerial efficiency. Webb had been using the rhetoric of management innovation and calling NASA the best managed agency in Washington. Many observers believed Webb. For Webb, the words were not mere rhetoric. They were gospel, and he invited management scholars to come into NASA and observe for themselves.

Many, including Johnson, also believed that the U.S. was catching up to the Soviet Union, maybe even surpassing the rival. Something had gone wrong in the Soviet program in the mid-1960s. Its leading technological genius had died and there was no one immediately able to pull the factions of the Soviet space enterprise together. The U.S.S.R. was no longer demonstrating a vitally active program. It seemed to be in trouble. In contrast, the U.S. lunar landing program was in full thrust. By 1966, Gemini was proving to be everything its initiators hoped it would be: a great technical learning experience, confidence builder, and public relations tool.

Also, in the same year, NASA sent its first surveyor spacecraft to "soft land" on the Moon. The Ranger program was providing photographs of the Moon. Surveyor built on this and engaged in televised digging into the lunar surface. There had been some scientific speculation about the risks of a lunar landing by a relatively heavy manned spacecraft. Surveyor was developed to precede astronauts to the Moon. It utilized unique machinery and had to function almost perfectly to succeed. On June 2, 1966, the first Surveyor landed on the lunar surface. Surveyor was a probe able to show "that lunar soil was the consistency of wet sand, firm enough to support lunar landings by the lunar module."[21] In 1967, the technical team responsible for Surveyor was awarded a Collier Trophy for their efforts.

The only drawback to these successes was that the Apollo program appeared to be going so well the Americans were getting complacent. This complacency—reflected by the President and Congress—did not impact negatively on Apollo funding. However, it did affect NASA's drive to acquire funds for long lead time items beyond landing on the Moon. Johnson, in particular, wanted to delay post-Apollo decisions. In late 1966, the President finally conceded to a modest effort in "Apollo Applications" that constituted an interim program to keep a production line of Saturn rockets and Apollo spacecraft going until a bigger decision (for a Mars trip or Space Station) was possible.

---

18. Roger E. Bilstein, *Orders of Magnitude: A History of the NACA and NASA, 1915–1990* (Washington, DC: NASA SP-4406, 1989), p. 78.
19. *Ibid.*
20. Barton C. Hacker and James M. Grimwood, *On the Shoulders of Titans: A History of Project Gemini* (Washington, DC: NASA SP-4203, 1977).
21. Bilstein, *Orders of Magnitude*, p. 82. Howard E. McCurdy, *Inside NASA: High Technology and Organizational Change in the US Space Program* (Baltimore, MD: Johns Hopkins University Press, 1993), p. 64.

The Apollo Fire

Thoughts about post–Apollo were put on hold in January 1967, when the entire Apollo effort was threatened.[22] A fire ignited in the Apollo spacecraft, while it sat on its platform at Kennedy Space Center, Florida. Three astronauts were killed. Under the NASA system of management this should not have happened. Indeed, there had been a warning about the fire danger issued to Joe Shea, Apollo spacecraft manager in Houston, by a technical support contractor. But the warning was not heeded. In the wake of the fire, a nationwide furor erupted. As the space program had become an icon because of its seeming perfection, it now was questioned unmercifully by the national media. The coalition that had brought NASA to this point was in danger of unraveling.

No one was more cognizant of Apollo's political vulnerability than Administrator Webb. He determined that NASA had to find out what went wrong, fix the problem, and get back into space. All of this had to be accomplished in a manner that kept the coalition of supporters together.

Johnson held firm to the coalition, granting Webb's request that NASA be allowed to investigate itself. Congress agreed to hold off its investigation until the NASA inquiry was complete. Between January and April 1967 the NASA investigating team did its work, completing a report that castigated NASA and the spacecraft's prime contractor, North American, for shoddy engineering and carelessness. The fire was most likely caused by an exposed wire. Once the fire started, it could not be stopped because the spacecraft was filled with highly combustible materials and an all-oxygen atmosphere. Moreover, the door of the capsule opened from the outside, adding to the difficulties of escape. The media and Congress accepted the report's credibility, but questioned the Apollo timetable and whether the haste to get to the Moon led to shortcuts and thus caused the deaths.

The subsequent congressional inquiry went beyond the NASA investigation, which was primarily technical, to probe the NASA-North American relationship. A "Phillips Report" from 1965 had surfaced in which NASA Apollo manager Sam Phillips had sharply criticized North American's work and left at least the implicit threat of going to another contractor if North American did not improve its performance. Congress wanted to know more about the Phillips Report and any other internal studies of NASA-contractor problems.

Webb himself had been unaware of the Phillips Report and blamed Seamans for not better alerting him to problems. Seamans had become Deputy Administrator in 1965 after Dryden's death, while retaining his position as "General Manager." Exceedingly busy, Seamans relied on OMSF Director Mueller for information, and Mueller did not want interference from above. Mueller thought he had taken care of the North American issue in 1965–66 and that the work was back on target. As information had moved up the line from Phillips to Mueller, to Seamans, to Webb, the basic message of the Phillips Report had become increasingly sedated. Angry at both Seamans and Mueller, Webb was taking firm control now. He made "surgical" changes in NASA—replacing a few key people who had clearly made mistakes, most notably Shea with George Low, Deputy Director of the Manned Spacecraft Center. He also "supplemented" Seamans with Harold Finger, in topside management, and moved the individual who worked on budgeting for Mueller under his own wing. Webb's intent at this time of crisis was to manage Apollo much more closely and personally. Finally, he forced North American to replace its principal space manager with another individual, and ordered the President of North American to take greater personal responsibility for Apollo himself. To show he meant business, Webb brought Boeing

---

22. For an account of how NASA coped with the Apollo fire, see Lambright, *Powering Apollo*, pp. 142–88.

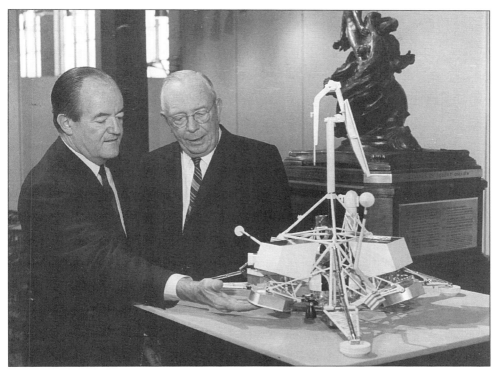

Lawrence A. Hyland received the Trophy in 1967 "representing the Surveyor Program Team at Hughes Aircraft Company, the Jet Propulsion Laboratory, and associated organizations that put the eyes and hands of the United States on the Moon." Hyland and Vice President Humphrey are shown with a model of Surveyor. (NASA photo).

aboard to supervise the integration of North American's work on the rocket and spacecraft with that of other contractors.

In the congressional hearings that followed the NASA investigation, however, Webb was evasive about the Phillips Report and his actions in connection with North American. Webb did not want Congress to get involved in his negotiations with North American. He felt that NASA should manage Apollo, not Congress, and that there were those in Congress who were anxious to use this moment to assert control over the space program.

Congress was suspicious of how North American got the lucrative spacecraft contract in the first place. Was "politics" involved? What about the rumors of Bobby Baker, Washington wheeler-dealer and one-time aide to Johnson, and his involvement in the deal? More and more, Webb became the target of the investigation, and long-time antagonists of Webb, including Senator Clinton Anderson, Chairman of the Senate Space Committee, looked for the smoking gun that would get Webb out of NASA.

No smoking gun could be found. Important elements of the NASA coalition held together. Frank Borman, speaking on behalf of the astronauts, said that they had confidence in NASA management. Webb's allies in Congress, especially the most influential Republican on the space committee, Senator Margaret Chase Smith, stuck with him. The congressional investigation petered out as the summer of 1967 came on, with NASA

The Surveyor program received the Collier Trophy in 1967 together with the Hughes Aircraft Company, the Jet Propulsion Laboratory, and associated organizations which put the successful series of spacecraft on the Moon, and facilitated the success, in 1969, of the Apollo lunar landing of humans on another planet. This mock-up of the Surveyor spacecraft was taken in 1966. (NASA photo no. 66-H-476).

promising to keep Congress better informed of emerging problems in the future. Webb did show Congress the Phillips Report and other internal documents—but on his terms, in closed session.

Of course, many legislators had already seen the report via leaks. A great deal of posturing and appeals to "principle" were involved. Webb interpreted his job as manager at this point to shield his agency and, if need be, even North American from excessive congressional strictures. He succeeded but at considerable cost to his own credibility with Congress and the media. He expended much of his political capital, and the coalition weakened—but it continued to function. Then, in November 1967, just 10 months after the Apollo fire, the first Saturn moon–rocket was launched. It was a great success, renewing confidence in NASA among doubters, and proving the wisdom of what was termed the "all–up" decision. This was a decision Mueller had earlier made, and Webb backed, to save considerable time by testing various components of the Saturn system all at once, in the first launch, rather than incrementally, as von Braun's team preferred. NASA had taken a large risk, and it paid off. The Apollo program could now recoup lost time.

Success

The next two Collier awards went to *Apollo 8* and *Apollo 11*. The former was the first circumlunar trip and the latter the actual lunar landing flight. Leading up to *Apollo 8* were a sequence of unmanned flights capped by the first manned flight, *Apollo 7*, in October 1968. Astronauts flew in a spacecraft that had been significantly redesigned to make safety changes in the wake of the Apollo fire. These included a new escape hatch, fireproof materials, and better distribution and protection of flammable materials. The spacesuits were made virtually fireproof and changes were made in the spacecraft's atmosphere to enhance safety.

A moving force in redesigning *Apollo 7* was George Low.[23] Low had been Deputy Director of the Manned Spacecraft Center in Houston. He was persuaded to head the spacecraft development office in Houston with the mission of redeeming NASA's reputation following the Apollo fire. In 1968, while working on *Apollo 7*, Low realized that the lunar landing module that was to be developed and tested on the scheduled *Apollo 8* flight later that year would not be ready. This could mean delays and ultimately missing the Kennedy deadline, unless a flight scheduled after that was moved ahead. The more advanced circumlunar flight would have to become *Apollo 8*. Flying men around the Moon was an extremely bold decision, not only technically, but psychologically. But Low believed the technical risks were acceptable since the rocket and basic spacecraft would be fully tested together in *Apollo 7*. Robert Gilruth, Director of the Manned Spacecraft Center, and von Braun agreed with Low.

Moreover, there was strong evidence, much of it classified, that the Soviet space program had been revived, and was pushing ahead again. The sense of competition burned deeply within NASA. The Soviet Union might well be gearing for a circumlunar flight and Low and others wanted to achieve this first.

What would Headquarters say? A key decision maker was Tom Paine, Deputy Administrator, replacement for Seamans, who had departed NASA at the beginning of the year, a casualty of the Apollo fire and deteriorating relations with Webb. Paine supported a circumlunar decision. Webb, abroad at the time, was contacted by telephone, and at first inclined to say no. Chastened by the fire, he saw the stakes as the nation's support for

---

23. See Murray and Cox, *Apollo: The Race to the Moon* for a description of the *Apollo 8* decision process.

The flight crew of the Apollo 8 *mission, Commander Frank Borman, Command Module Pilot James A. Lovell, Jr., and Lunar Module Pilot William A. Anders, hold a replica of the Collier Trophy awarded for their historic first flight to the Moon. (NASA photo no. 69-H-914).*

Apollo if anything went wrong. Never before had an attempt been made to send men the vast distance from near-Earth space to the Moon. On the other hand, he was aware that unless NASA took the risk, the Apollo goal might not be reached. Webb soon came around and gave a guarded decision to move ahead with the planning. Low had built a coalition within NASA for a major decision that was critical to NASA's Apollo schedule.

Apollo 7 proved successful and the stage was set for *Apollo 8*. When the *Apollo 8* flight took place, and astronauts went around the Moon on Christmas Eve 1968, the effect on the country was almost magical. This had been a dreadful year: Vietnam had taken a turn for the worse with the Tet Offensive; Martin Luther King and Robert Kennedy had been assassinated; there had been riots in Washington, DC; and Lyndon Johnson had gone on television to say he would not run for reelection. The country seemed to be coming apart. But on Christmas Eve, a quarter million miles from Earth, three brave men went around the Moon, and read from the book of Genesis. A divided country came together, at least for a while.

Next came a sequence of manned flights, equipment testing, and maneuvers in Earth orbit (*Apollo 9*) and lunar orbit (*Apollo 10*). Finally, *Apollo 11* was launched July 16, 1969. As *Apollo 8* had united the country, *Apollo 11* brought the world together—one fifth of the planet's population reportedly witnessed the moment, four days later, when Neil Armstrong took

A spectacular view of the rising Earth which greeted the Apollo 8 astronauts as they came from behind the Moon after the lunar orbit insertion burn. This view was also used for a United States postage stamp issued in 1969. Additionally, the views of Earth taken by Apollo 8 and subsequent astronauts are credited with giving visual stimulation to the environmental movement.

"one small step for [a] man, one giant leap for mankind."[24] As the three astronauts splashed down safely in the Pacific four days later, Kennedy's 1961 challenge was met.

At the helm of NASA at this point was Paine; Richard Nixon was in the White House. Webb left NASA the previous October, in part to give Paine a chance to show success in the remaining 1968 flights. Conscious of the disruption in momentum a presidential transition could cause, Webb wanted to keep as much of the NASA management team together after January as possible. He wished to enlist the new President in the Apollo coalition. Without Webb, that was more likely. The best way to get Johnson's successor aboard was to "depoliticize" the agency. Nixon inherited Paine and kept him as Administrator. The coalition behind Apollo—minus Webb and Johnson—carried out the remarkable feat of *Apollo 11*.

Voyages to the Moon

The final lunar landing program Collier award came for *Apollo 15*, described as "man's most prolonged and scientifically productive lunar mission," culminating a series of voyages whose intent increasingly differed from *Apollo 11*. The Apollo program had focused on technology development up to 1969. After the first lunar landing, the mission of succeeding flights shifted increasingly to acquiring scientific data about the lunar surface. NASA was trying to get scientists and engineers within the agency and outside to work in closer harmony. This part of the NASA "working coalition" was difficult to assemble, but it was essential that scientists and engineers cooperate to make the most of the lunar voyages.[25] The scientific

---

24. Bilstein, *Orders of Magnitude*, p. 91.
25. For voyages to the moon, see ibid., pp. 98–100; and William David Compton, *Where No Man Has Gone Before: A History of Lunar Exploration Missions* (Washington, DC: NASA SP-4274, 1989).

community had not been enthusiastic about Apollo, although a space science community had been built from the hundreds of millions of dollars NASA spent in universities in the name of Apollo in the 1960s. Now, however, NASA needed scientific support and help because scientists were in many ways users of the technological capability now in existence.

The voyages to the Moon lasted from 1969 to 1972 and were designed to learn more about the Moon. *Apollo 12* essentially repeated the *Apollo 11* journey but at a different lunar landing site. *Apollo 13*, launched April 11, 1970, was the flight that almost resulted in the first death in space when an oxygen tank ruptured, causing serious damage to the spacecraft. The trip to the moon was aborted and re-routed; the lunar module was used as a temporary "lifeboat." Through outstanding technical ingenuity on Earth and in space, the three astronauts made it safely back to Earth.

Apollo 14 lifted off January 31, 1971, and began more extensive scientific exploration of the Moon. A special cart was used to acquire rock samples and bring them back to Earth. Then came *Apollo 15*, launched July 26, 1971, which won the Collier award. *Apollo 15* demonstrated the introduction of the lunar rover, an electric-powered, four-wheel drive vehicle, developed at a cost of $60 million. Using the Rover, astronauts roamed far and wide beyond their immediate landing site, observing lunar features and collecting rock samples. They covered seventeen miles of lunar surface during their visit, taking photographs of the craters and ravines. Because of the Rover, they conserved their energy and doubled the amount of time astronauts were able to stay on the Moon.

Not surprisingly in 1969, Neil A. Armstrong, Michael Collins, and Edwin E. "Buzz" Aldrin, the crew of Apollo 11, received the Collier Trophy. The award honored the crew for their high courage and stunning success in the accomplishment of one of history's most spectacular adventures—the first manned Moon landing. Frederick B. Lee, then president of the National Aeronautic Association, presents the award to Michael Collins and "Buzz" Aldrin. (NASA photo no. 70-H-772).

A view of astronaut footprints in the lunar soil at Tranquillity Base, landing site of Apollo 11 *in July 1969. (NASA photo no. 69-H-1258).*

In April 1972 came *Apollo 16*, and on December 7 *Apollo 17* was launched, the last manned flight to the Moon. Both missions were scientifically productive, with *Apollo 17* being a harbinger of the future in having a professional geologist, Harrison Schmitt, as a member of the crew.

With the exception of *Apollo 13*, the other flights did not have the dramatic impact of *Apollo 11*. Americans had different priorities now. Great social change had taken place over the years since the Apollo mission began. Neither government nor technology were in favor. A conservative regime was in the White House, and an anti-technology counterculture in the universities. The Vietnam war soured everything. Hugely expensive programs like Apollo seemed to many an embarrassing frill.

The space race was over; America had won. With no post-Apollo decision comparable to the Kennedy choice politically possible, Nixon in 1972 selected Space Shuttle, his minimal manned space option, to keep NASA going. The space program, whose budget had begun declining in 1966, had only one-third the buying power in the 1970s it had in its heyday. The coalition behind Apollo had declined and was now disintegrating.

The last Apollo flight was the *Apollo–Soyuz* mission of 1975. As Cold War competition had launched Apollo, so a thaw in the Cold War brought the world's two space powers together for a meeting in Earth's orbit. Apollo ended and moved into history.

Conclusion

The lunar landing program was one of the great technological successes in history. It garnered five Collier awards, all of which were well deserved. Recognized were Gemini, Surveyor, and *Apollos 8, 11,* and *15.* While the awards cited specific individuals and achievements, they were really for an entire program and all those associated with it. This chapter has focused on the management of Apollo, specifically NASA. In the language of actor-network theory, NASA was the actor that established local and global networks (i.e., built and maintained both the working and political coalitions) to carry out the mission. Within NASA, the leaders of the agency did what actor-network theory suggests is essential for technical success, becoming an "obligatory point of passage" for decisions affecting the course of the program.[26]

What other factors were critical to success? How do these relate to central historical trends facing the agency, cited in this book's introduction, such as: (1) the growing web of bureaucratic and political obligations, (2) the increased complexity of R&D and disconnection between technology developers and users; and (3) changing attitudes towards funding?

Apollo did not succeed because of a mystical national "commitment." Such a consensus lasted but a moment in time, but it gave NASA leaders a year or two to procure major contracts, reorganize, found the Houston center, hire key managers, and launch a massive team of organizations capable of taking America to the Moon. The commitment to an ongoing coalition building and maintaining process was NASA's. Webb and his associates created a "lunar landing coalition" across Congress, the Executive Branch, and interest group constituents. Combined with working arrangements involving government, industry and universities, these political and administrative alliances were key to Apollo's technological success.

Strategies to mobilize such a huge coalition began in 1961 with the critical center and contractor decisions: choices not only of "who," but "where." Such strategies included the Sustaining University Program, which sought in part to neutralize scientific critics and win their favor for the space program. They continued in the mid-1960s, when NASA's Cold War rhetoric was supplemented by the rhetoric of the Great Society. Throughout, NASA management stayed in charge of the countless bureaucratic and political forces impinging on it, fending off challenges external and internal to its authority, thereby keeping an integrity to NASA decision-making and leadership in space policy. Webb created an "apolitical shield," using his political skills to insulate NASA's technical core from the political pressures of others.[27] This was a relative autonomy that did not survive much beyond Apollo.

Apollo engaged 400,000 people from government, industry, and universities at its apogee. With the Apollo deadline as a discipline on all parties, NASA leaders stressed management excellence and backed rhetoric with clear-cut technological success, gradually overtaking the Soviet Union in space feats. The technological coalition had outstanding personnel who worked with zeal, insulated against political disruptions. Internal struggles over R&D priorities among programs and Centers, and between engineering developers and scientific users, were minimized by the unmistakable primacy Apollo possessed.

Whatever else it was, Apollo was a giant technological development program. Developing the technology to go to the Moon took precedence over other aspects of the space program. Scientists might not have liked these priorities, but they knew what the priorities were and for a long time—until NASA reached the Moon—there was little ambiguity about NASA's mission. Having a clear goal was both a factor in success and a "connect" for the disparate parties of NASA's technical, political, and administrative system. Ranger

---

26. Law and Callon, "Engineering and Sociology in a Military Aircraft Project," p. 90.
27. The concept of "apolitical shield" is developed by Eugene Lewis, *Public Entrepreneurship: Toward a Theory of Bureaucratic Political Power* (Bloomington, IN: Indiana University Press, 1980).

and Surveyor were science in support of the Apollo goal. The SUP provided additional funds for space science. These, however, were possible only because of the larger support Apollo had, and many academic scientists realized this reality.

Then came the brief scientific use in the early 1970s of the technological capability that had been so arduously developed in the 1960s. The lunar voyages added enormously to the stock of scientific knowledge about the Moon, as users gained a measure of reward for their long wait. In succeeding years, however, the interests of developers and users would diverge sharply and eventually reveal outright competition.

NASA leaders were fortunate that they had moved quickly and adroitly enough to make visible progress toward the Moon by the mid-1960s. Gemini proved a political as well as technical link between Mercury and Apollo. There was thus an impetus in the latter 1960s to complete what had been started in 1961. However, the coalition supporting Apollo eroded steadily as the Great Society and then Vietnam changed national priorities and public attitudes toward funding large-scale science and technology. What was possible to launch at the outset of the decade was not possible at its conclusion as NASA learned when it sought to sell a post-Apollo program. This advocacy process wound up with the Space Shuttle decision in the early 1970s. Considerations of cost-benefit were influential in the shuttle decision, nonexistent when Kennedy decided to go to the Moon. It was a new era in terms of public and political attitudes toward funding R&D. Reaching consensus within NASA and among NASA and external forces would become steadily more difficult.

Nevertheless, what NASA demonstrated through Apollo was that great achievement by government in alliance with the private sector is feasible where leadership is present and political and technological conditions are ripe. Occasions that make an Apollo possible are rare, perhaps singular. But other opportunities can arise. When they do, and individuals and organizations coalesce around a clear goal, a nation can rise to awesome challenges.

The crew of Apollo 15, *Col. David R. Scott, USAF; Col. James B. Irwin, USAF; and Lt. Col. Alfred M. Worden, USAF, and Johnson Space Center director Robert T. Gilruth, received the Collier Trophy in 1971 for the most prolonged and scientifically productive mission of Project Apollo, and for demonstrating superb skill and courage. Astronaut Irwin is shown saluting the U.S. flag at the* Apollo 15 *landing site on the Moon, July 1971. (NASA photo no. 71-H-1414).*

Chapter 9

The Human Touch: The History of the Skylab Program

by Donald C. Elder

On February 8, 1974, astronauts Gerald P. Carr, Edward G. Gibson, and William R. Pogue, after an eighty-four-day mission in outer space aboard an orbital laboratory named *Skylab*, boarded an Apollo command module and returned to earth. Their splashdown marked the end of a venture involving three separate crews that had set twenty-five International Aeronautical Federation world records and had managed to complete ninety major scientific experiments.[1] It therefore surprised no one when two months later the Collier Committee announced that it had selected the *Skylab* program as the recipient of the 1973 Robert J. Collier Trophy. The official announcement of the award duly noted the importance of "the production of data of benefit to all the people on Earth," but had prefaced that praise by asserting that *Skylab* had proved "beyond question the value of man in future explorations of space."[2] The *Skylab* program, then, had a great immediate impact in aiding the expansion of scientific knowledge, but also offered the long-range benefit of demonstrating the importance of the continuance of the human component in the American space program.

This essay examines the history of *Skylab*. The story begins with the first suggestions for a laboratory in outer space, then turns to a discussion of how such an idea gained official acceptance from the National Aeronautics and Space Administration (NASA). Each of the missions will be considered, and the results they yielded will be assessed. Throughout the story, I will analyze how various groups competed with each other to control the design, timing, and function of the project. Although stressing the role technological innovation played in the eventual success of the *Skylab* program, I will demonstrate how the human element was crucial at every stage of this "exceedingly complex enterprise."[3] Finally, I will suggest the ways in which the program resonated with, and still influences, the goals and objectives of NASA.

The Promise of a New Day

The idea of placing a vessel with a human crew into Earth orbit first appeared in 1869 when an American, Edward Everett Hale, wrote a short story about launching a brick structure large enough to house a crew of thirty-seven into outer space. During the first half of the twentieth century individuals from the scientific community, including Konstantin Tsiolkovsky and Hermann Oberth, took up the subject. The possibility of placing such an object into orbit seemed remote at that time, however, due to the absence of a viable launch vehicle.[4]

---

1. W. David Compton and Charles D. Benson, *Living and Working in Space: A History of Skylab* (Washington, DC: NASA SP-4208, 1983), pp. 379–386.
2. *Astronautics and Aeronautics, 1974* (Washington, DC: NASA SP-4019, 1975), p. 95.
3. Ibid.
4. John M. Logsdon, "Space Stations: A Historical Perspective," in Mireille Gerard and Pamela W. Edwards, eds., *Space Station: Policy Planning and Utilization* (New York, NY: IEEE, 1983), p. 14.

The Skylab space station cluster in orbit. Note the solar shield, deployed by the second crew, which shades the Orbital Workshop. The OWS solar panel on the left side was lost on launch day. NASA's Skylab program, with special recognition to program director William C. Schneider and the Skylab astronauts, received the Collier Trophy in 1973, for the production of scientific data about long-term space flight. (NASA photo no. 74-H-98).

    The prospects for an orbiting Space Station improved remarkably with the success of the German V-2 rocket program during the second World War. In the postwar years, the United States Army combined German technology with work done during the war by the Jet Propulsion Laboratory to create successively more powerful rockets. Aware of the progress in this area, individuals from a number of countries soon made detailed proposals to take advantage of the launch capabilities rockets now offered. But in spite of the fact that launch vehicles by the mid-1950s had demonstrated great potential for lifting payloads into outer space, no government proved willing to commit itself to backing a venture to place an object into Earth orbit. A manned orbital laboratory therefore remained a distant goal.

    But official interest in the concept of a Space Station soon came in the wake of the launching of *Sputnik I* in October 1957. To develop and guide a systematic American response to the Soviet accomplishment, in 1958 Congress passed a law which created a civilian space agency, the National Aeronautics and Space Administration. Shortly after NASA became operational, T. Keith Glennan, the agency's first administrator, announced that the United States would launch a person into outer space as part of its program of operations.[5] Encouraged by this development, certain individuals at NASA soon began to explore the possibility of placing a Space Station and operations crew into Earth orbit.

---

    5.   For a full treatment of Project Mercury, see Loyd S. Swenson, Jr., James M. Grimwood, and Charles C. Alexander, *This New Ocean: A History of Project Mercury* (Washington, DC: NASA SP-4201, 1966).

This groundswell of interest soon found support from the higher echelons of the space agency. Appearing before the Senate Committee on Aeronautical and Space Sciences on February 20, 1959, NASA Deputy Administrator Hugh L. Dryden and Assistant to the Director of Space Flight Development DeMarquis Wyatt discussed the missions the agency eventually hoped to accomplish; a Space Station and an orbital manned laboratory figured prominently in their plans.[6] Indeed, NASA would request two million dollars for fiscal year 1960 to conduct a feasibility study regarding the building of a space laboratory. By the spring of 1959, NASA had definitely embraced a concept which two years earlier had seemed quite remote.

Other groups also began to express an interest in the idea. As part of a program to investigate the possibility of creating a military base on the moon, Wernher von Braun of the Army Ballistic Missile Agency, in June 1959, suggested that a necessary step in such a venture would involve first building an orbital Space Station. As a "quick fix" to accomplish this first step, von Braun proposed using the final stage of a launch vehicle which had achieved orbit as the foundation upon which to build such a vessel. Although the Army would never implement such a program, von Braun's idea would figure prominently in shaping the thinking about the configuration of *Skylab*.[7]

One month after von Braun submitted his proposal, a meeting took place at NASA's Langley Research Center to define the goals of the space agency for the orbital space laboratory that Dryden and Wyatt had testified about in April. At this conference, held on July 10, 1959, the participants agreed that such a venture should serve three purposes. First, it should allow scientists to study "the psychological and physiological reaction" of a human being in outer space over a long period of time. Second, it would permit technicians to analyze how materials, power sources, and control mechanisms would function "in a true space environment." And third, a space laboratory would provide a test of "communication, orbit control, and rendezvous" techniques, and would allow the evaluation of a person's ability to gather terrestrial and astronomical information while in orbit.[8]

The minutes of this conference clearly indicate that by the end of 1959 NASA had taken significant steps toward creating a program to place a laboratory and crew in orbit. But the record also shows the agency recognized even at that early date that landing a person on the Moon might rank as a higher priority program in the American space program. Indeed, the participants at the July 10 meeting noted that they envisioned the space laboratory "as one of the initial steps in the actual landing of a man on the moon in 10-15 years."[9] This focus suggests that from the beginning the agency saw an orbital laboratory as merely a component of a larger mission, rather than as a program which could stand on its own technical merit.

This evaluation of the potential priorities of the space agency seemed to be born out in the next three years. In 1960, NASA Administrator T. Keith Glennan favored making a lunar expedition the priority of his agency after the completion of Project Mercury; in May 1961, President John F. Kennedy announced his goal of placing a person on the Moon by the end of the decade. Wernher von Braun and others argued for adopting a

---

6. U.S. Congress, Senate, NASA Authorization Subcommittee of the Committee on Aeronautical and Space Sciences, *NASA Supplemental Authorization for Fiscal Year 1959; Hearings on S. 1096*, 86th Congress, 1st Session, 1959, pp. 46, 81.
7. Roland W. Newkirk, Ivan D. Ertel, with Courtney G. Brooks, *Skylab: A Chronology* (Washington, DC: NASA SP-4011, 1977) p. 9.
8. Beverly Z. Henry, Jr., Aeronautical Research Engineer, Memorandum for Associate Director, October 5, 1959, *Skylab*, Box 06,, File Skylab/AAP Documentation 1959, NASA Historical Reference Collection, NASA History Office, NASA Headquarters, Washington, DC.
9. *Ibid*. The prediction of the participants about the earliest possible date for a lunar landing was remarkably prescient.

mission configuration which would use an orbital vehicle as a staging base for a lunar mission; such a vessel, they reasoned, could also serve as a space laboratory between lunar missions.[10] In June 1962, however, NASA officials opted for a plan involving a command module placed in orbit around the Moon. Such a scheme did not require a vessel and crew as a way station in Earth orbit, thus effectively shelving von Braun's concept for at least the duration of the lunar program (known as Apollo).[11]

Although disappointed by the decision, proponents of a permanent laboratory in space remained hopeful that NASA would eventually embrace that concept, and continued to plan for such an eventuality. On October 17, 1962, Joseph F. Shea, Deputy Director for Systems, Office of Manned Space Flight, asked the agency's centers to submit opinions on how NASA could use and benefit from an orbital vessel capable of sustaining a crew. The Manned Spaceflight Center (MSC) quickly responded to Shea's request, putting together a proposal detailing the areas on which private contractors would have to do studies to determine the feasibility of such a venture. By March 4, 1963, work completed in this area allowed Hugh Dryden to testify before Congress that a manned laboratory in Earth orbit had become an "obvious candidate" for a place in the space agency's program (named the Apollo Extension System) after the completion of Apollo.[12]

Reflecting the optimism of the deputy administrator, the MSC in June 1963 engaged the Douglas Aircraft Company and the Boeing Company to do studies on possible configurations for such a vessel. Additional support for the concept of an orbital space laboratory came in October 1963, when officials from the Department of Defense and NASA agreed to explore the possibility of jointly developing plans for such a vessel. Although nothing ever came of this proposal, it did aid NASA officials in more clearly defining their aspirations regarding such a program.[13]

Up until this point, all of the work done on projects to succeed Apollo had a classification of "advanced study programs." But these feasibility studies and reports, suggesting the potential benefits from such efforts, soon began to change the perspective of the space agency. Moreover, the realization that without concrete plans in the very near future much of the Apollo workforce would be idle for a significant period of time prodded the NASA hierarchy to consider immediate steps to provide continued work for them. Accordingly, in August 1965 NASA officials decided to change the status of the post-Apollo program to that of "project definition." As part of this process, one month later the agency changed the name of the project to the Saturn-Apollo Applications Program (AAP).

George E. Mueller, NASA Director of the Office of Manned Space Flight (OMSF), had known for some time that the future of his Apollo workforce depended in large measure on the viability of the plans which would come out of this office. He also had seen the erosion of support in Congress for a continued space program on the scale of Apollo. Quickly recognizing that an orbital space laboratory represented the proposal of the AAP with the best chance of earning congressional approval, he began to accelerate work on finding an acceptable configuration for this idea.[14]

---

10. Ernst Stuhlinger and Frederick I. Ordway III, *Wernher von Braun: Crusader for Space*, (Malabar, FL: Krieger Pub. Co., 1994), pp. 172–79.

11. Compton and Benson, *Living and Working in Space*, p. 9.

12. U.S. Congress, House, Committee on Science and Astronautics, *1964 NASA Authorization: Hearings on H.R. 5466 (Superseded by H.R. 7500)*, 88th Congress, 1st Session, p. 20.

13. David S. Akins, *Skylab: Illustrated Chronology*, (Huntsville, AL MSFC, 1973), footnote, p. 105; Memorandum for the record, April 12, 1963, Box NASA, Administrator's Policy Meetings (1962–65), Folder Administrator's Policy Meetings—#2, Jan.–June 1963, NASA Historical Reference Collection; Compton and Benson, *Living and Working in Space*, pp. 15–18.

14. Memorandum on NASA future plans, from George Mueller, Associate Administrator for Manned Space Flight, April 28, 1964, Box 06, Skylab, Folder Skylab/AAP Documentation 1964, NASA Historical Reference Collection.

Mueller considered a number of possibilities, but soon chose an idea suggested jointly by the Douglas Aircraft Company and the Marshall Space Flight Center (MSFC). Ironically, it involved embracing the idea, first suggested by Wernher von Braun, of making a laboratory out of the final stage of a launch vehicle that had expended its fuel and had gone into orbit. A second launch vehicle would take a crew of astronauts into outer space to rendezvous with the spent stage; these individuals would then flush any residue out and put in the components of the laboratory. The concept, known as the "wet workshop," had technical merit, but also had an economical appeal as well. Mueller recognized that the rapid progress on the Apollo program meant that NASA would not need all the Saturn rockets it had originally ordered, therefore freeing him from the necessity of purchasing additional launch vehicles for the AAP. At a time when the financial demands of the Vietnam War had just started to compete with NASA's budgetary requests, such potential savings from AAP had great appeal to the agency. NASA Associate Administrator Robert C. Seamans gave Mueller his approval for the proposal in March 1966, and in November Mueller made a formal presentation to NASA Administrator James E. Webb.[15] Official backing came when Congress, acting upon Webb's recommendation, appropriated approximately $450 million for the AAP in the 1968 Fiscal Year budget.

With initial funding secured, Mueller then turned to the task of deciding what scientific experiments to conduct aboard the orbital laboratory. NASA officials saw a potential for investigations to yield information in seven areas: life sciences, solar physics, Earth observations, astrophysics, materials science and manufacturing, engineering and technology, and student experiments.[16] Mueller soon decided that one of the experiments should involve photographing solar activity. Originally NASA had intended to use a satellite called the Advanced Orbiting Solar Observatory for that purpose, but budgetary constraints had forced the cancellation of that project. Mueller felt that the crew of a space laboratory could complete those photographic tasks, and had directed AAP efforts to develop a telescopic mount for such a purpose. After extensive work at a number of NASA Centers, on September 19, 1966, NASA's Manned Space Flight Experiments Board (MSFEB) accepted the plan to include such a device, known as the Apollo Telescope Mount (ATM), in the space laboratory program.[17]

While the idea of the ATM had moved steadily from proposal to approval, the selection of other experiments took much longer. Although those in the AAP knew of over 100 experiments being considered by individuals at the various space centers, a lack of funds and supporting manpower prevented many of these potential contributors from moving quickly to solidify plans for such endeavors. But energetic work involving visits to the various centers by Douglas Lord, head of the Advanced Manned Missions Office Experiments Division, spurred interested individuals to complete their feasibility studies. By the end of 1966 the MSFEB had received and approved fifty-two such proposals.[18]

Individuals involved in preparations for a manned laboratory in Earth orbit felt quietly optimistic as 1967 began. They had seen the AAP receive formal approval from President Johnson, and knew that the program had become a line item of NASA's budget. Although some experts in the field questioned the timetable, the AAP team believed that

---

15. Memorandum from Robert C. Seamans, Jr., to Associate Administrator for Manned Space Flight, March 28, 1966, Box 09, *Skylab*, Folder S-IVB Workshop, NASA Historical Reference Collection; Compton and Benson, *Living and Working in Space*, p. 38.
16. Lee B. Summerlin, *Classroom in Space* (Washington, DC: NASA, 1977), pp. 11, 12.
17. Compton & Benson, *Living and Working in Space*, pp. 69–76. Astronomers, while initially distressed by the loss of the Advanced Orbiting Solar Observatory, would find the data gathered by the three *Skylab* crews had given them "the best observations ever obtained from space." *Ibid*, p. 344.
18. *Ibid*, pp. 76–79; George Mueller to Robert Gilruth, December 21, 1966, Box 09, *Skylab*, Folder S-IVB Workshop, NASA Historical Reference Collection.

NASA could place a space laboratory into orbit by the middle of 1968. Understandably, then, George Mueller considered the time right to reveal his plans to the national media; at a press conference on January 26, 1967, he spoke publicly, for the first time, about the ambitious space laboratory program of the agency. But an event which occurred the very next day put the hopes of those involved with the project on hold, and came close to scuttling the idea entirely.

Technicians at work on a piece of test hardware for the Apollo Telescope Mount for Skylab *at the NASA Marshall Space Flight Center. Called an ATM spar assembly, this piece of hardware was made for heat tests in a Marshall Center vacuum chamber. Replicas of the ATM solar experiments and other equipment were mounted on the spar and a metal outside covering added. The ATM canister was tested in MSFC's Sunspot II vacuum chamber. Working on the spar are, from the top left, Ronald Andrews and R.M. Freeman; J.B. Pendegraft, kneeling: and J.R. Clift, holding blueprint, all of the Manufacturing Engineering Laboratory. (NASA photo no. 69-H-1621).*

Battling Back From Adversity

In the late afternoon of January 27, 1967, a disastrous fire swept through an Apollo command module perched atop a Saturn launch vehicle at the Kennedy Space Center. The conflagration caused the deaths of astronauts Virgil "Gus" Grissom, Edward White, and Roger Chaffee, forcing NASA officials to reconsider many aspects of the Apollo program. This process of review meant that the agency would not move as quickly toward a lunar mission as originally planned, which caused the new director of the AAP, Charles W. Matthews, to recognize the impossibility of meeting the proposed orbital laboratory launch date.[19]

Unfortunately, the impact of the fatal fire did not end with the postponement. Congress, concerned about the efficacy of NASA programs and recognizing the increasing financial burden of the Vietnam War, appropriated only $300 million of the $457 million asked by NASA for the AAP in fiscal year 1968. This reduction forced officials in the AAP to further postpone the orbital laboratory schedule, which would involve three separate missions, back into the 1970 calendar year.

These setbacks put hurdles in the path of the space laboratory program, but paled in comparison to the challenge soon offered within the agency itself. Robert C. Seamans, Jr., visiting space centers in June 1967 to monitor activities in the wake of the Apollo disaster, found that many people in the agency wanted to express their strong reservations about the AAP mission design. Individuals at the MSC in particular believed that the idea of converting a spent booster stage into a laboratory in outer space posed too many potential difficulties. They asserted that building a workshop on Earth and then putting it into orbit offered a greater chance for success. Encouraged by the response of Seamans, personnel at the MSC decided to make a presentation on the subject at a conference to discuss plans for Earth-orbiting missions, scheduled for November 18.[20]

At the conference, both sides presented their views on the wet workshop proposal. George Mueller felt that the proponents sufficiently answered the objections to the AAP plans for an orbital laboratory, and reported favorably on the concept to Robert Seamans. But the top echelon of NASA management still had reservations based on the concerns voiced during Seamans' summer tour, and therefore NASA Administrator James Webb felt compelled to call for a total AAP review, which he scheduled for December 6. At that meeting Webb found adequate reasons for placing the wet workshop proposal on hold until the agency could conduct a total review.

To examine all the possibilities for NASA in the post-Apollo era, including the concept of a manned space laboratory, Webb convened a special committee in early 1968. This group, chaired by Langley Research Center Director Floyd L. Thompson, soon demonstrated agreement about the value of an orbital workshop. The members, however, recognized the limitations of the concept as then constituted. In March 1968, Thompson's group suggested that if the agency could not overcome the technical problems which beset the program it should consider "ground-assembling the workshop and launching it dry."[21]

This idea, which had circulated within the agency for some time, had great scientific merit. It would require the use of a powerful Saturn V launch vehicle, but would allow more flexibility in mission planning. Unfortunately, by the time of the Thompson committee report it had become apparent that Congress, beset by the increasing financial burdens of

---

19. Compton and Benson, *Living and Working in Space*, pp. 83–84.
20. Newkirk, Ertel, with Brooks, *Skylab*, p. 123.
21. Compton and Benson, *Living and Working in Space*, pp. 96–97.

the Vietnam War, Great Society efforts, and the space program, would not appropriate funds for such a modification. Moreover, NASA Administrator Webb, who at one time had strongly supported AAP, had by this time become a lukewarm proponent at best. Discussing AAP at a meeting with directors of the space Centers, Webb asserted that the program was merely "a surge tank for Apollo." Convinced that no other space agency program should demand funds which NASA could spend on Apollo, Webb authorized only those expenditures for the space laboratory regarded as absolutely necessary.[22] Many wondered if NASA would launch the workshop in any configuration, much less agree to a dry workshop program.

Just as the prospects for AAP seemed to have reached the lowest point, however, a series of events transpired which revived the flagging program. First, NASA Administrator James Webb announced his retirement in the fall of 1968. His successor, Deputy Director Thomas O. Paine, had demonstrated a much firmer commitment to the concept of a manned space laboratory, and would become a valuable asset to the project. Second, NASA successfully flew a manned Apollo mission in October 1968, restoring public confidence in the space program in general. Finally, the memorable December 1968 flight of *Apollo 8* seemed to have made Congress more amenable to NASA budgetary requests; by April 1969 NASA officials knew that they would have at least $252 million to spend on AAP. Thus reassured, they could plan for the space laboratory with confidence.

Revising the Configuration

With their hopes bolstered by the recent turn of events, NASA officials began to firm up plans for an orbital space laboratory. But as those involved in the project resumed their work, they noted a subtle change in the attitude of George Mueller. The man who had long championed the concept of a wet workshop now seemed determined to utilize the proposal of Floyd Thompson's committee as well. Indeed, at a conference on May 3 and 4, 1969, Mueller announced to those involved in the program his hope of developing a series of missions which would include both a wet and a dry workshop.

Mueller's plan struck many as impractical. Some pointed out the financial burdens of redesigning AAP missions, while others noted how adding a second type of workshop would delay the program. But it occurred to Leland F. Belew, the director of the Marshall Space Flight Center's Saturn—AAP Office, that NASA could save time and money by merely replacing the plan for a wet workshop with one for a dry workshop. Mueller immediately saw the logic of this suggestion, and called upon those involved in AAP to voice their opinions. Gradually, a consensus emerged in favor of launching a prefabricated laboratory on a Saturn V rocket as the sole AAP effort involving a manned space workshop. Thomas Paine made the decision complete when he gave his formal approval on July 18, 1969.

Public notice of this change had to wait, however, until NASA officials could officially authorize the use of a Saturn V by AAP. James Webb had insisted upon a commitment of those launch vehicles exclusively to Apollo; until a successful landing on the Moon, therefore, NASA could not guarantee the availability of a Saturn V for any other purpose.[23] But since the lunar module from *Apollo 11* would touch down on the lunar surface only two days after Paine's action, Mueller felt confident that such official notification would come

---

22. *Ibid*, pp. 99–104; W. Henry Lambright, *Powering Apollo: James E. Webb of NASA* (Baltimore, MD: Johns Hopkins University Press, 1995), pp. 193–196. While agreeing with Compton and Benson on Webb's attitude in 1968 towards AAP, Lambright disagrees with them about how Webb had felt about the program until that time. Compton and Benson assert that the NASA administrator "had never been an enthusiast for AAP," while Lambright maintains that Webb gave post-Apollo a high priority through at least 1966. For the respective arguments, see Compton and Benson, *Living and Working in Space*, p. 104, and Lambright, *Powering Apollo*, p. 139.

23. Lambright, *Powering Apollo*, pp. 195–96.

shortly. In fact, on July 22, 1969, those involved in AAP received the word to proceed with the dry workshop program.[24] With the goal of the Apollo program successfully attained, NASA could devote more of its attention to planning for the future. An indication of this shift came in February 1970, when the agency chose a name for the dry workshop proposal. After considering over 100 possible suggestions, the NASA Project Designation Committee selected the title *Skylab*, submitted by U.S. Air Force Lieutenant Colonel Donald Steelman. But it soon turned out that not everyone shared the optimism of the agency regarding the future of the American space program. Thomas Paine had hoped to use the success of the Apollo program as a springboard for funding more ambitious projects; much to his frustration, however, he found in September 1970 that the Nixon Administration hoped to cut the NASA budget substantially. On the fifteenth of that month, Paine resigned.

His acting replacement, George M. Low, tried to salvage as much funding as possible from the Nixon Administration. He succeeded in minimizing the cuts to the *Skylab* project, but Low knew that this would still necessitate a readjustment of that program's schedule. The Office of Manned Space Flight finally set the date for the first launch of *Skylab* for April 1973, and work on the program began in earnest.

In its final configuration, *Skylab* would consist of five components: a command module (which would join the other components in orbit), a service module, a docking adapter, an airlock, and a workshop. The workshop would itself have two separate sections. The upper compartment would house the work station and equipment for experiments, the frozen food locker, water containers, and film bins. The lower compartment would contain "the kitchen and dining room, bedrooms, an experimental work area, and the toilet." The workshop, forty-one feet long by twenty-two feet wide, would provide 10,426 cubic feet of work space for the astronaut crews. Satisfied with the design, NASA officials finalized contracts with private contractors for the various components.[25]

By this time, the *Skylab* program had lost one champion, but had gained another. George Mueller, the person who had guided the idea of an orbital laboratory through troubled times, had resigned from NASA in 1969. While he would be missed, William C. Schneider, who had become the director of the AAP in December 1968, had already demonstrated the same enthusiasm for the orbital workshop concept that Mueller had exhibited. Building on the foundation established by Mueller, Schneider worked steadily on ways to expedite the progress being made on the project. The former Apollo mission director found the NASA system of reviews, implemented during the Apollo program, of great help in keeping the program on schedule. The procedure, which involved formal assessments at seven different stages of development, insured that program officials could quickly identify potential difficulties.[26] By the end of 1970, Schneider had approved the last of the modifications necessitated by the change to the dry workshop configuration, and had authorized the testing of the components for the project.

While the various contractors began the preparation of the materials to be used, NASA officials finalized their assessment of the habitability of *Skylab*. In the fall of 1969, Schneider had already authorized a number of modifications in this regard; the orbital laboratory would now have a room for the astronauts to both sleep and dine in, and an observation window for the viewing pleasure of the astronauts. By late 1970, the only serious question remaining about living conditions involved food. The MSC wanted a more

---

24. Newkirk, Ertel, with Brooks, *Skylab*, p. 169.
25. Leland F. Belew, ed., *Skylab, Our First Space Station* (Washington, DC: NASA, 1977), pp. 16–34.
26. Compton and Benson, *Living and Working in Space*, pp. 118–23. For a discussion of the Apollo system of management see Robert C. Seamans, Jr., and Frederick I. Ordway, III, "Lessons of Apollo For Large-Scale Technology," in Frederick C. Durant III, ed., *Between Sputnik and the Shuttle: New Perspectives on American Astronautics* (San Diego, CA: Univelt, Inc., 1981), pp. 241–87.

elaborate system for storing and preparing meals than the Marshall Space Flight Center felt the laboratory could accommodate. In the end both sides compromised, and Schneider faced no further controversy involving the habitability of *Skylab*.

Confident that agency officials had designed a workable program, Schneider then turned his attention to the private companies supplying two major components for *Skylab*. He knew that in April 1969 workers at the McDonnell Douglas Corporation's Huntington Beach, California, facility had started the process of converting a Saturn IVB final stage into an orbital workshop. Schneider soon found that the work of the company had fallen behind schedule. The vast number of pieces, many supplied by other contractors, involved in assembling the craft proved especially daunting. But by utilizing more efficient management techniques, McDonnell Douglas proved able to increase the pace and improve the quality of its efforts. Indeed, in 510 hours of tests during the summer of 1972 company technicians found only minor technical flaws. Confident about the finished product, in a ceremony on September 7 of that year, McDonnell Douglas officially presented NASA administrators with the completed laboratory.[27]

One month later, NASA received the other major component of the *Skylab* vessel. Technicians at the space agency had designed a docking facility and airlock for inclusion in the project; these would allow a crew in an Apollo command module to rendezvous with *Skylab* in outer space and enter the orbital laboratory. In the summer of 1970, the Martin Marietta Company had built a multiple docking adapter and had sent in to the St. Louis facility of McDonnell Douglas where workers would connect it with the airlock they had constructed. On October 5, 1972, a NASA *Super Guppy* airplane flew this last component to the Kennedy Space Center (KSC).

As NASA technicians joined the various parts of *Skylab* together at KSC, agency officials finalized the process of preparing the astronauts who would eventually live and work in it. Initially, the selection of these individuals had created a controversy within NASA. In 1965 and again in 1967, the space agency had recruited a number of scientists for the astronaut program, and these individuals saw the *Skylab* program as a logical venue in which NASA could utilize their talents. They hoped that NASA would include two of their number on each mission. But Donald K. "Deke" Slayton, NASA Director of Flight Crew Operations, had chosen to include only one scientist in each of the three three-person crews. In spite of strong protests from the scientific community, Slayton held his ground, arguing that the missions might need two individuals trained primarily in piloting. Backed by Rober Gilruth of the MSC, Slayton won his case, and NASA finalized the composition of the three crews. An all-Navy team of Captain Charles "Pete" Conrad, Jr., Commander Joseph P. Kerwin, and Commander Paul J. Weitz would fly the first mission. In September 1972 all the crews started a program of mission simulations, preparing for the schedule of launches.[28]

Personnel at KSC began working longer hours in 1973, getting all systems ready for the long anticipated May launch date. Inclement weather threatened to disrupt the schedule, but by May 14, 1973, conditions had improved sufficiently to warrant initiation of the countdown sequence. At 1:30 p.m. the 334-foot Saturn V carrying *Skylab* lifted off from its launch pad, ending the years of planning and preparation. Now the question became whether the program could deliver on the promises its proponents had made for it.

---

27. Newkirk, Ertel, with Brooks, *Skylab*, p. 279.
28. Compton and Benson, *Living and Working in Space*, pp. 218–30.

Saving the Program

Telemetry received at the Johnson Space Center (JSC) indicated that the powerful Saturn V had taken only ten minutes to place its cargo into orbit at a height of 270 statute miles. At that point various electrical systems aboard *Skylab* began to come to life.[29] At first every indication suggested a normal process, but soon Flight Director Donald Puddy noticed readings that gave him pause for concern. After a few hours of investigation, NASA officials determined that two things had indeed gone wrong with the mission. First, the two solar panels, which would provide power for the laboratory, had failed to deploy properly. And second, it appeared that a shield designed to both protect against micrometeorites and provide shade for *Skylab* had been lost. Both of these developments posed serious, perhaps even mission-threatening, challenges for NASA officials to ponder.

Analysis by engineers at the Marshall Space Flight Center revealed that the loss of the micrometeorite shield would cause the temperature inside the orbital laboratory to rise to 190 degrees fahrenheit when bathed in sunlight. This heat, they realized, could damage the photographic film and food aboard *Skylab*; more ominous still, it might melt plastic inside the laboratory, releasing toxic fumes. The mission control team could change the attitude of the orbital laboratory to reduce the angle at which the rays would hit it, thus reducing temperatures, but that adjustment would also limit the amount of sunlight reaching the solar cells of the ATM –*Skylab's* major remaining source of power. The attitude finally chosen by NASA lowered the temperature inside *Skylab* to 130 degrees and allowed the generation of 2800 watts of power, but NASA officials believed that they would have to arrange for some sort of sunscreen to protect *Skylab* if they hoped to save the mission.[30]

After NASA postponed the launch of the first crew of astronauts for ten days, individuals at the JSC and Marshall immediately began to explore possible methods for providing a sunscreen. After discarding a number of suggestions, NASA officials finally narrowed the range of options to three. One came from Jack Kinzler of the JSC's Technical Services Division, who suggested creating a collapsible parasol out of aluminum-coated mylar supplied by the G.T. Schjeldahl Company. Kinzler felt that the astronaut crew could deploy such a device through the scientific airlock of *Skylab* after docking. A second suggestion involved having the astronauts affix a twenty-two-by-twenty-four-foot sunscreen to the outside of the laboratory while standing in the hatch of their Apollo command module. Finally, a third idea envisioned attaching a "twin boom" frame to the ATM and then hoisting a sail between the two rods. While finally favoring the Kinzler recommendation, NASA officials decided to have Conrad, Kerwin, and Weitz take all three devices with them aboard the Apollo command module.[31]

NASA officials also hoped that the astronauts could salvage some power capacity from the solar panels. Data received from *Skylab* indicated that one of the panels had remained attached to the laboratory, but had not deployed. If debris from the micrometeorite shield had trapped it, Conrad, Kerwin, and Weitz might have a chance to free the panel. To facilitate these repair efforts, the agency requisitioned specific tools from the A.B. Chance Company and included them in the cargo which the astronauts would take into outer space.

The astronauts lifted off on May 25, 1973. After achieving rendezvous with *Skylab* and

---

29. After the death of Lyndon Johnson in January 1973, the Senate passed a resolution in February to rename the Manned Space Center after the individual who had been in many ways "the father of the space program." Henry C. Dethloff, "*Suddenly, Tomorrow Came. . . . : A History of the Johnson Space Center*" (Washington, DC: NASA SP-4307, 1993), p. 214.
30. Belew, *Skylab*, pp. 41–57.
31. Interview of Mr. Don Arabian, Manager, Program Operations Office, Johnson Space Center, June 12, 1973, Box 05, *Skylab*, Folder *Skylab* Oral History Interviews, NASA Historical Reference Collection.

eating a meal, the crew began the planned repairs. Conrad took the Apollo command module away from the docking port and placed it next to the remaining solar panel which, as it turned out, had not deployed because of a strap from the micrometeorite shield which had become wrapped around it. Weitz attempted mightily to free the solar array panel, but had no success. Finally abandoning that endeavor, Conrad decided to dock the command module for the crew's rest period. But this time when he attempted the procedure, Conrad found that the latches of the respective entry ports would not engage. He tried a number of other docking techniques, but these too failed. As a last resort, the astronauts effected a manual modification of their craft's docking port, which allowed Conrad to finally couple the vessels. After a twenty-two-hour day, the crew went to sleep in the command module.[32]

When the crew awakened, Weitz entered *Skylab* to determine if the high temperatures had generated toxic fumes. Finding no such danger, the crew then entered the laboratory and prepared to deploy the parasol they had brought along. Working carefully, they managed to unfurl the device, and noticed an immediate cooling inside the laboratory. The temperature inside *Skylab* would soon stabilize at ninety degrees. The parasol did not cover as wide a surface area as originally hoped, but it would allow the crew to begin the schedule of experiments in relative comfort. Conrad, Kerwin, and Weitz would waste no time in initiating these efforts.

The "Can Do" Spirit

The crew soon established a regular schedule which involved carrying out experiments, exercising, eating, maintaining the vessel, and sleeping. They found no difficulty in adjusting to the sensation of living in outer space, experiencing no motion sickness, but found the pace of the tasks assigned to them very demanding. Within days Conrad suggested revising the time allotments for experiments and other duties, and his crew soon settled into a comfortable routine.[33] But the crew soon recognized that insufficient power levels might not allow the completion of a number of the proposed experiments. NASA officials concluded that the astronauts would have to make another attempt to free the remaining solar panel.

NASA officials had prepared Conrad, Kerwin, and Weitz for their extra-vehicular repair effort by having them train in a zero-gravity neutral buoyancy simulator. Kerwin, however, had not found the technique practiced on Earth any help to him in his first repair attempt. Therefore, NASA officials assigned a group headed by Russell "Rusty" Schweickart (the commander of the back-up crew for the first mission) to devise another method for deploying the solar panel. The team finally recommended using a twenty-five foot long cable cutter to free the panel and a tether to maneuver it, once freed, into proper alignment. Ground control relayed instructions on the procedure to the crew, and on June 7 Kerwin and Conrad left the workshop to make another attempt.

It became immediately apparent to the two astronauts that the actual repair efforts in outer space would differ significantly from those practiced on Earth. To begin with, Kerwin found that he could not establish a firm foothold on the solar observatory antenna as Schweickart had done; cables prevented him from getting close enough to secure his feet. Every time Kerwin attempted to position the cutter he simply drifted off until the

---

32. Belew, *Skylab*, pp. 61–63.
33. Compton and Benson, *Living and Working in Space*, pp. 286–94.

Skylab Ultra-violet Stellar Astronomy—A special airlock in the wall of the Workshop allowed the crew to perform several "outside" experiments. Here, a camera equipped with calcium fluoride transmission optics to pass ultra-violet light is aimed at stars in the Milky Way field. (NASA photo no. 72-H-891).

cord connecting him to the antenna restrained him. Moreover, when Conrad attempted to place the hooks of the tether into holes on the solar array panel he found them smaller than those on the panel on which Schweickart's team had practiced. He managed to fit one hook onto the restrained device, but wondered whether he could exert sufficient force to free it with only one clasp. Kerwin then began to cut the strap, but after thirty minutes of work the astronaut had not made any progress toward accomplishing his goal.[34]

At that moment, however, Kerwin had an idea. He took the line connecting him to the antenna and shortened the length until he had secured his position. Having thus gained leverage, within ten minutes Kerwin proved able to cut the strap. The solar array panel then swung open at an angle of about twenty degrees. At that point, Kerwin and Conrad began to pull on the tether connected to the device, but with no success. Finally Conrad stood up and slung the tether over his shoulder, while Kerwin continued to pull. This technique worked almost too well—the recalcitrant hinge broke loose suddenly, sending the astronauts flying off into space. When the restraining cords stopped their motion, Kerwin and Conrad saw the panel almost fully deployed. Within days the astronauts could report that power levels had almost doubled.

Having lost a significant portion of the four week duration of the mission to repair efforts, NASA officials hoped to salvage as much of the experimental schedule as possible. Forced to prioritize the investigations, they decided to give the highest ranking to medical experiments. During the time remaining in orbit, Conrad, Kerwin, and Weitz managed to achieve nearly 100 percent of the projected goals that medical officials had set forth for the mission.[35] Of particular interest to the medical community was the fact that weightlessness allowed the skeletal structure of the astronauts to fully extend, increasing their height by an inch. It also balanced fluids within the body more evenly, thus making the faces of Conrad, Kerwin, and Weitz appear fuller. With a regular regimen of exercise the astronauts remained in outstanding physical condition throughout the duration of the mission, proving the adaptability of humans to living in outer space.

The crew also managed to accomplish a high percentage of the assigned tasks involving the observation of solar activity, eventually taking over 25,000 pictures of the sun. This endeavor proved a highly challenging one for the crew, given the complicated nature of the control equipment. Located in the docking adapter, these instruments activated the ATM telescopes. Kerwin later asserted that the pace of mission experiments kept him from ever completely mastering the control system.[36] He found the procedure for observing solar flares especially frustrating. NASA technicians had developed a sensor which would alert the crew to solar flare activity, but the radiation field over the eastern section of South America would frequently trigger the alarm system. After many false warnings, the crew despaired of ever observing a solar flare. But after deciding to spend the entire work day of June 15 in hopes of witnessing such an occurrence, their patience paid off. Weitz proved able to record a flare of impressive magnitude for over two minutes. Although the crew may have felt that their work involving the observation of solar activity had not met up to expectations, astronomers proved genuinely pleased with the results, which they considered to have met eighty percent of pre-mission expectations.

The scientific aspect which suffered the most from the compression of the mission schedule involved earth resources experiments. NASA officials had included cameras and other devices to provide a more detailed analysis of geological, meteorological, and agricultural features than could otherwise be gathered by conventional methods. The equipment

---

34. Belew, *Skylab*, pp. 50–75; Compton and Benson, *Living and Working in Space*, pp. 268–76.
35. Compton and Benson, *Living and Working in Space*, p. 289.
36. *Ibid*, p. 290.

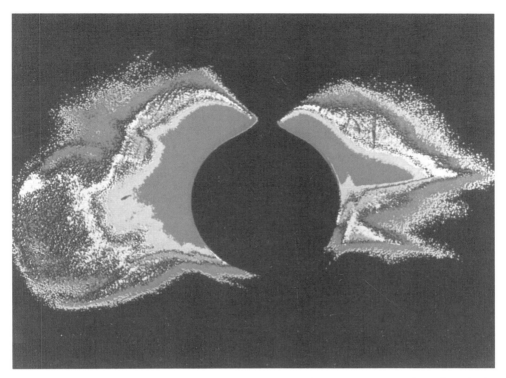

The Sun's hot outer atmosphere, or corona, color-coded to distinguish levels of brightness, reaches outward for millions of miles. A coronagraph, one of Skylab's eight telescopes, masked the Sun's disk, creating artificial eclipses. It permitted 8-1/2 months of corona observation, compared to less than eighty hours from all natural eclipses since use of photography began in 1839. (NASA photo no. 74-H-97).

which these experiments utilized required a significant amount of power and, because of this, NASA officials had to postpone a vast number of them at the beginning of the mission. The first crew never fully caught up, but still managed to complete an estimated 60 percent of the planned earth resources experiments.[37]

Ironically, the crew might not have done as many earth resource experiments as they did had the astronauts not completed an in-flight repair of the power system. During the mission, the device which collected power from the ATM panels failed to function properly, causing concern at JSC. NASA technicians, remembering that they had overcome the same difficulty in ground tests by delivering a sharp blow to the device, recommended a similar procedure to the astronauts. Conrad, wielding a hammer, soon had the power conditioner operating properly again, and mission experiments continued.[38]

As the end of their mission drew near, the astronauts began their preparations to depart from *Skylab*. On June 22 the crew boarded their command module to fly back to earth. After 404 orbits and twenty-eight days in space, the astronauts made a successful splashdown in the Pacific Ocean 800 miles west of San Diego. Picked up by the aircraft carrier *Ticonderoga*, the astronauts soon received a hero's welcome when they reached the

37. Newkirk, Ertel, with Brooks, *Skylab*, p. 312.
38. Belew, *Skylab*, p. 89.

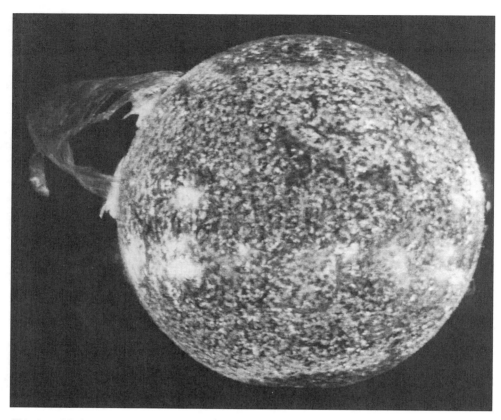

This photo of the Sun, taken from Skylab *on December 19, 1973, shows one of the most spectacular solar flares ever recorded, spanning more than 588,000 kilometers (367,000 miles) across the solar surface. The flare gave the distinctive impression of a twisted sheet of gas in the process of unwinding itself. The solar poles are distinguished by a relative absence of supergranulation network, and a much darker tone than the central portions of the disk. Several active regions are seen on the eastern side of the disk. This photo was taken in the light of ionized helium by the extreme ultraviolet spectroheliograph instrument of the U.S. Naval Research Laboratory. (NASA photo no. 74-N-434).*

United States. Encouraged by the success of the first mission, but concerned that further mechanical difficulties occurring with no crew aboard to rectify them might jeopardize the status of *Skylab*, NASA officials decided to advance the launch date of the second mission by three weeks. Attention would therefore turn again to KSC.

A Commitment to Excellence

As the second mission began on July 28, 1973, the crew—Navy Captain Alan L. Bean, Marine Major Jack R. Lousma, and civilian Owen K. Garriott—soon realized that their experience would differ in at least one respect from that of the first crew: Lousma reported nausea only one hour into the flight. The other two soon experienced the same malady. The crew attempted to work through their infirmities, but motion sickness slowed down the pace of activity during the first few days, prompting concern among medical officials at NASA. These officials, after reviewing the timing of the occurrences among the crew during the day, recommended that Bean, Lousma, and Garriott eat six small meals rather than three larger ones each day. Whether because of this procedure or simply through an adjustment to space, within days the crew reported that they felt back to normal.[39]

This return to good health came just in time. In addition to assigned tasks involved in making *Skylab* operational, the crew also had the responsibility of deploying another sunscreen over the laboratory. NASA officials had chosen the parasol concept for the first crew to use because of the speed with which the astronauts could deploy it, but had recognized that ultraviolet rays would soon cause the fabric of that device to deteriorate. They therefore decided to have the second crew deploy another device considered for use by NASA during the original mission: the twin boom frame sunscreen. Bean, Lousma, and Garriott had practiced extensively on the necessary construction techniques in the neutral buoyancy tank at Houston, and with their health vastly improved they felt ready to begin the task on August 6.

As frequently happened, this activity did not go exactly as planned. Garriott and Lousma soon recognized that the size of the components which they were using differed slightly from those they had practiced with on Earth, forcing them to make a series of adjustments while out of the vehicle. But patiently keeping at their task, the astronauts eventually managed to assemble the frame and raise the sunscreen shade. Lousma and Garriott then replaced film in the ATM, gathered data from previous experiments, and checked on a potentially threatening situation involving the rocket thrusters of the Apollo module. After six hours and thirty-one minutes the two returned to *Skylab*, having established a record for the longest period of extra-vehicular activity in outer space.[40]

After completing the necessary repair, the crew resumed the regular experimental schedule. As with the first mission, the observation of solar phenomena had received a high priority. From August 10-20 in particular, the astronauts devoted a significant portion of their work days to capturing solar activity on film. On August 21 the crew witnessed a gigantic bubble develop on one side of the Sun and watched it grow to a size nearly three-quarters that of the Sun itself. In all, the second crew would spend over 300 hours engaged in solar-related activity; their efforts received unqualified praise from the scientific community, and many had an opportunity to speak to the crew directly from Houston.

This approbation did not go unnoticed. Indeed, reporters covering the *Skylab* program asked NASA officials at a press conference on August 10 why the second crew seemed to find greater favor with the scientific community. Dr. Ernest Hindler of the High Altitude Observatory of Boulder, Colorado, asserted that the first crew had actually paved the way for the rapport with the second crew. By pointing out the flaws in the program, Conrad, Kerwin, and Weitz had helped the second team avoid problems on their mission. But, he conceded, the second crew also seemed more amenable to an interactive dialogue with scientists at ground control than the first had.[41]

While solar observations took much of the time of the astronauts during the first part of the mission, Earth resources observations also merited considerable attention. All three astronauts would participate simultaneously in these efforts, with each operating a particular piece of equipment—six major instruments in all. NASA officials planned to utilize the first and last segments of the mission for Earth resources observation, knowing that atmospheric conditions would limit visibility during the middle period. In spite of certain

---

39.   *Ibid*, pp. 104–106.
40.   Compton and Benson, *Living and Working in Space*, pp. 300–02. NASA officials had been troubled by readings indicating that leaks might have occurred in the rocket thruster systems, thus jeopardizing the mission. Lousma and Garriott confirmed during their space walk that the situation had stabilized.
41.   *Ibid*, pp. 302–03.

difficulties, the crew managed to take almost 16,000 photographs of Earth and load data onto 18 miles of magnetic tape.[42]

After the first period of earth resources activity ended, the crew then took up what NASA officials called "corollary experiments." These involved space technology, space physics, and observations of stars. Although the exploits of two spiders (named Arabella and Anita) overshadowed these experiments in the eyes of the American public, the crew in fact produced a wealth of important data.[43]

The endeavors of Arabella and Anita certainly certainly caught the fancy of the people on Earth, but the astronauts themselves proved able to capture their attention as well. Lousma provided amusing commentary about the mission on the twenty-four-hour-delay communication channel B, and Garriott conducted a number of experiments which gave vivid televised visual illustrations of scientific principles. But the crew's most famous exploit involved a mysterious voice. One day the ground control team at JSC heard a female voice speaking to them from *Skylab* over the public air-to-ground channel. Houston hesitantly responded, and the conversation with the unidentified person went on for a few minutes. Finally, Garriott aboard *Skylab* burst out in laughter; he had brought along a tape recording of his wife's voice for this bit of levity.

While these activities illustrate that the second crew certainly liked to have fun, they should not overshadow the fact that Lousma, Garriott, and Bean more than met the expectations set forth for them at the beginning of the mission. While illness and repair efforts had put them well behind schedule for the first ten days, the astronauts soon caught up and in fact began to ask for additional tasks. Indeed, Bean asked Houston to lengthen the mission by at least a week to give the crew more opportunities for meaningful research. JSC declined this request, but the astronauts still proved able to supercede expectations by fifty percent during their mission.[44]

On September 25, 1973, after fifty-nine days in space, the crew boarded the command module and began their return to Earth. At 6:20 p.m. EDT, the capsule splashed down in the Pacific Ocean 300 miles from San Diego. Picked up by the *U.S.S. New Orleans*, less than forty-five minutes later, the astronauts received accolades for their impressive accomplishments. Encouraged by their efforts, NASA officials eagerly began final preparations for the third, and last, *Skylab* mission.

A Series of Challenges

NASA officials had hoped to launch the third crew, consisting of Marine Lieutenant Colonel Gerald Carr, Air Force Lieutenant Colonel William Pogue, and civilian Edward Gibson, on November 11, 1973. Flaws in the launch vehicle temporarily postponed the mission, but on November 16 the Saturn IV blasted off, taking the crew to a rendezvous with *Skylab*. After initial difficulties, Carr managed to dock the two crafts. Under orders from NASA, the crew would wait until after a rest period to begin the activation of the laboratory.

To minimize the chances of having the crew suffer from space sickness, NASA medical personnel had ordered the astronauts to take anti-nausea pills. Unfortunately, this precaution did not help Pogue, who soon became quite ill. Gibson and Carr pondered what to do about this incident; if they reported the full extent of Pogue's distress (which included vomiting), NASA officials might alter the schedule of the mission. They chose, therefore, to reveal to Houston only that Pogue had experienced some nausea, but would withhold any mention of the vomiting.

---

42. Newkirk, Ertel, with Brooks, *Skylab*, p. 334.
43. Belew, *Skylab*, pp. 114–15.
44. Newkirk, Ertel, with Brooks, *Skylab*, pp. 333–35.

This information would not remain a secret for long, however. The crew had forgotten that NASA officials would hear their deliberations; an onboard taping device had recorded their discussion word for word, and when doing a routine review of the conversations ground control soon discovered the cover-up. Upset by the lack of candor displayed by the crew, NASA chief of astronauts Alan B. Shepard sternly warned them not to repeat their mistake. The crew admitted their error, and promised better behavior.[45]

Individuals at ground control soon found another reason to criticize the crew. After beginning the process of activating *Skylab*, during which they made a number of mistakes, the astronauts complained about the pace of the activities planned for them. It seemed to some at NASA, however, that the astronauts simply did not want to work as hard as they should. In defense of the crew, it should be noted that none of the astronauts had any space flight experience, thus making their adjustment to the mission more difficult. In addition, NASA officials had added a number of experiments to the mission in the period after the return of the second Skylab crew, which left little time for Pogue, Carr, and Gibson to learn the new procedures. Finally, the astronauts often could not find gear aboard *Skylab* where NASA officials had told them to expect to find it. In spite of these reasons, NASA flight planners continued to regard the third crew as not working up to maximum capacity.[46]

Matters came to a head on December 30. The crew initiated a frank discussion with capsule communicator Richard Truly, hoping to make NASA officials aware of their objections to the pace of activities set forth for them. Carr proved especially critical of the fact that NASA mission plans had allocated virtually no free time for the astronauts directly before and after the sleep periods. Truly attempted to mollify the astronauts by praising their efforts to date, but subtly suggested that NASA felt that the crew could work at a faster pace during the last half of the mission. After almost an hour of discussion (punctuated by a 20-minute break in the communications link) Truly ended the conversation by telling the crew that he was "very happy with the way you're doing business." Both sides seemed satisfied that they had clarified their respective points of view, and in the days to follow, ground control noted a new sense of responsiveness on the part of the astronauts.

This incident illustrates an inherent tension that ran through much of the history of the space program. Beginning with the confrontation between the original seven astronauts and the designers of the Mercury capsule over the inclusion of a window in the spacecraft, astronauts had repeatedly disagreed with positions taken by the scientific and engineering community regarding the subject of space flight. To the crew of the third mission, it seemed as though flight planners felt that every mission could be treated like a computer ready to be programmed. Pogue, Gibson, and Carr had asserted the astronauts' view that flexibility had to be built into a flight plan to provide for the diversity of each human put into outer space. In the wake of their conversation with Truly, the astronauts felt that they had regained their individuality.

Pleased with their new freedom, the astronauts threw themselves back into their work. Gibson in particular demonstrated almost a compulsion to record a solar flare in its entirety, spending extra time manning the solar observation controls. His diligence was rewarded on January 21, 1974, when he witnessed an impressive solar event.[47] Coupled with the observations already done on the comet Kahoutek, the third crew could therefore take great pride in their contributions to the field of astronomy.

Not much time remained for the third crew in space. The astronauts did a number of medical experiments in the days after January 20 and conducted their final televised press conference, but they also devoted an increasing amount of time to preparations for the

---

45. *Ibid*, pp. 312–16.
46. *Ibid*, pp. 316–24; Belew, *Skylab*, pp. 128–30.
47. Compton and Benson, *Living and Working in Space*, pp. 331–33.

eventual moment of deactivation. They experienced some difficulty in storing in the command module all the things they wanted to take back to Earth, but using ingenuity and a certain amount of force they succeeded in packing their craft. On February 8 the crew boarded the command module, but before disengaging from *Skylab* they used the thrusters of the Apollo spacecraft to lift the laboratory to a higher orbit. After accomplishing that task, Pogue, Carr, and Gibson headed for earth.

One final bit of adventure awaited this crew. When Carr attempted to maneuver the command module prior to reentry, he found that neither the pitch nor the yaw controllers worked. Momentarily nonplussed, he nevertheless quickly shifted to another system and resumed the landing sequence. Later, the astronauts revealed they had accidentally activated a set of circuit breakers, thus neutralizing the thrusters in question. Carr's rapid response had saved the day. As on the second mission, the Navy used the *U.S.S. New Orleans* to conduct the recovery operation for the final crew, thus bringing to an end the 171 days, 13 hours, and 12 minutes of space flight connected with the *Skylab* program.

The Mission in Retrospect

The February 8 splashdown marked the official termination of the *Skylab* program. Scientists would need a great deal of time to fully analyze all of the data that the three crews had gathered, but enough work in this regard had already been done to allow observers to appraise the value of the project. Of the seven types of experiments the nine astronauts had conducted, they surpassed expectations in all except the area of engineering and technology investigations.[48] Many scientists, hitherto skeptical of the allocation of financial resources to the *Skylab* program, spoke in glowing terms of its value. Leo Goldberg, Director of the Kitt Peak National Observatory, voiced the views of his profession when he asserted that the astronauts, "by their rigorous preparation and training and enthusiastic devotion to the scientific goals of the mission, . . . have proven the value of men in space as true scientific partners in space science research."[49]

Goldberg's observation provides a fitting epitaph for *Skylab* (which would disintegrate upon reentering the Earth's atmosphere in 1979). While the program unquestionably yielded valuable scientific information, its greatest value came from its demonstration of the importance of the human element in the space program. As John Disher, NASA Director of Advanced Programs in the Office of Space Transportation Systems, would later note, *Skylab* "turned around many people who thought men in space were a hindrance rather than a help."[50] Echoing this sentiment, Program Director William Schneider stated that *Skylab* had shown that, regarding the space program, "the limit is only our resolve, not the ability of men to work, and not our technical knowledge."

In his statement, Schneider succinctly captured the essence of a central debate regarding the American space program. Many individuals had argued that scientific inquiry in outer space did not depend on a human presence in that realm; in the words of Homer E. Newell in 1958, "all we need is a few thirty-pound satellites in Earth orbits [to] furnish enough observational data to keep our space scientists busy for decades to come."[51] But the majority of the personnel at NASA had from the beginning believed that "the ultimate objective [was] manned travel to and from the other planets," and often

---

48. Compton and Benson, *Living and Working in Space*, p. 340.
49. Belew, *Skylab*, p. 155.
50. John H. Disher, "Space Transportation: Reflections and Projections," in Durant, ed., *Between Sputnik and the Shuttle*, p. 213.
51. Stuhlinger and Ordway, *Wernher von Braun*, pp. 298–99.

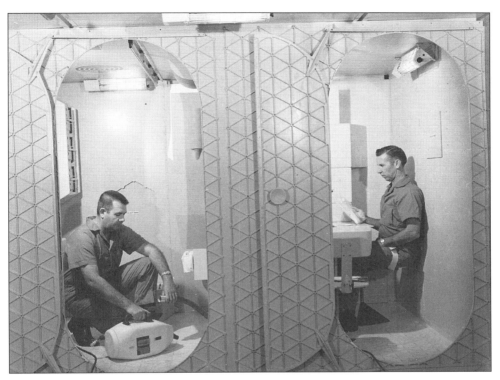

Ralph Geiger (left) and Ralph Murphy check out equipment in the waste management and food management compartments, respectively, in a full-scale mockup of Skylab *at the NASA Marshall Space Flight Center. The Workshop was developed by Marshall Center for three Earth orbit missions in 1972, the first lasting 28 days and the other two 56 days each. This view of the living area shows the "telephone booth" type doors used and the grid floor that separates the living area from the laboratory or working area making up the 10,000 cubic feet of space inside. Joined to the Workshop main body to form the "cluster" configuration was an airlock, multiple docking adapter, Apollo Telescope Mount and the Apollo command/service module. All of the components were to be launched by a Saturn V, with the exception of the command/service module, into an Earth orbit of about 250 statute miles. Several days later NASA planned for the astronauts in the command/service module to be launched by a Saturn IB rocket for rendezvous and docking with the multiple docking adapter. The cluster of spacecraft was used to conduct experiments, scientific and medical, and for photographic studies of the Sun. (NASA photo no. 69-H-1645).*

focused planning efforts in this direction.[52] The *Skylab* program, therefore, provided crucial and timely evidence to support the predominant NASA position.

Fittingly, the Collier Trophy Committee decided to present the 1973 trophy of their organization to the director of the *Skylab* program. Although the Committee noted the contributions of the over 26,000 individuals who had worked on the program, the members felt that William Schneider should receive the trophy on behalf of the three crews because of his leadership in the extraordinarily difficult venture. When Vice President Gerald Ford presented the award to Schneider on June 4, 1974, it provided an apt climax to an American space project which had proved without question the value of "the human touch" in space endeavors.

---

52. Howard E. McCurdy, *The Space Station Decision: Incremental Politics and Technological Choice* (Baltimore, MD: Johns Hopkins University Press, 1990), p. 8.

Chapter 10

LANDSAT and the Rise of Earth Resources Monitoring

by Pamela E. Mack

In 1974, Dr. John E. Clark, representing NASA, and Daniel J. Fink, representing General Electric, received a Collier Trophy for the Landsat program, for "proving in 1974 the value of U.S. space technology in the management of the Earth's resources and environment for the benefit of all mankind."[1] The Landsat program had proved its value in the eyes of the selection committee in the two years since the launch of the first satellite in 1972, but proving its value to potential users and to the Office of Management and Budget and to Congress turned out to be substantially more difficult. Somewhat improved Landsat satellites are still flying in the mid 1990s, back under government management after a failed effort at commercialization. But the project cannot be deemed a success; the United States maintains neither leadership in technology for civilian earth observation nor a robust operational program. Clearly the Collier Trophy award represented technical success and the hopes of the aerospace community for a new, more relevant mission to justify the space program. The larger story, however, lies in why a project which embodied such hopes came to so little.

Landsat was not a large project by the standards of the program to put human beings in space, but it involved broader concerns for NASA and a large-project management style. Landsat gained public attention, and a Collier Trophy, because it symbolized a wish that the space program would bring more obvious benefits on earth. NASA leaders sought to respond to such concerns from Congress and the general public by playing up the idea that Earth resources satellites could serve the public good, while at the same time promising quick commercialization.[2] But this commitment lacked stamina; NASA leaders still saw space exploration as the core mission of the agency, and the agency tended to further define that mission as research and development only, not operational data collection or promoting use of the resulting data. Landsat became a project intended to provide political or bureaucratic capital to NASA and its supporters, and those motivations further complicated the problem of balancing the needs of researchers and of potential operational users.

The project found itself repeatedly strangled in the budget process and by conflicts with the user agencies, even after it had (in the eyes of participants) "succeeded magnificently from a scientific and engineering sense."[3] Most of the scientists and engineers involved at the working level committed themselves wholeheartedly to developing the possibilities for a civilian earth-observation satellite to serve the public good. But funding for the project and approval of subsequent steps was repeatedly caught up in conflicts both between NASA, the Office of Management and Budget, and Congress over funding and between NASA and the agencies that would use the data over the future of the project.

---

1. Bill Robie, *For the Greatest Achievement: A History of the Aero Club of America and the National Aeronautic Association* (Washington, DC: Smithsonian Institution Press, 1993), p. 235.
2. For a passionate analysis of this contradiction see John H. McElroy, "Preface," in Kathleen M. Eisenbeis, *Privatizing Government Information: The Effects of Policy on Access to Landsat Satellite Data* (Metuchen, NJ: Scarecrow Press, 1995).
3. *Ibid.*, p. xi.

Disagreements among users made it difficult to design the satellite and resulted in only weak support from users when funding decisions were made.[4] These factors, plus a lack of strong leadership in the project after its earliest years, left it caught in limbo without sufficient funding to realize the potential it had demonstrated.

The 1974 Collier Trophy went to Dr. John F. Clark, NASA, and to Daniel J. Fink, General Electric Company, representing the NASA/Industry Team responsible for the Earth Resources Technology Satellite Program (later renamed Landsat). Landsat is shown in flight configuration with solar panels deployed after tests at the G.E. Valley Forge Plant. A Data Collection System, on board, gathered information from Earth based platforms then relayed data to a ground processing facility. (NASA photo no. 72-H-873).

---

4. Interagency conflicts were my main focus in Pamela E. Mack, *Viewing the Earth: The Social Construction of the Landsat Satellite System* (Cambridge, MA: MIT Press, 1990).

Building a Base of Support

In the early years of Landsat, advocates for Earth resources satellites built a complex web of political and bureaucratic support in order to make the project happen. Landsat developed in a period when NASA managers tried out the possibility of justifying the space program on the basis of its practical benefits, instead of simply appealing to the space race as a justification for popular and political support. Weather and communications satellites were already established on a firm footing by the mid-1960s; Landsat formed part of a new wave of interest in applications in the late 1960s, as space enthusiasts tried to limit post-Apollo cutbacks.[5] The trouble with practical applications, however, was that the benefits they brought inevitably fell within the responsibilities of some other agency or organization, and those agencies had their own interests (and usually much lower levels of research and development funding than NASA).

The use of satellites to observe the Earth for classified reconnaissance played a major role in the development of the U.S. space program, but civilian Earth observation satellites got off to a much slower start.[6] Geologists and geographers working in the classified reconnaissance satellite program saw that satellite data had potential value to civilian users, but the managers of the classified program attached great importance to keeping secret not only technological designs but also the capability of satellites. Therefore they not only prohibited civilian use of classified technology but also discouraged the development of a civilian Earth observation satellite program using unclassified technology.[7] However, pressure for a civilian program grew in the mid-1960s as scientists saw pictures of the Earth taken by astronauts with hand-held cameras and as studies of other planets provided examples of the potential of remote sensing.[8] In 1965, NASA started exploratory research on remote sensing of Earth resources with contracts to the U.S. Geological Survey (a branch of the Department of the Interior) and the Army Corps of Engineers for research using sensors flown in NASA aircraft.[9]

The Department of the Interior, and to a lesser extent the Corps of Engineers and the Department of Agriculture, quickly concluded that civilian Earth resources satellites had the potential to help them perform their assigned missions. However, each agency had different requirements and concerns about the future management of a satellite program, and NASA faced other needs and pressures as well. The Geological Survey wanted fairly fine resolution images that could easily be compared with maps and aerial photographs for studies of geology and natural resources (the Corps of Engineers had similar needs). The Department of Agriculture needed spectral accuracy; that is, agricultural scientists needed detailed information about the color of vegetation in order to differentiate crops

---

5. *Ibid.* pp. 18–27, provides a very brief history of communications and weather satellites. For surveys of the various applications programs see chapters in John Logsdon, et al., eds., *Exploring the Unknown: Selected Documents in the History of the U.S. Civil Space Program, Vol. 3*, to be published by the NASA History Office.

6. For the early history of classified reconnaissance satellites see Walter A. McDougall . . . *the Heavens and the Earth: A Political History of the Space Age*, (New York, NY: Basic Books, 1985); William E. Burrows, *Deep Black: Space Espionage and National Security* (New York, NY: Random House, 1986), and Robert A. McDonald, "Corona: Success for Space Reconnaissance, A Look into the Cold War, and a Revolution for Intelligence," *Photogrammetric Engineering and Remote Sensing* (1995): 689–720.

7. Mack, *Viewing the Earth*, pp. 33–49. For a particularly clear summary of the role of the intelligence agencies written by a participant see McElroy, "Preface," p. x.

8. Mack, *Viewing the Earth.* pp. 39–42.

9. Peter C. Badgley, "Current Status of NASA's Natural Resources Program," Proceedings of the Fourth Symposium on Remote Sensing of Environment, April 12–14, 1966, Willow Run Laboratories, University of Michigan, Ann Arbor, Michigan.

and detect disease.[10] NASA initially studied possibilities for an elaborate satellite carrying a number of sensors to meet a variety of requirements. However, the Department of Interior in particular wanted a relatively simple operational satellite quickly rather than an elaborate experimental program.[11]

In 1966, the Department of the Interior pulled off a public relations stunt in an effort to accelerate NASA's plans. The leaders of the U.S. Geological Survey persuaded Secretary of the Interior Stewart L. Udall to announce in September 1966 that the Department of the Interior was initiating its own operational satellite program. They proposed a simple satellite carrying just one kind of sensor to be launched in 1969.[12] The Department of the Interior did not have the necessary expertise (or a partner-agency with the necessary expertise) to start its own satellite program.[13] But even though the announcement did not reflect a realistic plan, it worked as a strategy. In the resulting controversy, NASA retained responsibility for experimental Earth resources satellites, but the space agency's leaders found themselves under pressure from the press and Congress to develop a satellite quickly.[14] NASA initiated a project initially called Earth Resources Technology Satellite (ERTS), then changed the name to Landsat in 1975.

Even with this pressure for a quicker and less ambitious experimental program, NASA planners and engineers sought to design a satellite more elaborate than the simple experiment proposed by the Department of the Interior. This decision reflected both the assumption of NASA program managers that their goal was to collect data of the maximum possible scientific sophistication and their interest in balancing the influence of the Department of the Interior by seeking to identify and satisfy the requirements of other users besides Interior. NASA had funded development of a multispectral scanner that could provide data more useful to agricultural scientists than the television-type camera (return beam vidicon) the Department of the Interior wanted. However, scanners had not yet been tested in space, and some engineers doubted that the sensor was ready for flight. The decision to include the scanner on Landsat resulted both from pressure from researchers at the Department of Agriculture and from the interest of NASA leaders in involving that department more fully in the project. NASA saw that support from the Department of Agriculture could provide a larger constituency for Landsat and could balance Interior's pressure for a quick transition to an operational program controlled by Interior rather than NASA.[15]

---

10. Mack, *Viewing the Earth*, pp. 66–79.
11. *Ibid*, pp. 52–60.
12. United States Department of the Interior, Office of the Secretary, Press Release, "Earth's Resources to be Studied From Space," September 21, 1966. The Department of the Interior called its program Earth Resources Observation Satellites (later Systems), or EROS.
13. The Weather Bureau played a similar game over management of the TIROS program, but in that case the Weather Bureau had made arrangements to cooperate with the Department of Defense instead of with NASA. See Richard LeRoy Chapman, "A Case Study of the U.S. Weather Satellite Program: The Interaction of Science and Politics," Syracuse University dissertation, 1967.
14. Charles F. Luce to Robert C. Seamans, Jr., October 21, 1966, with attached "Operational Requirements for Global Resource Survey by Earth-orbital Satellites." W. T. Pecora, Director, Geological Survey, to Under Secretary, Department of the Interior, "Status of EROS Program," Draft, June 15, 1967. For more details see Mack, pp. 56–65. Public pressure not only spurred NASA leaders to faster action than they had planned but also helped them overcome outside opposition to earth resources satellites. Resistance to the idea of a small earth resources satellite came not only from NASA leaders with an interest in a more sophisticated experiment but also from the intelligence community and their allies at the Bureau of the Budget who opposed any civilian earth resources satellite. See Mack, *Viewing the Earth*, pp. 58, 61.
15. Mack, *Viewing the Earth*, pp. 70–73.

This plan sounded good in theory, but lack of clarity about user needs. Meanwhile tensions between NASA and the user agencies made it almost impossible to build a base of support for the project. Different parts of the user community had different needs, and it was often not clear how satellite data could best serve agency missions. Users tended at first to expect the new technology to directly replace various older technologies. The agencies' own perceptions of their needs changed as they saw what the technology could do for them and as NASA sought to persuade them that data from a compromise satellite would have value to them even if it did not easily fit existing systems. Specific problems with requirements included a Department of the Interior expectation of map-like accuracy that required special correction of the data from either sensor. The Department of Agriculture, on the other hand, tended to expect the satellite data to directly replace aerial photographs, and therefore wanted fine resolution and frequent coverage of large areas, as well as spectral accuracy. In the fall of 1970, NASA project scientist William Nordberg described data specifications from the Department of Agriculture as "grossly overstated" and those from the Department of the Interior as "a bomb shell."[16] Limited by restrictions on resolution intended to prevent conclusions from being drawn about the capability of classified satellites, NASA had to persuade the users that data that did not meet these unrealistic requirements could still be useful.[17] Perhaps if funding had been plentiful project engineers could have designed a system more optimized for different users rather than trying to force compromises before users had time to learn from experience with satellite data. But in a tight funding situation, such disagreements simply became a justification for further cutbacks, and the project did not have an adequate chance to prove itself.

The Bureau of the Budget (later the Office of Management and Budget) opposed the project both because it did not appear that NASA and the users had their act together and for reasons entirely external to Landsat. The Budget Bureau refused approval for Landsat in late 1967, only to be overturned on appeal, and then in 1968 proposed cancellation of the project (and its replacement with an aircraft program). This early lack of support probably reflected most strongly opposition to Landsat from those involved in the classified reconnaissance satellite program. It also involved a general tendency for budget-makers to cut other parts of the NASA budget at a time when Apollo costs were high and policy-makers had moved on to other priorities, including the escalating cost of the war in Vietnam. In 1969, the Bureau of the Budget cut the Landsat budget from $41.5 to $10 million dollars, and eliminated funding for the Data Center proposed by the Department of the Interior, though NASA funding and minimal funds for the Data Center were restored on appeal.[18] By this point, cost-benefit analysis and lack of uniform enthusiasm from the users had become a key issue. The Bureau of the Budget required a whole series of cost-benefit studies, in which NASA had to justify Landsat not on the grounds of new benefits that would result but on the grounds of how the satellite project would save the government money by replacing old ways of doing things.[19]

---

16. William Nordberg to Wilfred E. Scull, "Review of User Agency Requirements for ERTS A in Response to September 17 Meeting at GSFC," October 14, 1970; William Nordberg to William Fischer, October 19, 1970.

17. Mack, *Viewing the Earth*, pp. 111–112. In fact, as in many cases of big technology advance predictions of usefulness of the data were often far from accurate; the multispectral scanner turned out to be more useful to the Department of the Interior than the Return Beam Vidicon and the Return Beam Vidicon was not included in later satellites.

18. The summary of BoB attempts to cut the program is based on Committee on Earth Studies, Space Science Board, National Research Council, *Earth Observations from Space: History, Promise, and Reality* (Washington, DC: National Academy Press, 1995), p. 113. For more explanation of this fight see Mack, *Viewing the Earth*, pp. 80–93 and Eisenbeis, *Privatizing Government Information*.

19. *Ibid.*, pp. 81–93.

Though NASA had to maintain the appearance of a united front within the Executive Branch, the agency sought to counter Bureau of the Budget opposition by trying to get stronger support from the user agencies and by playing to congressional interest in practical applications of the space program. Joseph E. Karth, chair of the House Subcommittee on Space Science and Applications, gave particularly strong support; he wanted to see the project move ahead and complained about the repeated cost-benefit studies that the Budget Bureau required for Landsat. He summed up his criticism of Executive Branch opposition to Landsat in 1969: "In looking at the history of ERS [Earth Resources Surveys], I come to the inescapable conclusion that there is a preponderance of evidence of foot-dragging, setting up of strawmen, and the assignment of unique and unusual and, I might say, ridiculous yardsticks, and so on and so forth."[20] In 1969, Congress actually restored funding to the Federal budget for the Landsat data processing system that had been requested by the Department of the Interior, but opposed by the Bureau of the Budget, and therefore not included in the budget that the President sent to Congress. However, the Bureau of the Budget refused to release the appropriated funds.[21]

While NASA did win enough funding to build and launch Landsat, the unending battles weakened the project. Even in 1971, just a year before the launch of the first satellite, the Budget Bureau proposed cancellation of the backup satellite.[22] Lack of funding and political support dramatically reduced NASA leaders' ability to respond to user needs, particularly as project managers sought increasingly to serve both experimental and operational needs. In addition, under pressure from the Bureau of the Budget, NASA had agreed to a conceptualization very different from weather satellites: "the program was permanently molded as a government R&D program that, once feasibility was demonstrated, would give way to a new commercial venture."[23]

Technology Development

Landsat showed both the strengths and the weakness of NASA's process for managing technological innovation. The project lacked strong leadership, but by the late 1960s NASA had a fairly standard pattern for research and development of satellite technology, and the Landsat satellite system fit into that pattern with few surprises. Management problems arose mostly in cooperation with the user agencies, an area in which NASA managers had less experience. The Landsat system involved a new combination of existing technologies; not significant technological innovation, though there were challenges in the data processing system and in development of applications. With more attempt to advance the state of the art, Landsat might have represented the kind of incremental development that had been successful at the NACA. However, an incremental approach did not work as well without a commitment to a long-term program and funding for continued improvements.

Landsat did not have consistent leadership or a strong institutional base even inside NASA. The first head of the Earth Resources Program Office, Peter Badgley, had been an early advocate for an Earth resources satellite, but he left NASA in 1968 and his successor left in early 1970.[24] In 1967, NASA leaders assigned Landsat development to the Goddard Space Flight Center, a Center with more expertise in space science than space applications. The team for the project came in significant part from a group that was finishing up

---

20. Joseph E. Karth, "Earth Resources Surveys–An Outlook on the Future," presented at an IEEE meeting, February 13, 1969, attachment to Bernard P. Miller to Tom Ragland, March 3, 1969, #002824, 81-416 (3), Record Group 255, Washington National Records Center, National Archives and Records Administration, Washington, D.C.
21. Mack, *Viewing the Earth*, pp. 86.
22. Committee on Earth Studies, *Earth Observations from Space*, p. 113.
23. McElroy, "Preface," p. xi.
24. Mack, *Viewing the Earth*, p. 95.

work on the Orbiting Geophysical Observatory satellite project. A reorganization of NASA management in 1972 introduced another level of complexity by establishing an Earth Resources Survey Program Office at the Johnson Space Center (abolished in 1977). NASA Headquarters assigned some of its coordination and evaluation duties to this office, which also strengthened the role of the Johnson Space Center in applications development and technology transfer for Landsat.[25]

A series of interagency committees provided coordination between NASA and the user agencies, with some success on the working level but limited support from higher levels. In 1968, an Earth Resources Survey Program Review Committee was formed to make policy for Landsat but it met only every few months and tended to provide mostly approval of decisions already made by NASA.[26] This committee did play a major role in the battle over Landsat specifications discussed above, but not in as much detail as the users would have liked. A cartographer at the Department of the Interior complained: "Specifications which vitally effect [sic] the users are being made without the users being properly informed. It is believed that this situation, if allowed to continue, will result in a satellite being flown the data from which cannot properly be utilized by the users."[27] That prediction was too dire, but certainly the users did not get everything they thought they needed. In 1972, the Program Review Committee was replaced with an Interagency Coordinating Committee: Earth Resources Survey Program, but NASA kept control of the new committee.

Landsat managers and engineers found that their task was not to develop new technology for Landsat, but only to decide between alternative approaches for the use of existing technology. When the Goddard group initiated a concept study for Landsat in 1967, they received unsolicited proposals from a number of aerospace companies for Earth resources satellites using technology that those companies had already developed (both for classified programs and for planetary probes). The Goddard group decided on a relatively small satellite using existing satellite technology; only the sensors required significant development and even they represented modifications of existing designs.[28] NASA selected General Electric as the satellite's contractor, with a plan to use the satellite bus (structure and common systems, such as power supply) already developed for the Nimbus experimental weather satellite. The program followed a standard NASA management system, called Phased Project Planning, and suffered only minor snags in coordinating contractors.

The data processing system provided more of a challenge, but again the research and development took place almost entirely in industry. The Bureau of the Budget wanted to keep the Landsat experiment as small as possible, and particularly did not want to see a de facto operational satellite created under the cover of an experiment. Therefore, the Budget Bureau repeatedly cut funding for the data distribution system to be provided by the Department of the Interior, and restricted the data processing system developed by NASA to the minimum size necessary to meet the requirements of a small number of scientific investigations.[29] Landsat data processing represented a new combination of challenges for NASA. The planetary program had provided agency engineers with experience with high resolution image data, which involved very large amounts of data for each image, but with Landsat that large data rate would be sustained for years (instead of for a

---

25. *Ibid.*, p. 97–98. For the larger issue of the role of the Johnson Space Center in applications development see pp. 146–58–users sometimes saw the JSC projects as an attempt to preserve jobs for people who had worked on Apollo.

26. *Ibid.*, p. 99.

27. Alden P. Colvocoresses to Research Coordinator, EROS program, "Liaison with Goddard on ERTS-A and -B," January 1970; "Landsat 1 Documentation" folder, NASA Historical Reference Collection, NASA History Office, NASA Headquarters, Washington, DC.

28. Mack, *Viewing the Earth*, p. 68–73, 101.

29. Committee on Earth Studies, *Earth Observations from Space*, p. 112.

few weeks at a time, as had been the case with planetary probes) and the data was much more useful if it was available quickly. In addition, project managers had to make decisions about the data processing system at a time when a new approach, digital image processing, was clearly the wave of the future but had not yet been proven for large-scale use. Given tight budgets and requirements that Landsat be designed only as a short-term experiment, Landsat managers decided to stick to the older technology (analog processing). The resulting data processing system was outdated before it was built and inadequate to meet user expectations. Planning for a major upgrade to a digital system began as early as a year after the launch of the first satellite in 1972, though lack of funding delayed the installation of an improved system until 1980.[30] In this case project leaders clearly chose (or were forced to chose) to use well-established technology rather than undertake research and development to meet a new challenge.

NASA took some role in the development of applications for Landsat data, but in most cases it simply provided funding to scientists and user agencies. NASA's largest in-house project was an effort to develop agricultural uses for Landsat data, conducted at the Johnson Space Center in partnership with the Department of Agriculture and the National Oceanic and Atmospheric Administration from 1974 to 1978. NASA engineers (some of whom had previously worked on Apollo) took a large-scale, brute force approach, setting up a computer system to measure the area planted in various crops and monitor and predict from weather data how well they were growing. The Department of Agriculture did not put the resulting system into use, in fact the Foreign Agricultural Service instead developed its own system using a different approach. Instead of calculating areas planted each year, the Foreign Agricultural Service system compared Landsat images from year to year, and analyzed only at the area where it detected differences.[31] Overall, NASA had some successes in developing applications for Landsat data, but potential users were often reluctant to consider applications developed by NASA because they assumed that any technology developed by the space agency would be too expensive and too sophisticated for everyday, practical use by resource managers.[32]

For a project like Landsat, NASA did almost none of the kind of original research and development that had characterized the NACA. The job of the space agency had become funding and managing research and development by industry, and in the case of a project with tight funding and important public relations implications like Landsat the space agency preferred to minimize technological risk. Landsat did not represent research that could only be done by the Federal government, and it only marginally fit the definition of a project too risky or too long-term for private investment. Landsat was a government enterprise because NASA leaders thought it would benefit the space program to show more practical results, and political issues such as the proper use of data collected over other countries made the government nervous of allowing private enterprise to get into the business in the early years.[33] An alternate model-that the government might provide Earth resources satellite data as a public good like weather satellite data-never gained official acceptance.

---

30. Mack, *Viewing the Earth*, pp. 107–18.
31. *Ibid.*, pp. 150–58.
32. *Ibid.*, pp. 159–70.
33. *Ibid.*, pp. 180–82, 185–88.

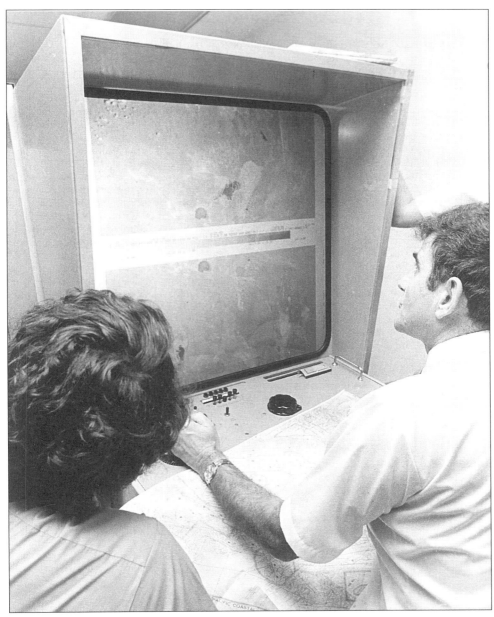

Scientists view a Landsat enlargement on a special machine in the control center. Federal agencies participating in the Landsat project included the Department of Agriculture, Commerce, Interior, Defense, and the Environmental Protection Agency. (NASA photo no. 72-H-1065).

Success and Failure

Landsat data proved its value to many users after the launch of the first satellite on July 23, 1972. However, success did not end the political travails of the project. In particular two problems resulted in continuing uncertainties about the future of the project. First, actual operational use of the data did not live up to the predictions that had been made by project supporters when they had campaigned for support of the project in the years before launch. Second, further development of satellites for civilian Earth observation became stalled in a fight over how an operational program should be conducted. In both cases difficulties arose from conflicting interests between NASA and the users. These were complicated by tensions about the proper relationship between research and practical applications and the proper role of the government as a technology moved along that spectrum.

In technical terms the project proved almost completely successful. The satellite functioned as planned and delivered the promised data, which provided information of value to scientists studying agriculture, geology, land use, and in many other fields. The sensor that had seemed more of a risk during development, the Multispectral Scanner, proved particularly valuable because the data from different spectral bands (in effect different colors) could be compared accurately. Satellite data proved most dramatically beneficial to developing countries; in many cases Landsat images provided the first adequate maps of remote regions.[34] Scientists found much useful information, however, even for well mapped areas. Landsat data could indeed provide information on everything from urban growth to ice cover in shipping lanes to the health of vegetation. The awarding of the Collier Trophy in 1974, two years after launch, reflected at least in part a large number of successful scientific experiments demonstrating that useful information of many different types could be extracted from Landsat data.[35] Given the continuing lack of political support for Landsat its continuing success was hardly assured in 1974, but the project had met its initial goals.

In 1974, supporters of Landsat would probably have recognized that the project was not ready for commercialization and hoped for a relatively quick transition to a government-controlled operational remote sensing system on the model of the weather satellite system, housed either in NASA or in the Department of the Interior. They would have expected an operational system to involve improvements in the satellites and the data processing system (particularly to deliver data more quickly after the satellite collected it), more user involvement, and, most important, a commitment to data continuity (that is, to launching another satellite before or quickly after the operating satellite failed). Initially, opponents of that vision argued that the satellite, while successful in the narrow sense, had not proved useful enough to establish the need for an operational system.

Indeed, operational use of Landsat data grew much more slowly than its proponents had predicted.[36] Partly this resulted from overly optimistic predictions. A 1985 study summed up Landsat's problem: "Large but unverifiable estimates of benefits from space remote sensing were used to 'sell' the program, within NASA and within the administration.

---

34. See Arnold W. Frutkin to distribution, "Some Recent International Reactions to ERTS-1," December 22, 1972, and Mack, pp. 189–92.
35. See for example, Stanley C. Freden and Enrico P. Mercanti, eds., *Symposium on Significant Results Obtained from Earth Resources Technology Satellite-1, Volume 3: Discipline Summary Reports* (Greenbelt, MD: Goddard Space Flight Center X-650-73-155, May 1973).
36. Mack, *Viewing the Earth*, pp. 139–41.

This photograph was taken from an altitude of 914 kilometers (568 statute miles) of the Santa Barbara, California, area about 11 a.m. on October 4, 1972. Some of the notable landmarks photographed include: Sierra Madre Mountains (center); Santa Barbara, California (lower center); Vandenberg AFB (left center); Santa Rosa Island (bottom left center); Santa Cruz Island (bottom right center); Ventura, California (lower right center); Bakersfield, California (upper right). (NASA photo no. 73-H-114).

Unfortunately, some of the early flamboyant and unrealizable projections of benefits later came back to haunt the program."[37] In addition, selling a new system for domestic operational needs proved difficult because of user resistance to changing existing systems. Potential users often found only marginal benefits from replacing existing data sources with data from Landsat, and the old ways of doing things often had strong reinforcement from constituencies.[38] More benefits came from using the new data in new ways, and NASA provided some funding and a lot of free data to researchers who studied more innova-

---

37. Committee on Practical Applications of Remote Sensing from Space, Space Applications Board, National Research Council, *Remote Sensing of the Earth from Space: A Program in Crisis* (Washington, DC: National Academy Press, 1985).
38. Mack, *Viewing the Earth*, pp. 141–45, 151–55. For example, the Department of Agriculture realized it would be politically unpopular to substitute satellite data for the employment of people all over the country to conduct agricultural surveys.

A four sequence image of the Rocky Mountains Wind Range in Wyoming, taken by Landsat, where the amount of snow cover changes from winter to summer. By the precise change of acreage in this snow cover, hydrologists have computed the water run-off from the mountains for the whole watershed and assessed the amount of water available for irrigation and human consumption. (NASA photo no. 74-H-504).

tive applications that made better use of the real advantages of Landsat data. Two NASA scientists summed up their view of the results in 1985: "The examples and the capability discussed here clearly illustrate the overall success of the program. Landsat data have resulted in totally new methodologies for resource inventory and environmental assessment for a worldwide community of users and as such have served an important role in bringing resource managers into the computer age."[39] However, the Bureau of the Budget had insisted that Landsat prove itself by replacing existing systems, and NASA leaders did not fully understand the difficulty of persuading users to adopt new techniques that supplanted, or greatly modified, existing systems.[40]

---

39. P. K. Conner and D. W. Mooneyhan, "Practical Applications of Landsat Data," In Abraham Schnapt, ed., *Monitoring Earth's Ocean, Land, and Atmosphere from Space–Sensors, Systems, and Applications,* Progress in Astronautics and Aeronautics vol. 97 (New York, NY: American Institute of Aeronautics and Astronautics, 1985), p. 391.

40. Mack, *Viewing the Earth,* pp. 123–29, 159–70. NASA struggled with only limited success to develop successful programs for technology transfer; see Samuel I. Doctors, *The NASA Technology Transfer Program: An Evaluation of the Dissemination System* (New York, NY: Praeger Publishers, 1971), and Granville W. Hough, *Technology Diffusion: Federal Programs and Procedures* (Mt. Airy, MD: Lombard Books, 1975).

This is one of the first photos taken by Landsat. Taken from an altitude of about 900 kilometers (560 statute miles), this photo shows several features in the Dallas/Fort Worth, Texas, area. The cities of Dallas and Fort Worth are at the bottom of the photo. (NASA photo no. 72-H-1044).

The continuing definition of the project as experimental also provided a major barrier to effective operational use. The Bureau of the Budget defined the project as strictly experimental: the Bureau "placed stringent limits on the throughout capability of the ground data processing system" and refused to fund a system that could process data quickly—"as a result, any exploratory use of Landsat data that required rapid access to processed data was precluded at the outset."[41] The Landsat data processing system could not meet all of the demands of operational use, and in any case potential users hesitated to invest in expensive new systems to use Landsat data while the project was still experimental and data continuity not guaranteed. In turn, the Office of Management and Budget would not approve the transformation of Landsat into an operational system until widespread use proved its value.[42] Without any separation between an experimental and an operational program, operational needs squeezed out innovation.[43] As delays in an

---

41. Committee on Earth Studies, *Earth Observations from Space*, p. 112.
42. Bruno Augenstein, Willis H. Shapley and Eugene Skolnikoff, "Earth Information From Space By Remote Sensing," Reported prepared for Dr. Frank Press, Director, Office of Science and Technology Policy, Executive Office of the President, June 2, 1978.
43. For example, a plan to launch a fourth Landsat satellite carrying only a new sensor–the thematic mapper–resulted in strong protests from users who wanted the satellite to carry the older sensor as well to provide data continuity. See for example M. Mitchell Waldrop, "Imaging the Earth (I): The Troubled First Decade of Landsat," *Science* 215 (March 26, 1982): 1600–03.

operational decision mounted, Landsat became increasingly out of date; in 1986, France launched an Earth resources satellite named SPOT (Systeme Probatoire d'Observation de la Terre) carrying more technologically-advanced sensors providing finer resolution.

The creation of an operational Earth resources satellite program in the United States became snagged not only in questions about effective use of the system but also in political interests, in particular a new emphasis on privatization. President Jimmy Carter made a priority of reducing the size of the Federal government, and his staff identified Earth resources satellites as one of the best candidates for transfer of a government function to private industry. Landsat appeared to be a perfect case because commercial success had been one of the promises of the early cost-benefit studies: "the original approval for Landsat was predicated on private markets growing to the point of having the capability to fully fund all system cost."[44] In October 1978, President Carter officially requested that NASA and the Department of Commerce investigate ways to encourage private industry participation in civilian remote sensing (including Landsat, weather satellites, and ocean observation satellites).[45] However, the disappointing demand for data gave private industry doubts about the profitability of Landsat. It quickly became clear that privatizing the project would not be an easy task.[46]

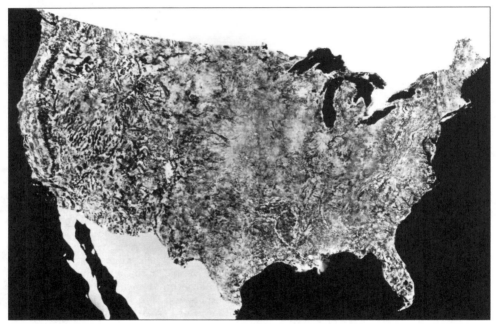

A giant photo map of the contiguous forty-eight states (1974) of the United States, the first ever assembled from satellite images, completed for NASA by the U.S. Department of Agriculture's Soil Conservation Service Cartographic Division. (NASA photo).

---

44. Committee on Earth Studies, *Earth Observations from Space*, p. 110.
45. Also in October, Senator Harrison Schmidt introduced a bill calling for the creation of an Earth Resources Information Satellite Corporation modeled on Comsat. No action was taken on the bill. Science Policy Research Division, Congressional Research Service, "United States Civilian Space Programs. Volume II: Applications Satellites," Prepared for the Subcommittee on Space Science and Applications of the Committee on Science and Technology, U.S. House of Representatives, May 1983, pp. 249–50.
46. "Private Sector Involvement in Civil Space Remote Sensing," prepared by an Interagency Task Force consisting of NASA, Dept. of Commerce/NOAA, Dept. of the Interior, Dept. of Agriculture, Dept. of Defense, Environmental Protection Agency, U.S. Army Corps of Engineers, and the Dept. of State, Draft, June 4, 1979.

Meanwhile, Landsat was stuck in place until the President made a decision on an operational system. NASA launched additional satellites that tested relatively minor improvements in technology and provided the data users needed, but the space agency had authorization neither for an ambitious research program to develop new generations of sensors nor for an operational program that would meet the needs of users for an assured supply of data. In addition, the wide range of users complicated the decision on an operational system: other user agencies did not want the Department of the Interior to take responsibility for an operational system because they feared that Interior would not serve their interests.[47] Faced with these constraints, Carter chose a short-term solution in November 1979. He gave the National Oceanic and Atmospheric Administration (NOAA) temporary responsibility for managing an operational Landsat system and asked it and its parent agency, the Department of Commerce, to study ways to encourage industry participation with the long-term goal of eventual operation by the private sector.[48]

President Reagan attached an even higher value to privatization than Carter, and in March 1983 he announced a decision to transfer Landsat, weather satellites, and future ocean observation satellites to private industry.[49] Congress strongly rejected the idea of privatizing weather satellites, but the Department of Commerce proceeded with a request for proposals from private industry to take over Landsat.[50] Congress passed a bill setting the terms for transfer, and the Earth Observation Satellite Company (a joint venture of Hughes and RCA) won the competition and took over the program.[51] The new company started out in a weak position; the Federal government provided only a small subsidy for the transition period and no guaranteed Federal data purchases, yet Congress continued oversight by holding hearings (in some cases leading to amendments to the original law) about the future of Landsat and the concerns of government agencies that used Landsat data.[52] By 1985, one Landsat supporter had concluded that "There seems to be little doubt that the present market cannot sustain the operating costs of a land observing system, to say nothing of the capital costs."[53]

---

47. Mack, *Viewing the Earth*, pp. 201–07.
48. "United States Civilian Space Programs. Volume II: Applications Satellites," pp. 238–42. For issues relating to the transition to an operational system see Richard D. Lamm to George S. Benton, NOAA, April 30, 1980, with attached "Recommendations of the National Governor's Association, National Conference of State Legislatures, Intergovernmental Science, Engineering and Technology Advisory Panel, National Resources and Environment Task Force, for the Final Transition Plan for the National Civil Operating Remote Sensing Program (first draft April 10, 1980)." For problems in cooperation between NASA and NOAA see National Research Council, R*emote Sensing of the Earth from Space: A Program in Crisis* (Washington, DC: National Academy Press, 1985).
49. "Statement by Dr. John V. Byrne, Administrator, National Oceanic and Atmospheric Administration, U.S. Department of Commerce," March 8, 1983. The presumption was that one company might take on all three programs, reflecting a proposal from Comsat to take over weather and earth resources satellites together because weather satellites were expected to be more profitable in the short term. Communications Satellite Corporation News Release, "Comsat President Proposes Bold Restructuring of Earth Sensing Satellite Systems," July 23, 1981. For a detailed survey of the debate over commercialization see Eisenbeis, *Privatizing Government Information*, ch. 1.
50. "Weather Satellites," *Congressional Record*, October 20, 1983, S 14367; "Transfer of Civil Meteorological Satellites," *Congressional Record*, November 14, 1983, H 9812–9822. See Hill p. 60. Dept. of Commerce, "Request for Proposals for Transfer of the United States Land Remote Sensing Program to the Private Sector," January 3, 1984.
51. Public Law 98-365, July 17, 1984.
52. Eisenbeis, *Privatizing Government Information*, pp. 49–52.
53. John H. McElroy, "Earthview–Remote Sensing of the Earth from Space," in Schnapt, ed., *Monitoring Earth's Ocean, Land, and Atmosphere*, p. 39. McElroy had been involved in the project in a number of different positions, but at that point was working at the National Oceanic and Atmospheric Administration. In the paragraphs following the quote he proposes an economic justification for a federally funded Landsat program to serve the public good.

The attempt at privatization failed in 1992. Because the corporate owner had never had significant new resources to invest in the system, little had been accomplished during the privatization period. The Land Remote Sensing Policy Act of 1992 ended "the 'experiment' which had so negatively affected the research use of remote sensing data acquired from the Landsat satellites."[54] The new law repealed the commercialization act of 1984 and transferred responsibility for Landsat from the Department of Commerce to NASA and the Department of Defense, which had found the broad coverage of Landsat data useful during Desert Storm.[55] After disagreements over funding the Department of Defense withdrew in 1994, and NASA resumed sole responsibility for Landsat, with plans to launch one more satellite.[56] Failure of Landsat 6 in October 1993, frequent changes in NASA's overall remote sensing plans, and increasing competition from other countries and possibly from private industry, left the future of the program uncertain.[57]

While Landsat commercialization had failed, interest in commercial remote sensing continued to grow.[58] Private industry could almost certainly sustain an Earth resources satellite that provided data similar to Landsat (though probably lacking some of the features scientists want) if the government would guarantee a significant purchase of data each year, or if the owner could offer commercial users exclusive use of certain data for a higher price. However, such a satellite would most likely not provide data of as much scientific value as that provided by Landsat. A private company would probably only collect data as ordered, rather than providing comprehensive coverage to build up a historical archive of data for later comparison, and would probably not invest as much in the precision of the sensors, since such precision is needed for only a few uses, mostly scientific.

The proper roles of the government and private industry became less and less clear as technology advanced. Landsat became less dauntingly "big technology," new innovations in the 1990s made it possible to design a much smaller and less expensive satellite with similar capabilities. Such a satellite was no longer too expensive for private companies to undertake without Federal subsidy.

Landsat was a relatively small project by NASA standards, but because of its practical goals it shows particularly clearly the problems of building a constituency for big science and technology projects and the complexities involved in determining the proper role of the government in the spectrum between research and practical applications. While the NACA had successfully served industry needs by providing background research rather than building whole new systems, NASA leaders found big projects with practical benefits much more problematic than projects oriented towards scientific research or exploration. NASA could justify a certain amount of basic science as worth doing for its own sake, but once a project was justified on the basis of its practical benefits then why was the government doing it rather than leaving it to private industry who presumably could make a profit by selling such beneficial data? At least for NASA, the public good has become increasingly difficult to define and use as a justification. One long-time participant in the program wrote in frustration: "One of the great conundrums of the Federal programs of the space age is that the more likely something is to be useful the more difficult it will be to sustain it."[59]

---

54. Eisenbeis, *Privatizing Government Information*, p. 157. Eisenbeis writes as a scholar of information management, but she had some experience inside the project.
55. *Ibid.*, pp. 156–58. W. Henry Lambright, "The Political Construction of Space Satellite Technology," *Science, Technology & Human Values* 19 (1994): 56.
56. Committee on Earth Studies, *Earth Observations From Space*, p. 114.
57. For a scathing review of the overall situation see Committee on Earth Studies, *Earth Observations from Space*.
58. William Stoney, "Landsat 8's World" briefing charts, February 2, 1996.
59. John McElroy, "Preface," p. vii.

Chapter 11

Voyager: The Grand Tour of Big Science

by Andrew J. Butrica

Of all the NASA missions, none has visited as many planets, rings, and satellites, nor has provided as many fresh insights into the outer planets, as Voyager, which was launched in 1977. On 19 May 1981, the National Aeronautic Association awarded its Collier Trophy to the "Voyager Mission Team, represented by its chief scientist Dr. Edward C. Stone, for the spectacular flyby of Saturn and the return of basic new knowledge of the solar system."[1] The awarding of the Collier Trophy was a fitting tribute to the science carried out by the Voyager spacecraft, which also received twice, in 1980 and 1981, respectively, the Dr. Robert H. Goddard Memorial Trophy, an aerospace industry prize awarded annually since 1958 by the National Space Club to recognize achievement in astronautics, for the Voyager encounters with Jupiter and Saturn.[2]

Neither the Goddard nor the Collier Trophy recognized completely the science accomplished by Voyager, for after flying by Uranus (1986) and Neptune (1989), it left the solar system to explore interstellar space until around 2020, when the spacecraft will lack sufficient power to operate the scientific instruments on board and to return data to Earth. By then, the two Voyager spacecraft will have operated longer, and returned data from greater distances, than any previous probe.

Voyager is planetary exploration on a grand scale. First conceived as a "Grand Tour" of the solar system from Jupiter to Pluto, then scaled back to a more modest mission called Mariner Jupiter-Saturn until its incarnation on the eve of launch as Voyager, the mission has been, and will remain well into the future, NASA's biggest planetary expedition. The two Voyagers have explored more planets (four), have discovered more moons (22), and have returned more photographic images, than any other space flight.[3] The original price tag of nearly a billion dollars made it the second most expensive planetary voyage, exceeded only by Viking, which landed on Mars in 1976.[4] Each Voyager spacecraft weighed more than any Surveyor or Ranger sent to the Moon and more than any Mariner or Pioneer probe (except for Pioneer Venus), though less than the combined weight of the Viking lander and orbiter.[5]

Its scientific, budgetary, and technological immensity makes Voyager archetypical big science. Born of what President Dwight D. Eisenhower called the military-industrial complex, and what historian Stuart Leslie more recently has called the military-industrial-

---

1. Wording cited from the Collier Trophy held at JPL and conveyed to the author in a memorandum from Edward Stone dated 28 November 1994. See also, Bill Robie, *For the Greatest Achievement: A History of the Aero Club of America and the National Aeronautic Association* (Washington, DC: Smithsonian Institution Press, 1993), p. 235.
2. Bruce Murray, *Journey into Space: The First Three Decades of Space Exploration* (New York, NY: W. W. Norton & Company, 1989), pp. 161-62; Gita Siegman, ed., *Awards, Honors & Prizes*, 6th ed. (Detroit, MI: Gale Research Company, 1985), 1: 440. The Trophy consists of a bronze sculpted bust of Goddard.
3. Magellan sent back more images per se, but those were range-Doppler images created by a synthetic aperture radar. Unless otherwise stated, information is from the NASA Headquarters web site: http://www.hq.nasa.gov/office/pao/History/
4. S. Ichtiaque Rasool, interview with author, Paris, December 12, 1994.
5. Fact sheet on missions prepared by JPL Archives.

The Voyager mission team, represented by Chief Scientist Dr. Edward C. Stone received the award in 1980, specifically for the spectacular fly-by of Saturn and the return of basic new knowledge of the solar system. This photo is a montage of images of the Saturnian system prepared from an assemblage of images taken by the Voyager spacecraft during its Saturn encounter in November 1980. (NASA photo no. 80-H-366).

academic complex,[6] big science quickly came to characterize the civilian enterprise to explore space, that is, what one might call the NASA-industrial-academic complex. Since its creation in 1958, the National Aeronautics and Space Administration has shaped American science to an extraordinary degree, namely by providing the financial and institutional aegis for the transformation of American planetary astronomy into big science, yet NASA's primary objective was (and whose budgetary bulk paid for) the designing, building, and launching of vessels for the exploration of the solar system.

Although the Voyager mission is inescapably an example of NASA big science, the actual scientific experiments were carried out by scientists employed by NASA Field Centers or by individual scientists who more appropriately fit the category of little science. The latter Voyager scientists worked individually or in small collaborative groups, often with graduate assistants, in university laboratories with relatively small budgets and limited laboratory equipment. In the case of Voyager, the management of decision making and the organization of scientists, just as much as the creation and utilization of monumental technology and mammoth technological networks, delineated big science.

Planetary astronomy has had a long existence as simultaneously both little science (astronomers working individually or in small groups) and big science (large expensive telescopes and observatories) that dates back to the sixteenth-century island observatory of Tycho Brahe. The number, sophistication, and expense of instruments have escalated over

---

6. Stuart W. Leslie, *The Cold War and American Science: The Military-Industrial-Academic Complex at MIT and Stanford* (New York, NY: Columbia University Press, 1993).

the centuries, particularly in the past 100 years. The interplanetary spacecraft has become the new observatory, carrying scientific instruments on trajectories independent of the Earth's course through space. Planetary astronomy's very dependence on instrument technology necessarily and inescapably has driven it in the direction of big science.

The Voyager mission, and NASA planetary missions in general, illustrate the amphibious life of planetary astronomy as both little science and big science. The Voyager project transformed geographically-dispersed individual scientists drawn from a spectrum of scientific disciplines and subdisciplines into members of a centralized, multidisciplinary big science team. As each Voyager spacecraft approached one of its target planets, the members of the mission's scientific teams arrived in Pasadena, the home of the Jet Propulsion Laboratory, to take up residence for the period of closest approach. The mission provided those scientists a set of instruments and a spacecraft observatory. Their role was not limited to using the spacecraft instruments, however; those scientists also played a critical role in shaping the mission even before it was funded principally through the Space Science Board and its summer studies. Conflict between the scientific community and the NASA Field Centers, in particular, served as the catalyst that brought about the demise of Voyager's predecessor, Grand Tour. This chapter examines the role of scientists in the shaping of Voyager before launch and their transformation into a big science project team through their participation in the Voyager mission, then considers the critical role of technology in the conduct and success of that mission's science, as well as the relationship between big science and little science and the role of technology in that relationship.

Voyager can be said to have begun in 1965 as Grand Tour, an extensive, if not grandiose, planetary mission planned in the midst of shrinking NASA and Federal budgets, at a time when NASA sought to define its mission in the post-Apollo era. The Apollo lunar program in 1965 was reaching its funding peak; NASA's annual overall budget declined from $5.2 billion in 1965 to slightly over $3 billion in 1972,[7] in response to social and political pressure on the Federal budget stemming largely from the Great Society programs and the Vietnam War, as well as the conservative fiscal policy of the Nixon administration.

In the summer of 1965, in order to define post-Apollo NASA missions, the National Academy of Sciences' Space Science Board[8] held a summer study of scientists at Woods Hole, Massachusetts. The scientists urged NASA to shift interest from the Moon to the planets, giving primary emphasis to Mars and Venus, more so than to the outer planets. As for the outer planets, the summer study recommended two directions: either reconnaissance flyby missions to each of the outer planets or an intensive study of Jupiter using orbiters and atmospheric entry probes.[9] These two exploration strategies dominated discussions of outer planet exploration over the following years. The 1965 Woods Hole summer study thus demonstrated that the congeries of scientists who made up the planetary scientific community already had ideas about how NASA ought to set about exploring the outer planets.

Most members of the planetary science community preferred smaller, tested spacecraft flying short missions over large, expensive, complex and lengthy projects. They feared that the government might cancel their smaller projects in times of tight budgets in favor of a few expensive high-profile missions. Moreover, with small inexpensive spacecraft launched at rela-

---

7. Craig B. Waff, "The Struggle for the Outer Planets," *Astronomy* 17 (1989): 44; David Rubashkin, "Who Killed Grand Tour?" ms., National Historical Reference Collection, NASA History Office, NASA Headquarters, Washington, DC. p. 15. Dollar amounts are not adjusted for inflation. The NASA budget decline in real dollars, therefore, was much more dramatic.

8. Established in June 1958 when the National Academy of Sciences combined the functions of the International Geophysical Year technical panels on rocketry and the Earth satellites into a single agency, the Space Science Board was the chief agency of the scientific community for advising NASA on space programs and for serving as its watchdog.

9. *Space Research: Directions for the Future* (Washington, DC: National Academy of Sciences, 1966).

tively short intervals, scientists could more easily follow up on new discoveries than they could with one large complicated spacecraft that took many years of preparation. Major missions to a large degree tended to solidify research into a specific line of investigation for a long time.[10]

Into the gelling consensus that emerged from the Woods Hole study came the idea for Grand Tour. The Grand Tour would take advantage of a once-every-175-year planetary alignment to send several spacecraft to all five of the outer planets, from Jupiter to Pluto. Launch windows were available relatively soon, between 1976 and 1980.[11] Despite its subsequent reputation as an exorbitant expenditure of public funds, a pair of Grand Tour spacecraft actually would have been far more economical than the several individual probes to the outer planets proposed by scientists at Woods Hole in 1965. Grand Tour could reduce costs further by surveying the outer planets in less time—in eight to thirteen years, depending on the trajectory, compared to thirty years for a direct flight to Neptune alone—by employing a maneuver called gravity assist,[12] in which the spacecraft exploited a planet's gravitational field to increase its velocity and alter its trajectory, thereby reducing both launch power requirements and flight time.[13] Grand Tour thus was intrinsically a money-saving concept.

Appearing to save money was critical to selling a large-scale project, even during the days of big NASA budgets, as illustrated by the recollection of Donald P. Hearth, NASA Planetary Programs Office director, when he learned about Grand Tour for the first time:[14]

> *You've got to remember selling a new start is a bitch. Even then-it's even worse today, but even then. It's almost as hard to sell a hundred million dollar project as it is a billion dollar project. And a hell of a lot more work to sell two $100-million projects than one $200-million project.*

Before NASA Headquarters considered Grand Tour, though, the Jet Propulsion Laboratory[15] started promoting it, beginning with a December 1966 article penned by

---

10. Homer E. Newell, *Beyond the Atmosphere: Early Years of Space Science*, (Washington, DC: NASA SP-4211, 1980), pp. 406–407.

11. James E. Long, "To the Outer Planets," *Astronautics & Aeronautics* 7 (June 1969): 32.

12. The origins of the gravity assist maneuver are lost in the many and conflicting attempts to determine those origins.

13. Gary A. Flandro, "Utilization of Energy Derived from the Gravitational Field of Jupiter for Reducing Flight Time to the Outer Solar System," pp. 12–23 in JPL, *Space Programs Summery No. 37–35, Volume IV, for the period August 1, 1965 to September 30, 1965* (Pasadena, CA: JPL, October 31, 1965); Gary A. Flandro, "Fast Reconnaissance Missions to the Outer Solar System Utilizing Energy Derived from the Gravitational Field of Jupiter," *Astronautica Acta* 12 (1966): 329–37; Michael A. Minovitch, *The Determination and Characteristics of Ballistic Interplanetary Trajectories under the Influence of Multiple Planetary Attractions*, Technical Report No. 32-464 (Pasadena, CA: JPL, October 31, 1963); Michael A. Minovitch, *Utilizing Large Planetary Perturbations for the Design of Deep-Space, Solar-Probe, and Out-of-Ecliptic Trajectories*, Technical Report No. 32-849 (Pasadena, CA: JPL, December 15, 1965).

14. Donald P. Hearth, interview with Craig B. Waff, Boulder, CO, August 7, 1988, cited in Craig B. Waff, ch. 3, "The Next Mission: Grand Tour or Jupiter-intensive?" in *Jovian Odyssey: A History of NASA's Project Galileo*, unpublished manuscript, NASA Historical Reference Collection, pp. 6–7.

15. JPL was unique among NASA Field Centers. It existed long before the creation of NASA in 1958, but not as a NACA laboratory. JPL was the child of California Institute of Technology faculty interested in rocket research. A Presidential order of December 1958 transferred JPL to NASA, but developing an effective relationship between NASA and JPL took time. The laboratory grounds, buildings, and equipment belonged to the Government, while the laboratory personnel originally came from Caltech. During the 1960s NASA management frequently debated the question of the NASA-JPL relationship: Should JPL be regarded as another NASA Field Center (an insider) or treated as a contractor (an outsider)? JPL was proud of its academic connection, despite the tenuous and often disregarded nature of that connection, and Caltech accorded the laboratory a good measure of independence to plan and execute its own research program. Clayton R. Koppes, *JPL and the American Space Program: A History of the Jet Propulsion Laboratory* (New Haven, CT: Yale University Press, 1982), esp. pp. ix, 4–5, 10–17, 20, 38, 45 and 65; Newell, *Beyond the Atmosphere*, pp. 258–63.

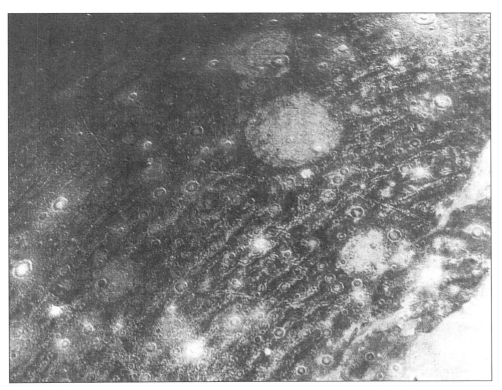

This reconstruction of part of the northern hemisphere of Ganymede was made from pictures taken by Voyager at a range of 313,000 kilometers (194,000 miles). The scene is approximately 1,300 kilometers (806 miles) across. It shows part of a dark, densely cratered block which is bound on the south by lighter and less cratered, grooved terrain. The dark blocks are believed to be the oldest parts of Ganymede's surface. Numerous craters are visible, many with central peaks. The large bright circular features have little relief and are probably the remnants of old, large craters that have been annealed by flow of the icy near-surface materials. The closely spaced arcuate, linear features are probably analogous to similar features of Ganymede which surround a large impact basin. The linear features may indicate the former presence of a large impact basin to the southwest. (NASA photo no. 79-H-393).

Homer Joe Stewart, head of JPL's advanced mission planning. In 1967, JPL used the project as a lure in its employee recruitment literature.[16] In short, although other NASA Field Centers competed, especially the Ames Research Center, JPL put forth a tremendous effort to make Grand Tour a JPL project.

The NASA Office of Space Science and Applications faced the task of establishing priorities among the various proposed missions to the outer planets. The agency called on its own scientific community to formulate outer planet exploration approaches and created the Outer Planets Working Group in 1969. Its creation was part of a larger agency reorganization initiated by Homer Newell, NASA associate administrator, in order to focus

---

16. Homer Joe Stewart, "New Possibilities for Solar-System Exploration," *Astronautics and Aeronautics* 4 (December 1966): 26–31; Craig B. Waff, "Searching for an Outer-Planet Exploration Strategy: NASA and its Science Advisory Groups, 1965–71," paper presented at the IAU General Assembly, Baltimore, August 6, 1988, pp. 5 and 31–32, copy at NASA Historical Reference Collection.

on the development of long-range plans, as opposed to the emphasis in preceding years on the budget year or on near-term plans. The reorganization resulted in the creation of twelve planning panels and six special study groups covering the gamut of NASA activities, with a Planning Steering Group chaired and coordinated by Newell himself.[17]

The Outer Planets Working Group consisted of two representatives (a scientist and an advanced mission planner) from each of the NASA Field Centers interested in Grand Tour and other outer planet missions (JPL, Ames, Goddard, and Marshall) and from the Illinois Institute of Technology Research Institute's Astro Sciences Center, a NASA think tank of sorts which had initiated a Jupiter mission study in the fall of 1968. The Working Group thus limited the decision-making process to NASA Field Centers that were vying to design spacecraft; the external scientific community was not part of that process.

Rather than favoring a single Grand Tour to the outer planets, the Working Group endorsed the concept of multiplanet flyby missions, preferably two three-planet voyages (Jupiter-Saturn-Pluto in 1977 and Jupiter-Uranus-Neptune in 1979), on the grounds that these would reduce the mission time from thirteen or more years to only seven and a half.[18] From June 1969, officials in the NASA Planetary Programs Office began to associate the phrase "Grand Tour" primarily with a pair of three-planet missions, rather than the original single tour concept.[19]

The Outer Planet Working Group also recommended that: "A new Mariner-class outer planets spacecraft appears adequate for accomplishing the more urgent scientific objectives."[20] Although NASA ultimately followed that recommendation by building Mariner Jupiter-Saturn, the space agency did not heed the advice until Grand Tour's demise. One of the chief activities of the NASA Field Centers was the design and construction of spacecraft. Not surprisingly, then, the Working Group's advice also called for the designing and building of a large number of spacecraft.

NASA next put the question of outer planet exploration to the twenty-three scientists of the Space Science Board summer study that met in June 1969. Those scientists recommended a specific schedule of five outer planet missions: one to Jupiter, one to Jupiter and the Sun, one to Jupiter and Uranus, and the two Grand Tour missions outlined by the Outer Planets Working Group (Jupiter-Saturn-Pluto in 1977 and Jupiter-Uranus-Neptune in 1979). The recommendations artfully combined Jupiter-intensive exploration and separate missions to the transjovian planets, that is, what the scientific community originally set out at Woods Hole in 1965, with the Grand Tour notion issuing from NASA's Jet Propulsion Laboratory. NASA headquarters planetary programs officials interpreted the findings of the 1969 summer studies as support from the scientific community for Grand Tour.[21] NASA now intended to request Grand Tour funding for fiscal 1971.

Although the opinions of scientists and NASA Field Center experts had played the greatest role in shaping outer planet exploration up to this point, a new, and ultimately more powerful, player took the stage: the recently elected Nixon administration. The Bureau of Budget under Nixon consistently reduced NASA's budget allocation. No longer

---

17. Arthur L. Levine, *The Future of the U.S. Space Program* (New York, NY: Praeger Publishers, 1975), pp. 120–22; Arnold S. Levine, *Managing NASA in the Apollo Era*, (Washington, DC: NASA SP-4102, 1982), pp. 256–63.
18. Memorandum, Advanced Program and Technology to Director of Planetary Programs, May 13, 1969, attachment 4, "Conclusions and Recommendations of Outer Planets Working Group," record no. 005148, NASA Historical Reference Collection.
19. See, for example, Robert S. Kraemer, "Impact of Deferring Grand Tour Launch," December 15, 1970, record no. 005148, NASA Historical Reference Collection.
20. Memorandum, Advanced Program and Technology to Director of Planetary Programs, May 13, 1969, attachment 4, "Conclusions and Recommendations of Outer Planets Working Group," record no. 005148, NASA Historical Reference Collection.
21. *Planetary Exploration 1968-1975* (Washington, DC: National Academy of Sciences, 1968); *The Outer Solar System: A Program for Exploration* (Washington, DC: National Academy of Sciences, 1969).

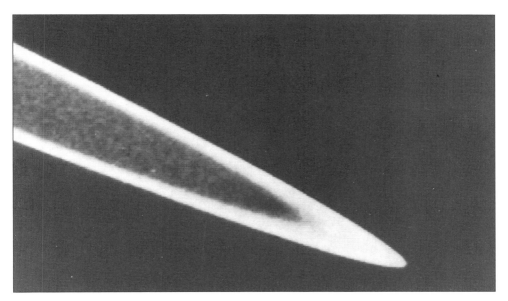

This view of Jupiter's ring was recorded by Voyager 2 on July 10, 1979, at a distance of 1.5 million kilometers (930,000 miles). The unexpected brightness is probably due to forward scattering of sunlight by small ring particles. Seen within the inner edge of the brighter ring is a fainter ring which may extend all the way down to Jupiter's cloud tops. The existence of the ring was first learned when photographed by Voyager 1 in March 1979. (NASA photo no. 79-H-507).

was space exploration a tool for competing with the Soviet Union. Nixon perceived the Apollo program in partisan terms, as a Kennedy program. Thus, for example, in December 1969, the Nixon administration quickly moved to shut down the only NASA laboratory ever closed, the Electronics Research Center in Cambridge, Massachusetts, which Nixon was said to have perceived as a Kennedy pork project.[22]

The Nixon budget cuts hit NASA's fiscal 1971 budget, in which the space agency requested funding for two three-planet "mini" Grand Tours scheduled for launch in 1977 and 1979. At the same time, NASA faced the cost overruns of the Viking orbiter and lander, whose dramatically escalating overall cost was earning Viking the title of NASA's most costly project after Apollo (rising from $364.1 million in March to $606 million in August 1969). The Nixon administration cut NASA's budget, which translated into a loss of $75 million to the $413.9 million budget for NASA's Office of Space Science and Applications, the budget portion that fed Grand Tour. The "pain" of NASA's fiscal 1971 budget was not confined to Grand Tour, though, and included suspending production of Saturn V launch vehicles, stretching out Apollo lunar missions to six-month intervals, and delaying the launch of Viking from 1973 to 1975.[23] This first postponement of Grand Tour thus did not

---

22. Rubashkin, "Who Killed Grand Tour?," pp. 9–10; Paul E. Green, Jr., interview with author, Hawthorne, New York, September 20, 1993. For the creation and demise of the NASA ERC, see Ken Hechler, *Toward the Endless Frontier: History of the Committee on Science and Technology, 1959–1979* (Washington, DC: U.S. Government Printing Office, 1980), pp. 219–31.

23. Attachment, Acting Associate Administrator for Advanced Research and Technology to Associate Administrator for Space Science and Applications, November 17, 1969, and "Office of Space Science and Applications FY 1971 Budget Issues," September 24, 1969, record no. 005148, NASA Historical Reference Collection; Rubashkin, "Who Killed Grand Tour?," p. 16; Waff, "The Struggle," p. 48; Waff, "The Next Mission," pp. 60–62.

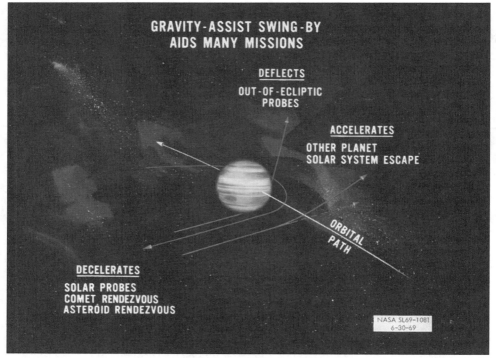

"*Gravity-Assist Swing-by Aids Many Missions*" chart shows the swing-by as it deflects, accelerates, and decelerates. (NASA photo no. 69-H-1521).

arise from any perception that the mission was too costly per se, but from a White House attempt to reduce NASA's, as well as the overall Federal, budget.

The severe and unprecedented reduction of NASA's Office of Space Science and Applications budget led Philip Handler, president of the National Academy of Sciences, to suggest to NASA administrator Thomas Paine in November 1969 that a Space Science Board panel evaluate and rank the disciplines supported by NASA, such as planetary and lunar exploration, astronomy, and Earth environmental sciences. Paine agreed. Subsequently, a summer study, involving nearly ninety scientists, took place at Woods Hole, Massachusetts, from July 26 to August 15, 1970. In addition, a fourteen-member executive committee, chaired by Space Science Board member Herbert Friedman of the Naval Research Laboratory, had the daunting task of combining the proposals of the working groups into an overall priority system.

What emerged was an ominous schism between the advice of the scientists of the Woods Hole Planetary Exploration Working Group and that of Friedman's executive committee. The Working Group urged that Grand Tour not be missed: it was a unique opportunity. The executive committee, on the other hand, favored Jupiter-intensive missions. The difference partly arose from concerns about the technological demands of the two types of missions. Jupiter-intensive missions required development of spacecraft lasting only five years; the real design challenge was in the probes, which had to withstand entry into the Jovian atmosphere. In contrast, while Grand Tour would not be entering any planetary atmospheres, it demanded spacecraft capable of enduring a much longer time period. In both cases, spacecraft design was intrinsically linked to mission cost. Even some Grand Tour advocates complained that JPL had not made any effort to design a more modest spacecraft with an estimated cost that would be more in line with the prevailing budgetary climate.[24]

The planetary scientists opposing Grand Tour fell into two camps. One camp preferred smaller, less costly, and shorter duration missions; the second feared that support of Grand Tour would divert funds from the building of a large space telescope. It was at this point that the perception that Grand Tour's price tag was too high emerged. Friedman led the contingent of astronomers who advocated building a large space telescope; they successfully placed the large space telescope in the highest priority category of the study. Friedman placed a higher priority on a large 45-inch orbiting telescope than on Grand Tour for reason of both its lower cost and its perceived higher scientific promise. Already, Grand Tour bore an estimated price tag of $700 million, and funding it, Friedman and others feared, would have a serious impact on other highly desirable scientific missions. The high cost of Grand Tour was being compared to Viking, which had become so costly that in early December 1969 an ad hoc Viking Review Panel set up by the Space Science Board almost recommended terminating the project.[25] The collision of opposing views among scientists that the Woods Hole summer study brought to light was to resound throughout the space exploration community and to have an impact on Grand Tour. By December 1970, members of the Space Science Board were raising questions about Grand Tour.[26] Elsewhere, in negotiations with the Office of Management and Budget (OMB) in December 1970, George M. Low, NASA Acting Administrator, suggested replacing Grand Tour with a mission to Jupiter-Uranus-Neptune in 1979 and a possible additional mission to Jupiter-Saturn-Pluto in 1977 or 1978, but which would require additional funding.[27]

In January 1971, months before the publication of the Friedman report on March 9, 1971, Friedman's report was leaked to the House Subcommittee on NASA Oversight, as well as to the Washington press. John Lannan, a reporter for the *Washington Evening Star*, made public Friedman's anti-Grand Tour views, and a *Science News* article reported the opposing views of some of the Working Group members. The contention over the funding of Grand Tour now spilled over from the space and astronomical communities to the public at large and even beyond the nation's borders.[28]

---

24. *Priorities for Space Research, 1971–1980* (Washington, DC: National Academy of Sciences, 1971); Space Science Board, *Outer Planets Exploration, 1972–1985* (Washington, DC: National Academy of Sciences, 1971); Waff, "Searching," p. 11; Waff, "The Struggle," pp. 48–49. As Homer E. Newell, *Beyond the Atmosphere*, p. 212 points out, the 1970 Wood's Hole summer study was thick with friction. The group chose to decrease support for magnetospheric and fields-and-particles research in favor of planetary research, thereby alienating fields-and-particles scientists, and by stressing high-energy astronomy over classical optical astronomy and solar physics, the group created more strife.

25. Waff, "The Next Mission," pp. 78 and 127; Rubashkin, "Who Killed Grand Tour?," pp. 14-15. In 1969, Grand Tour (four spacecraft–two flights to Jupiter, Saturn, and Pluto and two to Jupiter, Uranus, and Neptune–plus a test flight to Jupiter) was estimated to cost $660 million. Possible additional missions, such as a 1974 upper atmosphere probe of Jupiter, extra Jupiter-Saturn-Pluto or Jupiter-Uranus-Neptune launches, or two Uranus deep-entry probes in 1979, could raise the Grand Tour total price tag to $1.45 billion. "Office of Space Science and Applications FY 1971 Budget Issues," September 24, 1969, record no. 005148, NASA Historical Reference Collection.

26. Memorandum, Warren Keller to Elvira Haas and Julie Kertes, December 14, 1970, record no. 005148, NASA Historical Reference Collection.

27. George M. Low to Donald B. Rice, December 16, 1970, record no. 005148, NASA Historical Reference Collection.

28. John Lannan, "Space Tour Hopes Jolted," *The [Washington] Evening Star*, January 9, 1971,: p. A-9; "A Close Look at the Outer Planets: Scientists vying for Shares of NASA's Space Budget Disagree about this Once-in-a-lifetime Opportunity," *Science News* 99 (January 30, 1971): 77–78; Homer E. Newell, Daniel H. Herman, and Paul Tarver, "Potential Contributions of the United States Space Program to Exploration of the Solar System," pp. 285–314, esp. pp. 309 and 384–385 in Aina Elvius, ed., *From Plasma to Planet* (New York, NY: John Wiley & Sons, 1972).

At the heart of the contention was the JPL Grand Tour spacecraft called TOPS. Grand Tour consisted of four launches, two to Jupiter-Saturn-Pluto in 1976 and 1977, and two to Jupiter-Uranus-Neptune in 1979. NASA estimated the cost of the four missions to range from $750 to 900 million plus $106 million for launch vehicles.[29] One substantial portion of the cost of Grand Tour was development of a self-test and repair computer (STAR) that would operate for over ten years at a great distance from Earth. Another significant portion of the price tag represented development of the so-called Thermoelectric Outer Planets Spacecraft (TOPS) by JPL. The long lifetime of the TOPS spacecraft was to be achieved at the expense of increased vehicle weight and higher cost.[30]

Grand Tour TOPS and STAR development programs potentially represented a considerable fountain of paid employment for JPL employees, contractors, and subcontractors, as well as laboratory overhead, in the post-Apollo era. Contractor lobbying of the White House and Congress on behalf of the large space telescope helped to win congressional approval for it. Without that lobbying, historian Robert W. Smith has argued, Congress would not have approved funding the telescope.[31] But ultimately the bid to develop TOPS reduced potential political support for Grand Tour's other options.[32]

Further complicating matters was Senator Clinton P. Anderson (D-NM), champion of the Los Alamos nuclear weapons laboratories and an enthusiast, until his retirement in 1973, of the development of a nuclear rocket engine called NERVA. As chair of both the Senate Aeronautical and Space Sciences Committee and the Joint Atomic Energy Committee, Anderson provided NASA and the Atomic Energy Commission over $1.4 billion, about $500 million of which was spent in Los Alamos, for the development of the NERVA engine, which, Anderson held, was ideally suited for exploration of the outer planets, as well as for more advanced missions. Anderson worried that NASA and the OMB were shifting money from NERVA to fund Grand Tour. When the NASA budget came before Anderson's Aeronautical and Space Sciences Committee on May 12, 1971, his committee voted five to two to reduce Grand Tour's budget, while an amendment to increase NERVA funding passed. Werner von Braun worried that ardent congressional interest in NERVA would force a loss of Grand Tour in favor of a NERVA that had "no place to go."[33]

Meanwhile, NASA was trying to include Grand Tour as a new start in its 1972 fiscal budget. The Friedman report moved the Office of Management and Budget (OMB), in March 1971, to ask NASA to study simpler, less costly spacecraft alternatives to TOPS. The OMB also attempted to delay the Grand Tour start-up to fiscal 1973.[34]

---

29. House Committee on Appropriations, *HUD, Space, and Science Appropriations for 1972*, Part 1 Hearings before the subcommittee on HUD, Space and Science, 92nd Congress, 1st session, March 22, 1971, pp. 1234-1236; Robert S. Kraemer, "Impact of Deferring Grand Tour Launch," December 15, 1970, record no. 005148, NASA Historical Reference Collection.

30. Attachment, Acting Associate Administrator for Advanced Research and Technology to Associate Administrator for Space Science and Applications, November 17, 1969, record no. 005148, NASA Historical Reference Collection, indicates the distribution by RTOP of the $8,230,000 to be spent at JPL in fiscal 1970 on Grand Tour research and development work. The largest piece, $3,887,000, was to go just for the electric propulsion system, solar arrays, and batteries. At this point, Grand Tour had yet to be approved. J. K. Davies, "A Brief History of the Voyager Project," *Spaceflight* 23 (1981): 36-38; Rubashkin, "Who Killed Grand Tour?" pp. 2 and 11; Waff, "The Next Mission," p. 63. For a brief discussion of the development of the STAR computer at JPL, see P. J. Parker, "Grand Tour Spacecraft Computer," *Spaceflight* 13 (1971): 88 and 120.

31. Robert W. Smith, *The Space Telescope: A Study of NASA, Science, Technology, and Politics* (New York, NY: Cambridge University Press, 1989).

32. This is the interpretation of Rubashkin, "Who Killed Grand Tour?" p. 21.

33. Meeting notes, September 28, 1971, record no. 005148, NASA Historical Reference Collection; Rubashkin, "Who Killed Grand Tour?," pp. 17–18.

34. Waff, "The Struggle," pp. 50; Waff, "The Next Mission," pp. 96–97, 104, 110–12 and 114.

OMB and congressional pressure to cancel TOPS and to cut NASA's budget, combined with the debate induced by the Friedman report, left NASA management in a quandary. In order to energize support for Grand Tour, and to answer general questions about outer planet exploration, NASA administrators again turned to the scientific community at a Space Science Board summer study held at Woods Hole, August 8–14, 1971. Unlike previous summer studies, this one concerned itself solely with outer planet exploration.

This latest summer study concluded that both Grand Tour (four TOPS probes) and the intensive study of both Jupiter and Saturn ought to be supported. Although the summer study scientists supported Grand Tour by a vote of 12-1, they cautioned that if NASA funding levels fell too low, Grand Tour ought to be abandoned in favor of a Mariner spacecraft mission to Jupiter and Saturn.[35] The Mariner proposal was a return to the original 1965 Woods Hole idea of exploring the outer planets in piecemeal fashion.

As NASA prepared its fiscal 1973 budget, rumors spread that the "budget pinch" was going to affect planetary programs deeply and that the reduction of the Grand Tour payload from 205 to 130 pounds was "a likely fact of life."[36] Furthermore, Grand Tour now began to compete for funding with the latest NASA human program: the Space Shuttle. The fiscal 1973 budget request NASA submitted to the OMB on September 30, 1971 included both Grand Tour and the Space Shuttle. Throughout the autumn of 1971, several press reports presciently reported Grand Tour's vulnerability to a possible elimination or reduction.[37] On December 11, 1971, James Fletcher, NASA administrator since April 27, 1971, learned from White House officials that Nixon was prepared to approve the shuttle program and that Nixon would not let NASA simultaneously fund the shuttle and the full TOPS Grand Tour in the 1973 budget or in subsequent fiscal years.[38] Fletcher had to decide which was more important: Grand Tour or human flight.

By December 16, 1971, Fletcher had agreed to delete the TOPS version of Grand Tour from its fiscal 1973 budget request and to replace it with a pair of less expensive Mariner spacecraft to be known as Mariner Jupiter-Saturn to be launched in 1977.[39] The decision to kill Grand Tour was not made public immediately, and it was terrible Christmas Eve news at JPL.[40] Nixon, in his budget message of January 5, 1972, announced the development of the Space Shuttle, as well as the demise of TOPS Grand Tour and the substitution of the more modest Mariner Jupiter-Saturn mission.

Who killed Grand Tour? The demise of Grand Tour was less a simple case of its expensive price tag than its competition with other high-cost new starts (the shuttle and the space telescope) and Viking in a shrinking Federal and NASA budget. The smaller the budget became, and the more that costly programs competed for those shrinking funds,

---

35. Space Science Board, *Outer Planets Exploration, 1972-1985* (Washington, DC: National Academy of Sciences, 1971), esp. pp. 32–33.

36. Waff, "Searching," p. 25.

37. "Grand Tour, Shuttle Threatened: Cuts at NASA Possible," *Science News* 100 (September 18, 1971): 187; "A Question of Survival for Grand Tour of Planets," *Science News* 100 (October 9, 1971): 246; "Apollo 17 Threatened by OMB Budget Squeeze," *Space Business Daily* 58 (October 21, 1971): 182; John Noble Wilford, "Pressure is Reported on NASA To Make Heavy Cuts in Budget," *New York Times*, November 19, 1971; "President Approves Space Shuttle–$3.2 Billion NASA Budget," *Space Business Daily* 59 (December 14, 1971): 192; Waff, "The Next Mission," pp. 123–24.

38. Charles H. Townes to George M. Low, November 25, 1970, record no. 005148, NASA Historical Reference Collection; James C. Fletcher, interview with Craig B. Waff, Washington, August 5, 1988, cited in Waff, "The Next Mission," p. 125; John M. Logsdon, "The Decision to Develop the Space Shuttle," *Space Policy* 2 (May 1986): 103–119, esp. 115; and "The Space Shuttle Program: A Policy Failure?" *Science* 232 (May 30, 1986): 1099–105, esp. 1103.

39. For a description of the Mariner Jupiter-Saturn spacecraft, see H. M. Schurmeier, *The Mariner Jupiter Saturn Mission* (Pasadena, CA: JPL, 1974) and *Spaceflight* 19 (November 1977): 372.

40. James C. Fletcher to Caspar W. Weinberger, December 22, 1971, record no. 005148, NASA Historical Reference Collection; S. Ichtiaque Rasool, interview with author, Paris, December 12, 1994.

the more expensive each program appeared. To some extent, too, Grand Tour was a victim of the NASA preference for human space flight over scientific probes. The Space Shuttle was essential to continuing the U.S. human space flight program as Apollo wound down. The schism between how the planetary scientific community defined outer planet exploration—small, piecemeal ventures-and how JPL defined outer planet exploration—a large, expensive project to exploit a rare planetary alignment, and the public airing of that schism, certainly contributed to the pressure on NASA administrator James Fletcher to cancel Grand Tour. Thus, at NASA's fiscal 1973 budget briefing on January 22, 1972, NASA administrator Fletcher explained that Grand Tour was eliminated because of a "less than enthusiastic response from certain elements of the scientific community particularly, and to some extent, Congress."[41]

But was Grand Tour really dead? Even before the public announcement of Grand Tour's demise, planning had begun for Mariner Jupiter-Saturn, the reduced-cost, two-planet alternative to Grand Tour recommended by the most recent Woods Hole summer study. In December 1971, when NASA and the OMB agreed to delete fiscal 1973 funds for the TOPS Grand Tour, NASA informed the OMB that the JPL TOPS development group would be "retained and redirected into planning a new program to explore Jupiter and possibly Saturn with a three-axis stabilized Mariner-class spacecraft."[42] (Stabilization along three axes was a requisite for onboard cameras.)

NASA administrators next turned to the scientific community in the guise of the Space Science Board. The Board met February 8–9, 1972, and "unanimously and warmly endorsed" Mariner Jupiter-Saturn. The Space Science Board, through its chair Charles H. Townes, expressed the hope that the spacecraft would remain operational beyond Saturn "and return very significant data on cosmic particles and fields."[43]

Congress greeted with approval the replacement of the TOPS Grand Tour with Mariner Jupiter-Saturn and authorized funds for Mariner Jupiter-Saturn for fiscal 1973. The Mariner Jupiter-Saturn price tag, $360 million versus $1 billion for TOPS Grand Tour, could fit into a scaled back NASA budget that also financed development of the Space Shuttle. Although work on Mariner Jupiter-Saturn started at JPL as early as January 1972, the new project was not officially approved by NASA until the Contractual Task Order was signed on May 18, 1972.[44]

In order to reduce costs and overheads, NASA decided to leave design and construction of the Mariner Jupiter-Saturn spacecraft to JPL, rather than to Boeing, General Electric, Hughes, Martin Marietta, and North American Rockwell, all of which had some level of preparation for a Grand Tour proposal. The largest aerospace firms lobbied NASA Headquarters and Congress for the contracts. In order for expensive projects to pass congressional scrutiny as part of the NASA budget, they often had to include an intention to contract out much of the work. Thus, for example, Magellan, the radar imaging mission to Venus, although initially intended as a JPL in-house project for cost reasons, was let out to Martin Marietta (spacecraft contract) and Hughes (the radar contract).[45] The decision to go with JPL versus an industrial contractor was viewed at NASA Headquarters by John E. Naugle, Associate Administrator for Space Science, as a "many faceted problem" whose

---

41. Quoted in Waff, "The Next Mission," p. 125.
42. Quoted in Rubashkin, "Who Killed Grand Tour?" p. 25.
43. James C. Fletcher to Charles H. Townes, March 10, 1972, and Charles H. Townes to James C. Fletcher, February 22, 1972, record no. 005566, NASA Historical Reference Collection.
44. *Spaceflight* 7 (July 1972): 258; Davies, "Brief History," p. 38. The Voyager price tag fell further to $320 million. "Voyager costs, July 18, 1977" and "FY78 Senate Authorization, Part II," record no. 005556, NASA Historical Reference Collection.
45. Gordon H. Pettengill, interview with author, MIT, September 28, 1993; Gordon H. Pettengill, interview with author, MIT, May 4, 1994.

resolution was "of paramount importance to the future of NASA's Planetary Program as well as to the future of JPL.." In short, JPL needed the contract to maintain employment levels in the laboratory, and NASA Headquarters needed it to maintain the vitality of its planetary program. Therefore, he explained, "all of the various factors must be given careful and thoughtful consideration."[46]

Despite the limited aim of the Mariner Jupiter-Saturn, the mission had the Grand Tour launch window, that rare planetary alignment, and the engineers at JPL still had every intention of building a spacecraft that would last long enough to visit Uranus and Neptune. This intention was not emphasized; however, it was stated that a Mariner Jupiter-Saturn spacecraft might continue to Uranus if its mission at Saturn proved successful. The scientists working on the project knew that Mariner Jupiter-Saturn was going to go to Uranus and Neptune, too. As Bradford Smith, Leader of the Imaging Team, explained: "We understood at the time the enormous potential of this mission—that it could very well be one of the truly outstanding if not the most outstanding mission in the whole planetary exploration program."[47]

Grand Tour would rise from its own "death" as piecemeal additions to Mariner Jupiter-Saturn. As S. Ichtiaque Rasool, Deputy Director of Planetary Programs, Office of Space Science, reflected: "The lesson to be learned from Grand Tour cancellation was that you never fund such a big, long-term project at once. So we kept on adding piecemeal. And it's interesting that they always come out big. When you have less money, you can even do better sometimes."[48] The Mariner design and experience were used whenever possible and were supplemented with subsystems designed for the Viking orbiter to provide the required performance and reliability. NASA instructed the Atomic Energy Commission to upgrade the plutonium batteries so they might last more than ten years, enough time for Mariner Jupiter-Saturn to encounter Uranus and Neptune.[49] Despite the reliance on extant technology, some money was set aside to develop new technology. Congress and the OMB approved an additional $7 million to the Mariner Jupiter-Saturn appropriation for scientific and technological enhancements. Part of that appropriation went to develop a reprogrammable onboard computer,[50] which proved vital to maintaining *Voyager 2* as a functioning observatory in space. Without properly functioning hardware, no science could be conducted.

Just as scientists played a key role in shaping Voyager before it was funded, they collaborated actively with NASA in defining the mission's scientific objectives within orga-

---

46. John E. Naugle to William H. Pickering, April 13, 1972, and Memorandum, John E. Naugle to George M. Low, October 13, 1972, record no. 005556, NASA Historical Reference Collection. JPL already had campaigned for Grand Tour in 1969, but never gave cost reduction as a reason. Moreover, the reasons they gave for selecting JPL for Grand Tour did not compare JPL to industry contractors, but with other NASA centers, specifically Marshall and Goddard. D. P. Hearth, "Reasons for SL Recommendation of JPL for Grand Tour," September 15, 1969, record no. 005148, NASA Historical Reference Collection. The Ames Research Center, for example, had a contract with TRW under which TRW investigated the feasibility of conducting Grand Tour with a Pioneer spacecraft along the lines of Pioneers 10 and 11. Memorandum, Norri Sirri to distribution, May 15, 1970, record no. 005148, NASA Historical Reference Collection. In 1969, in fact, JPL was competing with other NASA centers, not with industry, to build Grand Tour. See, for example, Memorandum, Director of Planetary Programs to distribution, July 7, 1969, and Memorandum, Director of Planetary Programs, Office of Space Science and Applications, to distribution, May 13, 1969, and attachments, record no. 005148, NASA Historical Reference Collection.
47. "Interview: Bradford Smith," *Space World* n.v. (November 1985): 5, record no. 005586, NASA Historical Reference Collection; Eric Burgess, *Far Encounter: The Neptune System* (New York, NY: Columbia University Press, 1991), pp. 1 and 2.
48. S. Ichtiaque Rasool, interview with author, Paris, December 12, 1994.
49. Davies, "Brief History," p. 38; Rubashkin, "Who Killed Grand Tour?" p. 27.
50. Murray "Journey into Space," p. 174.

nizational frameworks established by NASA. On October 15, 1971, although Grand Tour had not yet been authorized, the space agency issued an "Invitation for Participation in Mission Definition for Grand Tour Missions to the Outer Solar System" to specify its scientific objectives, that is, typical payloads and scientific instruments requiring a long lead-time to develop. Among those primary objectives of Grand Tour (and Mariner Jupiter-saturn) were: 1. physical properties, dynamics, and compositions of atmospheres; 2. geological features; 3. thermal regimes and energy balances; 4. charged particles and electromagnetic environments; 5. periods of rotation, radii, figures, and other body properties; and 6. gravitational fields. While travelling between planets, both missions would study variations of the solar wind plasma and magnetic field, solar energetic particles, galactic cosmic rays, and interplanetary dust. Once the spacecraft left the solar system, they could make measurements of galactic cosmic rays unmodulated by the solar plasma.[51]

Regardless of which objectives or instruments the scientific community recommended, JPL insisted on including video cameras. At JPL, Harris M. "Bud" Schurmeier, JPL's Grand Tour and Mariner Jupiter-Saturn project manager, understood both the non-scientific and the scientific importance of imaging the planets and their satellites. In 1964, Ranger lunar-impact probes radioed back the first close-up pictures of the Moon, thanks to hardware designed at JPL under his guidance. Subsequently, in 1969, Schurmeier led the work on *Mariners 6* and *7* that achieved a hundredfold gain over tiny *Mariner 4* in the return of pictures from Mars, and in 1971, *Mariner 9* pictures of Mars, after waiting out a gargantuan dust storm. In addition to the imaging team, Mariner Jupiter-Saturn would have a Radio Science Team to exploit the scientific use of the spacecraft's radio systems.[52] In selecting members of the scientific teams, the first members chosen were those of the imaging and radio science teams, the teams using the video and radio equipment that JPL intended to put on board, regardless of whatever scientific instruments might be selected.[53] As the NASA Field Center in charge of the mission, JPL thus could exert a determining influence on the science to be conducted.

NASA, in April 1972, extended a formal request for experiment proposals and received over 200 replies. From those the space agency selected ninety scientists, mainly from the United States, but from France, Sweden, West Germany, and Great Britain, as well.[54] The selection process favored researchers in large institutional settings, but did not filter out little scientists entirely. NASA policy was to select scientists based on the merit of their research, as well as the "reputation and interest of the institution." The stated reason for this selection standard was to insure "scientific depth and breadth, and the availability of the resources to support the investigation." NASA assumed that selected scientists would be affiliated with an accredited academic institution, a private corporation "with sufficient contractual resources to provide the required scientific, technical, and

---

51. "Invitation for Participation in Mission Definition for Grand Tour Missions to the Outer Solar System," October 15, 1970, record no. 005148, NASA Historical Reference Collection.
52. Edward C. Stone, interview with author, JPL, November 23, 1994; Murray, *Journey into Space*, p. 139.
53. Milton A. Mitz to Chairmen and Headquarters Members of MJS77 Ad Hoc Subcommittee, July 26, 1972, record no. 005566, NASA Historical Reference Collection.
54. Memorandum, Milton A. Mitz to Bradford A. Smith, December 4, 1972, record no. 005566, NHO; NASA News, press release 72-239, December 6, 1972, record no. 005580, NASA Historical Reference Collection; Joseph P. Allen to Robert Griffin, June 23, 1977, record no. 005566, NASA Historical Reference Collection. For a discussion of the origins and evolution of this selection process, see John E. Naugle, *First Among Equals*, (Washington, DC: NASA SP-4215, 1991), pp. 32–39, 44–53, 57–59, 61–70, 79–105, and 119–23.

administrative resources and support" (e.g., TRW or The Rand Corporation), or a NASA or other government center or laboratory.[55]

Illustrating how the scientist selection process favored those in large institutional settings, such as NASA Field Centers, was the dominance of NASA's Goddard Space Flight Center scientists on the infrared spectroscopy and radiometry and the magnetic fields science teams. Eight of the eleven members of the first team were from Goddard, while seven investigators, only one of which came from outside Goddard (a German researcher) constituted the magnetic fields team.[56]

University planetary scientists populated most of the other science teams, although those scientists often counted on NASA funding for their research. NASA grants to university funding and NASA's use of university scientists drew them into the larger scientific enterprise of the NASA-industrial-academic complex, thereby weaving little science into the fabric of large-scale, big-budget science.[57] Such was the case of the radio science team, which was a mix of Stanford University and JPL researchers.

The Stanford investigators, Von R. Eshleman, Thomas A. Croft, and G. Leonard Tyler, came from that institution's Center for Radar Astronomy. Founded in 1962, initially in collaboration with SRI personnel, and underwritten by NASA, the Center for Radar Astronomy sought to conduct planetary atmospheric, ionospheric, and surface studies using the radio equipment ordinarily (and necessarily) included on each spacecraft, although special hardware often was developed to perform experiments. The Center was small, however, in terms of budget and personnel.[58] The remaining science team members, John D. Anderson, Gunnar Fjedlbo (now Lindal), Gerald S. Levy, and Gordon E. Wood were all JPL staff engineers and scientists. Fjeldbo, moreover, previously had been with the Stanford Center for Radar Astronomy.[59]

Stiff competition, and at times personality conflicts, reigned among the scientists submitting proposals.[60] Among other factors, the selection or rejection of instrument proposals hinged not as much on the qualifications of scientists or their research, but on the

---

55. Raymond L. Heacock to James W. Warwick, February 15, 1980, record no. 005566, NASA Historical Reference Collection. The matter was relevant to Warwick's case. He had begun as a Voyager science team Principal Investigator associated with the University of Colorado, Boulder, but later left and formed his own corporation, called Radiophysics, Inc., thereby raising the issue of his institutional affiliation. See additional related documents in record no. 005566, NASA Historical Reference Collection. The case of Allan F. Cook II, a member of the Imaging Team, raised the question of institutional affiliation of science team members who lost their institutional affiliation. Rodney A. Mills, Memorandum for the Record, February 7, 1978, record no. 005566, NASA Historical Reference Collection. The division of scientists into these three categories was not always unambiguous. For instance, David Morrison, a leading planetary astronomer, was on the faculty of the University of Hawaii, but also held a position for a period at NASA Headquarters in the Office of Space Science. David Morrison to Bradford A. Smith, April 19, 1978, record no. 005566, NASA Historical Reference Collection. The selection of foreign scientists not only required them to have an institutional affiliation, but to first be approved by their national space agency, which forwarded the proposal, after a screening process, to NASA. "Invitation for Participation in Mission Definition for Grand Tour Missions to the Outer Solar System," October 15, 1970, record no. 005148, NASA Historical Reference Collection.

56. *Voyager to Jupiter and Saturn*, (Washington, DC NASA SP-420, 1977), pp. 53–55.

57. Newell, pp. 223–42; Joseph N. Tatarewicz, *Space, Technology, and Planetary Astronomy* (Bloomington, IN: Indiana University Press, 1990), passim.

58. Andrew J. Butrica, *To See the Unseen: A History of Planetary Radar Astronomy*, (Washington, DC: NASA SP-4218, 1996), pp. 154–56.

59. *Voyager to Jupiter and Saturn*, pp. 53–55. Croft was an SRI staff member.

60. For example, one group of scientists attempted to control the process of selecting the magnetometer investigation team. Warren Keller, Program Manager of Outer Planets Missions, saw through the scheme: "We are dealing with problems arising from the personalities of some of the leading individuals in this field," he wrote. Program Manager, Outer Planets Missions, to Deputy Associate Administrator for Space Science, May 30, 1972, and Deputy Director, Physics and Astronomy Programs, to Deputy Associate Administrator for Space Science, May 17, 1972, record no. 005566, NASA Historical Reference Collection.

trajectory of the spacecraft and the discoveries of earlier missions. The assessment of the dangers of the asteroids, Saturn's rings, and Jupiter's electromagnetic environment was placed on a firmer foundation by the results beamed back by *Pioneer 10* and *11*, launched in 1972 and 1973, respectively. Although the asteroid hazard appeared less threatening, *Pioneer 10* encountered far more damaging radiation than had been expected. *Pioneer 11* reached Jupiter a year later (December 1974), then went on to Saturn, where the spacecraft passed within 21,000 km of Saturn's cloud tops in September 1979 and certified the safety of the narrow zone between Saturn and its rings.[61] These *Pioneer 10* and *11* results led to the dropping and adding of Mariner Jupiter-Saturn science experiments.

Once *Pioneer 10* discovered that the levels of radiation at Jupiter were a thousand times more intense then expected, NASA dropped an ultraviolet photopolarimeter experiment that had been selected on a provisional basis. In the place of that instrument, and at the urging of the concerned scientific community, S. Ichtiaque Rasool, NASA Office of Space Science, included on Mariner Jupiter-Saturn a plasma wave experiment which had been proposed but not selected until then.[62] On the other hand, other science experiments were selected or excluded on the basis of cost and spacecraft parameters. When drawing up the final list of investigators and instruments in September 1973, NASA dropped the micrometeorites experiment because of its development risk and cost, as well as the difficulty of integrating it into the spacecraft design.[63]

Perhaps the most unusual Voyager scientific experiment was that with no real Principal Investigator and essentially with no NASA budget for instrument construction or data analysis; it was the recording entitled "Sounds of Earth." On the chance that Voyager might encounter intelligent extraterrestrial life, NASA approved placement of a phonograph record on each of the two Voyager spacecraft. Recorded on a 12-inch copper disk, "Sounds of Earth" ran for nearly two hours. Its contents, assembled by a group of prominent scientists and educators led by Carl Sagan, who had placed extraterrestrial plaque messages on *Pioneers 10 and 11*, consisted of greetings from Earth in 60 languages, samples of music from different cultures and eras, and natural sounds of surf, wind, thunder, birds, whales, and other animals, as well as 115 photographs and diagrams in analog form, depicting human beings, the solar system, DNA, and various fundamental concepts from mathematics, chemistry, geology, and biology, and greetings from President Jimmy Carter and the Secretary General of the United Nations.[64]

The Voyager instruments and scientists selected, NASA then organized the scientists into twelve (later reduced to eleven) science teams. Except for the imaging and radio science teams, for which the project furnished the instrumentation, the individual science groups were responsible for designing and building the instruments associated with their

---

61. Memorandum, C. E. Kohlhase to J. R. Casani, April 27, 1976, accession number 90-61, box 2, folder 42, JPL Archives; "NASA Identifies 21 Promising New Missions," *Technology Week* 20 (June 12, 1967): 20; Long, "To the Outer Planets," p. 39; Davies, "Brief History," p. 35; David L. Roberts, comp., *The Multiple Outer Planet Mission (Grand Tour)*, Report no. M-16 (Chicago, IL: Astro Sciences Center, January 1969). For an overview discussion of Pioneer 10 and 11 science, see Richard O. Fimmel, James Van Allen, and Eric Burgess, *Pioneer: First to Jupiter, Saturn, and Beyond* (Washington, DC: NASA, 1980).
62. S. Ichtiaque Rasool, Memorandum for the Record, July 11, 1974, record no. 005566, NASA Historical Reference Collection; Noel Hinners to J. E. Blamont, July 25, 1974, record no. 005566, NASA Historical Reference Collection; S. Ichtiaque Rasool, interview with author, Paris, December 12, 1994.
63. Edward C. Stone, interview with author, JPL, November 23, 1994.
64. NASA press release no. 77-159, August 1, 1977, record no. 005554, NASA Historical Reference Collection. Among those members of Sagan's committee and others who played a major role in devising the Voyager record: Frank Drake, Cornell University; Philip Morrison, MIT; Bernard Oliver, Hewlett-Packard Corporation; Leslie Orgel, Salk Institute; Alan Lomax, Choreometrics Project, Columbia University; D. Robert Brown, Center for World Music, Berkeley; Murry Sidlin, National Symphony Orchestra, Washington; Jon Lomberg, Toronto. The record was prepared for NASA by Columbia Records.

investigation areas. The eleven investigation areas were: imaging, radio science, infrared and ultraviolet spectroscopy, magnetometry, charged particles, cosmic rays, photopolarimetry, planetary radio astronomy, plasma, and particulate matter. Specific scientific objectives included the study of the physical properties, surface features, periods of rotation, energy balances, and thermal regimes of the planets and moons and investigation of electromagnetic and gravitational fields throughout the mission. Items of special scientific interest included Jupiter's giant red spot and Saturn's rings and moons, Iapatus, Titan, and Rhea.[65]

Each science team had a leader, called a principal investigator, though the heads of the imaging and radio science groups were designated team leaders. The design and construction of the scientific instruments were the responsibility of the principal investigators, who either could have them built in their own laboratory or could contract for their construction. The team leaders and principal investigators formed the Science Steering Group, which had overall responsibility for advising NASA in the area of Mariner Jupiter-Saturn science. By the end of 1972, Ed Stone, a magnetospheric physicist from California Institute of Technology who had started on Grand Tour in 1970, during the preplanning stage, was appointed Project Scientist.[66] The Project Scientist stood at the interface between scientific needs and engineering and budgetary constraints, between the Science Steering Group and NASA, the public, the scientific community, and the press. In Stone's own words, the Project Scientist served "an impedance matching function between the engineering requirements and constraints and the science requirements and constraints to try to find a way to achieve the optimum match between these two different sets of requirements and desirements."[67] In short, the management of science and decision making were centralized in the Project Scientist.

Management of science included assuring that scientists' instruments were built on time and within budget and that they fit spacecraft parameters, especially payload weight, power requirements, physical and functional interface conditions, exposure to radiation, and the telemetry budget, that is, the allocation of down-link data bits without which data did not return to Earth.[68] The Project Scientist also was the ultimate arbiter in deciding which experiments and which observations would or would not be done. At times, the scientists lacked agreement on which observations to make, and the Project Scientist had to decide which of two equally good observations would be made. Rather than vote on the issue, Stone made the decision himself. "It turns out," Stone reflected, "that's a much more critical role than I had thought ahead of time, and that's because ultimately what science is all about is making discoveries. By deciding to make this observation rather than that one, you're effectively deciding that that group of scientists gets to make a discovery and this group doesn't."[69]

The most visible of the Project Scientist's activities as the interface between the Voyager scientists and NASA, the public, and the media was the press conference. The

---

65. Milton A. Mitz to Distribution, October 18, 1972, record no. 005566, NASA Historical Reference Collection; "Grand Tour Science Support Requests for Mission Definition," June 3, 1971, record no. 005148, NASA Historical Reference Collection; NASA News, press release 72-239, December 6, 1972, record no. 005580, NASA Historical Reference Collection; Edward C. Stone, interview with author, JPL, November 23, 1994.

66. "Invitation for Participation in Mission Definition for Grand Tour Missions to the Outer Solar System," October 15, 1970, record no. 005148, NASA Historical Reference Collection; S. Ichtiaque Rasool, interview with author, Paris, December 12, 1994; Edward C. Stone, interview with author, JPL, November 23, 1994.

67. Edward C. Stone, interview with author, JPL, November 23, 1994.

68. Milton A. Mitz to Distribution, October 30, 1972, record no. 005566, NASA Historical Reference Collection; "Invitation for Participation in Mission Definition for Grand Tour Missions to the Outer Solar System," October 15, 1970, record no. 005148, NASA Historical Reference Collection; S. Ichtiaque Rasool, interview with author, Paris, December 12, 1994; Edward C. Stone, interview with author, JPL, November 23, 1994.

69. Edward C. Stone, interview with author, JPL, November 23, 1994.

press conference was a keystone activity of the Project Scientist in his role as mediator among Voyager scientists. Press conferences, not scientific publications or conferences, were the venues where discoveries were first announced. Dealing with the media and the scientific process of discovery was the Project Scientist's major concern during the week around each encounter. Every day, working with the scientists of each investigation group, the Project Scientist had to determine what had been discovered, which discoveries were ready for release, and how they would be released in the press conferences.

The announcement of discoveries almost as they occurred, as well as the very aggregation of scientists into working groups, raised the question of intellectual property rights and priority of discovery. Ordinarily, scientists would hold their own data as proprietary, not sharing with any other scientist, so as to assure priority of discovery. However, not sharing data, Stone believed, "would have inhibited the total development of the scientific program."[70]

The idea of everyone sharing findings came to Stone from the need to communicate those findings to the media. He attended the press conference of *Pioneer 10* when it encountered Jupiter. Stone, who previously had worked only on Earth orbiting missions, was impressed by the scene: "Here was a room full of reporters wanting to know what the scientists had discovered. I mean, to me that was incredible. Normally there just isn't that interest in what you're doing as a scientist. And here they were day after day saying, 'Tell us what you've discovered. Tell us what you've discovered.' I realized that with Voyager we had both the opportunity and the obligation to communicate what we were discovering. To help the media tell the story. But we had to do it in a scientifically credible way."[71]

Having all Voyager scientists share their data made the scientists act less as individuals and more as members of a group, as they would on a typical big science project. The initial publication of results, too, followed this big science group approach. All of the initial publications resulting from a given encounter were published in the same issue of a given journal, such as *Science, Nature*, or the *Journal of Geophysical Research*. All scientists, therefore, published at the same time, but as a group, that is, one paper represented the discoveries of an entire science team. There was no question of priority, Stone explained, "everybody had equal priority, because everybody was there at the same time."[72]

The Mariner Jupiter-Saturn mission name persisted until March 1977, only a few months before launch. Many within the project and within NASA felt that the Mariner Jupiter-Saturn spacecraft departed enough from the Mariner family that a new name would be appropriate. As early as 1971, William H. Pickering, director of JPL, had suggested the name Navigator for the spacecraft pair.[73] NASA organized a name competition to choose the new name, and the winning nomination, "Voyager," was approved on

---

70. Edward C. Stone, interview with author, JPL, November 23, 1994.

71. *Ibid.*

72. *Ibid.* These measures accomplished their goal of protecting intellectual property and discovery rights, despite a memorable apparent exception, called the "Morabito incident," reported by Eric Chaisson. During the Voyager Jupiter encounter in 1979, a young technician named Linda Morabito accidentally discovered the first of nine active volcanoes on Io. The discovery, according to Eric Chaisson, "apparently embarrassed the doctoral scientists who had overlooked it" and affected the organization of science on the Hubble Space Telescope project. Eric Chaisson, *The Hubble Wars: Astrophysics Meets Astropolitics in the Two-Billion-Dollar Struggle over the Hubble Space Telescope* (New York, NY: Harper-Collins, 1994), esp. p. 102.

Morabito, however, was not an outsider; on the contrary, she was part of the JPL Voyager team, and her discovery was recounted in *Science* along with the other Voyager science reports. Moreover, neither Morabito nor anyone else outside the project "scooped" the Voyager science team. According to *Physics Abstracts*, the first announcement of the discovery of volcanism on Io was the IAU Circular of March 1979, issued by Torrence Johnson of the JPL Voyager team. Morabito, S. P. Synnott, P. N. Kupferman, Stewart A. Collins, "Discovery of Currently Active Extraterrestrial Volcanism," *Science* 204 (June 1, 1979): 972; T. Johnson, E. E. Becklin, C. O. Wynn-Williams, C. B. Pickett, and J. S. Morgan, IAU Circular 3338, March 16, 1979.

73. John E. Naugle to William H. Pickering, May 3, 1971, record no. 005148, NASA Historical Reference Collection.

March 4, 1977. The name change, however, coming so close to launch date, gave rise to a certain amount of confusion. References to *Mariners 11* and *12* and even *Voyagers 11* and *12* are a legacy of this last change of name.[74]

Despite the name change, Voyager remained in many ways the Grand Tour concept, though certainly not the Grand Tour (TOPS) spacecraft. *Voyager 2* was launched on August 20, 1977, followed by *Voyager 1* on September 5, 1977. The decision to reverse the order of launch had to do with keeping open the possibility of carrying out the Grand Tour mission to Uranus, Neptune, and beyond. *Voyager 2*, if boosted by the maximum performance from the Titan-Centaur, could just barely catch the old Grand Tour trajectory and encounter Uranus. Two weeks later, *Voyager 1* would leave on an easier and much faster trajectory, visiting Jupiter and Saturn only. *Voyager 1* would arrive at Jupiter four months ahead of *Voyager 2*, then arrive at Saturn nine months earlier. Hence, the second spacecraft launched was *Voyager 1*, not *Voyager 2*. The two Voyagers would arrive at Saturn nine months apart, so that if *Voyager 1* failed to achieve its Saturn objectives, for whatever reason, *Voyager 2* still could be retargeted to achieve them, though at the expense of any subsequent Uranus or Neptune encounter.

The taking of such precautions was normal for a venture where a certain number of spacecraft hardware breakdowns, called "anomalies" by NASA, are considered to be normal. Most are minor and have no impact on the ability of the spacecraft to carry out its scientific mission, such as the glitch that occurred during the launch of *Voyager 2*. Nonetheless, these anomalies emphasize the critical role that technology plays in the gathering of scientific measurements from a space-based observatory. Without that technology, no science is possible. The performance of the Voyager science mission from the moment of launch is a lesson in the critical role played by technology in the conduct of big science.

One serious anomaly that actually did limit the amount of Voyager science conducted was that of the scan platform. On February 23, 1978, before *Voyager 1* reached Jupiter, its scan platform became "stuck" during an azimuth scan. The platform turned on three axes in order to aim the cameras, spectrometers, and photopolarimeter in a scientifically useful direction. The platform jam thus threatened to compromise critical scientific observations. Luckily, command sequences transmitted to the spacecraft succeeded in moving the scan platform; the crisis subsided.[75]

As *Voyager 2* began to leave Saturn, and most of the scientific observations had been made on the planet, its 220-pound scan platform became stuck. The spacecraft cameras were sending back images of black space. The heavy workload during encounter, combined with an ineffective lubricant, likely caused the trouble, as engineers demonstrated on Earth-bound duplicate equipment. To help alleviate the platform problem at Uranus, the spacecraft was rolled when possible to perform large azimuth changes. The scan platform was moved only for smaller changes, and then only at slower speeds.[76]

The scan platform jam at Saturn occurred after most of the scientific observations had been made. Nonetheless, some science was lost. Whether one considered that science critical depended on one's interests. Certain project scientists wanted to play down the situation, and this annoyed those scientists who suffered real losses. The loss of two images of the moon Enceladus at a resolution of 1.6 km was perhaps not as great as the loss of the six

---

74. Davies, "Brief History," p. 39; *Spaceflight* 6 (June 1977): 19.
75. *Voyager Bulletin* no. 69 (June 20, 1985): 1–3, record no. 005586, NASA Historical Reference Collection; "Daily Activities Report," February 28, 1978, record no. 005566, NASA Historical Reference Collection; Murray, pp. 152–153.
76. JPL, Public Information Office, News Clips, May 22, 1985, record no. 005586, NASA Historical Reference Collection; Henry S. F. Cooper, Jr., *Imaging Saturn: The Voyager Flights to Saturn* (New York, NY: Holt, Rinehart and Winston, 1982), pp. 187, 200–01 and 209; Davies, "Brief History," vol. 24 (1982), pp. 256–57.

images of Tethys at 1.7 km, because the best pictures available of that moon had a resolution of no better than 5 km. The loss of coverage was serious in the case of both moons, however, for now neither of them could be measured all the way around with the precision that scientists would have liked, or mapped as comprehensively as the mission cartographers had expected. Other lost science included imaging the dark side of Saturn's rings, and non-imaging lost data included a further occultation experiment using the star Beta Tauri; infrared measurements of the ring material as it entered the planet's shadow; ultraviolet spectroscopy of ring material by observations of the Sun through the rings, as well as a field and particle maneuver. In the judgement of Ed Stone: "We were fortunate that the platform didn't stop a few hours earlier."[77]

The other major hardware failure was in the radio systems of *Voyager 2*. *Voyager 2*, which encountered more planets and moons than its double, seemed to suffer the greater number of serious hardware failures. No science was lost in this instance, although the potential was present, and an attempt to repair the situation raised the possibility of creating a spinoff ground facility for use in radio astronomy and ionospheric research.

In late November 1977, while the two Voyagers were still on route to Jupiter, one of *Voyager 2*'s two duplicate radio transmitters began to degrade. It was switched to low-power mode to nurse it along. Something was wrong, but there was no way to know exactly what. Months later, in April 1978, the Voyager team discovered that *Voyager 2*'s backup receiver had failed to detect signals sent from Earth because of a shorted capacitor. The primary radio receiver suddenly failed completely, as well. *Voyager 2* was silent. Continuing to Uranus and Neptune was no longer possible, unless a way could be found to communicate with the backup receiver. Moreover, the failure of the *Voyager 2* primary radio system had potential repercussions beyond the Voyager project. Its radio equipment was very similar to that on Pioneer Venus, which was launched the following month, in May 1978.[78]

Normally, the radio receiver automatically compensated for the Doppler shift of signals transmitted from Earth. The changing velocity and direction of the spacecraft relative to Earth caused this Doppler shift. Without the ability to compensate for the Doppler shift, the *Voyager 2* radio system could not detect any signals sent to it. The solution to *Voyager 2*'s radio problems came from NASA Deep Space Network engineers. They prepared computer tapes that slowly varied the frequency of the radio signals transmitted from Earth in order to compensate for the expected Doppler shift. The Deep Space Network station outside Madrid transmitted the first test signals on April 13, 1978. Fifty-three minutes later, *Voyager 2*'s acknowledgement returned. The trick worked. As a backup measure, in October 1978, *Voyager 2*'s memory banks were loaded to the brim with commands that would provide

---

77. Cooper, *Imaging Saturn*, p. 182.
78. Fred D. Kochendorfer, Memorandum for the Record, April 21, 1978, Rodney A. Mills to Distribution, April 21, 1978, and A. Gustaferro to Associate Administrator for Space Science, February 1, 1980, record no. 005566, NASA Historical Reference Collection; Murray, *Journey into Space*, pp. 152–54.

for a bare-minimum science encounter at both Jupiter and Saturn, should radio contact once again be lost. The same procedure was followed for subsequent encounters at Uranus and Neptune.[79]

The radio and scan platform breakdowns were not the only hardware failures that threatened or curtailed Voyager science. Some of the scientific instruments themselves experienced intermittent malfunctions and even complete breakdowns. The high radiation levels at Jupiter caused difficulties in transmitting commands, and the photopolarimeter instrument suffered radiation damage. Moreover, in November 1980, as *Voyager 1* was leaving Saturn, its plasma instrument stopped transmitting usable data. A similar fault had disabled the instrument for three months earlier in the year, as well as back in February 1978.[80]

In spite of the hardware problems that constantly threatened to diminish the mission's scientific returns, Voyager encountered Jupiter and Saturn, then continued on to Uranus and Neptune, following the Grand Tour route. Piece by piece, the Grand Tour itinerary came together. The continuation of Voyager to Uranus and beyond was made possible by reprogramming the onboard computers, creating new software, and building new ground facilities, new technologies and techniques without which the science could not be conducted. The expansion of Voyager into an even larger scientific enterprise also had spinoffs of value to the little science conducted on Earth. But first, funding the extension to Uranus had to be approved.

In 1975, the Space Science Board recommended a Mariner Jupiter-Uranus mission to be launched in November 1979, fly by Jupiter in April 1981, and proceed to Uranus arriving in mid-1985. Mariner Jupiter-Uranus was not the only mission under consideration by NASA. The space agency still was attempting to cobble together a program of exploration of the outer planets in the face of declining budgets. Other proposals included a Mariner Jupiter Orbiter (later developed into project Galileo) and Pioneer missions carrying atmospheric entry probes either directly to Saturn or to Uranus via Jupiter.[81]

Mariner Jupiter-Uranus was planned for NASA's fiscal 1977 budget and bore a price tag of $177 million, but it was in serious question because of the Administration's announced federal budget squeeze. In May 1975, NASA issued an announcement of opportunity for scientists to participate in Mariner Jupiter-Uranus. Facing budget restric-

---

79. Murray, *Journey into Space*, pp. 155; Davies, 23 (1981), pp. 72 and 73; Burgess, *Far Encounter*, 19 and 20. Deep Space Network engineers were not always successful at first. A slight change in temperature of the *Voyager 2* radio receiver (hardly more than one-tenth of a degree centigrade) was sufficient to wreck the improvised scheme. Also, changing amounts of electrical power consumption within the spacecraft gave off heat, and that heat, too, upset the scheme. NASA engineers then developed an elaborate computer model to account for all the temperature vagaries that fell on the spacecraft, as well as the heat created by changing patterns of electrical consumption within the spacecraft.

An alternate solution to *Voyager 2's* radio problem considered by NASA was to send commands through one of the science instruments, namely the planetary radio astronomy receiver. Tests conducted during September 1978 using the Stanford radio telescope indicated less received signal strength than had been anticipated, and the approach required both major changes in the onboard computer programs and the construction of a suitable ground transmitter facility. Implementation would cost an estimated $10 million ($7.5 million facility, $2.5 million project) and would require about twenty-four months to develop. Realizing that if the capability were never used, NASA would be open to criticism for having built an unnecessary facility, the Voyager Program Office decided against the planetary radio astronomy solution. A. Gustaferro to Associate Administrator for Space Science, February 1, 1980, and attachments, record no. 005566, NHO; Raymond L. Heacock to Rodney A. Mills, 29 March 1979, record no. 005566, NASA Historical Reference Collection; Von R. Eshleman, interview with author, Stanford University, May 9, 1994, JPL Archives.

80. Rodney A. Mills, Voyager Program Manager, to Distribution, February 24, 1978, record no. 005566, NASA Historical Reference Collection.

81. Burgess, *Far Encounter*, pp. 1 and 2; Davies, p. 39.

tions, still, NASA Headquarters dropped the project from its fiscal 1977 budget request, causing a severe manpower problem at JPL.[82] Adding Uranus to the Voyager project, on the other hand, bore a price tag far less than that for Mariner Jupiter-Uranus, about $100 million over five years, and it would bring money to JPL. Approval of the mission extension was received in November 1980 and was based on *Voyager 1* achieving adequate Titan and Saturn ring science and the health of *Voyager 2*. JPL had a long lead-time, five years, to prepare for Uranus: *Voyager 2* would not reach Uranus until January 1986.[83]

The extension of the mission to Uranus and beyond required re-engineering the spacecraft, which was already far from Earth, and upgrading Earth communication facilities. These changes were compelled by the vast distances over which the Deep Space Network had to communicate with Voyager, and by the dearth of sunlight needed for imaging and certain scientific experiments. The Sun at Uranus is only one fourth as bright as at Saturn and provides less than one four hundredths of its earthly illumination. Television exposures needed to be longer; camera shutter speeds reduced.

In upgrading the *Mariner 10* camera to image Mercury, JPL engineers developed a new electronic technique that read out the image signal three times more slowly when desired. When the Voyager cameras were operated in this slow mode, the lower radio transmission rate was adequate for real-time communications from Saturn, because the video signals could still flow directly from the camera to the radio transmitter and on to an attentive Earth. While Voyager would use the same slow camera mode and transmission rates at Uranus, additional techniques, namely compression and improved encoding, were demanded.

Part of the $7 million of additional appropriation granted Mariner Jupiter-Saturn in fiscal 1973 for technical improvements went toward design of an electronic means for transmitting error-free data to Earth, what is known as Reed-Solomon coding in honor of its inventors. Only basic coding hardware had been incorporated into Voyager's computer when it was launched. For the Uranus encounter, JPL engineers developed a special Reed-Solomon coding, which the Deep Space Network transmitted to Voyager's computer. The improved encoding worked, but it required more work on the ground.

Voyager engineers also used a technique called compression to obtain images from beyond Saturn. Normally, the full light-intensity value of each pixel of every image is transmitted back to Earth. Compression consists of sending back only the difference in light intensity between adjacent pixels on each line of each image. The technique reduced the communications rate needed by a factor of two and a half. But, in order to exploit compression, the spacecraft's computers had to be assigned new tasks, and that involved a certain risk. If a problem arose with the primary flight data computer, while the backup computer was tied up executing compression commands, key scientific observations, or even the entire mission, might be lost.[84]

Following the Grand Tour road to Uranus and Neptune also required revamping ground-based communication facilities. The distance to Uranus was over a billion kilometers. Signal strength was about one-fourth the level of the Saturn fly by in 1981, when Voyager was transmitting from a distance of 605 million kilometers. Existing ground-based Deep Space Network facilities were unable to adequately communicate with

---

82. *Defense/Space Daily*, December 3, 1975, p. 175, record no. 005556, NASA Historical Reference Collection; Burgess, *Far Encounter*, p. 66; Waff, "Jovian Odyssey: A History of Project Galileo," outline, December 9, 1987, n.p., NASA Historical Reference Collection.

83. Frank A. Carr to Edward C. Stone, December 18, 1980, and Frank A. Carr to Raymond L. Heacock, December 12, 1980, record no. 005566, NASA Historical Reference Collection; Edward C. Stone, interview with author, JPL, November 23, 1994; Davies, pp. 210, 253 and 257.

84. JPL, Public Information Office, News Clips, May 22, 1985, record no. 005586, NASA Historical Reference Collection; Murray, *Journey into Space*, pp. 173–175.

Voyager at those great distances. The solution was to array antennas together, a technique commonly used in radio astronomy. At the Deep Space Network site outside Canberra, Australia, two 34-meter and one 64-meter dish antennas were arrayed together. In addition, through an international agreement, NASA linked its Deep Space Network Canberra dish antennas with the 210-foot Parkes radio astronomy telescope located 200 km away via a microwave connection. Of the three Deep Space Network locations, that in Australia would have the best view of Uranus during *Voyager 2*'s ring plane crossing and closest encounter with the planet.[85]

A similar arrangement was put together at the Deep Space Network site at Goldstone, California, for the Voyager encounter with Neptune. Because Neptune is three times farther away from Earth than Saturn is, the Voyager X-band radio signal would be less than one-tenth as strong as during the Jupiter encounter in 1979 and less than one-half as strong as during the Uranus encounter in 1986. Part of the Voyager upgrade of the Goldstone 64-meter antenna involved enlarging the dish diameter 70 meters, increasing the surface accuracy, and improving the receiving system, as well as the installation of 34-meter antennas, to be used in an array formation, at the Goldstone and Canberra Deep Space Network sites.[86]

NASA approached the management of the Very Large Array (VLA), a radio telescope located in New Mexico, about participating in the formation of an antenna array with the Deep Space Network dishes at Goldstone, in order for NASA to communicate with Voyager at Neptune. The space agency installed low-noise X-band receivers on each of the 27 VLA antennas. Through the radio astronomy technique of arraying, and the installation of low-noise receivers on each VLA dish at NASA's expense, the echoes received from the VLA were combined with those received at the Goldstone 70-meter and 34-meter dishes to provide a data rate more than double that which would have been available with Goldstone's antennas alone. Just as with the Parkes radio telescope, a microwave link permitted NASA to array the VLA and Deep Space Network dishes at Goldstone.[87]

The Voyager upgrade of the VLA inadvertently created a state-of-the-art facility for planetary radar astronomy, a scientific activity that was, and remains, little science in terms of manpower, instruments, budget, and publications, but which took root within the interstices of big science.[88] When radar astronomers linked the Goldstone radar and the VLA in a bistatic mode, that is, with Goldstone transmitting and the VLA receiving, they created a radar with an extraordinary capacity for exploring the solar system. Duane O. Muhleman, California Institute of Technology, his graduate students Bryan Butler and Arie Grossman, and Martin A. Slade of JPL have used the Goldstone-VLA facility to explore Titan, Venus, Mars, and Mercury. Their exploration has led to a number of major discoveries, including the presence of polar ice on Mercury.[89]

---

85. JPL, Public Information Office, News Clips, May 22,1985, record no. 005586, NASA Historical Reference Collection; Bruce A. Smith, "NASA Reconfigures Voyager 2, Ground Stations for Uranus Flyby," *Aviation Week & Space Technology*, (May 20, 1965) pp. 65–66.

86. JPL, Public Information Office, News Clips, May 22, 1985, record no. 005586, NASA Historical Reference Collection; JPL Annual Report, 1973–1974, p. 15; *ibid.*, 1984, p. 13; *ibid.*, 1987, p. 41; and *ibid.*, 1988, p. 28, JPL Archives.

87. JPL, Public Information Office, News Clips, May 22, 1985, record no. 005586, NASA Historical Reference Collection; Murray to Morton S. Roberts, February 25, 1982, "Chron 1982 #1," and Memorandum, Associate Administrator for Space Tracking and Data Systems to Deputy Director, JPL, February 28, 1983, "NASA Correspondence, 1983, pt. #1," Peter Lyman Collection, JPL Archives; JPL Annual Report, 1984, p. 13, and ibid., 1987, p. 41, JPL Archives.

88. This is the argument developed in extenso in Butrica, *To See the Unseen*, passim.

89. For the discovery of polar ice on Mercury, see David A. Paige, "Chance for Snowballs in Hell," *Nature* 369 (1994): 182; Clark Chapman, "Ice Right Under the Sun," *Nature* 354 (1991): 504 and 505; J. Kelley Beatty, "Mercury's Cool Surprise," *Sky & Telescope* 83 (January 1992): 35–36. For a general discussion of the work done with the Goldstone-VLA, see Butrica, *To See the Unseen*, Chapter Nine.

As Voyager travelled from one planet to another, from one spectacular and unexpected discovery to the next, scientists and the public marvelled at the outcome of this scientific expedition. In the words of Project Scientist Ed Stone: "There's one lesson we learned from Voyager: Nature is much more inventive than our imaginations."[90] The Voyager mission truly deserved the honor of the 1980 Collier Trophy. Moreover, its subsequent accomplishments beyond Saturn in the face of hardware and budgetary hindrances have merited further recognition.

The crucial role played by technology in the success of the Voyager scientific mission allows us to draw some conclusions about the nature of big science. Obviously, Voyager science was entirely dependent on the availability of the spacecraft and its assemblage of scientific instruments. Hardware failures threatened the loss of science. That science depended, too, on the availability and proper functioning of an extensive network of telecommunication facilities on Earth. A similar dependence on technology is found in ground-based planetary astronomy.

The technologically driven nature of Voyager science raises questions about the epistemology of space-based science. In an Earth observatory, an astronomer can look through the lenses of a telescope and see the object of study. Using a space-based observatory, such as the Voyager spacecraft, scientists do not experience nature as directly as through a telescope. Instead, a scientific instrument makes the observation, then electronic circuitry aboard the space-based observatory converts the observation into strings of digital bits and transmits those bits to Earth, where a Deep Space Network facility acquires them. Through various signal-processing stages, which require extensive manipulation by large computers, the strings of digital bits transmute into data, which scientists then study. It is this data that scientists study and from which they draw conclusions about the phenomena that interacted with the scientific instrument in space. Data, rather than direct observation, has become the object of research, and that change has required inclusion of certain assumptions about the relationship between phenomena and the data. Thus, the instrument of scientific research is no longer just the spectrometer or the telescope (to use an Earth-bound analogy), but the observatory and the totality of electronic operations (both telecommunications and computing) required to turn the observation of the instrument into data. Historians of science need to explore how computers, signal processing, and other electronic techniques have come to mediate between the observer and the observed and to determine to what extent this transformation has been precipitated by the advent and growth of big science. Clearly, though, it is large-scale technology and techniques that make possible the science.

The Voyager project was an example of big science as measured by a number of yardsticks, such as the number of planets, satellites, and rings studied, mission longevity, and cost. At the same time, little science was an integral part of the project. The creation of the Goldstone-VLA array to receive Voyager images from Neptune also furnished radar astronomy's little science with a facility. More directly, university based scientists became part of Voyager big science through their organization into science teams and through the centralization of science and other decision making in the Project Scientist. The literature holds additional examples of big science as the centralization and management of little science.

James Watson, former head of the Human Genome Initiative, claims the project utilized a "little science approach" partly because only its management, and not the work, was centralized. In her study of fusion, Joan Lisa Bromberg argues that centralizing the research decision-making process, rather than centralized facilities, defined the institutional boundaries of big science. James H. Capshew and Karen A. Rader, furthermore,

---

90. Edward C. Stone, interview with author, JPL, November 23, 1994.

contend that activities that are broad in scope, scientific exploration being a specific example they cite, are big in the sense that they require coordination among geographically dispersed investigators or facilities. Consequently, the hallmark of such Big Science is horizontal integration and a reliance on extensive communication networks and centralized work processes.[91] The history of Voyager shows that yet another example of big science as horizontal integration of science management is the organization of the geographically dispersed Voyager scientists into teams and the concentration of decision-making in a single individual, the Project Scientist.

Within the NASA-industrial-academic complex, little science and big science do not always dovetail. The discussions of outer planet exploration within Space Science Board summer studies leading up to the decision to terminate Grand Tour illustrate this point. Planetary scientists wanted Jupiter-intensive studies and separate missions to the individual outer planets, while NASA (especially JPL) wanted to send numerous spacecraft to the outer planets, but each one taking advantage of the rare Grand Tour launch window. The division between JPL and the planetary science community stemmed largely from their divergent interests. The primary activity of JPL and NASA was the designing, building, and launching of vessels for the (preferably manned) exploration of the solar system. The planetary science community, on the other hand, wanted to do science, rather than build spacecraft.

Despite this division, NASA and the planetary science community had much in common. As Joseph Tatarewicz has shown, NASA has transformed American ground-based planetary astronomy into big science through its financing of the scientific enterprise.[92] By funding the construction and launching of spacecraft laden with scientific instruments, NASA also has positioned itself as the patron of space-based big science. NASA funding of both space-based and earth-bound planetary science is not the only way in which NASA has incorporated little science into big science. The organization of scientists into investigation areas and the centralization of the management of science into the Science Steering Group and the Project Scientist on the Voyager mission was another way in which NASA weaves little science into the larger fabric of big science.

This brief overview of Voyager stressed the critical role of properly functioning technology in the success of the scientific mission. The dependence of science on instrumentation for observation and the need for science funding is at the core of the relationship between big science and little science. Critical, too, is the inescapable fact that planetary science is based on observation. Without the Voyager observatory and its payload of instruments, planetary scientists would have been without data, without observations. NASA funding also paid for the scientists to participate in the project. To what extent could planetary science be conducted without NASA and the trappings of big science?

And by any estimation the planetary science conducted by Voyager was impressive. Just a partial list would include the following, and fully justify the recognition the mission has received:

- Discovery of the Uranian and Neptunian magnetospheres, both of them highly inclined and offset from the planets' rotational axes, suggesting their sources are significantly different from other magnetospheres.
- The Voyagers found twenty-two new satellites: three at Jupiter, three at Saturn, ten at Uranus, and six at Neptune.
- Io was found to have active volcanism, the only solar system body other than the Earth to be so confirmed.

---

91. James H. Capshew and Karen A. Rader, "Big Science: Price to the Present," *Osiris*, ser. 2, vol. 7 (1992): 14, 16, 20–22.

92. Joseph N. Tatarewicz, *Space, Technology, and Planetary Astronomy* (Bloomington, IN: Indiana University Press, 1990), passim.

- Triton was found to have active geyser-like structures and an atmosphere.
- Auroral zones were discovered at Jupiter, Saturn, and Neptune.
- Jupiter was found to have rings. Saturn's rings were found to contain spokes in the B-ring and a braided structure in the F-ring. Two new rings were discovered at Uranus and Neptune's rings, originally thought to be only ring arcs, were found to be complete, albeit composed of fine material.
- At Neptune, originally thought to be too cold to support such atmospheric disturbances, large-scale storms (notably the Great Dark Spot) were discovered.

As big science became the dominant way of doing science in the latter half of the twentieth century, what we call little science has become a necessary and integral part of Big Science. In the case of Voyager, spinoff facilities, summer studies, and, above all, the organization of scientists into the Science Steering Group, integrated little science into the overall big science undertaking. Many Earth observatories continue the tradition of blending little and big science. Individual scientists from universities request time on a large telescope, usually funded by public money, in order to make observations. The scientist might have funding from the National Science Foundation or NASA. There is no longer a distinction between big science and little science, but a single scientific enterprise in which the two are woven together in a set of interdependent relationships, each part of the same fabric.

Chapter 12

The Space Shuttle's First Flight: STS-1

by Henry C. Dethloff

The first mission of the space transport system (STS-1) or Space Shuttle, flew on April 12, 1981, ending a long hiatus in American space flight. The last Apollo lunar mission flew in December 1972, and the joint American Russian Apollo-Soyuz Earth orbital mission closed in July 1975. The National Aeronautics and Space Administration (NASA) intended that the shuttle make that permanent link between Earth and space, and that it should become part of "a total transportation system" including "vehicles, ground facilities, a communications net, trained crews, established freight rates and flight schedules—and the prospect of numerous important and exciting tasks to be done." It was to be "one element in a grand design that included a Space Station, unmanned planetary missions, and a manned flight to Mars."[1]

Awarded the Collier Trophy (in a tradition that began in 1911), the flight of STS-1 represented the greatest achievement in aviation for 1981. NASA, Rockwell International, Martin Marietta, Thiokol, and the entire government/industrial team responsible for the design, construction, and flight of the spacecraft, as well as the crew of the shuttle, John Young, Robert Crippen, Joe Engle, and Richard Truly, were all recipients of that award. Since 1962, NASA aerospace projects, including Mercury, Gemini, Apollo, Landsat, and Skylab, had received ten of the twenty Collier awards. Now, the eleventh in twenty years went to a NASA team that had designed and flown something remarkably different from those previous craft. For the Space Shuttle was a true aerospace craft, a reusable vehicle that could take off from the Earth, enter and operate in space, and return to an Earth landing. N. Wayne Hale, a missions flight director for the shuttle, likened it to a battleship, which while it may have only a few aboard, nevertheless had a crew of thousands stationed around the world and linked by Mission Control. Owen Morris, the Engineering and Systems Integration Division head for the shuttle Program Office, described the shuttle as a particularly complex, integrated machine and an enormous engineering challenge.[2]

Although it flew its maiden voyage only in 1981, NASA's shuttle program began many years earlier and predated Apollo. In the late 1950s, as human space flight began to be seriously considered and planned, most scientists and engineers projected that if space flight became a reality it would build upon logical building blocks. First, a human would be lofted into space as a passenger in a capsule (project Mercury). Second, the passengers would acquire some control over the space vehicle (project Gemini). Third, a reusable space vehicle would be developed that would take humans into Earth orbit and return them. Next, a permanent Space Station would be constructed in a near-Earth orbit through the utilization of the reusable space vehicle. Finally, planetary and lunar flights would be launched from the Space Station using relatively low-thrust and reusable (and thus lower cost) space vehicles. The perception of what became the shuttle as that reusable space vehicle associated with an orbiting space station held fast well into the vehicle's developmental stages.

---

1. Howard Allaway, *The Space Shuttle at Work* (Washington, DC: NASA, 1979), Foreword, pp. 21–27, 51–63.
2. Interview, Henry C. Dethloff with Owen Morris, Houston, Texas, August 8, 1990; Interview, Dethloff with N. Wayne Hale, Jr., Johnson Space Center, Houston, Texas, October 19, 1989; and see the author's "*Suddenly, Tomorrow Came . . . : A History of the Johnson Space Center*" (Washington, DC: NASA SP-4307, 1993), pp. 221–55, 285–305.

One of the known quantities in space flight was that the velocity required for a vehicle to escape earth's gravitational pull was only 1.41 times the velocity required to achieve earth orbit. The great costs associated with space flight included the cost of fuel used to achieve orbit, the cost of the expendable boosters and fuel tanks used to drive a space vehicle into orbit or into space, and the effective loss of the inhabited capsule or vehicle which, while it returned, could not be reused. Space quickly came to be an expensive business, and as it developed, the shuttle, more so than previous projects, was cost-driven, both in its incentives and in its construction. But because the nation's mission in space came to be to put an American on the Moon within the decade of the sixties, NASA's Apollo lunar program preempted both the Space Station and the shuttle. And, when the shuttle appeared without a Space Station to build and service, it appeared emasculated and detached from its intended purpose—to some extent an aerospace plane without a space mission.

When did the Space Shuttle begin? At what Point was it Created?

It could have been in March 1966, when a NASA planning team developed a statement of work for a "Reusable Ground Launch Vehicle Concept and Development Planning Study." Or it could have been at an Apollo applications conference held at the Manned Spacecraft Center (later the Lyndon B. Johnson Space Center) in Houston on October 27, 1966, when leaders of the Marshall Space Flight Center and the Manned Spacecraft Center agreed to pursue independent studies of a shuttle system along the lines of a March 1966 statement of work. Or most certainly a point of inception would be January 23, 1969, when George E. Mueller, NASA's Associate Administrator for Manned Space Flight, approved contract negotiations for initial shuttle design work.[3] Or it could have been even much earlier.

Under the authority of House Resolution 496, approved March 5, 1958, the House Committee on Science and Astronautics, chaired by Senator Overton Brooks, Democrat of Louisiana, convened hearings designed to provide direction and guidance for the creation of a new Federal agency that would head America's space program. During those hearings many "experts" described the development of space stations and "controlled space flight" as the prerequisites for expeditions to the Moon and beyond. Brigadier General A.H. Boushey, Air Force Director of Advanced Technology listed the development of spacecraft, piloted by humans, as "the most important" of the goals which must receive attention before there could be true exploration of space:

> *By piloted spacecraft, I refer to a vehicle wherein a pilot operates controls and directs the vehicle. This is quite a different concept from the so-called man-in-space proposal which merely takes a human 'along for the ride' to permit observation of his reactions and assess his capabilities.*[4]

Boushey believed that by the end of the decade of the 1960s, a large Space Station could be assembled by piloted "space tugs," that would remain in orbit throughout their useful life and operate only outside the atmosphere. "In addition to the 'tugs,' manned

---

3. Memorandum, Max Akridge (PD-RV), Space Shuttle History, January 8, 1970, MSFC Reports Subseries, JSC History Office Houston, TX.
4. Staff Report of the Select Committee on Astronautics and Space Exploration, *The Next Ten Years in Space, 1959–1969*, 86th Cong., 1st Sess., House Doc. No. 115 (Washington, DC: Government Printing Office, 1959), pp. 8–9.

resupply and maintenance spacecraft will shuttle from the Earth's equator to the orbiting satellites." Subsequently, a piloted spacecraft that would refuel at the Space Station in Earth orbit, "will land on the Moon."[5]

T.F. Morrow, vice president of Chrysler Corporation, thought that space stations or platforms might come in later decades, but that by 1969 one could expect "space trips encircling the Earth and the Moon." Dr. Walter R. Dornberger, rocket expert for Bell Aircraft, expected to see "manned and automatic space astronomical observatories; manned space laboratories; manned and automatic filling, storage, supply and assembly space facilities; manned space maintenance and supply and rescue ships-all climaxed by the first manned flight to the Moon."[6]

Roy K. Knutson, Chairman, Corporate Space Committee for North American Aviation, offered a much more exact definition for a "winged" space vehicle. While a piloted capsule (such as Mercury) would take a person into space and provide important physiological data, "Ultimately . . . consideration must be given to the problem of reentering the Earth's atmosphere from orbit in a winged vehicle capable of landing at a designated spot under control of a pilot."[7] He viewed North American Aviation's X-15 (then under development) as a forerunner of an aerospace craft, and believed solving the reentry problem would be the most crucial engineering task associated with developing a reusable shuttle. He offered, in 1958, a remarkably clear description of what would one day become the shuttle:

> *A large rocket booster would be used to boost the vehicle to high altitudes. Then a rocket engine installed in the ship itself would be ignited to provide further acceleration to the 25,000 miles per hour required for orbiting. In a low trajectory, the vehicle would pass halfway around the Earth in 45 minutes. A retrorocket would start the ship out of orbit at perhaps 10,000 miles from the landing point. As the vehicle enters the denser atmosphere, the nose and edges of the wing and tail will glow like iron in a blacksmith's forge. The structure will be built to withstand this extreme condition, however, and the pilot glide down to a dead stick landing.*[8]

If not a point of inception, there was at least in 1958 a sense of direction for the development of a reusable aerospace craft.

Even earlier, before the launch of the Soviet Sputnik satellite, scientists and engineers seriously discussed the construction and operation of space craft. Krafft A. Eriche, for example, presented "Calculations on a Manned Nuclear Propelled Space Vehicle" to the American Rocket Society in September 1957. In January 1957, NACA engineers on the staff of the Ames Aeronautical Laboratory at Moffett Field, California, filed a secret report on their "Preliminary Investigation of a New Research Airplane for Exploring the Problems of Efficient Hypersonic Flight." It was to be an aircraft considerably exceeding the performance levels of the X-15 with "a rocket boost . . . to Mach numbers of the order of 10 and altitudes of the order of 140,000 feet."[9]

---

5. *Ibid.*, p. 9.
6. *Ibid.*, pp. 9–10.
7. *Ibid.*, pp. 85–91.
8. *Ibid.*, pp. 91.
9. "Preliminary Investigation of a New Research Airplane for Exploring the Problems of Efficient Hypersonic Flight," by the Staff of the Ames Aeronautical Laboratory Moffett Field, California, NACA, Washington, DC, January 18, 1957; K.A. Eriche, "Calculations on a Manned Nuclear Propelled Space Vehicle," September 5, 1957, JSC History Office.

With the insight and direction provided by Congress, the experiences of National Advisory Committee for Aeronautics (NACA), and the American (and Canadian) aircraft industry, NASA set about after its inception in 1958 to provide the United States leadership in space exploration, space science, and space technology.[10] But American successes in space seemed painfully gained, and slowly realized.

Not only had the Soviet Union launched the first satellite into orbit on October 4, 1957, but in 1959 Soviet rocket scientists launched three successful interplanetary craft into space. The second, *Luna II* impacted on the Moon in September; *Luna III* flew behind the Moon in October 1959. On April 12, 1961, Major Yuri Gagarin became the first person to "leave this planet, enter the void of space, and return." By 1961, with the encouragement of the Democratic Party campaign for the presidency, Americans had begun agonizing over the "missile gap." After the elections and the inauguration, on May 25, 1961, President John F. Kennedy and Congress set a new course for NASA, preempting existing developmental programs and schedules. The United States, before the decade is out, should land "a man on the Moon" and return him safely to Earth.[11]

The Apollo program became the leading effort. An orbital Space Station and Earth-to-orbit spacecraft, while they might contribute to a continuing presence in space and provide a platform for further lunar or planetary exploration, did not contribute to the short term goal of an American lunar landing within the decade. NASA readjusted its schedules and priorities to accommodate Apollo. The Space Station and the reusable aerospace craft remained viable, but future, options. Marshall Space Flight Center (MSFC), in particular, continued to study the reusable vehicle concept and as early as January 1963, developed a statement of work for a fully reusable rocket-powered vehicle that could carry civilian passengers, and a sizable payload. Marshall awarded independent contracts to Lockheed Aircraft and North American Aviation for design and development studies. But the NASA focus continued to be on Mercury, Gemini, and Apollo. By the end of 1963, the Mercury program ended. The last Gemini mission flew on November 11, 1966. NASA scheduled the first Apollo flight for December 5, 1965. An Apollo with a Saturn booster, which was to send Apollo on its lunar voyages, flew an unpiloted test on February 26, 1966.[12] It appeared likely through most of 1966 that the Apollo-Saturn lunar program was on schedule. Should NASA complete its mission to land a man on the Moon within the decade, what would happen next?

NASA began to address that issue by establishing an Apollo Applications Office, in 1966, that would devise programs to utilize Apollo technology in non-lunar programs. In October 1966, the annual meeting of the American Institute of Aeronautics and Astronautics focused on the question, "After Apollo, What Next?" And, in 1966, just as the Apollo-Saturn program seemed on the verge of success, Congress and the American public began to divert attention and public funds from space and NASA to the more urgent business of a growing war in Vietnam. The war, and money, began, even in the midst of Apollo, to turn NASA's attention to the "more practical" approach to space.[13] More practical meant more efficient, less costly, more economic. Discussion of an orbital space platform or station, and a reusable Earth-to-orbit supply vehicle revived.

---

10. Memorandum for the Secretary of Defense and Chairman, the National Advisory Committee for Aeronautics, April 2, 1958, Special Committee on Space and Astronautics, Senate Papers, Box 357, Lyndon B. Johnson Library (hereinafter LBJ Library), Austin, Texas. A large contingent of Canadian and British aeronautical engineers were recruited by NASA following he Canadian governments decision in 1958 to halt development of the AVRO fighter plane.

11. Lyndon B. Johnson, *The Vantage Point: Perspectives of the Presidency, 1963–1969* (New York, NY: Holt, Rinehart & Winston, 1971), pp. 280–81.

12. See Dethloff, "*Suddenly Tomorrow Came . . .*", pp. 108–12, 221–22.

13. *Ibid.*, pp. 191–93.

Thus, in March 1966, a special NASA planning team developed a statement of work for a reusable ground launch vehicle, and in October Marshall Space Flight Center and the Manned Spacecraft Center agreed to pursue independent study and research on such a spacecraft. NASA budgets, however, were becoming increasingly constrained, and at a January conference at NASA Headquarters administrators reluctantly agreed that there should be no new launch vehicle development in order to reduce the budget problems. The year, 1967, passed without any real progress in the development of a reusable spacecraft, but financial pressures became greater rather than less. In January 1968, George Mueller rekindled sentiments for work on a reusable spacecraft as potentially a cost-saving measure:

> *Where we stand now is the feasibility generally has been established for reusability. And we have much data on many concepts. We have an uncertain market demand and operational requirements. The R&D costs for fully reusable systems, including incremental development approaches, appear high. Personnel and cargo spacecraft seem to dominate Earth-to-orbit logistics costs. R&D costs for new logistics systems are in competition with dollars to develop payloads and markets (dollars are scarce).*[14]

Nevertheless, NASA put a decision for the development of a reusable vehicle on hold.

Meanwhile, in collaborative sessions with the Air Force, which was independently studying orbiting laboratories and aerospace planes, NASA and Air Force engineers agreed on the need to develop a logistics space vehicle with a payload range of 5,000 to 50,000 pounds for use with a Space Station. Marshall and Manned Spacecraft administrators again conferred in October, and agreed to issue a request to NASA Headquarters for a joint Phase A (concept definition) study for a logistics space vehicle. Headquarters tentatively agreed to award a study contract, but withheld approval pending the results of the *Apollo 8* flight.[15]

Apollo 8 was the first Apollo flight carrying "human cargo" powered by the Saturn rocket. Its original flight plan was to go into Earth orbit, but again MSFC and MSC combined to convince leaders at NASA Headquarters that *Apollo 8* should be a circumlunar flight. Although perceived to be a "high risk" effort, *Apollo 8*, launched on December 28, 1968, put astronauts Frank Borman, James A. Lovell, Jr. and William A. Anders into ten orbits about the Moon, and returned them safely to Earth. That flight provided greater assurance of the probability of completing a lunar landing within the decade, and accelerated the need to commit to a post-Apollo program. On January 23, 1969, George Mueller approved contract negotiations for design work on what would become the Space Shuttle.[16] Touchdown by *Apollo 11* on the Moon's surface in July 1969 brought work on the shuttle into sharper focus. The question, "After Apollo, What Next?" needed to be answered soon.

President Richard M. Nixon appointed a Space Task Group to study the problem and offer options. Internal NASA studies complemented the work of the task group. On January 29, NASA awarded Phase A study contracts for elements of an "integral launch and reentry vehicle" (ILRV). Lockheed Missile & Space Company studied clustered or modular reusable flyback stages. General Dynamics/Convair examined expendable fuel tanks and solid propulsion stages. Both contracts were administered by Marshall. The Manned Spacecraft Center in Houston directed a study by North American Rockwell for expendable tank configurations coupled with a reusable spacecraft. McDonnell Douglas,

---

14. Akridge, *Space Shuttle History*, p. 36.
15. *Ibid.*, pp. 36–48.
16. *Ibid.*, 49; Linda Neuman Ezell, *NASA Historical Data Book*, III, *Programs and Projects, 1969–1978*, (Washington, DC: NASA SP-4012, 1988), pp. 113–18.

working under Langley Research Center supervision, examined tank, booster, and spacecraft ("triamese") configurations. Martin Marietta conducted an independent design study also submitted to NASA.[17] Concurrently, a joint DOD/NASA study began on space transportation which would also go to the President's Space Task Group.

In October 1969, Congressman Olin E. Teague, Chairman of the House Committee on Science and Astronautic's subcommittee for NASA oversight, asked the Director of each NASA Center involved directly in the manned space flight program to review various "levels of effort" as they might affect future programs when measured against the Space Task Group recommendations. He requested an evaluation of the Space Task Group's preliminary recommendations that NASA focus on a reusable space craft and a permanent space station. And he requested personal letters from Dale D. Myers (Associate Administrator for Manned Space Flight), Robert R. Gilruth (Director of the Manned Spacecraft Center), Kurt H. Debus (Director of Kennedy Space Center), Eberhard Rees (Director of Marshall Space Flight Center), and Wernher von Braun (Deputy Associate Administrator), "setting forth their views on the importance of moving forward with the Manned Space Flight Program at this time."[18]

Dale Myers described the changing focus of the mission in space from the single purpose pursued in the Apollo program, to a broader effort to use space technology for the benefit of man. "In earth orbit, a space station supplied by the reusable shuttle will provide additional economic gains and practical benefits." They would facilitate a considerable expansion in space activities and increase the number of visitors into space.[19]

Robert R. Gilruth, Director of the Manned Spacecraft Center, responded that he firmly believed "that the reusable Space Shuttle and the large Space Station are vital elements which must be developed." He described the "earth-to-orbit shuttle" as "the keystone to our post-Apollo activities." Kurt Debus described the broad technology advances required for the development of a shuttle and Space Station, and noted that one cannot always identify the total utility of an innovation. Throughout history, he noted, innovations have been made without identifying all the uses and applications-he named the wheel, the telephone, the car, and the airplane as good examples. He advised proceeding now with the development of a fully reusable Space Shuttle, and the initiation of Phase B studies. Eberhard Rees wrote that the answer to the high costs of space transportation is to develop a system "which operates much like the cargo and passenger airlines, namely a Space Shuttle System."[20]

Wernher von Braun reviewed the accomplishments of the past decade, noting that the space program thus far "brought renewed strength in national leadership, in security, in education, and in science and technology, and in the will of America to succeed."

> ... the key to our future accomplishments in space will be willingness to undertake the developments that will advance this nation to new plateaus of operational flexibility and will give us the technological advances needed to assure economical operations in space. No one would question the justification for a jet aircraft that can be flown over and over again instead of just once. With the Space Shuttle and the Space Station we will have the space age equivalent of the jet liner.[21]

---

17. Ezell, *NASA Historical Data Book*, 3:48; Akridge, *Space Shuttle History*, pp. 36–49; J.P. Loftus, Jr., S.M. Andrich, M.G. Goodhart, and R.C. Kennedy, The Evolution of the Space Shuttle Design, Personal Files, Joseph P. Loftus, Johnson Space Center, Houston, Texas.
18. Olin E. Teague, Washington, DC to Robert R. Gilruth, Houston, TX, October 3, 1969, Apollo Program Chronological Files, JSC History Office; Robert F. Freitag to Distribution (with letters attached), April 29, 1970, Apollo Applications File 072-44/45, JSC History Office.
19. Freitag to distribution, AA 072-44/45, JSC History Office.
20. *Ibid.*
21. *Ibid.*

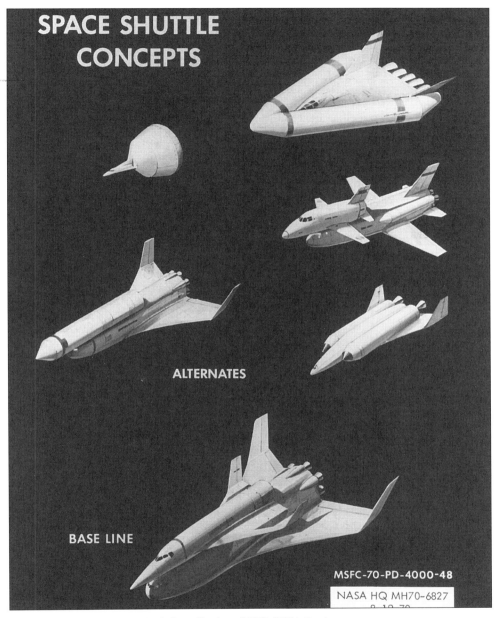

Possible configurations considered for the Space Shuttle as of 1970. (NASA photo).

Robert F. Thompson, who became the Manned Spacecraft Center's Space Shuttle Program Director in April 1970, explained that the emphasis in the initial Phase A and DOD studies was to develop a fully reusable system, which he perceived at the time as the most cost-effective configuration, because of anticipated lower operating costs. However, as early as May 1969, the costs of developing fully reusable systems became ominous. By the end of the year NASA Headquarters shifted the Phase A studies to an emphasis on a combination of expendable and recoverable boosters coupled with reusable spacecraft. The Phase A reports were received in November 1969, and the DOD/NASA joint studies were completed in December 1970. Both the NASA internal studies and the DOD/NASA study continued to support a fully reusable spacecraft.[22]

In May 1970, NASA awarded Phase B contracts to a North American Rockwell and General Dynamics team and to a McDonnell Douglas and Martin Marietta team for definition studies of a fully reusable shuttle. But in June, contracts were awarded to the Grumman Aerospace and Boeing partners for studies of various expendable and reusable booster and fuel tank designs, to Lockheed to examine an expendable fuel tank for the orbiter, and to Chrysler for design study of a single stage reusable orbiter. There were other contracts to study various assemblies through the remainder of 1970.[23] The year ended without a decision as to the design of the shuttle, but with a number of interesting options.

But the estimated costs of developing a fully reusable shuttle were rising, and costs soon became the decisive element, not only in the shuttle design, but in determining future NASA programs.

The development of a fully reusable shuttle was conservatively estimated to "require more than a doubling of NASA's budget, unrealistic at any time and particularly so in the light of increasing military expenditures in Southeast Asia." During congressional hearings on the FY 1971 NASA budget, NASA Comptroller Bill Lilly responded to questioning that if choices had to be made, the shuttle had to precede the Space Station because, "if they could not be developed concurrently, the shuttle in extended sortie, could act as a surrogate Station and the long term future of space flight lay in reducing the cost of all operations, but foremost in the cost of delivery to low Earth orbit."[24] As will be seen, funding was tenuous throughout the development program. The decision on a fully, or even a partially, reusable shuttle apparatus was still pending.

Finally, on April 1, 1971, NASA directed that the Phase B contracts shift the emphasis from "fully reusable" to consider an "orbiter" with external expendable hydrogen tanks. James C. Fletcher, who had replaced NASA Administrator Thomas O. Paine in April, believed that whatever the technical merits of a fully reusable space vehicle might be, the $10.5 billion price tag currently assigned shuttle development simply would "not fly" with Congress. In June 1971, Max Faget, who headed MSC's Advanced Missions Program Office, presented an alternate configuration, that is, a two-stage shuttle with a drop tank orbiter. Administrator Fletcher accepted the configuration as NASA's choice, and on June 16, 1971, sent Congress a letter of decision. Studies of the new configuration with a fully reusable orbiter, and expendable or reusable external booster rockets and tanks, subsequently lowered estimated R&D costs to about $5 billion, or one-half that of the fully reusable vehicle.[25]

The new partially reusable configuration involved the lowest development costs, but also enhanced the aerodynamics of the shuttle's orbiter, and safety. An internal tank

---

22. Ezell, *NASA Historical Data Book*, 3:48; Dethloff, *Suddenly Tomorrow Came . . .*, pp. 224–35; Akridge, *Space Shuttle History*, pp. 49–98.
23. Ezell, *NASA Historical Data Book*, 3:48.
24. Loftus, Andrich, Goodhart, and Kennedy, Evolution of the Space Shuttle Design, p. 8.
25. Eagle Engineering Inc., Shuttle Evolution Study, April 23, 1986, Loftus Historical Documents File, JSC History Office.

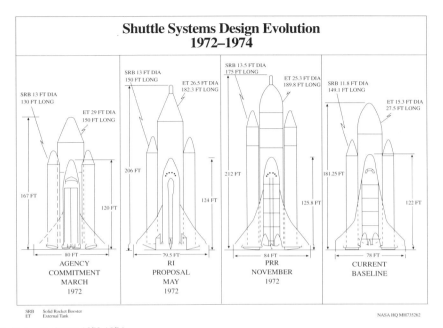

Shuttle Design Evolution 1972-1974.

design required heavy insulation of the spacecraft, much heavier launch weights, and flight difficulties resulting from tank torsion and "slosh." The very high pressure required in the fuel tanks also created higher risks and engineering and maintenance problems.[26] Refinement of the proposed new configuration took yet another two years. For the time, the solution seemed the best in terms of costs and technical development.

Despite NASA's June 1971 commitment to a reusable orbiter launched by an expendable or partially reusable propulsion system, there was no specific congressional funding for shuttle R&D. Shuttle funding came from general NASA spaceflight operations programs through FY 1973. Moreover, shuttle program expenditures had risen from $12.5 million in 1970 to $78.5 million in 1971.[27] Clearly, formal approval had to be secured or study on the shuttle project had to be terminated.

In June 1971, NASA's Associate Administrator for Manned Space Flight, Dale D. Myers, who had managed North American Rockwell's shuttle development work before he replaced George Mueller at NASA headquarters, assigned Marshall responsibility for development of the shuttle main engine and boosters, and the Manned Spacecraft Center responsibility for developing the orbiter. Throughout 1971 and into 1972, NASA extended the Phase B contracts, and awarded new ones to examine variously the use of existing Titan and Saturn rockets as shuttle launch vehicles, the feasibility of using liquid or solid propulsion boosters, and methods of recovering boosters and external tanks. In January 1972, Marshall Space Flight Center awarded contracts to Aerojet-General, Lockheed Propulsion Company, Thiokol Chemical, and United Technology Center to study the possibilities of using

---

26. *Ibid.*, p. 222.
27. Ezell, *NASA Historical Data Book*, 3:69.

existing 120-inch and 156-inch solid rocket motors as part of the shuttle booster system.[28] Preliminary and final reports confirmed the lower costs of the new shuttle configuration.

On January 5, 1972, Administrator Fletcher and Deputy Administrator George Low met with President Nixon and his staff assistant, John Erlichman, for a review of the shuttle program. Nixon approved the revised and less costly shuttle program, and wanted to stress both the civilian and the international aspects of shuttle development and future missions.[29]

Nixon's support for the shuttle, however, became hoisted on the petard of the growing difficulties in Vietnam, the proposed Air Force supersonic transport plane (SST) cancelled by Congress the previous year, and party politics. On January 7, Senator Edmund Muskie (D-ME), a Democratic candidate for the presidency, told Florida audiences while campaigning there that the Space Shuttle was an extravagance and should be shelved. Reflecting the sentiments of many Americans, the greater priorities of the nation, he said, were "hungry children, inadequate housing, decaying cities, and insecure old age." He accused President Nixon of practicing "pork barrel politics" by supporting the $5.5 billion space program.[30]

Senator Walter Mondale (D-MN), another aspirant for president, called the Space Shuttle program "ridiculous" on a nationally televised debate. "At the present and known levels of space activity, to produce the Space Shuttle would be like buying a fleet of gold-plated Cadillacs to go out and repair the tire of a Pinto. . . . It is not a new exploration weapon. It is simply a truck–a very expensive truck that is not worth the money."[31]

Senator William Proxmire (D-WI), who successfully led the fight against the SST in 1971, called Nixon's decision to go ahead with what he estimated to be the "$15.5" billion shuttle project, "an outrageous distortion of budgetary priorities." The President, Proxmire said, had chosen the Space Shuttle over schools, public health, housing, mass transit, open space, environmental needs and other vital programs.[32] The space program also had powerful advocates in Congress, including Texas Congressman Olin E. Teague (and the entire Texas delegation), Mississippi Senator John C. Stennis, and Senator Stuart Symington of Missouri, among others. Nevertheless, the administrative decision to proceed with shuttle development rested upon Congressional approval and budgets. The future of the Space Shuttle seemed particularly tenuous in 1972 as Congress began the budget debates near the end of January.

Meanwhile, NASA increased its allocation for shuttle spending from $78 million in 1971 to $100 million for 1972 from its internal operations funds. In March 1972, Myers assigned the Manned Spacecraft Center in Houston "lead center" authority for overall Space Shuttle Program Development management and control. Robert F. Thompson, a member of the original Space Task Group at Langley Research Center (which became the nucleus of the Manned Spacecraft Center in Houston, Texas) continued as manager for the NASA-wide Shuttle Program Office. Thompson previously headed the Manned Spacecraft Center's Apollo Applications Program Office, concerned with post-Apollo planning.[33]

During 1971 and 1972, the Manned Spacecraft Center and Marshall Space Flight Center began to fold personnel from Apollo offices into the shuttle program. Under the duress of budget cutbacks, and tenure, and with the successful close of the Apollo program, many NASA administrators and engineers began to leave NASA. Wernher von Braun relin-

---

28. *Ibid.*, p. 48.
29. George Low, "Meeting with the President on January 5, 1972," memo for the record, January 12, 1972, Shuttle Series, JSC History Office; Ezell, *NASA Historical Data Book*, 3:48.
30. *Miami Herald*, January 7, 1972; Typed memorandum, political roundup, January 7, 1972, Shuttle Papers, 007-24, JSC History Office.
31. Wirephoto WX2, January 16, 1972, Shuttle Papers, 007-24, JSC History Office.
32. *Houston Post*, January 9, 1972.
33. Ezell, *NASA Historical Data Book*, 3:48; Manned Spacecraft Center Announcement, Shuttle Files, 007-43, JSC History Office.

quished the post of Director of Marshall Space Flight Center to Eberhard Rees in 1970. Robert Gilruth stepped down as Director of the Manned Spacecraft Center in January 1972. Chris Kraft, formerly head of Apollo flight operations, replaced him.[34] At the very height of Apollo successes, NASA seemed to be imploding, while at the same time it redirected personnel and funds into the shuttle program. There were concurrent reductions in force and organizational realignments among NASA's aerospace contractors.

Although NASA had some 14 years of space flight experience behind it by 1972, the shuttle was something very new and very different from what had gone before. As Aaron Cohen, manager of the Orbiter Project Office in Houston explained, the "orbiter, although similar to Apollo in that it goes into space, is very different." The shuttle orbiter (which is usually identified in the public mind as the shuttle) is not simply a spacecraft, but a launch vehicle, a spacecraft, and an airplane combined. The transition from Apollo to shuttle, Cohen said, represented a transition of technology spanning ten years. There were major technological advances over Apollo in terms of materials, electronics, propulsion, and software. The launch configuration of the Space Shuttle was also different than had ever flown before. With Apollo the thrust was through the center of gravity, but with the shuttle the thrust was through the orbiter with an offset external tank. That configuration raised enormous problems with the structural dynamics of the assembly. In addition, whereas Apollo, Gemini, and Mercury launched from series burns, the shuttle utilized a parallel engine burn.[35] Most significantly, perhaps, the shuttle engines, unlike the Saturn or Titan engines, were "throttlable," having a controlled engine burn.

Cohen stressed that certain technical elements of the shuttle were so advanced they were "outside the existing state of the art." The controlled burn and the high pressures and temperatures at which the engines operated were an engineering challenge. Even to test the apparatus required innovative testing equipment and procedures. The thermal protection system involved the development of a heat-resistant tile that had never previously existed. Each individual tile fitted on the orbiter nose and underbody had to be individually designed and tested.[36] One of the most highly sophisticated and advanced systems was the avionics (guidance, navigation, and control) system which fused electronics with aviation (hence avionics) and made the guidance and control systems responsive and complementary to human direction.

"The avionics system synchronized four centralized computers and had a single computer independent of the other four." The fifth computer was on standby to step in should there be a software problem in one of the other computers. The four synchronized computers, the "heart and brains" of the shuttle, "communicated with each other 440 times per second." One computer was the lead computer, the other three "voted" on the input and output of each other. "Should the three other computers disagree with the lead computer, it was voted out of the system." Air data, microwave sensors, gyros, accelerometers, star trackers, and inputs from ground based laboratories all fed into the avionics system.[37] The shuttle avionics system represented revolutionary advances in electronics, computer technology, and guidance and control in the few short years since Apollo. Similarly, Apollo communications systems (using a unified S band) were inadequate to support shuttle missions.

Shuttle avionics systems were so advanced that special laboratories were required to design and develop them. NASA constructed a $630 million Shuttle Avionics Integration Laboratory (SAIL) at Johnson Space Center for the job. A special Shuttle Mission Simulator (SMS) trained crews to use the shuttle and fly missions in what is now popular-

---

34. See note above, and Dethloff, "*Suddenly Tomorrow Came . . .* ", pp. 209–10.
35. Aaron Cohen, "Progress of Manned Space Flight from Apollo to Space Shuttle", presented at AIAA 22nd Aerospace Sciences Meeting, January 9–12, 1984, Reno, Nevada, in Shuttle Papers, JSC History Office.
36. *Ibid.*
37. *Ibid.*

ly termed a "virtual reality" setting. Astronauts returning from shuttle missions reported that the simulations were so accurate they felt they had flown the mission many times.[38] Despite the advanced technologies used by the shuttle as compared to Apollo, Cohen believed that a permanent presence in space, that is the establishment of a Space Station, would require yet again major advances in new technologies.

New technologies were expensive. Research and development costs (R&D) grew rapidly. Inflation, which peaked at almost 13 percent in 1973, diminished appropriated funds and budgets proportionately. NASA and other government agencies were particularly affected by inflation because appropriations were approved in a previous year at fixed dollar levels. NASA found itself spending dollars that bought much less than anticipated. Congressional appropriations for NASA R&D declined by almost $450 million (15 percent) in 1971, and were reduced again in 1972 by another $40 million. R&D appropriations improved by about $80 million in 1973, but collapsed by over $400 million in 1974. During the most critical years of shuttle development, from 1971 through 1977, R&D appropriations remained remarkably stable. But the value of the dollars appropriated declined by about 50 percent in those five years. Budget stresses caused "slippage" and delays in development and production, and those in turn, raised the final costs of developing the shuttle.

Table I, below, provides an overview of total NASA R&D funding and designated shuttle funding during the developmental stage of the shuttle.[39]

Table I

NASA Appropriations, 1969-1978
(in thousands of dollars)

Fiscal Year	Research & Development	Space Shuttle Funding
1969	$ 3,530,200	$ -0-[a]
1970	2,991,600	9,000[b]
1971	2,630,400	160,000[c]
1972	2,623,200	115,000
1973	2,541,400	200,000
1974	2,421,600	475,000
1975	2,420,400	805,000
1976	2,748,800	1,206,000
1977	2,980,700	1,288,100
1978	2,988,700	1,348,800
1979	3,138,800	1,637,600
1980	3,701,400	1,870,300
1981	4,223,000	1,994,700

[a] the shuttle was funded as part of the spaceflight operations program through FY 1973.
[b] for a space station only.
[c] for shuttle and station; $6 million requested for station definition.
[Source: *NASA Pocket Statistics* (January 1994), and for shuttle funding, 1969-1977, see Linda Neuman Ezell, ed., *NASA Historical Data Book*, 3:69.]

---

38. *Ibid.*; and see Dethloff, "*Suddenly Tomorrow Came . . .* ", pp. 243, 247–51.
39. Ezell, *NASA Historical Data Book*, 3:12, 69.

That funding should be viewed in light of NASA's overall budget which, based on the value of 1992 dollars, dropped sharply from the FY 1965 peak in excess of $22 billion, to a 1974-1979 average of only $9 billion, as adjusted for inflation using 1994 constant dollars.

Although shuttle-specific funding by Congress did not begin until 1974, in 1972 and 1973 NASA began to move from the planning and study stage of shuttle development to the design and production stage. One of the great achievements of shuttle development had to do with the production (and business) management of complex disparate systems and integrating those systems or machines into one wholly integrated greater machine. There were many (in fact all) of the NASA centers involved in the creation of the Space Shuttle. There were far more, literally hundreds, of independent private manufacturers involved in its development. NASA, in effect, was the management team assembled for the production of a single machine by hundreds of diverse private manufacturers. NASA did not build the shuttle, private industry did. Thus, the Space Shuttle continued the peacetime mobilization of American science, engineering, and industry, begun at the inception of NASA and America's entry into the space age—albeit, perhaps, at a lower level.

The general NASA management structure was, of course, inherited from the Apollo and earlier programs, but there were important refinements. In 1971, NASA Headquarters assigned Marshall Space Flight Center responsibility for developing the booster stages and the shuttle main engines. Marshall, of course, had basic propulsion (engine) responsibilities from the beginning. Engine testing was assigned to Stennis Space Center, which had begun as Marshall's testing laboratory for the Apollo-Saturn engines. The Manned Spacecraft Center had responsibility for developing the orbiter, or piloted vehicle. Such had been Houston's basic responsibility since its establishment in 1961. Kennedy Space Center, formerly the Cape Canaveral Launch Operations Directorate under Marshall, had responsibility for launch and recovery of shuttle flights— as it had throughout the program. The technical, developmental work on the shuttle at all the NASA centers was coordinated through the shuttle Program Office located at MSC in Houston. (Under the Apollo program, many collaborative management decisions were reached informally between the Manned Spacecraft Center and Marshall, or were coordinated or passed through the Manned Space Flight Office in Washington.) The Shuttle Program Office, in turn, reported to the Office of Manned Space Flight at NASA Headquarters in Washington.[40]

The command and control management structure resembled the Apollo management systems, but there were some important differences. Production management was more decentralized than before, but control (integration) was more centralized. The shuttle program did rely (even more heavily) on Apollo-type Integration Panels which coordinated design and construction projects so that the pieces literally fit together and worked together. Integration was the critical element in shuttle production-which, as Owen Morris noted, was a so much more complex machine than Apollo. The Integration Panels reported to the Systems Integration Office in the Shuttle Program Office at the Manned Spacecraft Center and the Systems Integration Office reported to a Policy Review Control Board chaired by NASA Headquarters.[41]

Shuttle management became a "state-of-the-art" system for very large-scale industrial production. There were, of course, important precedents, such as the construction of the Panama Canal, a battleship, hypersonic aircraft, and Apollo. None of those systems, how-

---

40. *Ibid.*, pp. 121–22.
41. Memorandum from Joseph Loftus in response to an inquiry from Aaron Cohen [1990], "How did we manage Integration in the Apollo and Shuttle Programs?" Loftus personal files; MSC, *Roundup*, June 18, 1971; Dale D. Myers, "Space Shuttle Management Program," March 14, 1972, NASA Management Instruction Subseries, JSC History Office.

ever, involved the complexity of machinery, electronics, computers, and materials as entered into shuttle construction.

Within the three basic management levels for shuttle development technical engineering and management decisions flowed from the bottom up. The "bottom" consisted of the Level III project offices, such as the Orbiter Office at the Manned Spacecraft Center and the Booster Office at Marshall Space Flight Center. The Level III offices managed the production contracts. Level III offices maintained a Resident Office (or engineer) at the primary contractors production site, and often co-located a manager with the appropriate Level II division. The Level II office was the Shuttle Program Office. It had responsibility for systems engineering and integration, configuration, and overall design and development, or as Dale Myers stated: "program management responsibility for program control, overall systems engineering and system integration, and overall responsibility and authority for definition of those elements of the total system which interact with other elements." The Level II office established "lead center" authority for engineering and development management. Headquarters, or Level I, in turn had overall program responsibility and primary responsibility for the assignment of duties, basic performance requirements, the allocation of funds to the Centers, and control of major milestones.[42]

The management structure created a very decentralized, independent production system—very compatible, if not necessary, to the very diverse and autonomous private entities that made up the manufacturing or production base of the NASA program. One of the great achievements of the space program, contrary to the tendency in large scale bureaucratic enterprises, was to harness the basic strengths of American industry through decentralized management and production.

Although it was not designated "Level IV," the real production base of the shuttle program was private industry. The basic management tool was the NASA contract, and effectively, competition for the contract. It was the contract (and the primary contractor's subcontracts) that mobilized American industry in support of the space program.

The preliminary study, design, and feasibility contracts (Phases A & B), mentioned earlier, with in-house study and tests produced the technical parameters for issuing an RFP or Request for Proposal. NASA began issuing RFP's for shuttle procurement in the spring of 1971. Aerojet Liquid Rocket Company, Pratt & Whitney, and Rocketdyne were invited to submit proposals for the development of the shuttle main engines. Soon after, the Manned Spacecraft Center issued an RFP for a shuttle thermal protection system, to protect the orbiter and its occupants during the critical reentry phase. In July 1971, MSFC selected Rocketdyne as the primary contractor for the production of thirty-five shuttle main engines. Pratt & Whitney challenged the Rocketdyne award and during a GAO (General Accounting Office) review, Rocketdyne was given an interim contract. In March 1972, MSC issued an RFP for the development of containerized shuttle payload systems, and NASA issued an RFP for the development of the shuttle, with the design due in May.[43]

North American Rockwell (later Rockwell International), McDonnell Douglas, Grumman, and Lockheed submitted proposals for the shuttle. NASA approved an interim letter contract with Rockwell in August 1972, and issued a final contract on April 16, 1973. Rockwell, in turn, subcontracted major components of the shuttle orbiter to other aerospace firms. Fairchild Republic Division of Fairchild Industries constructed the vertical tail unit; Grumman, the delta wings; General Dynamics' Convair Aerospace Division subcontracted for the mid-fuselage section, and McDonnell Douglas had responsibility for

---

42. See note above; *Catalog of Center Roles* (Washington, DC: NASA, December 1976), pp. 1–30, Loftus Subseries, JSC History Office.

43. Ezell, *NASA Historical Data Book*, 3:122.

The Space Shuttle rises from Launch Pad 39A at Kennedy Space Center, Florida, a few seconds after 7 a.m., April 12, 1981. This first flight was flown by astronauts John Young, Commander, and Robert Crippen, Pilot. (NASA photo no. 81-H-285).

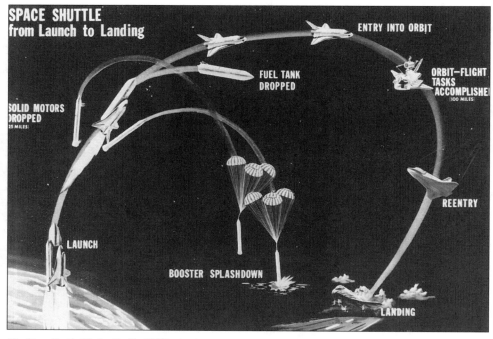

The Space Shuttle Mission Profile. (NASA photo).

the orbital maneuvering system.[44] The contractor and subcontractors, in turn, had subcontracts and suppliers from the very broad gamut of American industry. Electronics, ceramics, metal fabrications, plastics, and chemicals were all heavy contributors to the shuttle. The shuttle was to be a composite creation of American industry, technology, and labor.

The shuttle grew and changed even as it came into being. New problems, new concerns, and new technologies altered the configuration and the engineering as the shuttle took shape. Each new alteration, in turn, often affected the design, performance, and configuration of other systems. The shuttle offers a classic study of "systems engineering." For example, the decision to utilize a "returnable" external fuel tank rather than build the tank as part of a fully integrated reusable vehicle, did not solve the fuel tank problem. Similarly, although NASA opted for a fully reusable orbiter, the decision as to how to build or equip the orbiter to resist the extreme reentry temperatures came later. And while the major function of the shuttle was to carry "payloads" into space, the design of the payload bay continued to change. Changing payloads altered flight characteristics and changed flight plans. Building an aerospace craft unlike anything built before, and one that could never be "test" flown in an unmanned version (unlike Apollo), placed engineering and design work on the creative edge.

---

44. Ezell, *NASA Historical Data Book*, 3:122–23.

Robert F. Thompson, the Space Shuttle Program Manager from 1970 through 1981, credits "the decision to abandon the 'fully reusable' ground rule and employ expendable tankage for the orbiter main rocket engines propellant was perhaps the single most important configuration decision made in the shuttle program." And it occurred late in the definition stage of shuttle development. Through most of 1972, NASA intended to launch the shuttle into orbit with two solid rocket boosters fueled by an external propellant tank, which package would then be deorbited using smaller solid rocket motors, retrieved, and reused. On June 5, 1972, Howard W. (Bill) Tindall, John Mayer's deputy and data coordination chief for Apollo mission planning, flagged a critical problem in returning the fuel tank from orbit. "It's becoming increasingly evident that a probable major problem area and operations cost driver will be the HO tank separation and retrofire." It appeared, he said, that a very expensive, complex, and expendable attitude control system would be required for the tank to return it from orbit. The problem, he suggested, should be given high priority.[45] It was.

The problem was directed to a team from the Advanced Mission Design Branch of the Mission Planning and Analysis Division in the office of the Director of Flight Operations at the Manned Spacecraft Center. The team reported in August that the fuel tank could be "staged" (dropped) prior to orbit. That would solve the expensive and difficult tank reentry problem. The idea was rejected, however, because for the orbiter to achieve orbit, it would need to do so with its own engines, and that would require additional internal liquid oxygen/liquid hydrogen fuel tanks. That would mean a heavier lifting body, higher risks, and redesign of the entire shuttle configuration. The Advanced Mission Design Branch restudied the problem and discovered that the existing orbital maneuvering system could accelerate the orbiter into orbital velocity after separation of the external tank.[46]

Thompson rejected the idea because the orbital maneuvering system would require more fuel and larger tanks. This was September. In December, new studies and a "resizing exercise," revealed that orbital maneuvers could be accomplished on less fuel than originally planned–meaning that additional fuel would be available for the use of the orbital maneuvering system to achieve orbital velocity. The Advanced Mission Design Branch passed this information on in their Weekly Activity Report (January 29, 1973) and in March the Advanced Mission Design Branch team planned a launch to include suborbital staging of the external propellant tank with a recovery in the Indian Ocean. It also became apparent that not only could the suborbital staging work, but it would give the orbiter an additional 5,000 pound payload capacity. NASA elected, however, to retain the previous 32,000 pound payload requirement, and use the savings to reduce the thrust of the solid rocket boosters, and substantially lower flight costs. NASA subsequently estimated total program savings of $238 million.[47] Costs remained a compelling ingredient in shuttle design.

At almost every step design and development options constantly appeared. Thompson pointed out that NASA selected the more advanced, higher performance main liquid rocket engine over lower pressure but less costly engine as used in the upper stages of the Apollo program. Despite its higher developmental costs, the higher pressure engine could drive a larger orbiter, created maximum launch acceleration, and improved abort capabilities, and in total seemed to offer better capabilities at reasonable costs. Once the expendable tank design was accepted, NASA restacked the launch, enabling the use of the high

---

45. Memorandum, April 24, 1974, Development of Suborbital Staging for the Shuttle External Propellant Tank, Loftus Historical Documents File, JSC History Office; Robert F. Thompson, 1984 Von Karman Lecture, "The Space Shuttle—Some Key Program Decisions."
46. Development of Suborbital Staging for the Shuttle External Propellant Tank, p. 2.
47. *Ibid.*, pp.2–3; Members of the Advanced Missions Design Branch who developed the suborbital staging plan included Jack Funk, John T. McNeely, Burl G. Kirkland, Stewart F. McAdoo, Jr., and Victor R. Bond.

performance orbiter engines throughout the launch phase, and gained the protective margin of orbiter engine start and thrust verification before the main booster ignited. Another "developmental" decision had to do with attempting a crewless test flight. The guidance, navigation and control systems on the shuttle, however, were constructed for human control. Such a shuttle flight, if it could be accomplished, would not truly test the shuttle flight controls. The first flight of the shuttle then, would be a piloted flight.[48]

One problem that seemed to defy a wholly satisfactory solution had to do with insulating the orbiter adequately for its return into the atmosphere, a journey that generated temperatures on its outer body of 3,000° F (1,650° C). Designers recognized two basic approaches to the problem. One was to use conventional aircraft materials such as aluminum, titanium, and composites for the body and then insulate over the external skin with silicate materials. Another was to build a "hot structure" of metals that could withstand the high temperatures and absorb and disperse the temperatures throughout the external skin. This entailed the development of new metals. NASA chose the more known quantities-that is building the shuttle of basic aircraft metals, and overlaying the leading edges with thermal protective coatings.[49]

There were, however, no thermal protective materials in use that could adequately insulate against the high temperatures. Those had to be developed. A task group of NASA engineers, working with Lockheed, McDonnell Douglas, Battelle/Columbus laboratories and university scientists and engineers, developed a silicone type tile (high purity foamed silica coated with borosilicate glass) that could withstand the temperatures. But once developed, the tile created new problems. For one, it was extremely fragile. The tile was tested by simply firing missiles (such as a .22 slug) at the material to simulate an impact by a meteorite. The prototype tile crumbled. The tiles were then thickened and redesigned with a ludox (silicon-boron) base. That seemed to work. Then, the next problem involved attaching the tiles to the leading edges of the orbiter. That required the creation of new glues, several of them in fact, before a suitable adhesive could be found. Finally, 31,000 tiles, each independently cast to fit the appropriate location on the shuttle, had to be hand glued to the leading edges. The job required 670,000 hours of labor (or 335 person-years).[50] While tile development might euphemistically be called "leading edge" technology, the work did reflect the fact that building a Space Shuttle required invention and new technology ranging from flush toilets that would work in the environment of space and the development of adhesives and insulating materials, to the creation of intricate life support, avionics, and computer systems. One of the important and enduring elements of shuttle development relates to the inception of new technology and the application of that technology to other areas. Conventional airplane construction, air safety, navigation, and flight control have been rich recipients of NASA shuttle technology, as have human medicine, computers, plastics and metallurgy. The shuttle and space flight have had a much more pervasive and profound influence on Americans than is evidenced by the construction of the vehicle, or by its flights into space. Its greatest impact has been on Earth, rather than in space.

The significance of the Space Shuttle lay not in its flight per se, but in its payload, that is the freight, cargo, laboratory, or experiments delivered from the earth into space, and returned safely to earth. Shuttle payloads became one of NASA's most complex problems, as much in the social and political context as in the technical realm. Because of the chang-

---

48. Thompson, Von Karmen Lecture, pp. 5–9.
49. *Ibid.*, pp. 10–12.
50. See Roger E. Bilstein, *Orders of Magnitude: A History of the NACA and NASA, 1915–1990* (NASA: Washington, DC, NASA SP-4406, 1989), pp. 69–70; and Loftus, Andrich, Goodhart and Kennedy, "The Evolution of the Space Shuttle Design," p. 12.

ing payloads to be carried by the shuttle, each flight involved unique technical preparations and refitting. But the social and organizational structuring required for payload delivery proved most troublesome.

A special Ad Hoc Shuttle Payload Activities Team, headed by Charles J. Donlon, manager of the Shuttle Program Office, concluded that what would be needed in NASA would be "a radical change in thinking . . . to meet the vastly different "ferris wheel" mode of operation . . . required in the shuttle operational period." NASA must disassociate the transportation system from the hardware. Authorization for shuttle payloads within and without NASA must be carefully defined. The authority of the payload project manager and the transportation operator must be carefully delineated, and the flight people must be out of the "payload approval loop." Science payloads cannot be given lower priority than commercial payloads. Lead times for the development of payloads and the boarding of payloads need to be short in order to make the system work. And the committee particularly (and repeatedly) warned of the problem of competition among NASA Centers for control over payload operations and decisions. There was considerable skepticism that NASA could ever truly become a service organization, which would be required for effective shuttle operations, as opposed to its traditional mode of operation as a research and development agency.[51] Thus, the effort to build and launch the first shuttle involved some very basic social and philosophical re-evaluations, as well as technological innovation.

Despite the problems, and continuing financial constraints, NASA anticipated the first shuttle flight could occur in 1978. But budget pressures and technical problems continued to cause "slippages." As early as 1972, Dale Myers believed that cost overruns being experienced in the Skylab program would delay shuttle development and possibly cause it to be cancelled: "The Shuttle Program will live or die based on our capability to keep it reasonably on schedule, and this first schedule impact caused by funding limitations will cause an increase of cost at completion which cannot now be estimated." Delays did increase costs, and technical problems as with the tiles, the tanks, and the rocket motors did so as well.[52]

For example, Rockwell engineers working on the Orbiter's Thermal Protection System (the insulating tiles) complained that funding shortages caused work on the thermal protection system to be performed out of sequence and later than planned. Budget constraints often led to deferring quality testing. Problems were identified much later than they should have been. More work had to be done (at additional costs) simply to try to minimize the impact of performing tasks out of sequence. Design work on the thermal protection system originally required 18,750 drawings-by 1981 the required engineering drawings had increased to 25,456 (a 35 percent increase) because of delays and changes. Rockwell sought a "Program Adjustment," that is more money to compensate for the additional costs.[53]

Wayne Young, whose job was management integration in the Shuttle Program Office at the Johnson Space Center, explained that the shuttle came into being in "an austere budget environment." NASA had to first look at the budget, and then decide what could be done within that financial framework. Decisions sometimes had to be made on the basis of costs, rather than on the basis of engineering. As costs rose, scheduling and integration became even more critical.[54]

---

51. Minutes, Ad Hoc Shuttle Payload Activities Team, Center Series, Loftus Papers, Box 27, JSC History Office.
52. Dale D. Myers to James C. Fletcher, August 18, 1972, Shuttle Papers, 007-43, NASA History Office.
53. Memorandum, August 17, 1981, Rockwell Papers-Shuttle Series, JSC History Office.
54. Interview, Henry C. Dethloff with Wayne Young, Deputy Administrator, Johnson Space Center, July 18, 1990.

The orbiter Columbia *is seen in the final approach prior to landing on Rogers Drylake Runway 23 at NASA's Dryden Flight Research Center, April 14, 1981. (NASA photo no. 81-H-342).*

In 1977 the fuselage of orbiter 101, designated the *Enterprise* (which would not be the first shuttle to be launched), had been completed and the *Columbia* neared completion. Congress authorized, before the end of the decade, the construction of five shuttles (including the *Challenger, Discovery,* and *Atlantis*) estimated at a cost of $550 to $600 million each. Each finally exceeded $1 billion. During the year, NASA conducted five unpowered glide tests by dropping the craft from a Boeing 747. Rockwell's Rocketdyne Division began testing the Space Shuttle main engine at the National Space Testing Laboratory (formerly the Mississippi Test Facility, and soon to be Stennis Space Center) in March. Tests on the engine terminated after 70 seconds when a fire erupted in the engine causing damage to the A-1 test stand. Rockwell and NASA engineers conducted over 650 test firings between 1977 and 1980 before the first shuttle flight in 1981.[55] The problems most often encountered had to do with the use of conventional valves and fittings in a very unconventional 6.5 million pound thrust hydrogen-oxygen engine.

By the time the *Columbia* fired its engines on the launch pad at Kennedy Space Center in Florida, on April 12, 1981, the Space Shuttle already had experienced a long and diffi-

---

55. NSTL News Release, March 25, 1977; November 4, 1979; December 4, 1980; and see Neil McAleer, *The Omni Space Almanac* (New York, NY: World Almanac, 1987), pp. 72–91.

cult history. Simply being there, on the launch pad, was something of a triumph. The greater achievement lay ahead. The three main shuttle engines fired in rapid sequence. Then the twin solid rocket boosters, each generating 2.65 million pounds of thrust, ignited. *Columbia* lifted off. Just short of leaving the Earth's gravitational pull, the solid rocket boosters burned out, separated from the orbiter, and parachuted into the Atlantic where they were retrieved. The shuttle main engines continued to burn, taking fuel from the external tank. The main shuttle engine cut off, and the external tank detached and disintegrated as it reentered the atmosphere. The *Columbia* then fired its two orbital maneuvering system engines. The first burn put it into orbit, a second burn stabilized the circular orbit about the earth. Twelve minutes had elapsed since launch.[56]

The shuttle carried mission commander John W. Young, a Georgia Tech aeronautical engineer and a space veteran who made his first space flight aboard *Gemini 3*, and then was command module pilot for *Apollo 10* and commander of the *Apollo 16* flight. Robert L. Crippin, a native of Beaumont, Texas, and graduate of the University of Texas, had come into the astronaut training program by way of an aborted Air Force Manned Orbiting Laboratory Program. During the launch his heartbeat rose from 60 to 130 per minute. He described it as "one fantastic ride!"[57]

The *Columbia* changed orbits, and for most of the flight flew in a tail-forward upside-down position, relative to the Earth, giving the crew a better view of Earth and its horizon. Young and Crippin checked all systems, the computers, navigational jet thrusters, and huge cargo bay doors. The ship began the return at 12:22 EST on April 14. Young and Crippin fired the orbital maneuvering rockets for two minutes and twenty-seven seconds to reduce their speed to less than the orbital velocity of 17,500 miles per hour. Gravity would do the rest. They began an hour-long descent. They fired their attitude control thrusters to turn *Columbia* right side up and nose forward. Thrusters were fired again to keep the nose up so that the thermal protective tiles could absorb the heat of reentry. The *Columbia* lost speed as its altitude dropped, and over Rogers Dry Lake in the Mojave Desert, Crippen and Young banked the ship sharply, looped back into a landing pattern, and touched down at a speed of 215 miles per hour, about twice that of a commercial airliner. "The touchdown marked the successful conclusion of STS-1, 2 days, six hours, twenty minutes and fifty-two seconds after lift-off from Florida." President Ronald Reagan greeted the returning crewmen, "Today our friends and adversaries are reminded that we are a free people capable of great deeds. We are a free people in search of progress for mankind."[58] That search for progress, in the form of a reusable spacecraft, involved not only NASA, and the industries and astronauts who were identified as the recipients of the 1981 Collier Trophy, but reflected more fully the past and present energies, initiatives, technologies, aspirations, and capital investments of the American people.

---

56. See Michael Collins, *Liftoff: The Story of America's Adventure in Space* (New York, NY: NASA, Grove Press, 1988), pp. 201–22; NASA, Mission Report, MR-001.
57. NASA, Mission Report, MR-001.
58. *Ibid.*

Chapter 13

"More Favored than the Birds": The Manned Maneuvering Unit in Space

by Anne Millbrooke

In 1984 a Manned Maneuvering Unit (MMU) enabled a few astronauts to maneuver in outer space, outside of spacecraft, and free of tether lines. This manned maneuvering unit and its predecessors are, as the name implies, maneuvering devices. Flight is the function of the spacecraft. Life support is the function of the space suit. Maneuvering is an extravehicular activity independent of the protective and supportive space suit, yet integrated with the suit and even the spacecraft. The spacecraft and space suit are prerequisites to extravehicular activity, the craft to transport the astronaut into outer space, and the suit to protect and support life. The maneuvering unit is an optional aid. The maneuvering unit, spacecraft, and space suit are complementary components of the human space flight program. Whereas all such space flights have involved spacecraft and space suits, only a few have utilized manned maneuvering units.

Outer space is a micro- or zero-gravity environment that requires special techniques for moving inside the spacecraft as well as out. Based upon experience aboard the Space Shuttles *Columbia* and *Discovery*, astronaut Joseph P. Allen described the experience inside a spacecraft:

> *During the first few days in space, the act of simply moving from* here *to* there *looks so easy, yet is so challenging. The veteran of zero gravity moves effortlessly and with total control, pushing off from one location and arriving at his destination across the flight deck, his body in the proper position to insert his feet into Velcro toe loops and to grasp simultaneously the convenient handhold, all without missing a beat in his tight work schedule. In contrast, the rookies sail across the same path, usually too fast, trying to suppress the instinct to glide headfirst and with vague swimming motions. They stop by bumping into the far wall in precisely the wrong position to reach either the toe loops or the handholds.*[1]

Space writer Harry L. Shipman expressed this more directly: "Velcro takes the place of gravity" inside the spacecraft.[2] Outside the spacecraft, there is no Velcro and no enclosing walls. Civilian and military engineers thus explored various mechanism to aid astronauts outside the spacecraft. These aids included foot restraints, hand grips, tether lines, and self-propelled maneuvering units, yet few space missions required the technology and capability of manned maneuvering units.

Floating in space was a lesson learned by experience gained gradually during the Mercury, Gemini, Apollo, Skylab, and Space Shuttle missions. In fact, the Mercury,

---

1. Joseph P. Allen with Russell Martin, E*ntering Space: An Astronaut Odyssey* (1984; revised edition, New York: Stewart, Tabori, and Chang, 1985), p. 75.
2. Harry L. Shipman, *Humans in Space, 21st Century Frontiers* (New York, NY: Plenum Press, 1989), p. 97.

Astronaut Bruce McCandless on a spacewalk using the manned maneuvering unit (MMU) on STS-41B, February 1984. NASA and Martin Marietta Corporation were awarded the Collier Trophy in 1984 for the development of the MMU, and for being the NASA teams that rescued three disabled satellites, with special recognition to astronaut Bruce McCandless II, NASA's Charles E. Whitsett, Jr., and Martin Marietta's Walter W. Bollendonk. (NASA photo no. 84-H-71).

Gemini, and Apollo spacecraft were too small to allow astronauts much mobility within the craft. Project Gemini included the construction of two types of maneuvering units and the training of astronauts in their use. In 1965 Gemini astronaut Edward H. White made the first American spacewalk. Using a hand-held maneuvering unit, and wearing a space suit for life support, he spent twenty minutes outside of *Gemini 4*. He was tethered to the space capsule for safety. Project Gemini thus provided the first American experience with extravehicular activity in space; a Soviet cosmonaut had achieved the first extravehicular activity months before White ventured out of the Gemini capsule. Later Gemini astronauts also completed extravehicular activities in the harsh environment of outer space.

Apollo and later Skylab added to NASA's research and development experience with the concept and technology of maneuvering in space, though not with the operation of any maneuvering aids in free flight in outer space. Apollo's objective was the lunar surface, not outer space. The three Skylab missions in 1973 and 1974 provided astronauts experience with weightless floating in a relatively large open space within a spacecraft, but not outside. It was not until the Space Shuttle, a reusable transportation system, that astronauts acquired operational experience floating both inside and outside a spacecraft.

The award-winning manned maneuvering unit was designed for a specific type of mission: satellite rescue missions. All earlier maneuvering units had been designed for experimental missions, that is to test the technology, but with the reusable Space Shuttle NASA introduced an operational, mission-oriented maneuvering unit—the award-winning manned maneuvering unit (MMU). This operational unit was used three times—on the tenth, eleventh, and fourteenth flights of the Space Transportation System, more commonly known as the Space Shuttle. The year of these flights was 1984. The Collier Trophy for that year recognizes astronaut Bruce McCandless II, who first used the unit in space, NASA's Charles E. "Ed" Whitsett, Jr., and Martin Marietta's Walter W. "Bill" Bollendonk. These three men were instrumental in the development, and McCandless in the use, of the unit. Behind this award is a story of technological development involving a variety of institutions within the national infrastructure of the space program and amid the superpower rivalry known as the Cold War.

From the preliminary research and development in the 1950s to the achievements of 1984, civilian and military personnel—engineers, technicians, and astronauts—defined and redefined the technology of maneuverability in terms of perceived needs and capabilities, and within the limitations imposed by budgets and flight schedules. At each step reviews, tests, and experiments, as well as political decisions affecting the space program in general, influenced decisions about whether to continue development, in what direction, and by which next step. The identification and definition of applications for maneuvering units actually began in science fiction literature, which included earth-based as well as outer space missions. Civilian and military agencies and government contractors, that is industry, participated in the development of maneuvering units of several types, including finally the award-winning manned maneuvering unit.

Science Fiction

Before the "science fact" there was science fiction. From novels of the nineteenth century to moving pictures of this century, humans traveled in space—sometimes using maneuvering units outside the spaceships and sometimes not, mostly not. Early literary classics of space travel include Jules Verne's novel *From the Earth to the Moon* (1865) and H.G. Wells' *First Men in the Moon* (1901). Both of these books were made into movies of the same names, respectively; Verne's in 1958 and Wells' in 1964.[3] In both the print and film versions, the space travelers left Earth with neither space suits nor maneuvering units. In Verne's story the characters flew in a ballistic projectile, shot from a huge cannon, toward the Moon. They relied upon the probability that they would be able to survive in the rarified atmosphere of the Moon.

En route to the Moon in Verne's sequel, *Round the Moon* (1870), the French adventurer Michel Ardan asked his traveling companions, "Why cannot we walk outside like the meteor? Why cannot we launch into space through the scuttle? What enjoyment it would be to feel oneself thus suspended in ether, more favored than the birds who must use their wings to keep themselves up!"[4] Practical Impey Barbicane, president of the Gun Club that had shot their projectile into space, responded with two reasons. First, there was no air in the ether of outer space. Second, the density of a man being less than that of the projectile in which they traveled meant that a man outside the spacecraft would move at a speed not equal to that of the craft and thus the man would move apart from the craft.

Later Ardan exclaimed, "Ah! what I regret is not being able to take a walk outside. What voluptuousness to float amid this radiant ether, to bathe oneself in it, to wrap oneself in the sun's pure rays. If Barbicane had only thought of furnishing us with a diving apparatus and an air-pump, I could have ventured out."[5] Again practical Barbicane countered the proposal: a diving apparatus in space would burst like a balloon that had risen too high. Barbicane thereupon prohibited "all sentimental walks beyond the projectile," but his authority applied only to his fictional companions and not to writers of other science fiction works.

What Ardan missed both in print and on film, Buck Rogers and Flash Gordon achieved. These twentieth-century fictional heros provided inspiration—and humor—to the astronauts and engineers involved with maneuverability in space. Mission specialists Bruce McCandless and Robert Stewart even called each other Buck and Flash in the cabin of the Space Shuttle on that historic mission when the award-winning manned maneuvering unit was first used.[6] Originally the star of a comic strip, Flash Gordon entertained movie audiences via three serials: *Flash Gordon* (1936), *Flash Gordon's Trip to Mars* (1938), and *Flash Gordon Conquers the Universe* (1940). Created by Alex Raymond and played by Buster Crabbe, Flash Gordon fought the evil forces of Ming the Merciless. He traveled in Dr. Zarkov's rocketship and other spacecraft. The technology of life support and maneuverability did not clutter his adventures.

Buck Rogers similarly appeared in print and on film. Under the name Anthony Rogers, he made his debut on the pages of a pulp magazine in 1928. Using the name Buck, he moved into a comic strip the following year. A decade later he appeared on film in twelve episodes of *Buck Rogers* (1939); like Flash Gordon, he was played by Buster Crabbe. In the original story, written by Philip Francis Nowlan, Rogers awoke from suspended animation in a future time—year 2419—when Americans wore "inertron" belts, both "jumpers" and

---

3. *From the Earth to the Moon*, Waverley, 1958, and *First Men in the Moon*, Columbia/Ameran, 1964. Information about all movies mentioned in this article appears in Leslie Halliwell, *Halliwell's Film Guide*, 8th edition, edited by John Walker (New York, NY: Harper Perennial, 1991), pp. 384, 418. For Verne and Wells. The best overview of science fiction and spaceflight is Frederick I. Ordway III and Randy Liebermann, editors, *Blueprints for Space, Science Fiction to Science Fact* (Washington, DC: Smithsonian Institution Press, 1992).

4. Jules Verne, *From the Earth to the Moon and a Trip around It* (1865, 1870; Philadelphia: J.B. Lippincott, n.d.), part 2, p. 173.

5. Verne, *From the Earth to the Moon*, part 2, p. 177; Barbicane response is on the same page.

6. "'Steppin' out with Flash and Buck," NASA Lyndon B. Johnson Space Center, *Space News Roundup*, 23/4 (24 February 1984): 1–2.

"floaters" that increased their mobility.[7] A jumper made the wearer weigh "in effect" as little as desired and therefore able to jump considerable distances. Floaters were advanced jumpers equipped with rocket motors that enabled the wearers to float in air similar to a diver floating in water; directional control came through the wearer twisting his body and moving his arms and legs. These rocket-powered mobility units inspired science fiction writers and the recent Walt Disney-Silver Screen movie *The Rocketeer* (1991), which like the original Nowlan story involved maneuverability on Earth, not in outer space.

Generally, spaceships and space suits received more attention in science fiction stories than did technology for maneuvering. Robert A. Heinlein is an example. His *Rocket Ship Galileo* (1947) is about three boys, recent high school graduates, who accompany an atomic scientist to the Moon and while there defeat Nazis. The heroes wore pressurized stratosphere suits that look like diving suits. The helmets were bowl-shaped plexiglass, and the soles and seats of the suits were insulated with asbestos. The characters in print had no maneuvering devices. Heinlein's thin book became the loose basis of *Destination Moon* (1950), a good Cold War movie about combined American industry racing to get to the Moon before the Russians. In both book and movie the space vehicle is an atomic-powered rocketship; and Woody Woodpecker gives a delightful explanation of rocketry in the movie.

Traveling to the Moon was a race, not simply a space race, but also a military arms race. In *Destination Moon* the fictional General Thayer announced, "there is absolutely no way to stop an attack from outer space" and thus "the first country that can use the Moon for the launching of missiles will control the Earth."[8] In the movie private industry supported the lunar mission because government was unable to mobilize the necessary resources during peacetime; the movie script thus failed to anticipate federal appropriations in time of a Cold War. The film travelers performed extravehicular activities in space. Wearing space suits and tethered to the spacecraft, they unstuck an antenna. One man released hold of his safety line and drifted away from the ship. He was rescued by another man who carried a large oxygen bottle, released gas for propulsion, and steered by manually facing the nozzle away from the desired direction of travel.

Science Fact

Michel Ardan's wish for "a diving apparatus and an air-pump" was not far afield from the early development of special suits for high-altitude flight, the predecessors of early space suits. In the 1930s aviator Wiley Post attempted and achieved stratospheric flight. As he said, "The main objective of high-altitude flight is to increase the speed of

---

7. Philip Francis Nowlan, *Armageddon 2419 A.D.* (1928–1929; New York, NY: Ace Publishing Corporation, 1962), pp. 24–25. This version of the story contains both "Armageddon 2419 A.D." (1928) and its sequel "The Airlords of Han" (1929), both originally serialized in *Amazing Stories*, but with "a certain amount of revision and condensation" in this book, according to the foreword.

8. *Destination Moon*, Universal/George Pal, 1950. In Heinlein's book *Have Space Suit–Will Travel* (1958), alien flying saucers transport space travelers. The hero of this tale, Clifford Russell, won a space suit in a Skyway Soap slogan contest. Previously used at a satellite station, this obsolete pressure suit was a real–though fictional–space suit, not a toy. Made by Goodyear, air conditioned by York, and equipped with auxiliary systems by General Electric, the suit had a body of silicone, asbestos, and glass-fiber cloth. The helmet was chrome-plated with a bright reflecting surface so as not to absorb heat from the sun. Oxygen came from steel bottles carried in a backpack. The suit, named Oscar by its new owner, served Cliff as he was kidnapped and taken to the Moon, then to cold Pluto, and then beyond, and–of course–back to Earth. Cliff concluded his adventures by deciding to become a space suit engineer. Throughout his adventures, Cliff rode in spacecraft or lunar rovers, or he provided his own mobility without the aid of any maneuvering device.

travel."[9] He foresaw transcontinental and transoceanic flights for the transportation of passengers and freight. But to fly in the thin air of high altitudes, he needed both oxygen and sufficient pressure to protect the cells of the body. He recognized that the aircraft's cabin could be sealed and filled with air under pressure, but at the addition of "prohibitive weight." More specifically, it would be impossible to pressurize the plywood shell of his *Winnie Mae*, a Lockheed Vega that he had flown twice around the world and that he used in his high flying. He therefore approached the B.F. Goodyear Company with the idea of "a suit, something like a diver's outfit, which the pilot can wear, and which can be blown up with air or oxygen to the required pressure."[10]

In response to Post's request, Goodyear built him three pressurized suits. The first cost less than $75 (equivalent to about $800 in 1994 dollars). It ruptured during an unmanned test, before flight testing. Post got stuck in the second suit, which literally had to be cut off him, again before any flight test. The third suit not only passed tests in the Army's low-pressure chamber at Wright Field, but also met requirements during the 25 hours that Post logged in the suit. That suit proved compatible with his airplane and allowed sufficient mobility for him to operate the plane's controls; though when inflated, the suit allowed only very limited mobility. In a series of flights in 1934 and 1935 Post successfully demonstrated the utility of the pressure suit. To continue his experiments in high altitude flight, he acquired another airplane. He and the humorist Will Rogers died in a crash of that experimental plane, a crash from very low altitude.[11]

Post's successful pressure suit, however, influenced research programs of the Army and Navy, which contracted with manufacturers—Goodrich, Bell Aircraft, U.S. Rubber, National Carbon, and later others—for pressure suits, initially for experimental designs, later for production suits. Military contracts, that is military money and military specifications, defined technical progress in the development of pressure suits. One goal was to increase the mobility of the person inside the suit, to allow the pilot more range of movement. Although progress was slow, two key developments were achieved in the 1950s. One was the linknet restraint. This linknet-nylon restraining layer prevented a suit from ballooning under pressure. Introduced in 1956 by the David Clark Company and the Air Force, this feature helped make the A/P22S-2 pressure suit standard Air Force equipment. Test pilots flying the X-15 experimental plane wore this pressure suit in supersonic flight. In 1957 B.F. Goodrich and the Navy built swivel joints with airtight rotating bearings, and also fluted joints (semirigid accordion pleats), into the Mark II suit, made of a rubberized fabric. Improved versions appeared in 1958—Mark III—and 1959—Mark IV.

When the United States began its man-in-space program, high-altitude pressure suits were adapted into space suits. As stated in a Smithsonian publication, Goodyear engineer "Russell Colley is considered the father of the American space suit for constructing the first successful fully pressurized flying suit for Wiley Post."[12] Project Mercury provided the spacecraft in which astronauts first used space suits and proved the technological basis of human space flight. The Mercury space suits were adapted from the Goodyear-Navy Mark IV suit. The space suit was a protective system made of aluminized nylon, Neoprene-coated nylon, and vulcanized nylon. It would pressurize only in the event of an emergency. In that sense,

---

9. Wiley Post, "Flying the Stratosphere: Wiley Post Seeks New Record," *Popular Mechanics Magazine*, 62 (October 1934): 492–95; quote on page 492.

10. Post, "Flying the Stratosphere," p. 493; see also Stanley R. Mohler and Bobby H. Johnson, *Wiley Post, His Winnie Mae, and the World's First Pressure Suit* [Smithsonian Annals of Flight, Number 8] (Washington, DC: Smithsonian Institution Press, 1971).

11. Byran B. Sterling and Frances N. Sterling, *Will Rogers & Wiley Post: Death at Barrow* (New York, NY: M. Evans and Company, 1993).

12. Lillian D. Kozloski, *U.S. Space Gear, Outfitting the Astronaut* (Washington, DC: Smithsonian Institution Press, 1994), p. 15.

equipping the suit for pressurization was a precaution against the possibility that the spacecraft might decompress, a redundancy built into the Mercury program for the protection of the astronaut. Still the suit was specially adapted for ventilation of the astronaut, for waste removal, and for safety and comfort. Changes were made as experience warranted.

One astronaut at a time, and safely confined in the spacecraft, Mercury astronauts orbited the earth. No Mercury astronaut ventured outside the spacecraft into outer space. The importance of integration of all aspects of the space program became apparent during Project Mercury. As one historian concluded, "The greatest lesson learned from the Mercury flights was probably the unique importance of people to machines. The Mercury program began with a machine that had a man in it. And by the end of the program, it truly became a manned spacecraft."[13] Spacecraft, crew, space suit, and other mission equipment needed to be integrated. This lesson applied to maneuvering units then and later under development.

Although maneuvering units were not specifically a part of Project Mercury, the possibilities of maneuvering in space were explored concurrent with the Mercury flights. The Air Force, for example, began testing space propulsion units, hand-held, pistol-like, compressed-air devices, at its Aerospace Medical Laboratory in 1958. That work was done at the laboratory at Wright-Patterson Air Force Base, Ohio, with some testing also conducted at NASA's Air Bearing Facility in Houston. The main problem of a hand-held unit was "the difficulty of aligning the thrust vector with the center of mass of the man, causing rotation with translation resulting in unworkable flight paths";[14] in other words, the astronaut could not maintain control. The Air Force's Aerospace Medical Division issued a report on "Self-Maneuvering for the Orbital Worker" in 1960. That year the Rocket Propulsion Laboratory at Edwards Air Force Base, California, provided assistance in designing the propulsion system for an experimental maneuvering unit, a research device designed for testing under weightless conditions, but not for use in the environment of outer space.

This early work of the Air Force led to maneuvering units for Project Gemini, during which astronauts wore space suits and remained tethered to the spacecraft during extravehicular activity. Only briefly using maneuvering units, astronauts began maneuvering in space, outside the spacecraft, during Project Gemini.

Maneuvering in Space

Project Gemini provided NASA experience with extravehicular activity in space and with two maneuvering devices. One device was the Hand-Held Maneuvering Unit (HHMU) that White used in 1965, also known as the self maneuvering unit, pressure gun, or simply gun. The second device was a backpack called variously an Astronaut Maneuvering Unit (AMU), the Modular Maneuvering Unit (MMU), Modular Astronaut Maneuvering Unit (MAMU), and Department of Defense experiment D-12. Gemini crews accomplished extravehicular activities, six hours of tethered time and six hours of standing up in the open hatch of the spacecraft. The five crews who accumulated the tethered time were aboard *Gemini 4, 9A, 10, 11,* and *12.* Both the hand-held and backpack maneuvering

---

13. *Ibid.,* p. 50.
14. Julien M. Christensen, Aerospace Medical Laboratory, to A[ir] F[orce] F[light] T[est] C[enter] (Howard Barfield), Edwards Air Force Base, document #1 in "Support of the Gemini Program, 'A Document Collection,' Volume II—MMU, through December, 1964," by Research and Technology Division, Rocket Propulsion Laboratory, Air Force Systems Command, on microfilm roll #26225, Air Force Historical Research Agency, Maxwell Air Force Base, Alabama. See also John C. Simons and Melvin S. Gordner, *Self-Maneuvering for the Orbital Worker* [Wright Air Development Division Technical Report 60-648] (Wright-Patterson Air Force Base, OH: Aerospace Medical Division, Wright Air Development Division, 1960).

devices were scheduled for testing during tethered activity on these flights, but only the hand-held unit was ever used in space and only on *Gemini 4* and *10*. Both maneuvering units, however, provided experience and established precedents that contributed to the Collier-winning manned maneuvering unit of 1984. From the 1960s to 1984, however, the developers of maneuvering units explored several directions.

Civilian and military branches of the Federal government and contractors and subcontractors in industry participated in the design and development of maneuvering units for the Gemini program—and thereby established the pattern of collaboration that continued thereafter in the development of maneuvering units. NASA was the lead agency. Project Gemini was phase two of NASA's manned space flight program; Project Mercury had been phase one, and Project Apollo would be phase three. NASA's Manned Spacecraft Center in Houston managed the agency's manned space flight program. The prime contractor for Project Gemini was McDonnell Aircraft Corporation, headquartered in St. Louis. Other contractors designed, developed, built, or delivered a variety of products incorporated in the Gemini missions, including space suits, life support systems, and maneuvering units.

The Manned Spacecraft Center managed the development of the hand-held maneuvering unit. Per policy, this NASA center participated in research, design, and testing, but contracted out construction. In developing the Gemini hand-held unit, NASA balanced the advantages of tractor or tow thrusters and the pusher mode. It developed a proportional thrust system, allowing the astronaut more control than an on-off system. The unit accommodated the limited dexterity of the gloved hands of an astronaut. The initial unit used on *Gemini 4* had two one-pound tractor jets and one two-pound pusher jet. The gas was oxygen, deemed safe to store in the cabin of the spacecraft. This was a self-contained system. A later model, intended for use on *Gemini 8*, received its propellant, Freon 14 gas, from a tank packed on the astronaut's back. In a still later model used on *Gemini 10*, a hose bundled in the astronaut's umbilical cord transported nitrogen gas from the spacecraft to the hand-held unit. Refinements in the handle of the unit were also made as the Gemini program progressed. Equipment to train astronauts to use the hand-held units included air-bearing simulators in the Air Bearing Facility.[15]

The Air Force managed what it called the "modular maneuvering unit" (MMU) program, initiated in 1963, and the Air Force's Space Systems Division became the lead division for developing this maneuvering backpack unit. Why was the Air Force participating in the civil space program? First, NASA requested the Air Force's assistance because the Air Force had launch vehicles (like the Titan II rocket modified for use as the Gemini launch vehicle) and other resources.[16] Also, the Air Force had effectively supported NASA's Mercury program. The Air Force, in fact, had pursued its own human space flight program, Dyna-Soar, from 1957 into 1963. Canceled three years before the scheduled first flight, the Dyna-Soar program provided important technical information about hypersonic flight, reentry flight control, and heating problems. Secondly, the Air Force, and Department of Defense in general, approached space in terms of national security and military strategy. Dyna-Soar, for example, grew out of military interest in a piloted boost-glide bomber-missile (called Bomi), a reconnaissance system (called Brass Bell), and a hypersonic weapon and research and development system (HYWARDS). The three programs were consolidated into Dyna-Soar in response to the Soviet's successful orbiting of *Sputnik* in

---

15. Harold I. Johnson, William C. Huber, Edward H. White, and Michael Collins, "EVA Maneuvering about Space Vehicles," typescript report (without the referenced figures), no date, pp. 2–7 plus tables I-III, in Record Number 007189, NASA Historical Reference Collection, NASA History Office, Washington, DC.
16. Colonel Daniel D. McKee, "Gemini Program," pp. 6–15, in *The U.S. Air Force in Space*, edited by Lieutenant Colonel Eldon W. Downs (New York, NY: Frederick A. Praeger, 1966), p. 6.

October 1957. Even after the cancellation of Dyna-Soar, the Air Force retained military objectives for a space program.[17]

In the United States, civilian and military objectives became interwoven in national policy. Congress responded to Sputnik by establishing three space organizations in 1958: the civilian National Aeronautics and Space Administration (NASA), the military Advanced Research Projects Agency (ARPA), and the executive National Aeronautics and Space Council (an advisory panel reporting to the President). At that time the country publicly entered a technological and scientific, as well as political, space race. Of the early years of that space race, NASA historian Roger D. Launius concluded, "First, NASA's projects were clearly cold war propaganda weapons that national leaders wanted to use to sway world opinion about the relative merits of democracy versus the communism of the Soviet Union. . . . Second, NASA's civilian effort served as an excellent smoke-screen for the DOD's [Department of Defense] military space efforts."[18]

From the military perspective, General Bernard A. Schriever explained the nature of civil-military cooperation:

> NASA programs by themselves will not build a military capability. That is not their purpose, nor should it be their purpose. A military capability can be created only by a military organization which possesses a combination of technical knowledge and operational experience with suitable military equipment. Both NASA and the Department of Defense have valid and distinctive roles in the national space program. Their efforts are complementary, not competitive; their programs are cooperative, not conflicting.[19]

The Air Force had particular interest in launch vehicles, operation of spacecraft, communications systems, and—in General Schriever's words—"techniques needed to transport and support man in space and to permit him to function effectively there." To function effectively in space implied maneuverability, and thus the Air Force's research and development of maneuvering units, including experiment D-12 in the Gemini flight program.

Among the companies involved in the early maneuvering work was Aero-Jet General, which prepared an influential report entitled "A Rocket System for Limited Manned Flight" (1959) and proposed an "AeroPak Flight Vehicle." As early as 1953, Wendell F. Moore of Bell Aerosystems had begun designing a rocket belt. He continued his effort, and in 1960 Bell obtained an Army contract to produce the Army's A-1 prototype rocket belt. In 1962 President John F. Kennedy viewed a Bell rocket belt in flight demonstration at Fort Bragg, North Carolina.[20] The next year Ling-Temco-Vought (LTV) prepared for the Department of Defense preliminary designs of a Remote Maneuvering Unit (RMU) to be ejected from the spacecraft and then remotely moved, by an astronaut inside the spacecraft, toward a target that had also been ejected. The unit was to be man-rated so that

---

17. See chapter two, "Dynamic Soaring," pp. 11–19, in Dennis R. Jenkins, *The History of Developing the National Space Transportation System, the Beginning through STS-50* (Marceline, MO: Walsworth Publishing Company, 1992); and Roy Franklin Houchin II, "The Rise and Fall of Dyna-Soar: a History of Air Force Hypersonic R&D, 1944–1963," Ph.D. dissertation, Auburn University, 1995.

18. Roger D. Launius, *NASA: A History of the U.S. Civil Space Program* [Anvil series] (Malabar, FL: Krieger Publishing Company, 1994), pp. 34–35. Regarding the space race in the context of the Cold War, see particularly Walter A. McDougall, . . . *The Heavens and the Earth, a Political History of the Space Age* (New York, NY: Basic Books, 1985). Regarding the immediate response to Sputnik, see Robert A. Divine, *The Sputnik Challenge, Eisenhower Response to the Soviet Satellite* (New York, NY: Oxford University Press, 1993).

19. General Bernard A. Schriever, "Does the Military Have a Role in Space?" pages 59–68 in *Space: Its Impact on Man and Society*, edited by Lillian Levy [Essay Index Reprint Series] (1965; Freeport, NY: Books for Libraries Press, 1973), p. 63; the phrase quoted in the same paragraph is from p. 62.

20. Regarding the history of the Bell rocket belt program, see Barry E. DiGregorio, "The Rocket Belt," *American Heritage of Invention & Technology*, 11 (Spring 1996): 46–50.

manual operation would also be possible. Bota Reaction Motors also did relevant small rocket work under government contract.

In November 1963 the Air Force's Aero Propulsion Laboratory proposed to develop "an individual back-pack experiment which would permit the astronaut to maneuver independently around the Gemini vehicle."[21] This was the beginning of the Gemini astronaut or modular maneuvering unit. The Air Force's proposed extravehicular experiments for Gemini had priority ".05A, equivalent to that of the Ballistic Missile Program," and thus required White House approval.[22] The Rocket Propulsion Laboratory in California accepted technical responsibility for the rocket propulsion system, which used hydrogen peroxide as the monopropellant. For design and fabrication, the Air Force initially planned to grant a sole-source contract to Ling-Temco-Vought, but soon issued a request for technical proposals. LTV and Bell responded.

The Air Force evaluated these technical proposals on nine points: propulsion, environmental control system, electronics, flight controller, power, aerospace ground equipment, overall system, reliability and quality control, and program plan. Bell was "very strong" in propulsion and scored a ninety percent overall for its proposal, and LTV scored only sixty percent, but "both were considered acceptable."[23] LTV then thoroughly amended its technical proposal into something "greatly improved" though still a bit weaker than the Bell proposal, but LTV's cost proposal was $750,000 less than Bell's. The Air Force negotiated a cost-plus-incentive-fee contract with LTV for the construction and delivery of three backpack maneuvering units.

As the prime Gemini contractor, McDonnell Aircraft Corporation integrated the maneuvering units and related experiments into the Gemini spacecraft. Although B.F. Goodrich had received the first contract to design Gemini space suits, even delivered two prototypes, the David Clark Company won the contract to produce the Gemini space suits.[24] Clark thus participated in integrating the space suits with both the spacecraft and the maneuvering units and maneuvering experiments.

Clark produced three models of space suits actually worn on Gemini flights and continually modified the models in response to the astronauts' comments. Like their Air Force predecessor (the A/P22S-2 pressure suit), these space suits had linknet Dacron woven throughout one layer of the suit. The linknet held the pressurized containment layer to the contours of the body and thereby aided mobility. The basic extravehicular model G4C weighed thirty-five pounds, ten pounds more than model G3C worn only inside the spacecraft and only on *Gemini 3*. G4C weighed more mostly because of additional outer layers of nylon, aluminized Mylar, unwoven Dacron insulation, and Nomex heat-resistant material that formed a protective hazardous-environment shield. The G4C helmet similarly had additional protective layers: visual, thermal, and impact shields. Like the rest of the suit, the helmet was continually modified. The lightest Gemini suit, just sixteen pounds, was the G5C, worn inside the spacecraft on only the *Gemini 7* mission. Variants of G4C were worn by both members of crews of *Gemini 4, 5, 6A, 8, 9A, 10, 11*, and *12*, whether or not extravehicular activities were planned, as any opening of the spacecraft's hatch exposed the crew to the space environment.

---

21. Colonel J.M. Silk, Director, Air Force Rocket Propulsion Laboratory, to Major Saavedra, November 20, 1963, document #4 in "Support of the Gemini Program, 'A Document Collection,' Volume II."
22. Paul C. Erickson, Memo for the Record, "Program 631A," December 5, 1963, document #6 in "Support of the Gemini Program, 'A Document Collection,' Volume II."
23. P.C. Erickson to PRPRE, regarding trip to the Aero Propulsion Laboratory of March 30–April 9, 1964, document #10 in "Support of the Gemini Program, 'A Document Collection,' Volume II." The "greatly improved" quote later in the paragraph came from document #11 about a May review, in the same volume.
24. Chapter 4, "Gemini Space Suits," pp. 51–72 in Kozloski, *U.S. Space Gear*; and Gregory P. Kennedy, "Development of the Gemini Space Suit and Extravehicular Equipment," pp. 97–109 in *National Air and Space Museum Research Report for 1984* (Washington, DC: Smithsonian Institution Press, 1984).

Integration of the backpack maneuvering unit and the space suit posed a particular challenge to Gemini engineers within government and industry. In 1964, for example, LTV uncovered a problem while testing the modular maneuvering unit. The company's exhaust plume analyses revealed that rocket exhaust plumes impinged on the space suit. The exhaust heated the suit, particularly the helmet and the legs. NASA opposed adding insulation patches to the space suit as a solution, so LTV proposed other solutions that involved modifying either the maneuvering unit or the space suit or both. One way to avoid overheating the suit was to extend all thruster nozzles far enough to avoid impingement. A second method was to extend the upper forward nozzles beyond the helmet impingement and to add a leg restraint device to prevent the astronaut's leg from moving into a lower plume. Third, LTV proposed modifying the space suit, extending the upper forward nozzles beyond the helmet, and rebuilding the lower suit of materials to withstand higher temperatures. A thermal skirt, a fourth idea, was proposed to cover the astronaut's legs. All the proposals posed their own problem—"delays of varying length to the MMU delivery schedule."[25]

The decision was to extend two forward nozzles on the maneuvering unit and to rebuild the lower section of the space suit, but without altering the delivery schedule, something "not possible if we are to meet NASA flight dates," according to the director of the Air Force Aero Propulsion Laboratory.[26] This colonel explained, "The only other alternative is termination of the program." He urged close monitoring of contractors and subcontractors, also simultaneous qualification testing, reliability testing, and hardware fabrication, yet "there must be no compromise with the astronaut's safety during space flight." This approach worked. It meant, however, additional weight to the special suit the astronauts wore outside the spacecraft, doubling the weight of the fabric in the legs over that of other G4C suits. As modified for the first in-space test of the backpack maneuvering unit, the suit's legs included neoprene-coated nylon, uncoated nylon, fiberglass cloth, aluminized high-temperature film, and Chromel-R cloth (stainless steel).[27]

As the modular maneuvering units neared completion in February 1966, an accident occurred. There was an explosion during a reliability test of one unit—at hour 96 of the planned 100-hour operating time. A quick investigation revealed the problem to be in LTV's now damaged test equipment, in the company's Space Environment Simulator (also known as the SES). There was no problem with the maneuvering unit. With the design and development phases complete, and the final verification tests in progress or on schedule, the explosion merely delayed tests conducted in that one facility.

By mid-April 1966 all the testing had been completed, and the three experimental modular maneuvering units had been delivered to the Air Force. Gemini astronauts were in final training for using the units. Wearing training packs, they experienced brief periods of zero gravity aboard a KC-135 aircraft.[28] To obtain zero gravity in flight, the pilot pushed the jet airplane into a dive, pulled the nose up, and flew over a parabolic arc; the weightless condition occurred going "over the hump."

---

25. J.F. Kephart, Trip Report [Dallas trip, March 24, 1965], document #12 in "Support of the Gemini Program, 'A Document Collection,' Volume III—MMU, January 1–June 30, 1965," by Research and Technology Division, Rocket Propulsion Laboratory, Air Force Systems Command, on microfilm roll #26225, Air Force Historical Research Agency, Maxwell Air Force Base, AL.

26. Colonel Richard T. Hemsley to A[in] F[orce] R[ocket] P[ropulsion] L[aboratory] (Colonel [J.M.] Silk), April 5, 1965, document #14 in "Support of the Gemini Program, 'A Document Collection,' Volume III."

27. Kozloski, *U.S. Space Gear*, p. 66.

28. Regarding the conclusion of the pre-flight activities, see "Support of the Gemini Program, 'A Document Collection,' Volume V—MMU, October 1, 1965–June 30, 1966," by Research and Technology Division, Rocket Propulsion Laboratory, Air Force Systems Command, on microfilm roll #26225, Air Force Historical Research Agency.

Gemini flights had already begun. In fact, Ed White had accomplished the first extravehicular activity on June 3, 1965, as part of *Gemini 4*.[29] He used the Hand-Held Maneuvering Unit, the pressure gun. In case of emergency, that is in case White dropped the gun, it too was tethered. Command pilot James A. McDivitt kept the spacecraft in a stable attitude while White maneuvered outside the vehicle, and McDivitt took pictures of White's space walk. After White used all the gas in the hand-held maneuvering unit, he could still maneuver with the aid of the tether line, but that gave McDivitt problems controlling the spacecraft. White confirmed the earlier Soviet finding: Man can maneuver in space.

For extravehicular activity, the astronauts wore the Clark G4C suit and a life-support chest pack, either a Ventilation Control Module (*Gemini 4*) or an Extravehicular Life Support System (ELSS, *Gemini 9A, 10, 11,* and *12*). In the spacecraft, astronauts connected their suits to the craft's life support system. Outside the craft an oxygen hose, electrical and communication wires, and a safety tether connected the astronaut to the spacecraft. These were bundled in the umbilical cord between the chest pack, astronaut, and spacecraft. Outside the spacecraft, according to plans for six Gemini flights, the astronauts would carry either the hand-held maneuvering device or wear the LTV-made astronaut maneuvering unit, the backpack.

Gemini 4, 8, 10, and *11* included among their missions experimenting with the Hand-Held Maneuvering Unit. White walked in space during *Gemini 4*. For reasons unrelated to the maneuvering unit, the *Gemini 8* extravehicular activity was canceled. *Gemini 10* and *11* carried an improved maneuvering unit, one supplied nitrogen through a hose within the umbilical cord. The hose connected the gun to two tanks aboard the spacecraft. During a *Gemini 10* docking exercise, Michael Collins successfully recovered a package from a target vehicle. In the process he lost hold of and drifted away from the target vehicle. He used the hand-held unit to maneuver back to place. This was an unscheduled use of the device; the scheduled use was canceled. Before using the maneuvering aid outside *Gemini 11*, Richard F. Gordon managed to tether the spacecraft and target vehicle together. Due to exhaustion from the physical effort involved in such early extravehicular activities, the crew halted the extravehicular experiment before using the Hand Held Maneuvering Unit. Evaluation of the mission focused on the workload and body restraints rather than the maneuvering unit.[30]

Gemini 9A provided the first opportunity to test the modular or astronaut maneuvering unit. That was June 1966. Astronaut Eugene A. Cernan experienced difficulty donning the maneuvering unit due to the problem of maintaining body position in zero gravity and the necessity of holding on to hand and foot bars. Outside the craft, he discovered that extravehicular tasks required both more time and more effort than ground simulations. Also, his visor fogged, reducing visibility—the result of his exceeding the design limits of the Extravehicular Life-Support System, the chest pack. Due to these problems, particularly the reduced vision, the extravehicular activity ended before any operational evaluation of the maneuvering unit. As a result of *Gemini 9A*, NASA changed the foot restraints on future Gemini craft, added underwater simulation of the weightless environment (this proved more effective than the brief periods of zero-gravity training aboard the KC-135 aircraft), and supplied astronauts an anti-fog solution to be applied to their visors before extravehicular activity. Such changes in equipment and technique were made after each mission.[31] The unused maneuvering unit required no modification.

---

29. *A Walk in Space, Gemini 4 Extravehicular Activity,* a 25-page commemorative brochure probably published by NASA in late 1965, in Record Number 007188, NASA Historical Reference Collection.
30. D. Owen Coons and G.F. Kelly, "Medical Aspects of Gemini Extravehicular Activities," typescript report (without the referenced figures), no date, in Record Number 007189, NASA Historical Reference Collection.
31. R.M. Machell, L.D. Bell, N.P. Shyken, and J.W. Prim, III, "Summary of Gemini Extravehicular Operation," typescript report (without the referenced figures), no date, in Record Number 007189, NASA Historical Reference Collection.

Gemini 12 again featured the backpack modular maneuvering unit in a plan that changed before flight. The unit was not even carried aboard the spacecraft. Both astronauts, James A. Lovell and Edwin E. Aldrin, completed extravehicular activities, aided by body restraints like waist tethers, foot restraints, and portable handholds that had been added to this mission based on previously identified need. These spacecraft-based aids to extravehicular activity had taken precedence over the rocket-powered units—foreshadowing the fate of the award-winning MMU.

The development, delivery, and integration of flight-ready maneuvering units, not the operation of the units in space, proved to be main accomplishments of the Gemini maneuvering programs.[32] The hand-held unit was tested in space twice, briefly during *Gemini 4* and *10*. Problems with the spacecraft or the environmental control systems occurred before the unit could be evaluated on other Gemini flights. Similarly, an environmental control problem caused the cancellation of the backpack experiment planned for *Gemini 9A*, and plans to test it on the *Gemini 12* mission were dropped before launch. No operational test of the backpack unit was achieved during Project Gemini.

Before Project Gemini drew to a close in 1966, the Manned Spacecraft Center awarded Rocket Research Corporation a contract to improve the Gemini hand-held maneuvering unit; improvements included using hydrazine as the propellant.[33] Under Air Force contracts, both LTV and Bell Aerosystems designed maneuvering units that could be operated remotely or controlled by an astronaut wearing the unit.[34] LTV's Remote Maneuvering Unit (RMU) could be worn on the back, whereas Bell's Dual-Purpose Maneuvering Unit (DMU) was to be mounted in front of the astronaut. Both units incorporated television cameras, stabilization and control systems, electronic sensors, and communications equipment. As unmanned units, they were intended for work too hazardous for a man, such as inspecting an enemy satellite, as well as for rescue, repair, and transfer operations. As manned units, they could be used during any extravehicular activity.

NASA explored these and other maneuvering technologies. Some units were considered for later Gemini missions, for the Apollo Applications Program (AAP), for even Project Apollo, and for Skylab. Project Apollo accumulated a total of 170 hours of extravehicular activity, mostly lunar surface time—walking on the Moon or riding the lunar rover. None of the Apollo extravehicular time involved the use of a maneuvering unit in free or tethered flight; none was needed to fulfill Apollo's lunar missions. Yet Apollo and the other early space programs provided opportunity not only to experiment with maneuvering units, but also for many companies and individuals to acquire space contracts and experience. Ed Whitsett, then with the Air Force, for example, worked on several extravehicular activity support devices for Apollo, including a hand-held, self-propulsion gun used in the low-gravity environment of the Moon's surface. One group of space scientists had recommended a "Lunar Flying Unit (LFU)" to increase lunar surface mobility and thereby to increase the scientific return from lunar missions;[35] the lunar rover, a wheeled vehicle, provided the increased surface mobility on the Moon. Whitsett also worked on the AAP that became Skylab.

Initially part of the AAP, Skylab experiment T020 consisted of a foot-controlled maneuvering unit (FCMU). Donald E. Hewes of NASA's Langley Research Center was the principal investigator; he built on the earlier work of John D. Bird, also of Langley. Some engineers at

---

32. Johnson and others, "EVA Maneuvering about Space Vehicles."

33. Memorandum from Charles W. Mathews regarding "MSC Contract with Rocket Research Corp. for Two Hand-Held Maneuvering Units," June 14, 1967, Record No. 008093, NASA History Office.

34. "Space Unit IQ Is Second Only to Man," *Denver Post*, August 4, 1965, p. 9; and "Bell Designs DMU to Propel Astronauts in Orbital Tasks," *Aerospace*, January 1967, p. 9.

35. William David Compton, *Where No Man Has Gone Before, a History of Apollo Lunar Exploration Missions* (Washington, DC: NASA SP-4212, 1989), pp. 338, see also 97–99.

the Manned Spacecraft Center opposed the Langley foot-control experiment and expressed "skepticism about the worth of the experiment's objective and concern over the monetary and manpower expenditures connected with its implementation."[36] The program continued. The main purpose of foot control was to free the astronaut's hands. The experimental unit used a cold gas, high-pressure nitrogen, supplied from a tank mount on the astronaut's back. Called "jet shoes" because the thrusters were mounted under the astronaut's feet, four thrusters per foot, the maneuvering unit was pedal operated. Astronauts tested the jet shoes inside Skylab's Orbital Workshop, a "shirt-sleeve" environment as astronauts no longer needed to wear space suits inside the protective environment of the spacecraft. They also conducted a space suit test of the maneuvering unit since the ultimate goal was a unit to be used outside of a spacecraft and thus by a suited astronaut.

Building upon the Gemini example and experience, Skylab's experiment M509 included both a backpack maneuvering unit called the Automatically Stabilized Maneuvering Unit (ASMU) and an improved hand-held maneuvering unit. Both units were propelled by high-pressure nitrogen drawn from a tank on the astronaut's back. In fact, both units could be used at the same time. These units were tested inside Skylab, which contained almost 12,000 cubic feet of living space. Skylab allowed the comparative testing of the jet shoes, the improved hand-held unit, and the backpack. On the three Skylab missions five astronauts flew the M509 experimental units on eleven sorties that totaled fourteen hours of orbital flight testing—all inside Skylab, some in shirt sleeves and some in space suits. The backpack proved superior in flight qualities and precision control. As an experimental unit designed for testing inside Skylab, the backpack lacked the system redundancy deemed necessary for safety outside a spacecraft, and it required a second person to assist the astronaut into the unit. Under NASA contract, the Martin Marietta company built and supported the M509 backpack maneuvering unit; North American Rockwell had also been a contender for the contract.

On assignments to the Manned Spacecraft Center and the Air Force Space and Missile Systems Organization (SAMSO), Major Whitsett headed the M509 experimental program. NASA admired his ability to balance experiment objectives, hardware development cost, and schedule constraints, and his efforts toward consolidating Air Force and NASA research into a single national program, important during that period of limited funding for space programs. Captain Bruce McCandless of the Navy and David C. Schultz of the Manned Spacecraft Center (renamed the Johnson Space Center in 1973) were co-investigators for the M509 experiment, generically labeled "astronaut maneuvering equipment." Whitsett later summarized the backpack program: "An experimental MMU tested onboard the NASA Skylab Program orbital workshop established key piloting characteristics and capability base for future MMU systems" and contributed to the "operational MMU" used on

---

36. Donald K. Slayton quoted in Gregory P. Kennedy, "HHMU, AMU, and MMU, the Development of Astronaut Maneuvering Units," pp. 471–82 in *Skylab, Space Platforms and the Future* [Advances in the Astronautical Sciences, Volume 49] (San Diego, CA: American Astronautical Society, 1982), 475. See also Donald E. Hewes and Kenneth E. Glover, *Development of Skylab Experiment T020 Employing a Foot Controlled Maneuvering Unit* (Washington, DC: NASA TN D-3809, 1972); and David F. Thomas, John D. Bird, and Richard F. Hellbaum, *Jet Shoes, an Extravehicular Space Locomotion Device* (Washington, DC: NASA TN D-3809, 1967).

the Space Shuttle.[37] Only ten years after Skylab, NASA called the Skylab backpack unit the "ancestor" of the Space Shuttle MMU.[38] In turn, Gemini's modular maneuvering unit was the "ancestor" of Skylab's automatically stabilized maneuvering unit.

Manned Maneuvering Unit

Engineer and historian Walter G. Vincenti wrote in *What Engineers Know and How They Know It*, "Engineering knowledge reflects the fact that design does not take place for its own sake and in isolation. Artifactual design is a social activity directed at a practical set of goals intended to serve human beings in some direct way. As such, it is intimately bound up with economic, military, social, personal, and environmental needs and constraints."[39] That is true not only of the experimental MMU—modular maneuvering unit of Project Gemini, but also of the operational MMU—manned maneuvering unit of the Space Shuttle program. Spacecraft, like maneuvering technology, made the transition from experimental (Gemini and Skylab) or exploratory (Apollo) to operational (Space Shuttle). The reusable and operational nature of the Space Shuttle influenced the design and fabrication of the Shuttle's MMU, as did the experimental experience with earlier maneuvering technology. In the post-Apollo environment of reduced NASA budgets, both the Space Shuttle, like a commercial truck, and the MMU, like a worker's tool, were expected to pay for themselves.[40] Neither would.

NASA, Rockwell International, Martin Marietta, Thiokol, "and the entire government/industrial team that improved the concept of manned reusable spacecraft" won the Collier Trophy for 1981.[41] That was the year that the Space Shuttle made its maiden flight (in April) and made the first flight of a reused spacecraft (in November)—both in the orbiter vehicle named *Columbia* and designated OV-102. Then officially called the Space Transportation System (STS), the Space Shuttle program was in fact a small fleet of orbiter vehicles. Other space shuttles built by the end of 1984 were *Challenger* (OV-099), *Enterprise* (OV-101), and *Discovery* (OV-103); *Atlantis* (OV-104) was under construction.[42] MMUs were used on two flights of *Challenger* and one of *Discovery*, all three flights in 1984.

---

37. C.E. Whitsett, *Role of the Manned Maneuvering Unit for the Space Station*, [SAE Technical Paper Series No. 861012] (Warrendale, PA: Society of Automotive Engineers, 1986), p. 1. Regarding Whitsett duties, see memorandum from William C. Schneider, Director of the Apollo Applications Program, to Assistant Administrator, Office of DOD and Interagency Affairs, April 9, 1969, copy in Record No. 002228, NASA History Office, Washington, DC. See also C.E. Whitsett and B. McCandless II, "Skylab Experiment M509 Astronaut Maneuvering Unit Orbital Test Results and Future Applications," *The Skylab Results* [Advances in Astronautical Science Sciences, Volume 31, Part 1] (Tarzana, CA: American Astronautical Society, 1975); Leland F. Belew and Ernst Stuhlinger, *Skylab, a Guidebook* (Washington, DC: NASA EP-107, 1973), pp. 197–98, 201–02; and W. David Compton and Charles D. Benson, *Living and Working in Space, a History of Skylab* (Washington, DC: NASA SP-4208, 1983). Regarding Whitsett, see Les Quiocho, "The Remarkable Flying Machine: a Tribute to Ed Whitsett," clipping from an unidentified Johnson Space Center employee publication, and Whitsett obituary [died 14 October 1993], from a Houston newspaper–both clippings obtained from the Johnson Space Center, Houston. Whitsett was educated at Auburn University and the Air Force Institute of Technology. His Navy colleague, McCandless, held degrees from the Naval Academy and Stanford University.

38. NASA, "41-B Tenth Space Shuttle Mission Press Kit" (Release No. 84-4, February 1984), p. 15.

39. Walter G. Vincenti, *What Engineers Know and How They Know It, Analytical Studies from Aeronautical History* (Baltimore, MD: Johns Hopkins University Press), p. 11.

40. Alex Roland, "The Shuttle, Triumph or Turkey?" *Discover* 6 (November 1985): 29–49; and Daniel A. Bland, Jr., Johnson Space Center, *Space Shuttle EVA Opportunities* [JSC-11391] (Houston, TX: Johnson Space Center, n.d. [pre-1984]).

41. Quotation from the base of the Collier Trophy, as quote in Bill Robie, *For Greatest Achievement, a History of the Aero Club of America and the National Aeronautic Association* (Washington, DC: Smithsonian Institution Press, 1993), p. 235.

42. Regarding the flight vehicle, see Jenkins, *History of Developing the National Space Transportation System*; Space Shuttle officially replaced Space Transportation System as the program name in 1990.

With the Space Shuttle, NASA introduced a new type of spacecraft, but the shuttle's MMU represented an evolutionary development of maneuvering technology, based heavily upon the M509 backpack tested on Skylab and the earlier Gemini backpack. What was the award-winning manned maneuvering unit? Whitsett defined it as "a self-contained propulsive backpack" and "a miniature spacecraft which an astronaut straps on for space walking."[43] Shuttle astronaut Joseph P. Allen called it "this spaceship's special dinghy," which "resembles a backpack with armrests, or some kind of overstuffed rocket chair."[44] In a brochure for the "payload community," NASA advertised the MMU: "Since the Manned Maneuvering Unit has a six-degree-of-freedom control authority, an automatic attitude-hold capability and electrical outlets for such ancillary equipment as power tools, a portable light, cameras and instrument monitoring devices, the unit is quite versatile and adaptable to many payload task requirements."[45]

Although approved for development in 1975, the shuttle maneuvering unit remained in the design definition stage until funding became available in 1979. Under preliminary design contract NAS9-14593, Martin Marietta established the operational MMU design definition and developed subsystems hardware.[46] During that period Martin Marietta also worked with Rockwell International on the MMU/shuttle interface and with NASA on the MMU's interface with the astronaut's space suit and life support system. Finally in 1979, NASA let the MMU fabrication contract, number NAS9-17018, to Martin Marietta. Preliminary designs and specifications were updated, technical changes were adopted, parts were procured, verification requirements were defined, components and then the MMUs were assembled, the units were qualified, mission profiles were drawn, training requirements were defined, and finally flight hardware was delivered. The astronaut representative for the MMU was Bruce McCandless, who as a member of the astronauts corps had served a CapCom or capsule communicator transmitting voice messages to Apollo spacecraft *10, 11,* and *14,* and who had participated in Skylab Experiment M509. Whitsett, who had moved from the Air Force to NASA, also brought experience with Apollo and Skylab. He worked in the Crew Systems Division of the Johnson Space Center. Walter W. Bollendonk managed the Martin Marietta program that built the manned maneuvering units. Martin Marietta delivered the two operational units to the Johnson Space Center in September 1983. Each MMU was valued at $10 million.[47]

New features of the shuttle maneuvering units included fingertip control (rather than the tiring hand-grip control of the Skylab unit), and storage in the cargo bay. Once in the MMU, an astronaut controlled position (forward/backward, left/right, up/down) with the left hand and rotation with the right hand. Tolerance of extreme temperatures was achieved in part by painting the MMU white to keep the temperature below 150° Fahrenheit and by using electrical heaters to keep components above their minimum temperature limits. An astronaut could recharge the propulsion system at the shuttle's cargo bay from airborne support equipment called the flight support station; this support station also provided storage of the MMU when not in use.

The shuttle MMU system had redundancy. Two silver-zinc batteries provided electricity. The propellant was gaseous nitrogen, GN2, stored in two tanks. The propulsion systems

---

43. Whitsett, *Role of the Manned Maneuvering Unit for the Space Station,* pp. 1, 5. Regarding the evolution of the award-winning MMU, see Kennedy, "HHMU, AMU, and MMU, the Development of Astronaut Maneuvering Units;" and D.J. Shayler, "The Shuttle MMU," *Spaceflight* 27 (June 1985): 263.
44. Allen, *Entering Space,* p. 113.
45. Bland, *Space Shuttle EVA Opportunities,* p. 7.
46. Martin Marietta, "Briefing: Manned Maneuvering Unit," at Lyndon B. Johnson Space Center, September 27, 1979, copy obtained from Johnson Space Center Library, Houston.
47. Craig Covault, "Maneuvering Unit Keyed to Simplicity," *Aviation Week & Space Technology* 120 (January 23, 1984): 43.

were arranged in two parallel sets, each set operating twelve thrusters; usually both sets—twenty-four thrusters—were operational at once, but the MMU was capable of full operations on only one set. Furthermore, the Space Shuttle normally carried two MMUs, the second in case of emergency, and the shuttle could be maneuvered into position to rescue an astronaut should an MMU fail. In conjunction with the propulsion system, three gyros—one each for the yaw, pitch, and roll axes—provided an attitude hold capability. Constructed mostly of aluminum, an MMU weighed 340 pounds—massive, though weightless in space. The operating time was six hours, and the operating range, 450 feet from the Space Shuttle.

Martin Marietta trained astronauts to fly the MMU at its Space Operations Simulator in Denver. A magazine editor who flew the MMU in that simulator reported, "The minimal training and precision flying features were demonstrated by my ability, with only a few minutes practice, to maneuver the unit safely in close proximity to fixed objects."[48] Astronauts received eighteen hours, not a few minutes, of training in the simulator. The two main features in the simulator were a six-degree-of-freedom moving-base carriage and a large-screen television display. NASA of course provided the standard astronaut and extravehicular-activity training.

NASA carried two MMUs (serial numbers 002 and 003) aboard three Space Shuttle flights in 1984: 41-B in February, 41-C in April, and 51-A in November. Six astronauts—Bruce McCandless II, Robert L. Stewart, George D. Nelson, James D. van Hoften, Joseph P. Allen, and Dale A. Gardner—flew the MMU. These mission specialists flew the MMU on a total of nine sorties for a total of ten hours and 22 minutes. Each astronaut donned and doffed the maneuvering unit in the open cargo bay.

Before exiting the pressurized spacecraft, the astronaut donned an extravehicular mobility unit (EMU) that consisted of the spacesuit and a portable life support system. A NASA brochure explained, "The Extravehicular Mobility Unit consists of a self-contained (no umbilicals) life support system and an anthropomorphic pressure garment with thermal and micrometeoroid protection."[49] The Hamilton Standard division of United Technologies Corporation, aided by subcontractor ILC (formerly International Latex Corporation), produced the space suit. The suit consisted of modular parts; the torso, for example, available in five sizes. Gloves were still custom-made for a particular astronaut. The life support system, also supplied by Hamilton Standard, was in a backpack that could attach to the MMU, which became in effect an outer backpack. The EMU was essential to extravehicular activity, but the MMU was one of several extravehicular aids available for a mission; the remote manipulator system, tools, tethers and other restraints, and portable workstations were the other aids. The astronaut and the extravehicular mobility unit, and any tools needed for an assignment, comprised the MMU's payload.

The shuttle manned maneuvering unit was a tool with a specific mission. That mission was the recovery of satellites. The astronaut using the manned maneuvering unit was a "serviceman" who serviced satellites. NASA offered this recovery service to civilian agencies, the military services, and commercial customers, all of which had satellites in orbit. To retrieve a satellite meant reaching the satellite, grabbing it, stopping its rotation, and moving it into the Space Shuttle's cargo bay. Although weightless in space, the satellite still had inertia, against which the maneuvering unit needed power to stop the rotation. Retrieving the *Solar Maximum* (*Solar Max*) satellite was to be the first operational assignment of the MMU.

---

48. Covault, "Maneuvering Unit Keyed to Simplicity," p. 43; and Craig Hartley, Dave Owynar, and Lex Ray, "Manned Maneuvering Unit Simulations on the Space Operations Simulator," typescript [1984], obtained from the Johnson Space Center, Houston.

49. Bland, *Space Shuttle EVA Opportunities*, p. 3. Regarding the Shuttle space suits, see Kozloski, *U.S. Space Gear*, pp. 123–44.

Launched in 1980, the *Solar Max* solar observatory had experienced electrical failures within six months, and NASA planned to repair the satellite in the cargo bay of a Space Shuttle.

Other missions were considered for the MMU, including inspection and repair of thermal tiles on shuttle orbiters, handling and transferring payload, construction of space structures (like a Space Station), and rescuing loose material or personnel floating in space. Martin Marietta promoted the MMU as support of shuttle extravehicular activities like inspection of the shuttle orbiter and like deploying, retrieving, and servicing payloads. But satellite retrieval was the primary mission in plans and in practice.[50]

The first use of the MMU occurred on the tenth flight of a Space Shuttle, mission 41-B, during which the MMU was flown on demonstration flights. These MMU flights demonstrated capabilities deemed appropriate for use in the planned retrieval of the *Solar Max* satellite on a later shuttle mission. Courtesy of the manned maneuvering unit, McCandless, then a Navy captain, became the first person to fly free, untethered in space; the date was February 7, 1984. While orbiting around the Earth at a speed of 17,500 miles per hour, McCandless floated from the cargo bay into outer space, 150 nautical miles above Earth, an experience he described as "a heck of a big leap."[51] Mission specialist Robert L. Stewart, an Army lieutenant colonel, also flew the MMU on shuttle mission 41-B. While flying the MMU, these men were in a journalistic phrase of the time "human satellites."[52] They checked out the equipment, maneuvered within the cargo bay, flew away from and back to the orbiter, performed docking exercises, recharged the MMU nitrogen tanks, and collected engineering data. The MMU, according to Martin Marietta's post mission report, "performed as expected and no anomalies were reported."[53]

The main purpose of flight 41-B, the fourth using the orbiter *Challenger*, was the deployment of two commercial communication satellites, Western Union's *Westar VI* and the Indonesian *Palapa-B2*. These satellites were released, but failed to reach geostationary orbit due to problems with the commercial upper-stage technology designed to lift the satellites from the low orbit of the Space Shuttle to the higher geosynchronous orbit—justifying a later rescue mission using MMUs. Also, in scheduled extravehicular activity during flight 41-B, astronauts demonstrated the shuttle orbiter's manipulator arm. One man at a time rode on the manipulator foot restraint work platform (a Grumman product) attached to the remote manipulator arm (a Spar Aerospace product), while mission specialist Ronald E. McNair inside the spacecraft controlled the movement of the arm. On this mission the arm developed a little problem with its wrist joint yaw motion capability, but on a later mission the manipulator arm would achieve a satellite rescue after MMU-retrieval attempts failed.

In April NASA launched the eleventh Space Shuttle mission, 41-C, which again used the orbiter *Challenger*. In response to the previous mission, Martin Marietta had made only two minor changes to the MMU hardware: new, adjustable lap belts installed on the MMU itself and a modification of the flight support station in the cargo bay. The main purpose of the 41-C mission was repairing *Solar Max*, and the main purpose of the MMU on the mission was retrieving the satellite. If successful, NASA predicted, this "Shuttle mission could launch an era of satellites with replaceable parts," satellites repairable in space.[54]

---

50. Martin Marietta, *Manned Maneuvering Unit, Users Guide* (MMU-SE-17-46, April 1982), p. 12.

51. Bruce McCandless II as quoted in Craig Covault, "Astronauts Evaluate Maneuvering Backpacks," *Aviation Week & Space Technology*, 120 (February 13, 1984): 16. Before the free flight in space, one magazine had predicted "Even Buck Rogers would envy Bruce McCandless;" "Rehearsal for a Space Rescue," *Discover* 4 (September 1983): 24–27, quote from p. 25.

52. Craig Covault, "Astronauts to Perform Untethered EVAs from Shuttle," *Aviation Week & Space Technology* 120 (January 9, 1984): 44.

53. Martin Marietta, *Manned Maneuvering Unit, Space Shuttle Program: Manned Maneuvering Unit Post Mission Summary Report for STS 41B* (Technical Report MMU-SE-17-1011, April 1984), p. 2.

54. Robert G. Nichols, "Repairing Satellites in Space," *High Technology* 4 (January 1984): 15–16.

Again in the post-mission report, Martin Marietta concluded that its "hardware performed as expected with no anomalies" for both mission specialists, George Nelson and James D. van Hoften.[55] But the astronauts using the MMU failed to retrieve the satellite.

Wearing the MMU, Nelson performed the equipment checkout flight, moved 150 feet to *Solar Max*, matched rates with the satellite, and attempted to dock three times. He was unable to stabilize the satellite, to stop its spinning. The failure was later attributed not to the MMU but to the trunnion pin attachment device mounted on the arms of Nelson's MMU, in front of him; this device was supposed to lock onto a trunnion on the satellite. Once Nelson had docked, he was to use the MMU thrusters to halt the satellite's rotation. With the satellite stabilized, the manipulator arm would grasp the satellite and move it into the cargo bay for repair. The MMU rather than the manipulator arm was to capture *Solar Max* in order to avoid the possibility of the rotating satellite snapping the manipulator arm. But Nelson in the MMU failed to stabilize the satellite, so NASA personnel in space and on Earth improvised.

Engineers at NASA's Goddard Space Flight Center in Maryland managed through radio commands to exert some control over the spinning satellite and by reprogramming the satellite's computer to slow the spin. Shuttle commander Robert L. Crippen flew the orbiter for a precision rendezvous with the satellite. Then astronaut Terry J. Hart operated the manipulator arm to capture the slowly rotating satellite. As an extravehicular activity in the open cargo bay, Nelson and van Hoften repaired *Solar Max*, and the *Challenger* crew released the repaired satellite back into orbit. The mission, though not the MMU's role in it, was a success. "Hart's small grab," not Nelson's free flight, quickly became the symbol of the utility of human space flight.[56]

Despite the docking problem experienced during mission 41-C and the use of the manipulator arm to achieve capture, NASA personnel still believed that the MMU could "provide an extra measure of control in the retrieval process" of future satellite recovery operations.[57] NASA scheduled the MMU for its next recovery mission for November. Mission 51-A, using the orbiter *Discovery*, was to rescue the *Westar* and *Palapa* satellites that mission 41-B had deployed in February. This time mission specialist Joseph Allen in MMU serial number three captured the *Palapa* satellite, and Dale A. Gardner in MMU serial number two recovered the *Westar* satellite. They used a new, improved capture device, a stinger, in the successful recoveries. The capture mechanism worked, and the MMU's automatic attitude hold function stopped the satellite rotation. Again, the MMUs "performed as expected with no anomalies."[58] And again, the recovery operations did not proceed as planned; the retrieval equipment did not fit one of the satellites, and the men had to hold the satellites and manually move them into the payload instead of using the manipulator arm. Despite the problems, Allen concluded, "the capture had been far easier than rodeo calf-roping."[59] Both satellites were secured aboard the *Discovery* and returned to earth for refurbishment and resale by insurance companies that had acquired the salvage rights.[60]

---

55. Martin Marietta, *Manned Maneuvering Unit, Space Shuttle Program: Manned Maneuvering Unit Post Mission Summary Report STS-41C* (Technical Report MMU-SE-17-107, August 1984), pp. 2–3.

56. Dennis Overbye, "Putting the Arm on Solar Max," *Discover* 5 (June 1984): 16–21. How the *Solar Max* rescue mission was supposed to go is detailed in Craig Covault, "Tight Pace Challenges Solar Max Repair," *Aviation Week & Space Technology* 120(March 26, 1984): 42–51; and how the mission actually went is reported in Craig Covault, "Obiter Crew Restores Solar Max," *Aviation Week & Space Technology* 120 (April 16, 1984): 18–20.

57. "NASA Believes EVAs Valid Despite Recovery Problem," *Aviation Week & Space Technology* 120 (April 16, 1984): 21–24; regarding the docking failure, see this article and Alton K. Marsh, "NASA Seeks Cause of Docking Failure," on page 25 of the same issue.

58. Martin Marietta, *Manned Maneuvering Unit, Space Shuttle Program: Manned Maneuvering Unit Post Mission Summary Report STS 51-A* (Technical Report MMU-SE-17-111, February 1985), p. 2.

59. Joseph Allen as quoted on page 294 in Henry C. Dethloff, "*Suddenly, Tomorrow Came . . .*", see also "Satellite Retrieval Succeeds Despite Equipment Problem," *Aviation Week & Space Technology*, 121 (November 19, 1984): 16–19.

60. The insurance underwriters had paid NASA $5.5 million recovery fee in advance of the mission; see "Underwriters Pay for Satellite Recovery Try," *Aviation Week & Space Technology* 121 (October 1, 1984): 28.

On three missions in 1984, the Manned Maneuvering Unit performed as expected and with precision and versatility. Humans could safely maneuver in outer space free of both spacecraft and tether. In recognition of the development of the MMU and the NASA-industry satellite rescue team, the National Aeronautic Association awarded the Robert J. Collier Trophy for 1984 to NASA and Martin Marietta, with special recognition of astronaut Bruce McCandless II, NASA's Charles E. Whitsett, Jr., and Martin Marietta's Walter W. Bollendonk.

Conclusion

The MMU was only one piece of space news in 1984. President Ronald Reagan had opened the year with a State of the Union address reminiscent in part of John F. Kennedy's 1961 "goal, before this decade is out, of landing a man on the moon."[61] Reagan directed "NASA to develop a permanently manned Space Station–and to do it within a decade."[62] The Air Force and Navy claimed no military requirement for a Space Station, which was seen as competition for funds the Department of Defense sought for military space operations. The Defense Advanced Research Projects Agency, for example, was studying a manned space cruiser, a light spacecraft in contrast to the heavy-cargo Space Shuttle. General James V. Hartinger, Commander of the Air Force Space Command, claimed the Soviets had "the world's only space weapon," an orbital anti-satellite system that threatened the low orbiting satellites of the United States.[63] This country needed, according to Hartinger, "to protect our assets in space." Regarding the arms race in space, a defense contractor declared that "the Soviets have taken the high ground on the technology, and we're left with the high ground on the debate."[64] Reagan's Strategic Defense Initiative, including controversial laser systems, addressed these military concerns.

In 1984 Congress appropriated funds for both the Strategic Defense Initiative and the Space Station, and the government's civilian and military agencies continued their routine cooperation in space matters. The Air Force, for example, had provided contingency support for Space Shuttle flights since the beginning, and it increased that contingency support in 1984. Furthermore, in August, the United States adopted a new National Space Strategy that delineated roles for NASA and the Department of Defense.[65] The civil-military-commercial infrastructure adapted to the changing space environment, which remained in part a political environment shaped by the international competition known as the Cold War. The Strategic Arms Limitation Talks (SALT) and Strategic Arms Reduction Talks (START), for example, had increased the importance of reconnaissance satellites, which were increasingly needed for verifying compliance with disarmament agreements. NASA and its hundreds of contractors began development of the civilian Space Station. In this context Whitsett forecasted ample roles for the manned maneuvering unit in the Space Station program: assembly, transportation, inspection, contingency, and rescue.[66]

---

61. John F. Kennedy, "Special Message to the Congress on Urgent National Needs," May 25, 1961, in *Public Papers of the Presidents of the United States: John F. Kennedy, 1961* (Washington, DC: Government Printing Office, 1962), p. 404.
62. Ronald Reagan, "State of the Union Address," January 25, 1984, quoted in Craig Covault, "President Orders Start on Space Station," *Aviation Week & Space Technology* 120 (January 30, 1984): 16.
63. General James V. Hartinger as quoted on page 128 in Edgar Ulsamer, "The Threat in Space and Challenges at Lesser Altitudes," *Air Force Magazine* 67 (March 1984): 128–31; see also Craig Covault, "DARPA Studying Manned Space Cruiser," *Aviation Week & Space Technology* 120 (March 26, 1984): 20–21.
64. Robert L. Kirk, president of LTV Aerospace, as quoted in "U.S. Urged to Negotiate Treaty Based upon Freedom of Space," *Aviation Week & Space Technology* 120/22 (May 28, 1984): 118.
65. Craig Covault, "U.S. Adopts New Space Strategy," *Aviation Week & Space Technology* 121 (August 27, 1984): 14–16.
66. Whitsett, *Role of the Manned Maneuvering Unit for the Space Station*, pp. 10–12.

Yet the MMU has not been used since 1984. There are several reasons for this. First, most extravehicular activities were effective without use of the MMU. Tethers, safety grips, hand bars, and other restraints allowed astronauts to work in the open cargo bay. Furthermore, the maneuverability of the Space Shuttle itself and the utility of the shuttle's robotic manipulator arm had proved capable of rescuing satellites—the primary function for which the MMU had been designed. The orbiter could be piloted with such accuracy that on mission 41-B, for example, commander Vance D. Brand piloted the *Challenger* into position so that McCandless on the manipulator arm could grab a foot restraint that had broken loose and floated away from the orbiter. On flight 41-C, the MMU failed to achieve mechanical mating to the *Solar Max* satellite, but the orbiter and manipulator arm recovered the satellite. On the *Discovery* mission, 51-A, commander Henry W. Hartsfield operated the remote manipulator arm to knock ice off a waste-water port, the ice being a reentry hazard. This sort of contingency was a potential MMU activity, but the manipulator arm solved the problem.

Another reason for lack of use of the MMU was the *Challenger* accident. In January 1986 the *Challenger* exploded 73 seconds after launch. The crew of seven, the spacecraft, and the payload were lost. That accident initially prompted a suspension of space flights that lasted into September 1988. The accident and resulting investigations also prompted new safety rules that would require expensive changes to the existing MMU, changes pending both a customer and a mission for the MMU. Still another reason for not using the MMU has been the lack of a new user with adequate funding and appropriate mission. Finally, since the Space Station is still under discussion, the Space Shuttle remains the main space human flight program of the United States. The MMU is not necessary to its operations.

Thus today, as Robert Frost observed in 1959:

> But outer Space,
> At least this far,
> For all the fuss
> Of the populace,
> Stays more popular,
> Than populous.[67]

---

67. *The Poetry of Robert Frost*, edited by Edward Connery Lathem © 1959, 1962 by Robert Frost, © 1969 by Henry Holt & Co., Inc. Reprinted by permission of Henry Holt & Co., Inc. (New York, NY: Holt, Rinehart, and Winston, 1969), p. 469, quoted with permission.

Chapter 14

The Advanced Turboprop Project: Radical Innovation in a Conservative Environment

by Mark D. Bowles and Virginia P. Dawson

In 1987, a *Washington Post* headline read, "The aircraft engine of the future has propellers on it."[1] To many this statement was something like heralding "the reincarnation of silent movies."[2] Why would an "old technology" ever be chosen over a modern, new, advanced alternative? How could propeller technology ever supplant the turbojet revolution? How could the "jet set mind-set" of corporate executives, who demanded the prestige of speed and "image and status with a jet," ever be satisfied with a slow, noisy, propeller-driven aircraft?[3] A *Washington Times* correspondent predicted that the turbojet would not be the propulsion system of the future. Instead, the future would witness more propellers than jets and if "Star Wars hero Luke Skywalker ever became chairman of a Fortune 500 company, he would replace the corporate jet with a . . . turboprop."[4] It appeared that a turboprop revolution was underway.

NASA Lewis Research Center's Advanced Turboprop Project (1976–1987) was the source of this optimism. The energy crisis of the early 1970s served as the catalyst for renewed government interest in aeronautics and NASA launched this ambitious project to return to fuel saving, propeller-driven aircraft. The Arab oil embargo brought difficult times to all of America, but the airlines industry, in particular, suffered and feared for its future in the wake of a steep rise in fuel prices. NASA responded to these fears by creating a program to improve aircraft fuel efficiency. Of the six projects NASA funded through this program, the Advanced Turboprop Project promised the greatest payoffs in terms of fuel savings, but it was also the most conceptually radical and technically demanding.

The project began in the early 1970s with the collaboration of two engineers, Daniel Mikkelson from NASA Lewis, and Carl Rohrbach of Hamilton Standard, the nation's last major propeller manufacturer. Mikkelson, then a young aeronautical research engineer, went back to the old NACA wind tunnel reports where he found a "glimmer of hope" that propellers could be redesigned to make propeller-powered aircraft fly faster and higher than those of the mid to late-1950s.[5] Mikkelson and Rohrbach came up with the concept of sweeping the propeller blades to reduce noise and increase efficiency and NASA received a joint patent with Hamilton Standard for the development of this technology. At Lewis, Mikkelson sparked the interest of a small cadre of engineers and managers. They solved key technical problems essential for the creation of the turboprop, while at the same time they attracted support for the project. After a project office was established, they became political advocates, using technical gains and increasing acceptance to fight for continued funding. This involved winning government, industry, and public support

---

1. Martha M. Hamilton, "Firms Give Propellers a New Spin," *Washington Post*, February 8, 1987.
2. Robert J. Serling, "Back to the Future with Propfans," *USAIR* (June 1987).
3. R.S. Stahr, Oral report on the RECAT study contract at NASA, April 22, 1976, Nored papers, NASA, Lewis Research Center, box 224.
4. Hugh Vickery, "Turboprops are Back!," *Washington Times*, November 1, 1984, p. 5B.
5. Interview with Daniel Mikkelson, by Virginia Dawson and Mark Bowles, September 6, 1995.

An advanced propeller swirl recovery model is shown in the NASA Lewis Research Center's 8 x 6 foot supersonic wind tunnel. Propeller efficiencies and noise are measured at cruise mach numbers up to 0.80 and at takeoff and approach conditions. Vane pitch angles and propfan-to-vane axial spacings are varied. The testing was part of the Advanced Turboprop Project, with the goal of providing the technology base to enable the U.S. development of quieter, fuel efficient turboprop engines with a comfortable aircraft interior environment. (NASA photo no. 90-H-78).

for the new propeller technology. Initially the project involved only Hamilton Standard, but the aircraft engine manufacturers, Pratt & Whitney, Allison, and General Electric, and the giants of the airframe industry, Boeing, Lockheed, and McDonnell Douglas joined the bandwagon as the turboprop appeared to become more and more technically and socially feasible. The turboprop project became a large, well-funded, "heterogeneous collection of human and material resources" that contemporary historians refer to as "big science."[6] At its height it involved over forty industrial contracts, fifteen university grants, and work at the four NASA research centers, Lewis, Langley, Dryden, and Ames. The progress of the advanced turboprop development seemed to foreshadow its future dominance of commercial flight.

The project had four technical stages: "concept development" from 1976 to 1978; "enabling technology" from 1978 to 1980; "large-scale integration" from 1981 to 1987; and finally "flight research" in 1987.[7] During each of these stages, NASA's engineers confronted and solved specific technical problems that were necessary for the advanced turboprop project to meet the defined government objectives concerning safety, efficiency at high speeds, and environmental protection. NASA Lewis marshaled the resources and support of the United States aeronautical community to bring the development of the new technology to the point of successful flight testing. In 1987, these NASA engineers, along with a wide-ranging industry team, won the coveted Collier Trophy for developing a new fuel efficient turboprop propulsion system.[8] The winning team included Hamilton Standard, General Electric, Lockheed, the Allison Gas Turbine Division of General Motors, Pratt & Whitney, Rohr Industries, Gulfstream, McDonnell Douglas, and Boeing—certainly the largest, most diverse group, to be so honored in the history of the prize.

Despite this technical success, the predicted turboprop revolution never came, and no commercial or military air fleet replaced their jets with propellers. The reason for this failure was socio-economic, not technical. Throughout the project, social issues influenced and defined the status of the advanced turboprop. From the beginning it was the perception of an energy crisis, not a technological innovation, that spurred the idea of the project itself. The Cold War and the existence of Soviet high-speed turboprops played a key role in convincing Congress to fund the project. As the project progressed, within each technological stage, the engineers used distinctive and creative approaches to deal with the complex web of government, industry, and academic contractors. More often than not, the main question was not does the technology work, but how can we get government, industry, and the public to accept this technology? In the end it was a socioeconomic issue again which shelved the program. The reduction of fuel prices ended the necessity for fuel conservation in the skies and today the advanced turboprop remains a neglected, or "archived" technology.

This is not to imply that the technical achievements were unimportant. Each distinct technical stage of the project determined a corresponding social action. During the concept development stage, creative advocacy was necessary to sell the government and industry on this radical idea. During the enabling technology stage, engineers used complex project management skills to ensure that this massive team would function effectively. During the large-scale integration stage, NASA had to deal with a competitor that surprised them by introducing its own high-speed turboprop. Finally, during the flight research stage, NASA became aware that no current airlines would adopt the advanced turboprop and thus the

---

6. See James H. Capshew and Karen A. Rader, "Big Science: Price to the Present," *Osiris*, 2nd ser., 7 (1992): 3–25.

7. Roy D. Hager and Deborah Vrabel, *Advanced Turboprop Project* (Washington, DC: NASA SP-495, 1988), p. 610.

8. Citation for the Collier Trophy in Roy D. Hager and Deborah Vrabel, p. vi.

engineers waged a battle to win the Collier Trophy to try and gain positive status and recognition for their technical achievement.

The relationship between these technical and social spheres was never either a simplistic story of social construction or technological determinism. Rather, the relationship was one of interdependence. At times the project advanced on its technical merits; at others, it progressed through political persuasion. At each stage, only after NASA engineers and their industrial and academic partners solved both the social and technical problems holding it back, was the advanced turboprop project able to obtain funding and move forward. But ultimately, the socio-economic issue of petroleum price and availability managed to scuttle NASA's technical success.

Thomas Hughes, a prominent historian of technology, has argued that the research and development organizations of the twentieth century, no matter whether they are run by a government, industry, or members of a university community, stifle technical creativity.[9] In these organizations there can be found "no trace of a flash of genius."[10] In contrast, the late 19th century for Hughes was the "golden era" of invention—a time when the independent inventor flourished without institutional constraints. Recently, David Hounshell has challenged Hughes's contention that industrial research laboratories "exploit creative, inventive geniuses; they neither produce nor nurture them."[11] Not only can the industrial research laboratory nurture a creative individual, but collectively, people engaged in research and development contribute to making an invention a commercial reality. In his study of the organization of research at Du Pont, Hounshell paid tribute to the individual brilliance of the organic chemist Wallace H. Carothers, but he argued that the real "genius of nylon was in the organization that developed it into one of the most successful and profitable materials of the twentieth century."[12] In our view, the NASA Advanced Turboprop Project represents another case in which organizational capabilities, not individual genius alone, create the opportunity for significant innovation. The organization that supported the development of the turboprop was far more complex than the research laboratory of an industrial firm, yet it responded to the energy crisis to advance a radical idea. As Donald Nored, who headed the office at NASA Lewis Research Center that managed the three aircraft energy efficiency projects remarked, "The climate made people do things that normally they'd be too conservative to do."[13] The history of the advanced turboprop demonstrates how a radical innovation can emerge from a dense, conservative web of bureaucracy to nearly revolutionize the world's aircraft propulsion systems.

The Conservative Team Environment

Although NASA won several Collier trophies for innovations related to the space program, it had produced no winners in aeronautics since the founding of the agency in 1958. NASA's predecessor organization, the National Advisory Committee for Aeronautics (NACA), had received five Collier trophies for contributions to aeronautics between 1929 and 1958. These trophies paid tribute to the individual creativity and the unique research environment of the NACA's research laboratories. James R. Hansen has described in this volume how engineer Fred E. Weick used the NACA's unique wind tunnel facilities to develop

---

9. Thomas P. Hughes, *American Genesis*: p. 54.
10. *Ibid.*, p. 183.
11. David A. Hounshell, "Hughesian History of Technology and Chandlerian Business History: Parallels, Departures, and Critics," *History and Technology* 12 (1995): 217.
12. *Ibid.* See also, David A. Hounshell and John Kenly Smith, Jr., *Science and Corporate Strategy: Du Pont R&D, 1902–1980* (Cambridge, MA: Cambridge University Press, 1988).
13. Interview with Donald Nored at Case Western Reserve University by Virginia Dawson and Mark Bowles, August 15, 1995.

the NACA low-drag cowling. Succeeding Collier trophies awarded under the institutional aegis of the NACA followed a similar pattern. Lewis A. Rodert won it for developing a thermal ice prevention system for aircraft (see the essay by Glenn E. Bugos, this volume), John Stack won it twice for his contributions to supersonic theory and the development of the transonic wind tunnel, and Richard Whitcomb carried off the prize for his discovery and empirical validation of the area rule. What made the award garnered by the NASA/industry team in 1987 different was that it recognized the collective talents of government engineers from four NASA research centers, academic researchers, and contractors from the propeller, engine, airframe, and airline industries.

The history of the turboprop project is interesting from an institutional standpoint because it took root and flourished within NASA's conservative, bureaucratic environment. It was modeled, not on NASA's small-scale aeronautical research projects (typically carried on by former NACA laboratories), but on the large-scale projects of the space program. The NASA Lewis Research Center adopted an administratively complex team approach that depended on input not simply from other NASA Centers, but also from numerous industrial and university contractors. Essentially, NASA Lewis Research Center became the center of an extensive government-industry-academic complex. At each stage in the project, the management team determined what needed to be done and sought the appropriate help both from within and outside NASA.

With its expertise in propulsion technology, the NASA Lewis Research Center was ideally suited to manage the turboprop project. Set up in Cleveland, Ohio, during World War II as an aircraft engine research laboratory, Lewis became the third laboratory of the National Advisory Committee for Aeronautics. Lewis engineers pursued aircraft engine research in the national interest—often over the objection of the engine companies who perceived the government as interfering with the normal forces of supply and demand. During the early years of the Cold War, the laboratory participated in engine research and testing to assist the engine companies in developing the turbojet engine. After the launch of Sputnik, the laboratory focused on a new national priority–rocket propulsion research and development. Almost all work on air-breathing engines ceased for nearly ten years.

The return to aircraft engine research coincided with drastic reductions in staff, mandated by cuts in NASA's large-scale space programs.[14] The mass exodus of nearly 800 personnel in 1972 sparked an effort to redefine the center's mission and find new sources of funding. The following year, OPEC's oil embargo galvanized the Center's director, Bruce Lundin, to look for ways to use its propulsion expertise to help solve the energy crisis. In 1974, Lewis received $1.5 million for a wind-energy program from the National Science Foundation and the Energy Research and Development Administration (ERDA). A program in solar cell technology development followed on its heels with increasing funding of various energy-related programs by ERDA and its successor, the Department of Energy. The changing focus of the Center's activities prompted rumors-emphatically denied-that it would become part of ERDA. The new emphasis on energy efficient aircraft, unlike the ERDA projects, promised to keep Lewis strongly in NASA's fold.[15] Moreover, it brought high visibility to the aeronautics side of NASA, long overshadowed by the enormous budgets and prestige of the space program.

Although it shared similarities in management with NASA's space projects, the turboprop project differed in significant ways. First, although the advanced turboprop was the reincarnation of an old idea, it involved the creation of cutting-edge technology. Space

---

14. Virginia P. Dawson, *Engines and Innovation: Lewis Laboratory and American Propulsion Technology* (Washington, DC: NASA SP-4306, 1991).

15. *Ibid.*

projects involved rigorous oversight, but generally relied on existing technology. When necessary, NASA contracted with industry to produce whatever new technology was needed for a particular mission. The turboprop project tapped the creative talents of engineers at NASA in ways that were reminiscent of the NACA tradition of in-house research, though in management scope it transcended the narrow institutional boundaries of NASA's research centers. Second, though all NASA projects of the early 1970s needed to be "sold" to an increasingly tight-fisted Congress, the controversial nature of the turboprop meant that NASA Lewis had to build support both at Headquarters and within the aviation community. What NASA referred to as "advocacy" needed to be vigorous and continuous throughout the life of the project.

The Energy Crisis and the Politics of Funding

The OPEC oil embargo of 1973 awakened the United States to the degree of control outside nations had over the lives of every American. The increased price of oil affected all areas of the economy, but none more than the airlines industry.[16] Earl Cook, noted geographer and geologist, has argued, "Whoever controls the energy systems can dominate the society."[17] An extension of this argument is, whoever possesses the fuel supply controls the energy systems. Five sources of energy, including petroleum, natural gas, coal, hydropower, and nuclear, accounted for all fuel consumption in the United States during 1973. Of these five sources, America was most dependent upon petroleum, consuming approximately seventeen million barrels of oil a day.[18] At no other time in American history was Cook's aphorism more evident than in 1973 when the United States imported six million barrels of oil a day, 64 percent of which came from the Organization of Petroleum Exporting Countries (OPEC).[19] The concern in the United States was that since OPEC controlled the petroleum, could they dominate American society?

In response to the energy crisis, in 1973 the airlines industry initiated its own fuel-saving program which reduced fuel consumption by over one billion gallons per year.[20] But these measures were not enough. Jet fuel prices jumped from twelve cents to over one dollar per gallon and total yearly fuel expenditures increased by one billion dollars, or triple the earnings of the airlines. Prior to 1972, fuel accounted for one-quarter of the commercial airlines' total direct operating costs.[21] During the crisis, fuel represented over half of the airlines' operating costs. The result was a reduction in the number of flights, the grounding of some aircraft, and the "furloughing" of some 10,000 employees. If the situation in the early 1970s seemed bad, prospects for the future appeared even worse. Linking the fate of the airlines, the cost of jet fuel and the prosperity of the nation as a whole, airlines industry lobbyists rushed to their congressmen. The politicians, in turn, appealed to NASA.

---

16. The Israeli victory during the Six-Day War in 1967 resulted in retaliation by OPEC. Seeking to force a pro-Arab stance from the United States (Israel's ally), Saudi Arabia imposed an American oil embargo concurrent with the quadrupling of oil prices from the other OPEC nations. See Don Peretz, *The Middle East Today* 5th ed. (New York, NY: Praeger, 1988), 154. Gary B. Nash, et al. *The American People: Creating a Nation and a Society* 2d ed. (New York, NY: Harper & Row, Publishers, 1990), p. 971.

17. Earl Cook, *Man, Energy, Society* (San Francisco, CA: W. H. Freeman, 1976), p. 208.

18. A barrel contains 42 gallons.

19. In 1973, total U.S. crude oil imports totaled 1,184 million barrels, 765 of which came from OPEC. The OPEC nations at that time included Algeria, Ecuador, Indonesia, Iran, Iraq, Kuwait, Libya, Nigeria, Qatar, Saudi Arabia, United Arab Emirates, and Venezuela. *Statistical Abstract of the United States* (Washington, DC: U.S. Department of Commerce, 1994), p. 593.

20. Clifton F. Von Kann, testimony before the U.S. Senate, Committee on Aeronautical and Space Sciences, September 10, 1975, p. 4.

21. Donald R. Nored, John B. Whitlow, Jr., William C. Strack, "Status Update of the NASA Advanced Turboprop Project," unpublished report, Nored private papers.

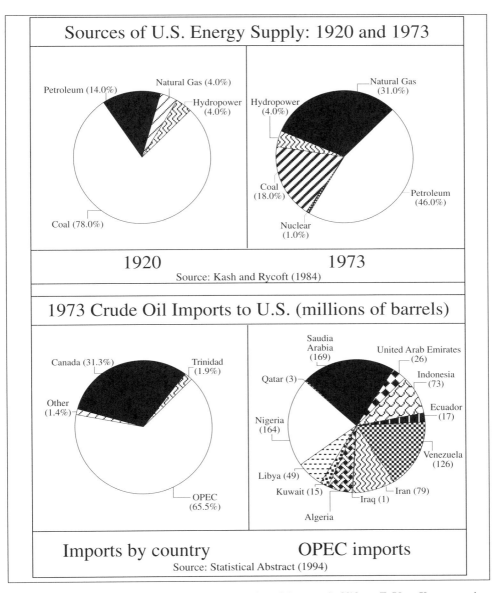

Why was jet fuel so important to our national interest? Clifton F. Von Kann, senior vice-president of the Air Transport Association of America, pointed out in a 1975 Senate statement that airlines were "more than just another means of transportation."[22] He asserted they played a major part in the economic and military success of the nation. They also

---

22. Kann testimony, p. 3.

provided the infrastructure for the mail system, the national export system, and the $60 billion tourist industry. Jet fuel was the "life-blood" of the airlines, but it was also their Achilles heel. He warned a failure to control the rising cost of fuel might result in either the nationalization or the withering away of the "basic building block in the structure of the U.S. economy."[23] Senator Barry Goldwater linked this crisis to the possible "loss of a large part of our world supremacy."[24] The fuel crisis created an opportunity for NASA at a time when Congress had drastically cut funding for the space program. Aeronautics, the first "A" in NASA, had long taken a back seat to the spectacular space missions of the Apollo years. Now the agency was ready to reassert its role as the nation's premier institution for research and development in civil aeronautics.

In January 1975, James Fletcher, the NASA Administrator, received a letter from Senators Barry Goldwater and Frank Moss.[25] The letter suggested a massive technology project involving NASA and industry to help ease the burden on the airlines caused by the energy crisis. Its goal was the realization of a new generation of fuel-efficient aircraft. Goldwater and Moss asked NASA to propose a plan, develop the technology, and facilitate the "technology transfer process" to industry.[26] Technology transfer later became a particularly thorny issue in the debate over whether the government should carry development to the point of costly flight testing, or leave that phase to the manufacturers who stood to benefit handsomely from this government-generated technology.

In February 1975, NASA formed the Intercenter Aircraft Fuel Conservation Technology Task Force to explore all potential options.[27] Sixteen government scientists and engineers from NASA, the Department of Transportation, the Federal Aviation Administration, and the Department of Defense took part in the seven-month study.[28] James Kramer, the task force leader, called for any new ideas that would satisfy government criteria, even those that might be considered "unusual." The task force defined six major areas with the potential for significant impact on aircraft fuel efficiency. It recommended the creation within NASA of the Aircraft Energy Efficiency (ACEE) Program, the administrative umbrella for six new aeronautics projects—three related to the airframe and three to the propulsion system.[29]

NASA assigned management of the three propulsion projects to the NASA Lewis Research Center. The first of these propulsion projects focused on improving existing turbofan engines through the redesign of selected engine components. It was the least technically challenging of the three projects and aimed for a five percent increase in fuel efficiency within a few years. The second project, the Energy Efficient Engine (E^3), involved building "a brand new engine from scratch" and offered a far greater payoff—an increase in fuel efficiency of ten to fifteen percent. In essence, NASA proposed to assume the risk for developing an "all new technology in an all up engine."[30] With a new "recoupment program" in place, the government expected to get back some of its investment out of the profits of the engine manufacturers, General Electric and Pratt & Whitney.

---

23. *Ibid.*, p. 5.
24. Senator Goldwater's response to Kahn, *ibid.*, p. 8.
25. It was likely that the NASA staff drafted the letter.
26. Barry Goldwater and Frank Moss to James C. Fletcher, as found in, Aircraft Fuel Conservation Technology, task force report, September 10, 1975, pp. 138–39.
27. Roy D. Hager and Deborah Vrabel, *Advanced Turboprop Project* (Washington, DC: NASA, 1988), p. 4.
28. Aircraft Fuel Conservation Technology, task force report, September 10, 1975, pp. 1 and 2.
29. The Aircraft Energy Efficiency (ACEE) Program airframe projects included: the Fuel Conservative Transport to improve on aerodynamic design and potentially save fifteen to twenty percent in fuel use; the Composite Primary Structures which would decrease the weight of aircraft through the use of composite materials and save 10 to 15 percent in fuel costs compared to an all metal airplane; and Laminar Flow Control to allow an aircraft to maintain low drag, thus creating a potential fuel savings of twenty to forty percent.
30. Nored interview.

In contrast to these two relatively conservative projects, the advanced turboprop offered dramatic increases in fuel efficiency. NASA planners believed that an advanced turboprop could reduce fuel consumption by twenty to thirty percent over existing turbofan engines with comparable performance and passenger comfort at speeds up to Mach 0.8 and altitudes up to 30,000 feet. (It should be noted that commuter turboprop-powered aircraft in current use fly at far slower speeds and lower altitudes.) The ambitious goals of the turboprop project made it controversial and challenging both from a technical and social point of view. Technically, studies by Boeing, McDonnell Douglas, and Lockheed pointed to four areas of concern: propeller efficiency at cruise speeds, both internal and external noise problems, installation aerodynamics, and maintenance costs.[31] Socially, the turboprop also presented daunting problems. Because of the "perception of turboprops as an old-fashioned, troublesome device with no passenger appeal," the task force report noted, "the airlines and the manufacturers have little motivation to work on this engine type."[32] Clifton Von Kann succinctly summed up these concerns to Barry Goldwater during his Senate testimony when he said that of all the proposed projects, "the propeller is the real controversial one."[33]

What made the government willing to assume the risk for such a difficult project? Proposed fuel savings was one important factor. However, the task force report indicated another significant and related issue—the Soviet Union had a high speed "turboprop which could fly from Moscow to Havana."[34] The continuing Cold War prompted the United States to view any Soviet technical breakthrough as a potential threat to American security. During the energy crisis, the knowledge that Soviet turboprop transports had already achieved high propeller fuel efficiency at speeds approaching those of jet-powered planes seemed grave indeed and gave impetus to the NASA program. During the government hearings, NASA representatives displayed several photos of Russian turboprop planes to win congressional backing for the project.[35] The Cold War helped to define the turboprop debate. No extensive speculation on the implications of Russian air superiority for American national security seemed necessary. The Soviet Union could not be allowed to maintain technical superiority in an area as vital as aircraft fuel efficiency. Thus, the report included the demanding Advanced Turboprop Project as part of the ten-year, $670 million Aircraft Energy Efficiency Program to improve fuel efficiency.

Concept Development and Early Advocacy

Industry resistance and NASA Headquarters' sensitivity to the public relations aspect of this opposition were among the key reasons that of the six projects within the Aircraft Energy Efficiency (ACEE) program, only the advanced turboprop failed to receive funding in 1976. John Klineberg, later director of Lewis Research Center, recalled that it was delayed "because it was considered too high risk and too revolutionary to be accepted by the airlines."[36]

---

31. Hager and Vrabel, *Advanced Turboprop Project*, p. 5.
32. Aircraft Fuel Conservation Technology, task force report, September 10, 1975, p. 44.
33. Clifton F. Von Kann, testimony before the U.S. Senate, Committee on Aeronautical and Space Sciences, September 10, 1975, p. 9.
34. Mikkelson interview.
35. Aircraft Fuel Conservation Technology, task force report, September 10, 1975, p. 48. These Soviet long-range turboprops included the Tupolev TU-95 "Bear" (which weighed 340,000 pounds, had a maximum range of 7,800 miles, a propeller diameter of 18.4 feet, and operated at a .75 mach cruise speed) and the Antonov AN-22 "Cock" (which weighed 550,000 pounds, had a maximum range of 6,800 miles, a propeller diameter of 20.3 feet, and operated at a .69 mach cruise speed).
36. John Klineberg, quoted in "How the ATP Project Originated," *Lewis News*, July 22, 1988.

If the advanced turboprop was so important to the national welfare, why did it encounter such opposition from the airframe and aircraft engine manufacturers? Donald Nored, the division chief in charge of the three propulsion projects at Lewis, remarked that his engineering peers in industry were "very conservative and they had to be." They were "against propellers" because they had "completely switched over to jets." Because of their commitment to the turbojet, they continually cited problems that they believed resulted from propellers. This included noise, maintenance, and the fear that the "blades would come apart." Nored recalled each problem had to be "taken up one at a time and dealt with."[37] It appears the government's revolutionary vision of the future frightened the aircraft industry with its large investment in turbofan technology. Aircraft structures and engines are improved in slow, conservative, incremental steps. To change the propulsion system of the nation's entire commercial fleet represented an investment of mind-boggling proportions. Even if the government put several hundred million dollars into developing an advanced turboprop, the airframe and aircraft engine industries would still need to invest several billion dollars to commercialize it. Revolutionary change did not come easily to an established industry so vital to the nation's economy.

Turboprop advocates encountered not only the opposition of industry representatives, but the hesitation and timidity of NASA Headquarters. By default, the advocacy role fell to NASA Lewis engineers, though the public relations aspect of technology funding had never been the Cleveland laboratory's strong suit. Lewis had a reputation for being more conservative and technical than the other NASA Centers.[38] One Lewis engineer remarked that when other Centers sent five representatives to important meetings, Lewis sent one. Moreover, research engineers from the aeronautics side of NASA had little experience managing major contracts. Yet the energy crisis and the need for projects to sustain the Center's viability within NASA galvanized a small cadre of Lewis engineers into action. They used their technical and new-found managerial creativity to sell NASA Headquarters and industry on a revolutionary new propulsion system—one that might forever ground all existing subsonic turbojets.

Technically, the entire future of the advanced turboprop project initially depended on proving whether a model propfan could achieve the predicted fuel efficiency rates.[39] If this model yielded successful results, then project advocates would be able to lobby for increased funding for a large research and development program. Thus, even during its earliest phase, the technical and social aspects of the project worked in tandem.

Lewis project managers awarded a small group of researchers at Lewis and Hamilton Standard a contract for the development of a two-foot diameter model propfan, called the SR-1 or single-rotating propfan. Single-rotating meant that the propfan had only one row of blades, as opposed to a counter-rotating design with two rows of blades, each moving in opposite directions. This model achieved high efficiency rates and provided technical data that the small group of engineers could use as ammunition in the fight to continue the program.

At the same time that they proved the technology using small-scale models, Lewis engineers built a consensus for the project, defending it against objections of skeptical segments of industry and government advisory committees. Advocacy is essentially "marketing" or "selling" to gain government funding and industry backing for new programs like the advanced turboprop. Funding government programs is neither scientific nor entirely rational, but depends on people and how they navigate a complex bureaucracy,

---

37. Nored interview.
38. Ibid.
39. Aircraft Fuel Conservation Technology, task force report, September 10, 1975, p. 46.

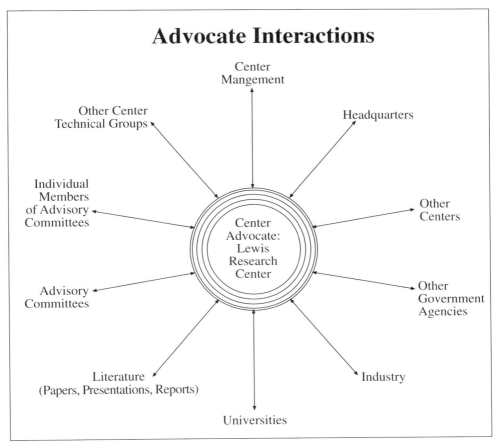

while avoiding numerous political entanglements. During the Apollo years, NASA had what amounted to a blank check to land a human being on the Moon within a decade. Not needing to spend time and energy fighting for funding, engineers had greater freedom to focus on building and testing hardware and managing space missions. But to keep the programs of the 1970s alive, even those that responded to a national crisis, required effort in non-technological spheres of activity.

Lewis was fortunate that Donald Nored, a maestro of project management, played a strong role in building a constituency in support of the project. Unlike most of the other Lewis engineers involved in advanced turboprop development, he hailed from the space side of NASA's house. He had worked on chemical rockets and high power lasers prior to taking up his post as head of the Aircraft Energy Efficiency Program Office at Lewis in 1975. He helped to show aeronautical engineers, more at ease with in-house research, how to negotiate the system to win funding. In 1981, with Frank Berkopec, Nored attempted to demystify the advocacy process by laying down guidelines for others within the Aeronautics Directorate. They disabused their order-seeking engineering colleagues of the notion that advocacy could be compressed into a series of well-defined steps. Rather, they wrote, it is

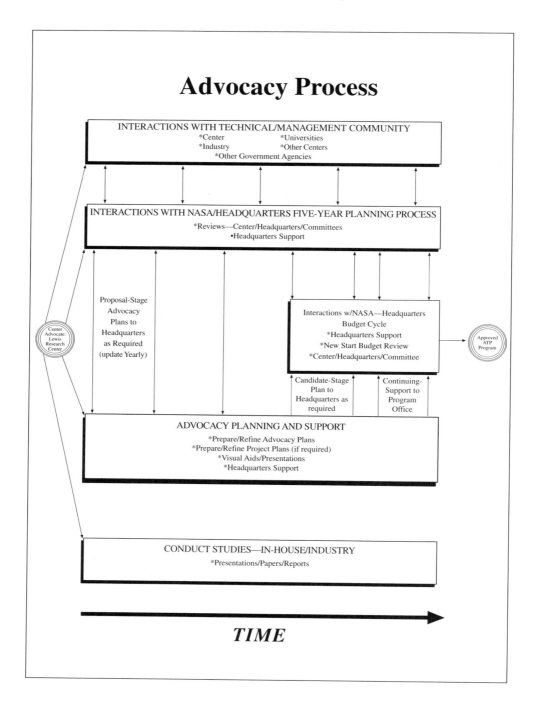

"basically informal, unstructured, and quite often confusing."[40] Since only a few of the proposed NASA programs received funding each year, they argued, the advocacy process had become essential and activities related to it should receive a "high priority."[41]

The advocacy guidelines indicated that the interactions with "industry, advisory groups, and especially Headquarters will often require rapid, comprehensive, and in-depth respondents [sic] to requests."[42] One early request of the turboprop project centered on the aircraft industry's concern over the safety of propellers. An aircraft accident advisor raised a question during a meeting of the Industrial Advisory Board at NASA Headquarters concerning the "safety aspect of propellers breaking away from the engine and the damage caused by their impingement into the fuselage."[43] Lewis engineers quickly launched their own study into propeller safety and commissioned similar studies at Hamilton Standard and Detroit Diesel Allison. The results were overwhelmingly positive. Lewis examined over 12,000 accident reports from 1973 to 1975 and found no instance where a propeller blade broke away from its engine.[44] Hamilton Standard reported that after fifty million hours of propeller flight time there had never been an instance of structural failure.[45] While after twenty million hours, Detroit Diesel Allison found one structural failure; they were quick to point out that "the aircraft landed routinely without further incident and no one was injured in the aircraft or on the ground."[46] This example typifies not only the early skepticism and resistance by industry to the idea of returning to propeller aircraft but also the "rapid, comprehensive, and in-depth responses" of NASA to industry's concerns. The advocacy process required to "market" and "sell" the radical turboprop project was in full swing. It continued to effectively diffuse the concerns of skeptics.

Enabling Technology and Project Management

Successful advocacy brought the formal establishment of the Advanced Turboprop Project in 1978 and initiation of the enabling technology phase. As the lead Center for the project, NASA Lewis had full responsibility for the management of its increasingly far-flung and complicated pieces. Before this phase began, NASA engineers devised a detailed "management approach" and the plan was approved in 1977. Officially, Lewis was to have "responsibility to execute all detailed project planning documentation, develop and implement the procurement of components and systems, provide technical direction to contractors, perform contract administration, perform engineering functions, coordinate the related in-house research and technology programs, and exercise the usual project review reporting and control functions."[47] These interrelated activities put Lewis in the middle of an intricate web of government (other NASA Centers), industry, and academic contracts. Project managers were responsible for assigning the technology contracts. They also had the equally important function of ensuring that both the public and the government viewed the ATP positively.

---

40. Donald L. Nored and Frank D. Berkopec, "Guidelines for Advocacy of the New Programs in the Aeronautics Directorate," unpublished report, January 1981, Nored papers, NASA-Lewis Research Center, file Nored/Berkopec, box 238, p. 1.
41. Ibid., p. 10.
42. Ibid.
43. J. E. Wikete, Aircraft Accident Information, August 4, 1976, Nored papers, NASA, box 224.
44. Paul J. McKenna (Lewis Research Center) to Wikete, July 12, 1976, Nored papers, NASA, box 224.
45. R.M. Levintan (Hamilton Standard) to Wikete, July 27, 1976, Nored papers, NASA, box 224.
46. P.C. Stolp (Detroit Diesel Allison) to P. Christman, July 14, 1976, Nored papers, NASA, box 224.
47. Project Plan for the Advanced Turboprop Program, September 1977, NASA, Nored papers, box 229, p. 26.

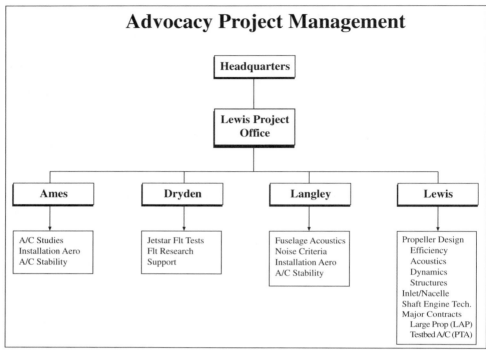

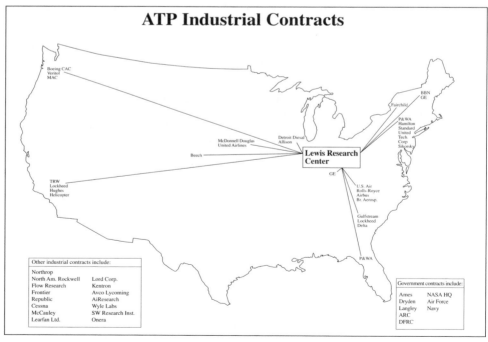

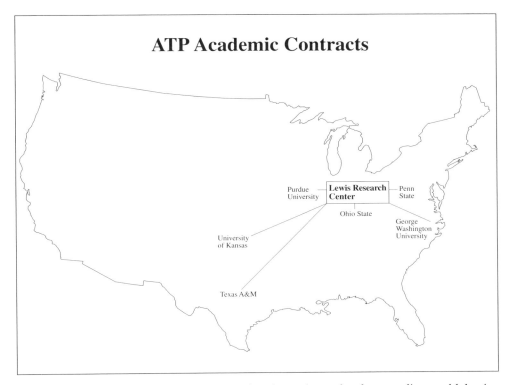

Once the management structure was in place, the technology studies could begin. Technically, this phase dealt with four critical problems: modification of propeller aerodynamics, cabin and community noise, installation aerodynamics, and drive systems.[48] Propeller aerodynamic work included extensive investigations of blade sweep, twist, and thickness. The late 1970s was the first time that engineers used a high speed computer to analyze the design of a propeller. Computers were not yet in widespread use when the turbofan replaced propeller-powered planes in the 1950s. Lewis programmers used their Cray supercomputers to develop the first three-dimensional propeller aerodynamic analysis. A further structural and aerodynamic achievement was to use thinner titanium blades to reduce the flutter problems associated with the steel propeller blades used in the 1940s and 1950s.

The advantage of propellers to save fuel had to be balanced against the potential harm to the environment their noise caused.[49] New computer-generated design codes not only contributed to improved propeller efficiency, but contributed to solving problems associated with noise. Engineers closely monitored the effect of propeller noise on both cabin occupants and people on the ground. To study propeller acoustics, they mounted propeller models on a JetStar aircraft fuselage at the NASA Dryden facility. Microphones located on the airframe and also on a Learjet chase plane provided data at close range and at a distance. After reviewing the sound pattern data, they concluded that substantial

---

48. Donald L. Nored, John B. Whitlow, Jr., William C. Strack, "Status Update of the NASA Advanced Turboprop Project," unpublished report, Nored papers, pp. 4–10.

49. Aircraft Fuel Conservation Technology, task force report, September 10, 1975, p. 18.

noise reduction technology was necessary to meet the established goals. Eventually, they achieved a reduction of sixty to sixty-five decibels of noise through a combination of structural advances and flight path modifications.

The final two technical problems of the enabling phase dealt with installation aerodynamics and the drive system. Numerous installation arrangements were possible for mounting the turboprop on the wing. Should the propeller operate by "pushing" or "pulling" the aircraft? How should the propeller, nacelle, and the wing be most effectively integrated to reduce drag and increase fuel efficiency? Wind tunnel tests were able to reduce drag significantly by determining the most advantageous wing placement for the propeller. Engineers also examined various drive train problems, including the gearboxes.

Solutions to all the enabling phase technical problems was still not enough to guarantee the continued funding of the program. Key social questions were still associated with this controversial technology. A vital concern for the advanced turboprop project managers was the social question concerning passengers: how receptive would they be to propeller-driven aircraft? In 1975, a government panel reported that they were "generally opposed to the turboprop aircraft, primarily because they felt that there would be little or no public acceptance."[50] If the public would not fly in a turboprop plane, all the potential fuel savings would be lost flying empty planes across the country.

In response to this concern, NASA and United Airlines initiated an in-flight questionnaire to determine customer reaction to propellers. Both NASA and industry were aware of the disastrous consequences for the future of the program if this study found that the public was against the return of propeller planes. As a result, the questionnaire de-emphasized the propeller as old technology and emphasized the turboprop as the continuation and advancement of flight technology. The first page of the survey consisted of a letter from the United Airlines vice president of marketing to the passenger asking for cooperation in a "joint industry-government study concerning the application of new technology to future aircraft."[51] This opening letter did not mention the new turboprops. The turboprop, inconspicuously renamed the "prop-fan" to give it a more positive connotation, did not make its well-disguised appearance until page four of the survey where the passenger is finally told that "'prop-fan' planes could fly as high, as safely, and almost as fast and smooth as jet aircraft." This was a conscious rhetorical shift from the term "propeller" to "prop-fan" to disassociate it in peoples' minds from the old piston engine technology of the pre-jet propulsion era. Brian Rowe, a General Electric vice president with oversight of the advanced propeller projects, explained this new labeling strategy. He said, "They're not propellers. They're fans. People felt that modern was fans, and old technology was propellers. So now we've got this modern propeller which we want to call a fan."[52] The questionnaire explained to the passenger that not only did the "'prop-fans' . . . look more like fan blades than propellers," they would also use twenty to thirty percent less fuel than jet aircraft.

The questionnaire then displayed three sketches of planes-two were propeller driven and the third was a turbofan. The passenger had to choose which one he or she would "prefer to travel in." Despite all the planes being in-flight, the sketches depicted the propellers as simple circles (no blades present), while the individual blades of the turbofan were visible. These were all subtle and effective hints to the passenger that the "prop-fan" was nothing new and that they were already flying in planes powered by engines with fan blades.

---

50. George M. Low to Alan Lovelace, April 28, 1975.
51. United Airlines Passenger Survey, NASA, Nored papers, box 224.
52. Quoted by Martha Hamilton, "Firms Give Propellers a New Spin: GE leads high-stakes competition for aircraft engineers with its 'fan,'" *The Washington Post*, February 8, 1987, p. H4, column 1.

Not surprisingly, the survey yielded favorable results for the turboprop. Of 4,069 passengers surveyed, fifty percent said that they "would fly prop-fan," thirty-eight percent had "no preference," and only twelve percent preferred a jet.[53] If the airlines could avoid fare increases due to the implementation of the turboprop, eighty-seven percent of the respondents stated they would prefer to fly in the new turboprop. Relieved and buoyed by the results, NASA engineers liked to point out that most of the passengers did not even know what propulsion system was currently on the wing of their aircraft.[54] According to Mikkelson, all the passengers wanted to know was "how much were the drinks, and how much was the ticket."[55] Equally relieved was Robert Collins, vice president of engineering for United Airlines, who concluded that this "*carefully constructed* passenger survey . . . indicated that a prop-fan with equivalent passenger comfort levels would not be negatively viewed, especially if it were recognized for its efficiency in reducing fuel consumption and holding fares down."[56]

At times project management also involved informing and changing government opinion. Aeronautics programs within NASA, because of the low levels at which they were traditionally funded, had never required close oversight by the General Accounting Office (GAO). The large budget and greater visibility of the Aircraft Energy Efficiency Program (ACEE) suddenly brought it unwanted attention. The first draft of the General Accounting Office's 1979 review, though generally favorable toward the ACEE program, was highly critical of the advanced turboprop project. It concluded with the statement that the "GAO believes that much of the fuel savings under ACEE attributed to the turboprop will not be realized."[57]

The draft's "negative tone" and "misleading and distorted view of the program" deeply concerned NASA Lewis project managers who feared the repercussions it would have on funding decisions.[58] They quickly went on the attack. Center Director John Klineberg heatedly responded that the GAO had treated the turboprop project unfairly in comparison with the other aircraft efficiency projects, calling the GAO ignorant of the project's "inherent uncertainties."[59]

NASA Lewis project managers prevailed in the battle against the negativity of the GAO draft report. The final publication specifically contained a retraction. The "GAO carefully reevaluated its presentation and made appropriate adjustments where it might be construed that the tone was unnecessarily negative or the data misleading." An example of these "appropriate adjustments" is apparent in a comparison of how one sentence changed from the draft to the final version. In the draft, the sentence appeared as: "The Task Force Report shows that in 1975 there was considerable disagreement on the ultimate likelihood of a turboprop engine being used on commercial airliners."[60] In the final publication, the GAO amended the same sentence to: "The possible use of turboprop

---

53. Prop-Fan, survey results, December 1978, NASA, Nored papers, box 231.
54. Interview with Keith Sievers, August 17, 1995, and telephone interview with Raymond Colladay, August 17, 1995, by Virginia Dawson and Mark Bowles.
55. Mikkelson interview.
56. Authors' italics. Robert C. Collins statement submitted to subcommittee on transportation, aviation, and materials, House of Representatives Committee on Science and Technology, February 26, 1981.
57. Preliminary draft of a proposed report, review of NASA's Aircraft Energy Efficiency Project, GAO office, August 1979, Nored papers, box 182, p. 36.
58. Unknown NASA Headquarters administrator to J. H. Stolarow, January 24, 1980, NASA, Nored papers, box 182, file GAO report.
59. John M. Klineberg to NASA Headquarters, December 21, 1979, NASA-Lewis Research Center, Nored papers, box 182, file GAO report.
60. Preliminary draft of a proposed report, review of NASA's Aircraft Energy Efficiency Project, GAO office, August 1979, Nored papers, box 182, p. 37.

engines on 1995 commercial aircraft is still uncertain, but has gained support since 1975."[61] These editorial changes giving the report a positive spin indicate the effectiveness of project managers in changing public opinion. Everyone, it seemed, had begun to associate the advanced turboprop technology with the possibility of bringing about an aeronautical "revolution," a paradigm shift, or as Forbes magazine headlined in 1984, "The Next Step." As surely as "jets drove propellers from the skies," the new "radical designs" could bring a new propeller age to the world.[62]

It is important to underscore how important the interpersonal skills of the project managers were to continuation of the program throughout this enabling technology phase. They were responsible not only for managing the project's technology, but also for enabling, proving, maintaining, and adjusting support for the turboprop. They continued to push this controversial technology against the conservative interests of the government, industry, and the public. Their consistent success paved the way for the third stage.

Large-Scale Integration and Competition

After two years of work, the advanced turboprop idea began to attract greater commercial interest. As a result of NASA's advocacy efforts, news articles began to predict the coming propeller "revolution." All indicators pointed to the introduction of the new turboprops on commercial aircraft by the 1990s. With the small-scale model testing complete, a data base, and an acceptable design methodology established, the project moved into its most labor and cost intensive phase—that of large-scale integration. The project still had serious uncertainties and problems associated with transferring the designs from a small-scale model to a large-scale prop-fan. Could engineers maintain propulsion efficiency, low noise levels, and structural integrity with an increase in size? The Large-Scale Advanced Prop-fan (LAP) project initiated in 1980 would answer these scalability questions and provide a database for the development and production of full-size turbofans.

As a first step, NASA had to establish the structural integrity of the advanced turboprop.[63] Project managers initially believed that in the development hierarchy performance came first, then noise, and finally structure. As the project advanced, it became clear that structural integrity was the key technical problem.[64] Without the correct blade structure, performance could never achieve predicted fuel savings. NASA awarded Hamilton Standard the contract for the structural blade studies that were so crucial to the success of the whole program. In 1981, they began to design a large-scale, single-rotating prop-fan made of composite material. Five years later they completed construction on a 9-foot-diameter design very close to the size of a commercial model. The model was so large that no wind tunnel in the United States could accommodate it. The turboprop managers decided to risk the possibility that the European aviation community might benefit from the technology that NASA had so arduously perfected. They shipped the large-scale propeller, called the SR-7L, to a wind tunnel in Modane, France, for testing. In early 1986, researchers subjected the model to speeds up to Mach 0.8 with simulated altitudes of 12,000 feet. The results confirmed the data obtained from the small model propeller designs. The large-scale model was a success.

---

61. "A look at NASA's Aircraft Energy Efficiency Program," by the Comptroller General of the United States, July 28, 1980, Nored papers, box 182, file GAO report, p. 45.
62. Howard Banks, "The Next Step," *Forbes*, May 7, 1984, p. 31.
63. "Large-Scale Advanced Prop-Fan Program (LAP)," technical proposal by Lewis Research Center, January 11, 1982, NASA, Nored papers, box 229.
64. Nored interview.

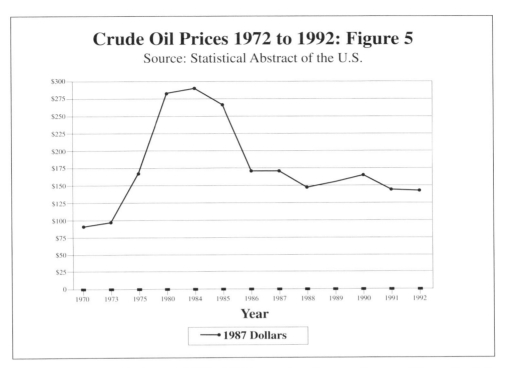

Crude Oil Prices 1972 to 1992: Figure 5
Source: Statistical Abstract of the U.S.

Success spawns imitators. While NASA continued to work with Allison, Pratt & Whitney—and Hamilton Standard to develop its advanced turboprop, General Electric (GE)—Pratt & Whitney's main competitor—was quietly developing an alternative propeller system. A feature of radical inventions is that competitors often introduce alternative forms of a similar technology before one form can prevail over another. Historians of technology have shown many cases of "interpretative flexibility" when "two or even more social groups with clearly developed technological frames [artifacts] are striving for dominance in the field."[65] This happened when General Electric introduced its own radical alternative to NASA's advanced turboprop project-the Unducted Fan (UDF). GE sprang the unducted fan on NASA completely by surprise.

In NASA's design, the propeller rotated in one direction. This was called a single rotation tractor system and included a relatively complicated gearbox. Since one of the criticisms against the turboprop planes of the 1950s (the Electra, for example), was that their gearboxes required heavy maintenance, GE took a different approach to prop-fan design. Beginning in 1982, GE engineers spent five years developing a gearless, counter-rotating, pusher system. They mounted two propellers (or fans) on the rear of the plane that literally pushed it in flight, as opposed to the "pulling" of conventional propellers. In 1983, the aircraft engine division of General Electric released the unducted fan design to NASA shortly before flight tests of the NASA industry design were scheduled. Suddenly there were two turboprop projects competing for the same funds. Nored recalled: "They wanted us to

---

65. Wiebe E. Bijker, "The Social Construction of Bakelite: Toward a Theory of Invention," *The Social Construction of Technological Systems*, 182.

drop everything and give them all our money and we couldn't do that."⁶⁶ NASA Headquarters endorsed the "novel" unducted fan proposal and told NASA Lewis to cooperate with General Electric on the unducted fan development and testing.

Despite NASA's initial reluctance to support two projects, the unducted fan proved highly successful. In 1985, ground tests demonstrated a fuel conservation rate of twenty percent.⁶⁷ Development of the unducted fan leapt ahead of NASA's original geared design. One year later, on August 20, 1986, GE installed its unducted fan on the right wing of a Boeing 727. Thus, to many NASA engineers' dismay, the first flight of an advanced turboprop system demonstrated the technical feasibility of the unducted fan system—a proprietary engine belonging entirely to General Electric, rather than the product of the joint NASA/industry team. Nevertheless, the competition between the two systems, and the willingness of private industry to invest its own development funds, helped build even greater momentum for acceptance of the turboprop concept.

NASA engineers continued to perfect their single-rotating turboprop system through preliminary stationary flight testing.⁶⁸ The first step was to take the Hamilton Standard SR-7A prop-fan and combine it with the Allison turboshaft engine and gearbox housed within a special tilt nacelle. NASA engineers conducted a static or stationary test at Rohr's Brown Field at Chula Vista, California, where they mounted the nacelle, gearbox, engine, and propeller on a small tower. The stationary test met all performance objectives after fifty hours of testing in May and June 1986. This success cleared the way for an actual flight test of the turboprop system. In July 1986 engineers dismantled the static assembly and shipped the parts to Savannah, Georgia, for reassembly on a modified Gulfstream II with an eight-blade, single-rotation, turboprop on its left wing.⁶⁹ The radical dreams of the NASA engineers for fuel efficient propellers were finally close to becoming reality. The plane contained over 600 sensors to monitor everything from acoustics to vibration. Flight testing-the final stage of advanced turboprop development-took place in 1987 when a modified Gulfstream II took flight in the Georgia skies. These flight tests proved the predictions of a twenty to thirty percent fuel savings (made by NASA in the early 1970s) were indeed correct.

On the heels of the successful tests, of both the GE and the NASA-industry team designs, came not only increasing support for propeller systems themselves, but also high visibility from media reports forecasting the next propulsion revolution. The *New York Times* predicted the "Return of the propellers" while a *Washington Times* headline read, "Turboprops are back!"⁷⁰ Further testing indicated that this propulsion technology was ready for commercial development. As late as 1989, the U.S. aviation industry was "considering the development of several new engines and aircraft that may incorporate advanced turboprop propulsion systems."⁷¹ But the economic realities of 1987 were far different from those predicted in the early 1970s. Though all the technology and social problems standing in the way of commercialization were resolved, the advanced turboprop never reached production, a casualty of the one contingency that NASA engineers never anticipated—that fuel prices would go down. (See figure 5) Once the energy crisis passed, the need for the advanced turboprop vanished.

---

66. Nored interview.
67. James, J. Haggerty, "Propfan Update," p. 11.
68. Hagar and Vrabel, *Advanced Turboprop Project*, pp. 49–74. This stage was called the Propfan Test Assessment (PTA) Project.
69. Mary Sandy and Linda S. Ellis, "NASA Final Propfan Program Flight Tests Conducted," *NASA News* May 1, 1989.
70. Andrew Pollack, "The Return of Propellers," *The New York Times*, October 10, 1985, D2. Hugh Vickery, "Turboprops are back!," *The Washington Times*, November 1, 1984, p. 5B.
71. Mary Sandy and Linda S. Ellis, "NASA Final Propfan Program Flight Tests Conducted," *NASA News* May 1, 1989.

Environmental Contingency and Insufficient Momentum

One of the main difficulties in the development of a radical new technology is the potential project threatening problems that arise. If they are left unsolved they can destroy an entire project. Historian of technology Thomas Hughes called these problems "reverse salients." Hughes argues that all large technological systems (of which the turboprop is an example) include political, economic, social, and technological components.[72] These system components are interrelated so that if one of the components is changed or altered in any way, the rest of the system will also be affected. The systems themselves grow and gain momentum by the process of removing "reverse salients," which arise and could potentially cause the system to fail. An example will help clarify the importance of solving these critical problems. In 1878, Thomas Edison encountered a technological reverse salient in his attempt to develop his electric-lighting system. This problem was the short-lived filament of the incandescent bulb. Edison realized that even if he solved this problem, a further economic reverse salient remained. The expense of the copper wire needed to link the entire system together was cost prohibitive for potential wide-scale acceptance. If Edison could not reduce the amount of copper needed for his electric system, then gas-lighting systems would become the more attractive alternative to the problem of street lighting. What is important to understand is that either the technological or the economic reverse salient could have caused the Edison system of electric-lighting to fail.[73]

Like Edison, the managers of the turboprop project also confronted a variety of critical problems. These problems included economic (the necessity of maintaining a favorable ratio of cost to implement turboprop technology versus savings in fuel efficiency), political (how to receive funding for a long-term project), social (how to implement a technology which the public could perceive as a "step backward"), institutional (how to successfully manage the government, industry, and academic relations), and technical (how to actually build a turboprop that improved fuel efficiency by twenty to thirty percent). Each of these problems had the potential to sabotage the entire system. NASA engineers had their own, more practical and direct term for "reverse salient"—a "showstopper." In 1984, engineers listed a number of technical show-stoppers that threatened to derail the project if left unsolved—for example, unacceptable levels of cabin noise.[74]

As system-builders solve critical problems, the system itself generates momentum. This momentum continues to increase and build until, according to Hughes, either a conversion, a catastrophe, or a contingency occurs. Conversions and catastrophes break momentum through either a change in societal belief, like a religious conversion, or a massive technological failure, like a nuclear-reactor catastrophe. But, it is the role of contingency which interests us here as the key factor in the current neglect of the advanced turboprop technology. Hughes identified one particular "contingent environmental change" that altered the course of the entire automobile industry—the energy crisis. He argues, "The oil embargo of 1973 and the subsequent rise in gasoline prices ultimately compelled U.S. automobile manufacturers to change substantially an automobile design that had been singularly appropriate to a low-cost-energy environment."[75]

---

72. Thomas P. Hughes, *Networks of Power: Electrification in Western Society, 1880-1930* (Baltimore, MD: The Johns Hopkins University Press, 1983).
73. Hughes, *American Genesis: A Century of Invention and Technological Enthusiasm* (London, England: Penguin Books, 1989), pp. 71–74.
74. Chart entitled, "Potential show-stoppers," February 6, 1984, NASA-Lewis Research Center, Nored papers, box 239, file ATP memos.
75. *Ibid.*, p. 462.

The development and subsequent neglect of advanced turboprop technology is the result of this same environmental contingency. In the early 1970s, the energy crisis created a situation which made it a national necessity for the government to explore new ways to conserve fuel. What the managers of the Advanced Turboprop Project (ATP) did not anticipate and could not control was a decrease in the cost of fuel. As the energy crisis subsided in the 1980s and the fuel prices decreased, there was no longer a favorable ratio of cost to implement turboprop technology versus savings in fuel efficiency. As John R. Facey, advanced turboprop program manager at NASA Headquarters, wrote, "An all new aircraft with advanced avionics, structures, and aerodynamics along with high-speed turboprops would be much more expensive than current turbofan-powered aircraft, and fuel savings would not be enough to offset the higher initial cost."[76] In the case of the ATP, its managers overcame all of their critical problems. However, when contingent economic conditions changed so that fuel cost was no longer a critical problem, regardless of the technical success of the project, the advanced turboprop lost its potential market in the industrial world.

Yet Keith Sievers, at that time the manager of the ATP, along with a handful of project staff, was convinced that the NASA industry team had made a significant contribution to aviation that ought to receive recognition. To win the Collier Trophy, he again summoned up the advocacy skills that had proved so valuable in bringing the controversial advanced turboprop to the point of technical feasibility. He used them to lobby for the prestigious Collier Trophy among the wide aeronautical constituency that had participated in advanced turboprop development. NASA Headquarters initially expressed some reluctance to lobby for awarding a prize for technology that was unlikely to be used—at least in the near future. But the timing was perfect. There was little competition from NASA's space endeavors since staff in the space directorate were still in the midst of recovering from the tragic *Challenger* explosion. As a result, the National Aeronautic Association awarded NASA Lewis and the NASA Industry Advanced Turboprop Team the Collier Trophy at ceremonies in Washington, DC. Today, the technology remains "on the shelf," or "archived," awaiting the time when fuel conservation again becomes a necessity.[77]

Despite the current neglect of the advanced turboprop, this case study demonstrates how radical innovation can emerge from within a conservative, bureaucratic government agency. The government—not industry—assumed the risk for developing the new technology. It used taxpayers' money to advance a radical idea to the point of technical feasibility. Engineers involved in the project used advocacy to build a consensus among the members of the aeronautical community that the advanced turboprop would prove a viable alternative to the far less energy efficient turbofan technology. Indeed, the technical and social achievements of the project were convincing enough to drive General Electric to invest its own funds to develop a competing design. This competition was evidence of wide acceptance for the turboprop concept.

The Collier Trophy in 1987 was presented to the "Lewis and the NASA Industry Advanced Turboprop *Team*." The team, defined in its widest possible context, included General Electric's independent contribution of the UDF and its subsequent flight testing by NASA. In contrast to previous Collier trophies in aeronautics won by the NACA, no individual received special mention. Certainly, throughout the eleven years of its existence the project had encouraged inventiveness of individuals in a variety of disciplines, from highly theoretical contributions in blade design and acoustics to more routine testing. Participants

---

76. John R. Facey, "Return of the Turboprops," *Aerospace America* (October 1988): 15.
77. Some specific technologies generated by the ATP project are in use today. These include noise reduction advances, gearboxes that use the ATP design, and certain structural advancements, for example, how to keep the blades stable, used in engine designs today, including large, high by-pass ratio turbofans of the 1990s. Sievers interview.

in the project ran the gamut from government, university, and industry researchers. But what the prize recognized above all was the project's management genius. NASA Lewis managers did not simply manage contracts. They kept the project alive. They used advocacy to win industry participation and cooperation, as well as stimulate competition. They pushed both the technical and the social aspects of the project to create the system's momentum. Yet once the energy crisis passed, this momentum was insufficient to dislodge the massive technological momentum of the existing turbofan system.

NASA engineers involved in the ATP project still remain confident that the future economic conditions will make the turboprop attractive again. When fuel becomes scarce and fuel prices begin to rise, the turboprop's designs will be "on the shelf" ready to respond with tremendous fuel-efficient savings. But, technological neglect is not the enthusiastic success on which NASA engineers built their careers. Donald Nored wistfully reflected on the project and said, "We almost made it. Almost made it."[78]

---

78. Nored interview.

Chapter 15

Return to Flight: Richard H. Truly and the Recovery from the *Challenger* Accident[1]

by John M. Logsdon

Seventy-three seconds after its 11:37 a.m. liftoff on September 29, 1988, those watching the launch of the Space Shuttle *Discovery* and its five-man crew breathed a collective sigh of relief. *Discovery* had passed the point in its mission at which, on January 28, 1986, thirty-two months earlier, *Challenger* had exploded, killing its seven-person crew and bringing the U.S. civilian space program to an abrupt halt.[2] After almost three years without a launch of the Space Shuttle,[3] the United States had returned to flight.

Presiding over the return-to-flight effort for all but one of those thirty-two months was Rear Admiral Richard H. Truly, United States Navy. Truly was named Associate Administrator for Space Flight of the National Aeronautics and Space Administration (NASA) on February 20, 1986. In that position, he was responsible not only for overseeing the process of returning the Space Shuttle to flight, but also for broader policy issues such as whether the *Challenger* would be replaced by a new orbiter, what role the shuttle would play in launching future commercial and national security payloads, and what mixture of expendable and shuttle launches NASA would use to launch its own missions. He served as the link between the many entities external to NASA—the White House, Congress, external advisory panels, the aerospace industry, the media, and the general public—with conflicting interests in the shuttle's return to flight. In addition, he had the tasks of restructuring the way NASA managed the Space Shuttle program and restoring the badly shaken morale of the NASA-industry shuttle team.

The citation on the 1988 Collier Trophy presented to Admiral Richard H. Truly read: "for outstanding leadership in the direction of the recovery of the nation's manned space program." This essay recounts the managerial and technological challenges of the return-to-flight effort, with particular attention to Richard Truly's role in it. However, as Truly himself

---

    1.   This essay's findings and conclusions are the responsibility of the author, and do not necessarily reflect the views of NASA or the George Washington University. The author wishes to acknowledge with gratitude the dogged research assistance of Nathan Rich; without his efforts, the task would have been much more difficult.
    2.   This essay is not an account of the *Challenger* accident, but rather the process of recovering from that mishap. For such an account, see Malcolm McConnell, *Challenger: A Major Malfunction* (Garden City, NY: Doubleday, 1987).
    3.   The formal name for the combined Shuttle orbiter, Space Shuttle main engines, external tank, and solid rocket boosters, plus any additional Spacelab equipment mounted in the orbiter's payload bay, is the Space Transportation System (STS). In this essay, the terms *Shuttle* or *Space Shuttle* are often used as an alternate way of identifying the STS.

It's thumbs up for the shuttle as the STS-26 Discovery crew celebrate their return to Earth with Vice President George Bush. The orbiter completed a successful four-day mission with a perfect touch down on October 3, 1988, on Rogers Dry Lake Runway 17. In this picture, from left to right, are: Mission Specialist David C. Hilmers, Commander Frederick H. (Rick) Hauck, Vice President George Bush, Pilot Richard O. Covey, and Mission Specialists George D. Nelson and John M. Lounge. (NASA photo no. 88-H-497).

recognized, the recovery program was a comprehensive team effort;[4] as the first post-*Challenger* flight approached, he sent a memorandum to the "NASA Space Shuttle Team," saying:

> *As I reflect over the challenges presented to us, and our responses to them, my overriding emotion is one of pride in association. You—the men and women who compose and support this unique organization—should take great pride in having renewed the foundation for a stronger, safer American space program. I am proud to have been a part of this effort; I am proud to have witnessed your extraordinary accomplishments.*[5]

Immediate Post-Accident Events[6]

When Truly was named NASA Associate Administrator for Space Flight, he told the press that in the three weeks since the *Challenger* accident he had not "had one moment" to review information about the mishap.[7] Whether he realized it or not, Truly was entering a very chaotic situation. At the time of the accident, NASA Administrator James Beggs was on a leave of absence to deal with a Federal indictment unrelated to his NASA duties. (Beggs was later completely exonerated of any wrong doing and even received a letter of apology from the Attorney General for being mistakenly indicted.) Acting as Administrator was NASA Deputy Administrator William Graham, a physicist with close ties to conservative White House staff members but no experience in civilian space matters prior to being proposed for the NASA job. A few weeks earlier, Graham had been named Deputy Administrator, a White House political appointment, over the objections of Beggs and other senior staff at NASA; in his short time on the job he had remained largely isolated from career NASA employees. When *Challenger* exploded, NASA was thus bereft of experienced and trusted leadership.

Graham was in Washington when the accident occurred. Later in the day, he flew to the Kennedy Space Center with Vice President George Bush and Senators John Glenn and Jake Garn. The latter three flew back to Washington after consoling the families of *Challenger* crew members and meeting with the Shuttle launch team. Graham stayed behind; in a series of phone calls to the White House during the night, a decision was made to have the President appoint an external review commission to oversee the accident investigation. Although Graham had been briefed by his NASA staff on how the investigation after the 1967 *Apollo 1* fire had been handled, he apparently did not argue that the NASA Mishap Investigation Board, set up immediately after the accident, should continue to lead the inquiry.

This naming of an external review panel was in marked contrast to what had happened nineteen years earlier, on January 27, 1967. When he learned that a fire during a launch pad test had killed the three *Apollo 1* astronauts, NASA Administrator James Webb immediately notified President Lyndon Johnson, and told him that NASA was best qualified to conduct the accident investigation. Webb later that evening told his associates that

---

4. Of those who worked closely with him in the return-to-flight effort, Truly singles out for particular praise Arnold Aldrich, Richard Kohrs, and Gerald Smith. Each of them, he notes "deserve an arm or a leg of the Collier Trophy." Personal communication to the author, August 14, 1995.

5. NASA, Memorandum from M/Associate Administrator for Space Flight to NASA's Space Shuttle Team, "Return to Flight," June 10, 1988.

6. Unless otherwise cited, this narrative of the return-to-flight effort is based on accounts in the leading trade journal *Aviation Week & Space Technology* (hereafter *AW&ST*), *New York Times*, and the *Washington Post*. All three gave detailed coverage to the effort.

7. *New York Times*, February 21, 1986, p. A12.

"this is an event that we have to control. . . .We will conduct the investigation. We will get answers. There will be no holds barred. We'll issue a report that can stand up to scrutiny by anybody." Meeting with the President the next day, Webb told him "They're calling for investigations. . . . A lot of people think it's a real issue for the future, and that you ought to have a presidential commission to be clear of all influences." But, argued Webb, "NASA is the best organization [to do the investigation]."[8] Johnson concurred in Webb's approach; NASA had already selected the initial members of the accident review panel, and they set to work immediately. Certainly there were external reviews of the Apollo fire, particularly by NASA's congressional oversight committees. However, their starting point was the NASA-led investigation.

By not even attempting to retain control of the *Challenger* accident inquiry at the start, NASA found itself subject to searching external scrutiny and criticism, and the space agency had to share decision-making power during the return-to-flight effort with a variety of external advisory groups overseeing its actions. Dealing with, on one hand, the desire to get the Shuttle back into operation as quickly as possible and, on the other, the recommendations of advisory groups who gave overriding priority to safety concerns and organizational restructuring, was one of Richard Truly's greatest challenges between February 1986 and September 1988. This was particularly the case as the accident investigation quickly changed from one focused on the technical causes of the *Challenger* mishap to one broadly concerned with NASA's organization and decision-making procedures.

On February 3, President Ronald Reagan announced that the investigation would be carried out by a thirteen-person panel chaired by former Secretary of State William P. Rogers; the group quickly became known as the Rogers Commission. Reagan asked the Commission to "review the circumstances surrounding the accident, determine the probable cause or causes, recommend corrective action, and report back to me within 120 days."[9]

Within a few days after the accident, NASA investigators had pinpointed a rupture in a field joint[10] of the shuttle's right Solid Rocket Motor(SRM) as the proximate cause of the *Challenger* explosion. As the Rogers Commission began its work, there appeared to be little controversy on this issue. However, in a closed meeting at the Kennedy Space Center on February 14, Commission members were "visibly disturbed" to learn that engineers from the firm that manufactured the SRM, Morton Thiokol Inc., had the night before recommended against launching *Challenger* in the cold temperatures predicted for the next morning; that their managers, at the apparent urging of NASA officials from the Marshall Space Flight Center, had overruled their recommendation; and that more senior NASA managers responsible for the launch commit decision were unaware of this contentious interaction. This was a "turning point" in the investigation; the Commission immediately went into executive session. It decided that the NASA team working with the Commission should not include any individual who had been involved in the decision to launch *Challenger*. It decided to broaden the scope of its investigation to include NASA's management practices, Center-Headquarters relationships, and the chain of command for launch decisions-in effect, shifting the focus of the inquiry from a technical failure to NASA itself. At the end of its executive session, the Commission issued a damning statement suggesting that NASA's "decision-making process may have been flawed."[11]

---

8. Webb is quoted in W. Henry Lambright, *Powering Apollo: James E. Webb of NASA* (Baltimore, MD: Johns Hopkins University Press, 1995), pp. 144 and 146. Lambright provides an account of the Apollo fire investigation on pp. 142–88 of his book.
9. *AW&ST*, February 10, 1986, p. 24.
10. So-called because it was assembled at a NASA field center (Kennedy Space Center) rather than at the manufacturer's plant.
11. *AW&ST*, February 24, 1986, pp. 22–25, and Boyce Rensberger, "Shuttle Probe Shifted Course Early," *Washington Post*, March 17, 1986, pp. A1 and A8. After a public hearing a week later in which much the same testimony took place, William Rogers told the press that in his opinion the decision-making process definitely "was flawed."

This indictment of shuttle management provided the backdrop against which Richard Truly would work in succeeding months. As the Rogers Commission tried to fix responsibility for the "flawed" decision to launch *Challenger*, the agency was rampant with internal conflicts and finger-pointing. The *New York Times* reported on its front page that the Marshall Space Flight Center, the key organization for diagnosing and fixing the SRM problem, was "seething with resentment, hostility, depression, and exhaustion."[12] *Aviation Week* described the U.S. space program as being "in a crisis situation."[13] Truly remarked in his first press conference "I have a lot to do"; he was certainly not overstating the situation.

Truly Takes Charge

While he may have been unfamiliar with the details of the *Challenger* mishap, Richard Truly was no stranger to the space agency; he had been a NASA astronaut from 1969 to 1983, had piloted several of the early unpowered tests of the shuttle, and had flown as pilot on the second shuttle mission in November 1981 and as commander of the eighth shuttle mission in August-September 1983. He left NASA on October 1, 1983, to become the first head of the Naval Space Command; it was from that position that he returned to NASA to assume control of the Office of Space Flight. Truly was an engineering graduate of the Georgia Institute of Technology and an experienced Naval aviator. To most, the combination of his technical background and astronaut experience—and his absence from NASA for the period preceding the accident—made him well qualified to head the return-to-flight effort.

Truly spent his first weeks as Associate Administrator becoming familiar with the situation he had inherited, organizing his immediate office, and establishing a close working relationship with the Rogers Commission. As soon as he entered office, Truly became chair of the "STS 51-L Data and Design Analysis Task Force,"[14] which had been set up by Acting Administrator Graham to provide NASA support to the Rogers Commission. One of Truly's crucial early decisions was to bring in J.R. Thompson as vice-chair and day-to-day head of this task force; in effect, this put Thompson in charge of NASA's part in the accident investigation. Like Truly, Thompson had been a long-time NASA employee, but had been in another job in the years preceding the *Challenger* mishap.[15] Other members of the task force were astronaut Robert Crippen; Col. Nathan Lindsay, Commander, Eastern Space and Missile Center; Joseph Kerwin, Director, Space and Life Sciences, Johnson Space Center; Walter Williams, Special Assistant to the NASA Administrator; and the leaders and deputies of the six task force teams on development and production, pre-launch activities, accident analysis, mission planning and operations, search, recovery and reconstruction, and photo and television support that had been set up to parallel the organization of the Rogers Commission investigation. The task force in turn drew on all relevant resources of NASA.

Between intensive task force efforts during March and April 1986 and the equally intense activities of the fifteen-person investigative staff of the Rogers Commission (plus a parallel investigation by the staff of the Committee on Science and Technology of the

---

12. *New York Times*, March 16, 1986, p. A1.
13. *AW&ST*, February 24, 1986, p. 22.
14. The *Challenger* mission had been designated 51-L; as noted above, STS was the acronym for the Space Transportation System, the official name for the Space Shuttle.
15. Thompson had spent twenty years at the Marshall Space Flight Center as an engineer and manager, but at the time of the accident had been for three years the deputy director of the Princeton Plasma Physics Laboratory in Princeton, New Jersey.

House of Representatives),[16] it was unlikely that any aspect of the accident would go unexamined. This was especially the case given the constant media scrutiny of the investigation.

By the end of March, Richard Truly was ready to go public with his return-to-flight strategy. In a March 24 memorandum which he later described as a "turning point" in the recovery effort,[17] he listed the "actions required prior to the next flight":

- *Reassess Entire Program Management Structure and Operation*
- *Solid Rocket Motor (SRM) Joint Redesign*
- *Design Requirements Reverification*
- *Complete CIL/OMI Review*
- *Complete OMRSD Review*
- *Launch/Abort Reassessment*

Truly also spelled out the "orderly, conservative, safe" criteria for the first post-accident Shuttle launch. These included: a daylight launch and landing, a conservative flight profile and mission rules, conservative criteria for acceptable weather, a NASA-only flight crew, engine thrust within the experience base, and a landing at Edwards Air Force Base in California. He closed the memo by noting that "our nation's future in space is dependent on the individuals who must carry this strategy out safely and successfully. . . . It is they who must understand it, and they who must do it."[18]

Truly reviewed his strategy before an audience of over 1,000 at the Johnson Space Center; his remarks were televised to other NASA Centers. He argued that "the business of flying in space is a bold business. We cannot print enough money to make it totally risk-free. But we are certainly going to correct any mistakes we may have made in the past, and we are going to get it going again just as soon as we can under these guidelines." The *New York Times* reported that "his upbeat words appeared to be meant to lift spirits at the beleaguered agency and to turn the staff's eyes forward to the shuttle's future. . . ."[19]

In just over a month after taking office, and well in advance of any recommendations from the Rogers Commission and the Congress, Richard Truly had set out the general outlines of the strategy he would follow over the following two and one half years. However, that it would take that long to return the Space Shuttle to flight was likely inconceivable to him and his associates at the end of March 1986. NASA planning at the time called for at worst an 18-month delay to July 1987 in launching the next shuttle. Left to its own devices, it is possible that NASA and its industrial contractors could have met this schedule. NASA was no longer a free agent, however; the *Challenger* accident and the resulting external scrutiny of NASA's decisions had changed the agency's freedom of

---

16. The report of the House investigation did not appear until October and, with some differences in emphasis, basically reiterated the major criticisms of the Rogers Commission. See House of Representatives, Committee on Science and Technology, *Investigation of the Challenger Accident*, House Report 99-1016, October 29, 1986.

17. Personal communication from Richard Truly to author, August 14, 1995. In this communication, Truly noted that "in my view, the strategy outlined in this memo (and in my JSC speech about it) was the turning point in the recovery. Although I had taken great care to brief the strategy to both Bill Graham and Bill Rogers, it significantly preceded any conclusions of either the Rogers Commission or the Congress . . ., and therefore did much to give NASA the leeway to implement it. Time and again, it was used by me and others to keep the people, the program and the budgets on track; in 1989, after the first year of successful flights were under our belt, I went back and reviewed it carefully. Despite all that happened in the interim, we had done almost precisely what was laid out in the March 24, 1986 memorandum."

18. Memorandum from M/Associate Administrator for Space Flight to Distribution, "Strategy for Safely Returning the Space Shuttle to Flight Status," March 24, 1986. With respect to the acronyms used in Truly's memo: CIL=Critical Items List; OMI=Operations and Maintenance Instructions; and OMRSD=Operational Maintenance Requirements Specification Documents.

19. *New York Times*, March 26, 1986, p. D24.

action forever. Over the coming months, Truly would have the almost impossible task of balancing the pressure to fly as soon as possible in order to get crucial national security and scientific payloads into space while convincing the agency's watchdogs that a return to flight was adequately safe. It was not to be an easy assignment.

Trying to Get Flying Soon

As mentioned earlier, it was clear within a few days of the accident that the direct cause of the mishap had been a failure in the joint between two segments of one of the shuttle's two solid rocket motors. That failure was in turn quickly traced to the failure of the "O-rings" designed to prevent the escape, through the joint, of the hot gasses generated during SRM firing. On March 11, Acting NASA Administrator Graham told a congressional committee that a redesign of the SRM joint and seals would be needed, and estimated the cost of the redesign at $350 million.[20]

Responsibility within NASA for overseeing the SRM lay with the Marshall Space Flight Center in Huntsville, Alabama. On March 25, Truly, acting on his memorandum of the previous day, announced the creation of a Solid Rocket Motor Team "to recommend and oversee the implementation of a plan to requalify the solid rocket motor (SRM) for flight, including the generation of design concepts, analysis of the design, planning of test programs and analysis of results, and any other initiatives necessary to certify flight readiness." The following day, Truly named James Kingsbury, Director of Science and Engineering at Marshall, to head the team on an interim basis.[21]

Within a few days, Kingsbury told The *New York Times* that he believed a redesigned solid rocket motor could be ready for flight within twelve months, and would not require ordering substantial new hardware. "We can use everything we have, and just modify it," he told the *Times*. In particular (though it was not publicly acknowledged at the time), NASA hoped to be able to use 72 steel casings for the SRM that had been ordered six months before the *Challenger* accident. As would become evident in the course of the accident investigation, NASA had been aware for some time of problems with the original design of the field joint; these casings had been planned to accommodate a new joint design incorporating a "capture fixture" that had been suggested as an improvement on the original joint design as early as 1981.[22]

In its eagerness to get started on the return-to-flight process, NASA appeared to be getting ahead of the findings and recommendations of the Rogers Commission, which was not scheduled to report to the President until early June. For example, Truly had said on March 25 that it was probably infeasible to add a crew escape pod to the shuttle orbiter, but "certainly if the Presidential Commission concludes we should do that, we will do it."[23] Particularly troubling to the Commission was the speed with which a redesign of the SRM field joint was being proposed. On May 7, the Orlando *Sentinel*, in an article headlined "Red Flags Fly Over Joint Redesign," reported that "engineers redesigning the shuttle's flawed booster joint will submit a preliminary plan to NASA today, but members of the Challenger

---

20. *Washington Post*, March 12, 1986, p. A1.
21. NASA, "STS 51-L Data and Design Analysis Task Force: Historical Summary," June 1985, pp. 3–75 and 3–76. In this essay, the term Solid Rocket Motor(SRM) is used except when the context is clearly one that deals with the overall Solid Rocket Booster(SRB), which incorporates not only the SRM but other elements such as parachute recovery systems and an aft skirt that contains the bolts which hold the Shuttle to the launch pad until the time of launch.
22. *New York Times*, March 30, 1986, p. A1, September 22, 1986, p. A1, and September 23, 1986, p. A1.
23. *Ibid*. Truly, in an August 14, 1995 personal communication to the author, notes "I don't remember making a public comment like that about a crew escape pod, and if I did, it was certainly an ill-advised statement, since a pod was totally out of the question for several technical, budgetary, and schedule reasons."

Commission say the agency is moving too fast on the project and could repeat its mistakes." Some Commission members, the article claimed, "are so concerned about Marshall botching the redesign that they want an independent panel of experts to approve the new joint."[24]

NASA had little choice but to respond to the Commission's concerns, particularly once they had become public; the agency in the wake of the *Challenger* accident had lost the ability to act counter to those reviewing it from the outside. The commission's concerns were communicated in a private meeting with NASA's top officials, and a response followed quickly. On May 9, Truly announced that James Kingsbury would be replaced as head of the solid rocket motor redesign team by John Thomas, who had been Spacelab Program Office manager at Marshall before being assigned to the 51-L Data and Design Analysis Task Force in March. This was a switch that had been in the works for some time, but it may have been accelerated by Kingsbury's bullish approach to SRM re-design. Truly also announced that "an independent group of senior experts will be formed to oversee the motor redesign" and that this group would be involved in all phases of the redesign effort, "will report directly to the Administrator of NASA, and will thoroughly review and integrate the findings and recommendations" of the Rogers Commission in carrying out its responsibilities.[25] The interactions between this external panel, which was appointed by the National Research Council(NRC) in June, and NASA during the redesign and testing of the SRM would be a key determinant of the pace of the return-to-flight process.

On May 12, Richard Truly got a new boss. James Beggs had long since resigned as NASA Administrator. The White House, in March, had nominated James C. Fletcher as his replacement. NASA Administrator from 1971-1977, the period during which the Space Shuttle had been approved and developed, Fletcher was quite familiar with the program. It took two months for Fletcher's nomination to be approved by the Senate. After being sworn in by Vice President Bush, Fletcher told the press that, if necessary changes to make the shuttle safe were not completed by the July 1987 target date for the next launch, "we just won't fly."[26]

In effect, any chance of a next launch before early 1988 had vanished with NASA's acceptance of the oversight role of an external advisory group, though it took several months before the agency fully recognized that reality. If there had been any prior doubt, it was now clear that the recommendations of the Rogers Commission, due out in early June, would be the defining context for NASA's return-to-flight effort, at least in the public mind. It was clear, moreover, that those recommendations would go well beyond the need for a redesign of the SRM to many other suggestions on how the Space Shuttle should be operated and managed; The *New York Times* commented that, with such a broad set of recommendations combined with White House and congressional pressure for full compliance with them, "the complexity of NASA's [and thus Richard Truly's] task appears to have been greatly magnified."[27]

---

24. Mike Thomas, "Red Flags Fly Over Joint Redesign," Orlando *Sentinel*, May 7, 1986, p. 1.
25. NASA Release 86-58, "Thomas Assumes Responsibility for SRM Redesign," May 9, 1986.
26. *Washington Post*, May 13, 1986, p. A10. Fletcher brought with him to NASA some baggage that was to complicate matters in subsequent months. Before coming to NASA for his first term as Administrator, Fletcher, a Mormon, had been President of the University of Utah. Congressional critics, particularly Senator Albert Gore, charged that there was a "Utah conspiracy" that had resulted, both in the original 1973 choice of the contractor for the SRB and in the plans for its redesign, in favoritism towards the Utah-based facilities of Morton Thiokol Inc. This bias, they claimed, was leading NASA to give limited attention to SRB re-design proposals from contractors other than Morton Thiokol. In particular, Aerojet had proposed a SRB cast in one piece, without field joints, that would eliminate the need for a joint redesign altogether. See coverage of this issue in *The New York Times*, July 19, 1986, p. A1; September 23, 1986, p. A23; December 7, 1986, p. A1; December 8, 1986, p. A1; and in a December 9, 1996 editorial, p. A20. According to Richard Truly, these attacks "deeply and personally" troubled Administrator James Fletcher, "but they really had zero effect either on the recovery program or the redesign." Personal communication to author, August 14, 1995.
27. *New York Times*, June 12, 1986, p. A1.

The Rogers Commission Report

The Presidential Commission on the Space Shuttle Challenger Accident (the official name of the Rogers Commission) submitted its report to President Ronald Reagan on Friday, June 6; the report was released to the public the following Monday. The over 200-page document, which contained detailed assessments of the causes of the accident and of NASA's overall failings related to the mishap, culminated in nine recommendations. Among them were:

> *Recommendation I* - *"The faulty Solid Rocket Motor joint and seal must be changed. This could be a new design eliminating the joint or a redesign of the current joint and seal."* Also, *"the Administrator of NASA should request the National Research Council to form an independent Solid Rocket Motor design oversight committee to implement the Commission's design recommendations and oversee the design effort."*
> *Recommendation II* - *"The Shuttle Program Structure should be reviewed."* Also, *"NASA should encourage the transition of qualified astronauts into agency management positions."*[28]
> *Recommendation III* - *"NASA and the primary shuttle contractors should review all Criticality 1, 1R, 2, and 2R items and hazard analyses."*
> *Recommendation IV* - *"NASA should establish an Office of Safety, Reliability and Quality Assurance to be headed by an Associate Administrator, reporting directly to the NASA Administrator."*
> *Recommendation VI* - *"NASA must take actions to improve landing safety. The tire, brake and nosewheel system must be improved."*
> *Recommendation VII* - *"Make all efforts to provide a crew escape system for use during controlled gliding flight."*
> *Recommendation VIII* - *"The nation's reliance on the shuttle as its principal space launch capability created a relentless pressure on NASA to increase the flight rate . . . NASA must establish a flight rate that is consistent with its resources."*[29]

In carrying out its mandate, the Rogers Commission had interviewed more than 160 people and held more than 35 formal investigative sessions, generating more than 12,000 pages of transcripts. The full-time staff grew to 43, plus some 140 part-time support specialists. In the end, the report toned down any strong criticism of NASA's overall performance and responsiveness; such a harsh approach had been proposed by Commissioner Richard Feynman.[30] Rather, the report's recommendations were followed by a conciliatory "concluding thought": "the Commission urges that NASA continue to receive the support of the Administration and the nation. . . . The findings and recommendations presented in this

---

28. Criticality 1 items were those where a failure could cause loss of life or vehicle; Criticality 1R, where a failure of all redundant hardware items could have the same effect; Criticality 2, where failure could cause loss of mission; Criticality 2R, where failure of all redundant hardware items could have the same effect.

29. Presidential Commission on the Space Shuttle Challenger Accident, *Report to the President*, June 6, 1986, pp. 198–201. Other recommendations dealt with the need to improve internal communications within NASA, particularly at the Marshall Space Flight Center, and improving maintenance procedures for Shuttle parts.

30. *Washington Post*, June 8, 1986, p. A1 and *New York Times*, June 8, 1986, p. A1. See Richard P. Feynman, "An Outsider's Inside View of the Challenger Inquiry," *Physics Today*, February 1988, pp. 26–37 for Feynman's views of the investigation and report. Feynman's critical views of NASA were published as an appendix to the full Rogers Commission report, but the volume of the report in which they appeared was not printed until well after the release of the main text of the report itself.

report are intended to contribute to the future NASA successes that the nation both expects and requires as the twenty-first century approaches."[31]

On June 13, President Ronald Reagan directed NASA Administrator Fletcher to implement the Rogers commission recommendations "as soon as possible," and asked for a report within thirty days on a plan for doing so.[32] NASA's response came on July 14; Administrator Fletcher told the President that "NASA agrees with the [Rogers Commission] recommendations and is vigorously implementing them." On June 20, in a memorandum to Richard Truly, Fletcher said that he would take direct responsibility for implementing recommendation IV on a new safety organization to replace what the Rogers Commission had characterized as NASA's "silent safety program."[33] Fletcher told Truly that "the Office of Space Flight is directed to take the action for all other Commission recommendations." Fletcher asked him to "status me on your progress on a weekly basis."[34]

While submitting its report to the President, NASA released a schedule for the return-to-flight effort that slipped the earliest possible date for the first launch by 6-8 months, to early 1988. Administrator Fletcher noted that some within and outside of NASA were urging that the three remaining Space Shuttles be returned to flight immediately, with constraints on the conditions under which they could be launched, but that, although he was "uneasy" and disappointed "about the additional delay," in view of the large visibility of the accident . . . when we start flying again we want to make sure that it is really safe."[35]

Implementing the recommendations of the Rogers Commission, and modifying them when justified, would occupy much of the time of Richard Truly and his Space Shuttle team for the next twenty-six months. They worked in the glare of constant congressional and media scrutiny and outside reviews of their actions. There was little margin for error in their task. This was in marked contrast to the situation in the months following the Apollo accident, where, after one round of congressional hearings on the NASA accident report, the space agency made the required technical and management fixes without anyone looking over its shoulder. Indeed, NASA in August 1968 even secretly made a decision to send the second post-accident mission, *Apollo 8*, around the moon. This decision came before the modified Apollo capsule had been tested on the October 1968 *Apollo 7* flight.

Fixing the Solid Rocket Motor

As mentioned earlier, a Solid Rocket Motor Team based at Marshall (but including personnel from other NASA centers, particularly Johnson), and led since May by John Thomas, had gotten an early start on SRM redesign. Sharing leadership with Thomas was Royce Mitchell, another Marshall engineer. Working with the NASA team was a parallel group of engineers from the SRM manufacturer, Morton Thiokol.

This group was headed by Allan J. McDonald, who had been one of those vociferously opposing the launch of *Challenger* on the night of January 27. McDonald's testimony to

---

31. Presidential Commission, *Report to the President*, p. 201.
32. *Washington Post*, June 14, 1986, p. A2.
33. Fletcher announced on 8 July that he was establishing a new Office of Safety, Reliability, and Quality Assurance, reporting directly to the NASA Administrator. This office would be an internal watchdog with respect to the actions of Truly's Office of Space Flight. *Washington Post*, July 9, 1986, p. A10. Because the operation of this office was outside of Richard Truly's responsibility during the return-to-flight effort, it is not discussed in detail here. However, the inputs of the Office of Safety, Reliability, and Quality Assurance into Truly's management decisions were clearly an important consideration in that effort.
34. NASA, *Actions to Implement the Recommendations of The Presidential Commission on the Space Shuttle Challenger Accident*, July 14, 1986, pp. v, 43.
35. *New York Times*, July 15, 1986, p. A1.

the Rogers Commission about the events of that night had brought him much positive media attention. Following that testimony, however, Morton Thiokol had reassigned McDonald and another senior engineer who had opposed the launch, Roger Boisjoly, to jobs not related to the SRM. Congressional outrage at such a reassignment and NASA pressure had led the firm to restore McDonald to a central role in the SRM effort.[36]

The Marshall and Morton Thiokol teams played the central role in developing an approach to SRM redesign and testing; from late 1986, the team worked out of temporary quarters near the Morton Thiokol facility in Brigham City, Utah, north of Salt Lake City. The SRM redesign effort received two overall directives from Truly's office: most fundamentally, "to provide a solid rocket motor that is safe to fly," and, secondarily, "to minimize the impact of the schedule by using existing hardware if it can be done without compromising safety."[37]

Thomas revealed on July 2 that the redesign effort was focusing on two alternatives for fixing the field joint, both of them based on using the previously ordered castings.[38] On August 12, he announced an overall plan for SRM redesign, which included not only changes in the field joint but also fixes to the SRM nozzle-to-case joint and to the nozzle itself. The redesign proposed for the field joint incorporated the capture feature that had been discussed since before the *Challenger* accident, added a third O-ring, and made other modifications.[39]

NASA's plan was controversial. For example, the front page of The *New York Times*, on September 23, reported "rising concerns that it [NASA] may be discarding more reliable designs in an effort to save time and hundreds of millions of dollars."[40]

Among those with reservations about the path NASA was taking were members of the NRC Panel on Technical Evaluation of NASA's Redesign of the Space Shuttle Solid Rocket Booster. This was the external review group that had been established in June at the urging of the Rogers Commission; the eleven-man Panel was chaired by H. Guyford Stever, a highly respected engineer who had been Director of the National Science Foundation and Science Adviser to President Gerald Ford.

The Stever Panel's first report was submitted to James Fletcher on August 1. It acknowledged that, of the factors driving SRM redesign, "safety is the prime consideration," but that "the critical national need for the launch capability of the shuttle makes time a close second." The Panel expressed early concern that the test program for the redesigned motor "meets only a minimal requirement."[41]

Over the next two years, the Stever panel would keep constant pressure on NASA to explore alternative designs and to conduct an extensive test program.[42] The panel's next report was submitted on October 10, after NASA had announced its choice for the redesign of the field joint. The Panel gave only a tepid endorsement to NASA's plans, noting that "if this approach is successful, i.e., if the test program succeeds and the level

---

36. *Washington Post*, May 4, 1986, p. A4; The *New York Times*, June 4, 1986, p. A23.
37. NASA, *Report to the President: Implementing the Recommendations of the Presidential Commission on the Space Shuttle Challenger Accident*, June 1987, p. 13.
38. *New York Times*, July 3, 1986, p. A1.
39. *AW&ST*, August 18, 1986, pp. 20–21. For a detailed description of the SRM redesign, see NASA's June 1987 report of how it was implementing the recommendations of the Rogers Commission cited above.
40. *New York Times*, September 23, p. A1. See also the Washington Post, November 10, 1986, p. A1 and November 29, 1986, p. A3.
41. Commission on Engineering and Technical Systems, National Research Council, *Collected Reports of the Panel on Technical Evaluation of NASA's Redesign of the Space Shuttle Solid Rocket Booster* (Washington, DC: National Academy Press, 1988), pp. 2, 5. This document is hereafter referred to as NRC, *Collected Reports*.
42. Richard Truly remarks that: "Guy Stever and his NRC group were without doubt the most helpful outside advisors" of "any commission, council, group, or Congressional committee. They stayed with NASA all the way to the end, and were constructively critical every time they needed to be." Personal communication to the author, August 14, 1995.

of safety is judged acceptable, the shuttle flight program can resume at the earliest time." The Panel expressed some skepticism about the likelihood of such success, however, urging that "NASA maintain a program to explore and develop original, possibly quite different designs . . . for the contingency that the baseline design may not offer sufficiently good performance and margin of safety." It noted that if the design competition had not been constrained by the desire to use the previously-ordered castings, "we believe that more basic alternatives to the basic design would probably be preferred once thoroughly analyzed." The Panel also told NASA "we believe that the planned test program requires significant augmentation with additional facilities and tests."[43]

NASA, after spirited internal debate, concluded that the panel's suggestions were well-founded, and added a number of partial and full-scale tests to its plans. On October 16, NASA also announced that it would follow the Panel's recommendation and build a second facility for full-scale tests of the SRM.[44] NASA did get Panel endorsement of its decision not to follow one of the Rogers Commission recommendations. At the urging of member Joseph Sutter of Boeing Aircraft, the Commission had suggested that the redesigned SRM be tested in a vertical position, since that was thought to more closely simulate the various conditions during actual SRM use. Constructing a stand for such a test would have cost twenty million dollars and added at least a year to the time before the next shuttle launch. Both the NASA Marshall team under John Thomas and Allan McDonald at Morton Thiokol argued that a horizontal test could be conducted in a way that better simulated flight stresses than would a vertical test. The Stever Panel concurred "that horizontal testing can be appropriate."[45]

Between 1986 and August 1988, the NASA-Morton Thiokol team conducted a test program that included eighteen full-scale but "short burn" tests of SRM joints; seventy-six tests of subscale motors; fourteen SRM assembly tests; and five full-duration tests of the redesigned SRM. Flaws in SRM insulation and seals in joint areas were deliberately introduced in a number of tests; particularly severe flaws were created for the last full-scale SRM firing before return to flight, in August 1988.[46]

The test program did not always go smoothly, and on occasion produced results that forced the team to revise their baseline design. As a result, the date for the first launch slipped twice from a February 1988 target, to June 1988 and then to the August-September period. Early subscale tests convinced the team to stay with the original O-ring material, rather than introduce a substitute. The first full-scale firing was delayed from February to May 1987. The redesigned joint was first tested in a subscale firing in early August 1987; the full scale test came on August 30. (Richard Truly's reaction to the successful test was "a couple of grins.")[47] A December 23 test of the new design at temperatures close to those at the time of the *Challenger* launch was at first called a success, but a few days later engineers discovered that the redesigned outer boot ring at the junction between the SRM nozzle and the rest of the motor had failed.[48] After this test, even though it had not identified the specific cause of the failure, in order to save time the redesign team abandoned the new design and returned to one that was a modification of the pre-*Challenger* design and had performed well in the August test. A successful fourth full-scale test on the

---

43. NRC, *Collected Reports*, pp. 7, 13, 12, and 14.
44. NASA Release 86-146, October 16, 1986.
45. *Washington Post*, October 3, 1986; NRC, *Collected Reports*, p. 10.
46. Allan McDonald, "Return to Flight with the Redesigned Solid Rocket Motor," AIAA paper 89-2404, July 1989, p. 13.
47. *New York Times*, August 31, 1987, p. A1.
48. *Washington Post*, December 30, 1987, p. A1 and January 5, 1988, p. A1.

new test stand that had been suggested by the Stever Panel came in June 1988; it simulated the bending, vibrations, and other stresses of an actual liftoff.

The final full-scale test came on August 18; it was the most demanding and controversial of the series. The need for such a test, introducing the "worst credible" flaw, had been urged on NASA by the Stever Panel as "essential."[49] The redesign team used a putty knife and shoelaces, among other means, to introduce holes in the primary SRM seals; these flaws allowed the seepage of gases in order to check whether backup seals would actually work. Such deliberately induced major flaws were unprecedented in the history of solid rockets, and "months of internal debate" within NASA and Morton Thiokol had preceded Richard Truly's decision to accept the NRC recommendation and approve the politically very risky $20 million test. (If there had been a failure during the test, NASA certainly could not have launched *Discovery* a month later, even though the test motor contained flaws well beyond anything likely to appear in *Discovery's* SRMs.) Although there were some within NASA who favored the test, most did not; that Truly approved it suggests the power the Stever Panel had over the character and pace of the return-to-flight effort.[50]

As the test ended, Allan McDonald and Royce Mitchell, the NASA engineer who had shared leadership of the SRM redesign effort with John Thomas, leapt on the still smoking booster to check for joint failure. There was no evidence of it. In the crowd watching the test, Truly shouted "we did it!"[51]

A few weeks later, a Morton Thiokol spokesman announced that the test had been "as near perfect . . . as you can imagine."[52] With that outcome, NASA judged the redesigned SRM ready for use. In its September 9 report to the NASA Administrator, the Stever Panel concurred, noting that "risks remain. . . . Whether the level of risk is acceptable is a matter that NASA must judge. Based on the Panel's assessment and observations . . ., we have no basis for objection to the current launch schedule for STS-26."[53]

To its great relief, NASA was now felt both technically and politically ready to return the Space Shuttle to flight. Successfully redesigning the solid rocket motor had been the "long pole in the tent" of the return-to-flight effort; with the muted endorsement by the Stever Panel of the redesign effort, the last obstacle to an initial post-*Challenger* flight had been removed.

One person close to the program suggested that the redesign and testing work between early 1986 and August 1988 "exceeded, by four or five times, the amount of work put into original motor work in the mid-1970s."[54] While Richard Truly was necessarily removed from the day-to-day engineering details of the enterprise, he at its outset focused efforts on only those redesign activities that were mandatory for requalifying the SRM for use on the first post-accident flight, and resisted pressures from many fronts to introduce changes, including new designs, additional tests, and different contractors, that would

---

49. The recommendation came in the June 22, 1987 panel report to Administrator Fletcher. See NRC, *Collected Reports*, p. 27.

50. *Washington Post*, August 19, 1988, p. A3. Truly remarks that, after the "fierce" internal debate, he decided that the Stever panel was correct, and that the risk of the test was "worth taking." He also suggests that "I wouldn't have hesitated to go the other way had I believed that they were wrong." Personal communication to author, August 14, 1995.

51. *Ibid.*

52. *New York Times*, August 31, 1988.

53. NRC, *Collected Reports*, p. 58.

54. Morton Thiokol assistant general manager for space operations Richard Davis, quoted in *AW&ST*, September 26, 1988, p. 17.

have delayed resumption of shuttle flights even more.[55] Truly defended the NASA-Morton Thiokol effort to a sometimes hostile Congress. He accepted the risk that the proposed "minimum necessary change" approach to redesign would not be successful, and authorized ordering SRMs incorporating the baseline design changes for the first post-*Challenger* flights at the time the redesign reviews were completed, but before major tests of the redesign had begun. If there had been a major design failure in the test program, NASA would have had to go back to square one, and those SRMs redesigned or scrapped.[56] When the pre-launch test program concluded with the August 18 success, Richard Truly had reason to be excited.

A New Management Structure

Putting a new management structure in place was second in importance to redesigning the SRM as a prerequisite to clearing the Space Shuttle for its return to flight. Richard Truly made a reassessment of the entire shuttle program management structure the first item in his return-to-flight strategy in March 1986, and the Rogers Commission listed such a review as its second recommendation. In May 1986, newly reinstalled NASA Administrator Fletcher had charged the former manager of the Apollo program, retired General Samuel Phillips, with conducting an overall review of NASA organization and management. On June 25, Truly directed astronaut Robert Crippen to form a fact-finding group specifically responsible for assessing the National Space Transportation System (NSTS) management structure.

A first step in reforming program management was the departure or transfer of a number of those who had been in key management positions at the time of the *Challenger* accident. By October 1986, there were new directors at the Johnson, Marshall, and Kennedy Centers, and several other individuals at Marshall who participated in the decision to launch *Challenger* had left NASA.

The Crippen group submitted its findings in August. They were consistent with the views of the Phillips review, and so on November 5, after extensive consultations within NASA, Truly announced a new shuttle management structure.[57] *Aviation Week* described it as "resembling that of the Apollo program, with the aim of preventing communication deficiencies that contributed to the *Challenger* accident."[58]

The key management change was moving lead responsibility for the shuttle from the Johnson Space Center to NASA Headquarters in Washington. Arnold Aldrich, who had

---

55. NASA diverted some of the pressure for involving firms other than Morton Thiokol in the SRM redesign effort by announcing on 18 July 1986 that it would seek to develop a second-generation Advanced Solid Rocket Motor for use beginning in the early 1990s, and that the competition to build this booster would be an open one. *The New York Times*, July 19, 1986, p. A1. NASA also asked other solid rocket manufacturers to critique the Morton Thiokol redesign, but this did not totally relieve Congressional and industry pressure for a more broadly-based redesign effort. *AW&ST*, February 9, 1987, pp. 116–17.

56. As it was, NASA had to retrofit the SRMs intended for use in the STS-26 mission with the design for the SRM nozzle outer boot ring that had been tested in the August 1987 full-scale firing; the boosters had been built with the design that had failed in the December test. This change took almost three months and was a primary reason why the STS-26 launch had to be delayed until August or September 1988. NASA did not know whether the December failure was due to a faulty design or to a demanding test that had been performed at the end of the test firing. Rather than wait for the results of an analysis to determine which was the case, NASA, wanting to launch the Shuttle as soon as possible, chose to go with a modification of the pre-*Challenger* design. *AW&ST*, January 4, 1988, p. 22 and January 11, 1988, p. 24.

57. Memorandum to Distribution from M/Associate Administrator for Space Flight, "Organization and Operation of the National Space Transportation System(NSTS) Program," November 5, 1986.

58. *AW&ST*, November 10, 1986, p. 30.

been NSTS manager in Houston, was asked by Truly to come to Washington as Director, NSTS—in effect, the single director of the Space Shuttle Program, with all shuttle-related activities at the Johnson, Marshall, and Kennedy Centers reporting to him. He in turn would report directly to Truly. Aldrich, who was the only top-level shuttle manager who retained his position after the *Challenger* accident, would have two deputy directors, one for the NSTS Program based at Johnson, and one for NSTS Operations, based at Kennedy. Richard Kohrs was named to the first deputy position; Robert Crippen, the second. The Director, NSTS would have "approval authority for top-level program requirements, critical hardware waivers, and for budget authorization adjustments. . . ."[59]

Truly in his memorandum also noted that "a key element in the ultimate success of the Office of Space Flight is a revitalization of the OSF Management Council."[60] This body included the Associate Administrator for Space Flight and the Directors of Johnson, Kennedy, and Marshall (and the much smaller National Space Technology Laboratories). It had not been very active in the pre-*Challenger* period. This top-level group, lead by Truly, began to meet on a monthly basis, and served as the forum for overseeing the return-to-flight effort in the months following. Its meetings were described as "free-wheeling, no-holds-barred," at which "programme issues are flushed into the open and relentlessly pursued to resolution."[61]

A secondary aspect of the Rogers Commission recommendation on management changes was that "NASA should encourage the transition of qualified astronauts into agency management positions." Richard Truly was himself a former astronaut, and it might have been expected that implementing this recommendation would have been a straightforward matter.

The reality turned out to be somewhat different. In the wake of the *Challenger* accident, the public discovered that the image of the astronaut corps was very much at odds with reality, and that the group was racked with "longstanding strains and resentments," and with "low morale, internal divisions, and a management style that uses flight assignments as a tool to suppress discussion and dissent."[62] Chief astronaut John Young, who had commanded the first shuttle mission, was particularly critical of NASA's approach to flight safety.[63]

Truly's first challenge, then, was rebuilding a positive attitude among his former astronaut colleagues. He met with them privately in March 1986, and made sure that Crippen considered astronaut views as he reviewed shuttle program management. He was not totally successful; some in the astronaut office believed he was too ambitious in trying to return the shuttle to flight by February 1988, and was planning on too many launches per year once the shuttle was back in operation. They were critical of the measured pace of the recovery effort, given a launch target only sixteen months in the future, pointing out that after the *Apollo 1* fire, the command module was redesigned in only eighteen months and suggesting that "management has either got to cut back what they want to do before restarting flights, or get a 'tiger team' approach to pick up momentum."[64]

By July 1987, NASA noted that "ten current or former astronauts hold key agency management positions."[65] One of them had been Rick Hauck, who served from August 1986 to January 1987 as NASA's Associate Administrator for External Relations before he returned to Houston to train for the STS-26 mission. It was rather well known that Hauck was likely to command the first post-accident shuttle flight; he was thus a convincing spokesman for the safety

---

59. Richard Truly memorandum, November 5, 1986.
60. *Ibid.*
61. L.J. Lawrence, "Space Shuttle–Return to Flight," *Spaceflight*, September 1988, p. 352.
62. *New York Times*, April 3, 1986, p. B9 and *Washington Post*, April 1, 1986, p. A1.
63. See, for example, Memorandum from CB/Chief, Astronaut Office to CA/Director, Flight Crew Operations, "One Part of the 51-L Accident-Space Shuttle Program Flight Safety," March 6, 1986.
64. *AW&ST*, October 20, 1986, pp. 34–35.
65. NASA, *Implementation of Recommendations*, July 1987, p. 32.

aspects of the return-to-flight effort. Other astronauts brought into management positions had "some difficulties in adjusting to the realities of bureaucratic life," but felt that "their presence had made a difference, pointing with pride to influence on key policy issues."[66]

Other Changes to the Shuttle

Even before the Rogers Commission submitted its report, Richard Truly made one key decision related to reducing the risks of future shuttle operation. Some in NASA, even before the accident, were concerned about the wisdom of using a modified Centaur rocket, fueled by highly combustible liquid hydrogen, as an upper stage to carry satellites from the shuttle's payload bay to other orbits. Among the payloads for which the Centaur was to be used were two solar system exploration missions, Ulysses to explore the Sun's polar regions and Galileo to orbit Jupiter; several classified Department of Defense payloads were also scheduled to employ the Centaur upper stage.

A combination of congressional pressure and the more stringent safety criteria being applied to the shuttle after the accident led to a NASA reassessment of Centaur. Although over $700 million had already been spent on modifying the Centaur for shuttle use, and its unavailability would cause major delays in the solar system exploration program, Truly recommended cancelling the Shuttle Centaur program. Administrator Fletcher agreed and announced the decision on June 19, 1986.[67]

Another key decision was to terminate planning for launching the shuttle into polar orbit from the Vandenberg Air Force Base in California. This decision meant that the very expensive Shuttle Launch Complex 6 at Vandenberg would be mothballed and that the number of overall Department of Defense (DoD) flights on the shuttle reduced (DOD would use a Titan IV expendable launch vehicle for payloads originally scheduled for a shuttle launch from Vandenberg). This decision reduced overall schedule pressure on a four-orbiter shuttle fleet, and eliminated the need for a lighter, filament-wound SRM case.[68]

The third recommendation of the Rogers Commission had directed NASA and its industrial partners to review, in terms of safety and mission success, all Criticality 1, 1R, 2, and 2R items and hazard analyses. Richard Truly had called for an even more extensive risk review in his March 1986 return-to-flight strategy. The Rogers Commission had also separately recommended a series of actions to improve landing safety.

That the shuttle had been flying with a number of less-than-optimum systems and components was well known to those close to the program, but the pressures of maintaining an ambitious launch schedule and budget constraints had blocked any extensive review and upgrading of the shuttle before the accident. When it became clear that the shuttle would be grounded for some time, Arnold Aldrich, at the time still in charge of the shuttle program at the Johnson Space Center, had on March 13, 1986, initiated a comprehensive review aimed at identifying possible shuttle upgrades. By the end of May, this review had identified "44 potentially [critically] flawed components of the space shuttle . . . that may have to be fixed before shuttle flights can resume."[69]

The conduct of a comprehensive Shuttle Failure Modes and Criticality Analysis and the audit of the resulting Criticality 1 and 2 items recommended by the Rogers Commission was an extensive and complex process. In its July 1986 report on implementation of the

---

66. *New York Times*, June 2, 1987, p. C2.
67. *New York Times*, June 20, 1986, p. A1.
68. Comment on draft of this essay by Richard Kohrs, July 19, 1995.
69. *Washington Post*, May 28, 1986, p. A5.

Rogers Commission recommendation, NASA indicated that "the overall reevaluation is planned to occur incrementally and is scheduled to continue through mid-1987."[70] By the time *Discovery* was ready for launch, the list of Criticality 1 items had grown from the 617 items at the time of *Challenger* to 1,568; each of those items had to pass particularly rigorous review before *Discovery* was cleared for flight. The number of Criticality 1R items had also grown dramatically, from 787 to 2,106.[71]

Similar to his situation with respect to SRM redesign, Richard Truly found an external review committee assessing NASA's actions with respect to risk assessment and management. The National Research Council created a Committee on Shuttle Criticality and Hazards Analysis Audit in September 1986; the Committee was chaired by retired Air Force General Alton Slay. In its initial report, submitted to James Fletcher on January 13, 1987, the Slay Committee noted that it had "been favorably impressed by the dedicated effort and extremely beneficial results obtained thus far." The Committee raised a point that recurred throughout its work, that "the present decision-making process within NASA . . . appears to be based on the judgment of experienced practitioners and has received very little contribution from quantitative analysis." The Committee also questioned the timing of the risk review in terms of incorporating any resulting design changes in the shuttle before its scheduled return to flight (then February 1988), noting that there may not be "time to incorporate any substantial design changes that may be indicated by the outcome" of the review.[72]

The Slay Committee continued its work throughout 1987 and submitted its final report to Administrator Fletcher in January 1988, although the report was not made public for two months. While generally positive in tone, it criticized NASA's risk assessment activities as still too "fragmented" and "subjective," and for not taking advantage of widely used quantitative techniques such as probabilistic risk assessment.[73] But, most important to Richard Truly and his associates, the Committee found "absolutely no show-stoppers" from a risk assessment perspective in terms of NASA's return-to-flight plans.[74]

Richard Truly had relieved much of the pressure of implementing the separate Rogers Commission recommendation on improving landing safety by mandating in his March 24, 1986, return-to-flight strategy that the first flight would land on one of the extremely long runways at Edwards Air Force Base in the California desert. In its 1987 report to the President, NASA said that it had identified several design improvements "to improve the margins of safety for the landing/deceleration system. Some of these improvements are modifications to existing designs and will be completed prior to the next flight." But, added NASA, improvements involving more extensive design changes would have to be certified for flight and then introduced "later in the program."[75]

In fact, this was the philosophy followed for almost all design changes to the shuttle in the aftermath of the *Challenger* accident which were not related to SRM redesign. The first post-accident shuttle flight was launched as soon as possible after the requalification

---

70. NASA, *Actions to Implement the Recommendations*, July 14, 1986, p. 19.
71. *Washington Post*, August 23, 1988, p. A3; NASA, "NSTS SR&QA Assessment," September 13–14, 1988.
72. National Research Council, Committee on Shuttle Criticality Review and Hazards Analysis Audit, *Post-Challenger Evaluation of Space Shuttle Risk Assessment and Management*, (Washington, DC: National Academy Press, January 1988), pp. 98–100.
73. This was not a new criticism of NASA. Staff of the White House Office of Science and Technology had made similar criticisms in 1962 as NASA evaluated various ways of carrying out a manned mission to the moon. See John M. Logsdon, "Selecting the Way to the Moon: The Choice of the Lunar Orbital Rendezvous Mode," *Aerospace Historian*, June 1971.
74. *Washington Post*, March 5, 1988, p. A6; *New York Times*, March 5, 1988, p. B2; *Science*, March 11, 1988, p. 1233.
75. NASA, *Implementing the Recommendations*, July 1987, pp. 55–56.

of the SRM for flight; the introduction of other redesigned shuttle elements as a result of the risk reviews or of Arnold Aldrich's examination of desirable shuttle improvements did not have significant influence on the shuttle launch schedule. However, the post-*Challenger* reviews did have other important impacts, both before and after return to flight. The system was overall much safer and reliable on September 29, 1988, than it had been in the 1981–1986 period. The shuttle's main engines were upgraded, its brakes improved, and the valves in the orbiter that controlled the flow of fuel to the orbiter's engines modified to prevent accidental closure. But the result was a "shuttle in transition"; "the hard truth," said Aldrich, " is that the really major changes take years."[76]

Adding an Escape System

As a former astronaut, Richard Truly gave particular, personal attention to the Rogers Commission recommendation that an escape system be added to the shuttle to allow its crew to leave the vehicle in an emergency while it was in controlled gliding flight (i.e., after the SRMs had finished firing and been jettisoned and the shuttle's main engines shut down). In fact, a search for a viable escape system had begun in March 1986; as the search progressed astronaut Bryan O'Connor played a key role in assessing various options. Alternatives considered included ejection seats, "tractor rocket" extraction of seated crew members, bottom bail out, and tractor rocket extraction through the side hatch. All but the last alternative were eliminated by the end of 1986, but in its July 1987 report to the President on how it was implementing the Rogers Commission recommendations, NASA said that a decision to implement the side hatch, rocket-powered escape approach "had not been made."[77]

NASA in December 1986 had in fact made a tentative decision to go forward with this approach, if it could be shown satisfactory in tests and installed in time for the next launch.[78] By September 1987, due to delays in the testing program and the possibility that an adequate supply of parts for the system might not be available on a timely basis, NASA began to consider a simpler alternative—one using a telescoping metal pole extending nine feet beyond the shuttle escape hatch. In an emergency, crew members would attach themselves to the pole and slide away from the shuttle orbiter's wing before they parachuted to Earth.[79]

Based on tests of the two systems, Truly in April 1988 selected the pole escape approach. This was perhaps the last major pre-launch choice stemming from a Rogers Commission recommendation. One factor in the decision was avoiding the additional risks created by installing the pyrotechnic tractor rockets in the shuttle cabin; also, the STS-26 crew preferred the pole system. The escape system could be used only with the shuttle in controlled flight at a less than 20,000 foot altitude, with landing on a primary or emergency runway impossible. (Whether in an emergency to push the shuttle's main engines beyond their design limits to enable the orbiter to reach a trans-Atlantic abort site, or to bail out was a controversial issue up almost to the time of the *Discovery* launch. Astronauts and mission controllers favored a bail out option, but they were overruled by Truly who wanted to avoid losing another orbiter in an ocean ditching.)[80] Bailing out of the shuttle was considered far preferable to trying to survive a water landing; one individual responsible for the escape system commented, "the orbiter doesn't survive ditching very well."[81]

---

76. *New York Times*, December 28, 1986, p. 1.
77. NASA, *Implementing the Recommendations*, p. 67.
78. *AW&ST*, January 5, 1987, p. 27 and July 6, 1987, p. 28.
79. *AW&ST*, September 7, 1987, p. 125.
80. *AW&ST*, September 26, 1988, p. 63.
81. *Ibid.*, April 11, 1988, p. 31.

Setting a Flight Rate

The Rogers Commission had identified "the relentless pressure to increase the flight rate" as a major contributing factor to the *Challenger* accident. Though not directly related to getting the shuttle ready for its first post-accident flight, determining the appropriate schedule for shuttle launches after the STS returned to flight occupied much of the time of Richard Truly and his staff at NASA Headquarters while the shuttle was grounded.

A first consideration was what payloads the shuttle would carry as the launch rate was reduced; it was clear that critical national security payloads would have first priority. After a series of intense debates within the Reagan administration–over NASA's objections–the President announced on August 15, 1986, that, except in situations where there were overriding national security, foreign policy or other reasons, the shuttle would no longer be used to launch commercial communication satellites.[82] This decision and plans for its implementation announced two months later removed a major category of payloads from the shuttle manifest; prior to the accident, eleven of the twenty-four earlier shuttle missions had carried one or more commercial communication satellites.

In October 1986, NASA released a shuttle launch schedule that called for a buildup to fourteen or sixteen launches per year, four years after the STS returned to flight, and after a replacement orbiter had entered service.[83] This was more ambitious than the launch rate thought reasonable by yet another National Research Council review committee. At the request of NASA's House Appropriations Subcommittee, the NRC created a panel to carry out a "post-*Challenger* assessment of Space Shuttle flight rates and utilization." In its October 1986 report, the panel concluded that with a four-orbiter fleet NASA could sustain a launch rate of eleven to thirteen launches per year, but only if there were significant improvements in various aspects of the shuttle program. Without such improvements, the panel estimated, the maximum rate was eight to ten launches per year. The panel noted that only "under special conditions" might the launch rate surge to fifteen launches per year.[84]

Balancing the desire to get flying again on a regular basis, the pressure to launch critical national security and scientific payloads as soon as possible, and the need to ensure continued safe and reliable operation of the Space Shuttle was a constant challenge for Richard Truly. He recognized that "we will always have to treat it [the shuttle] like an R&D test program, even many years into the future. I don't think calling it operational fooled anybody within the program. . . . It was a signal to the public that shouldn't have been sent and I'm sorry it was."[85] Media watchdogs were quick to report perceptions that NASA was "putting schedule over safety."[86] But, as Truly had said on many occasions, "the only way to operate the shuttle with zero risk is to keep it on the ground." That was not his intent.

Return to Flight

The Space Shuttle *Discovery* was rolled out from the Vehicle Assembly Building to launch pad 39B on July 4, 1988; as a morale-boosting measure, throughout the day Kennedy Space Center workers and their families were allowed to drive around the pad.

---

82. *Ibid.*, August 18, 1986, pp. 18–19.
83. *Ibid.*, October 13, 1986, pp. 22–23.
84. National Research Council, Committee on NASA Scientific and Technical Program Reviews, *Post-Challenger Assessment of Space Shuttle Flight Rates and Utilization* (Washington, DC: National Academy Press, October 1986), pp. 7–8.
85. *AW&ST*, September 26, 1988, p. 16.
86. *Time*, February 1, 1988, p. 20.

There were no waivers (permissions to launch even though specifications were not met) on any hardware element, and an internal NASA committee had found a "positive change in attitude" with respect to safety considerations and a "healthy redundancy of safety reviews and oversights." The group found no safety issues that would adversely affect the launch of STS-26, then set for September 6.[87]

There were a few minor delays before the shuttle was ready for launch, however, slipping the launch date to late September. A 21.8-second Flight Readiness Firing of *Discovery's* main engines was conducted on August 10, and a two-day Flight Readiness Review in early September. The final launch date of September 29 was set when it was determined that Hurricane Gilbert would not affect operations at mission control at the Johnson Space Center.

When *Discovery* roared off of the launch pad after a 98 minute weather-caused delay on the morning of September 29, a great weight was lifted off of not just Richard Truly, but the whole NASA organization. Truly would later say that "the time when the Space Shuttle did not fly was time well spent by NASA. When we look back at 1986–1988, we will see it as a time when NASA and the country took an unwanted, but necessary, breather in the space program. During this time, we took a hard look at ourselves and at what we hoped to accomplish in space. What we saw was solid. Some things needed changing and changes were made. It was a time of introspection, not without pain, but mostly it was a time when we rechartered our course and rededicated ourselves to space exploration."[88]

Richard Truly brought a perhaps unique set of attributes to his job as NASA Associate Administrator for Space Flight. Though admitting frustration at the inefficiencies of the political process and impatience with the need to testify so frequently to Congress and to participate in frequent executive branch meetings, he was skilled at charting a course through the political process. His status as a former astronaut gained him credibility on Capitol Hill and with the public, and legitimacy within the space flight community inside and outside of NASA. He was able to gain the support of the many external groups overseeing the accident recovery effort for most, if not all, of NASA's actions and decisions. He had enough technical background to understand the issues under debate during the recovery process. He surrounded himself with a team as committed as he was to the shuttle as the centerpiece of the U.S. effort in space.

As he reflected on his experience a few months before the shuttle returned to flight, Richard Truly suggested that during the preceding months "the high and low points have been very high and very low" and that "there have been great frustrations," particularly in dealing with the criticisms of NASA and its employees. He admitted that NASA deserved "some" of the criticism, and so his approach "has simply been to try to build a team that will win our credibility back." The high point in his experience during the return-to-flight effort, according to Truly, "has simply been watching this team come back together."[89]

Richard Truly accepted the 1988 Collier Trophy on behalf of all those in government and industry that had participated in the return-to-flight effort. It was an honor well earned.

---

87. *AW&ST*, July 11, 1988, pp. 34–35.
88. RADM Richard H. Truly, USN, *Space Shuttle—the Journey Continues*, NASA NP-117, September 1988, p. 19.
89. NASA, *NASA Activities: Special Edition*, Winter 1988, p. 4.

Chapter 16

The Hubble Space Telescope Servicing Mission

by Joseph N. Tatarewicz

Prelude
Big Science, Hubble, and Historical Trends

In March of 1994, the National Aeronautic Association announced that its 1993 Robert J. Collier Trophy would be awarded to the NASA Hubble Space Telescope Recovery Team "for outstanding leadership, intrepidity, and the renewal of public faith in America's space program by the successful orbital recovery and repair of the Hubble Space Telescope." Representing the more than 1,200 men and women directly involved in the mission, the seven-person astronaut crew of Space Shuttle Mission STS-61 and four ground managers were named as the recipients. It was truly, by all assessments of the participants and the observers and in the language of the Collier award, "the greatest achievement."

The Hubble Space Telescope had taken longer to build and launch than any other NASA spacecraft, including Apollo. It had cost more than any other scientific space program, and more than nearly any other space mission. Deployed in the Spring of 1990 with the express mission of addressing the most enigmatic and exciting questions of astronomy, it promised a revolution in understanding the origin and evolution of the universe and myriad other astronomical and cosmological questions. Within weeks, a horrible realization gradually emerged: the instrument bore a seemingly fatal and irreparable manufacturing flaw that would severely degrade or even scuttle its fifteen-year mission. Worse, various other systems and components began to act erratically or to fail. Soon the Hubble Space Telescope and NASA itself were the objects of anger, scorn, and ridicule. What began even before launch as a planned and routine servicing mission grew into a bold and comprehensive overhaul, a "rescue mission in space." The Hubble Space Telescope Recovery Team rescued more than just the telescope and its mission, however. By all accounts they rescued NASA and the U.S. space program as well.[1]

---

1. National Aeronautic Association, "Hubble Space Telescope Recovery Team Wins 1994 Collier Trophy," News Release, Arlington, Virginia, March 1, 1994; Loren S. Aguirre, (producer, writer, and director), *Rescue Mission in Space*, video cassette recording (Boston, MA: WGBH / NOVA, 1994); specific citations to this program will be given in minutes:seconds from start of the tape.
  This article is dedicated to Bob Bless, leader of the High Speed Photometer Team at the University of Wisconsin, who worked tirelessly for decades and in many ways to see the Space Telescope achieved, and then gracefully made room for COSTAR.
  I am grateful to a very large number of people who graciously gave freely of their time by sending or giving me information, oral history interviews, and informal discussions. The Hubble Servicing Mission was a massive effort that drew a dozen institutions and thousands of people into its cast. I have tried to give representative weight to the various contributors, but it would have been impossible within the limits of this brief article even to mention all their names. My own prior knowledge of the program from having worked on the Space Telescope History Project and the restoration and exhibit of the Space Telescope Structural Dynamic Test Vehicle was an important foundation for writing this article, and I am grateful to colleagues and friends at the Smithsonian Institution's National Air and Space Museum, especially Robert W. Smith. Finally, as ever I am grateful to the NASA History Office, especially Director Roger Launius, for supporting this article. Needless to say, I am solely responsible for all conclusions and interpretations herein.

Astronaut F. Story Musgrave, anchored on the end of the Remote Manipulator System (RMS) arm, prepares to be elevated to the top of the towering Hubble Space Telescope (HST) to install protective covers on magnetometers. Astronaut Jeffrey A. Hoffman (bottom of frame) assisted Musgrave with final servicing tasks on the telescope, wrapping up five days of space walks. (NASA photo no. 94-H-16).

In the introduction to this volume, Pamela E. Mack discerns a number of trends that altered the character of air and space flight over the eight decades of Collier awards:

> *(1)* *"NACA and then NASA [became] increasingly caught in a web of bureaucratic and political obligations"; individuals came to matter less and the planning process itself became more important.*
>
> *(2)* *"Research and development projects have become more complicated in fundamental ways,"* requiring more diverse expertise than any lone inventor could muster and separating developers of technology from users.
>
> *(3)* *"Attitudes towards funding research have changed"; after Apollo there arose a "new emphasis on cost-benefit calculations but also more willingness to fund projects on the basis of popular support."* [2]

The Hubble servicing mission represents these trends in full bloom, as well as other characteristics which have been identified as the classical indicators of big science: "money, manpower, machines, media, and the military." The Hubble Space Telescope program generally presents these characteristics in their more benign forms, as well as the pathological variants identified by Alvin Weinberg of "journalitis, moneyitis, administratitis."[3] In what follows, I shall first discuss these characteristics as they apply to the Hubble servicing mission. Then I shall present a basic historical narrative of planning and executing the mission. Finally, I shall conclude with some of the scientific results that ensued during 1994, and present a few conclusions addressing broader issues.

(1) *Bureaucratic and political obligations:* Between the initial planning in the 1970s for frequent routine shuttle maintenance of the telescope and the December 1993 servicing mission, replacing components came to acquire a heavy load of significance and implications. The servicing mission became a way to repay bureaucratic and political obligations that extended far beyond the telescope program itself, even beyond NASA. In late 1993, the then-tarnished reputation of NASA as a whole, the capability of its shuttle system and astronaut corps, and the viability of its most important next program, the Space Station, all rode with the crew of the Endeavour. It seemed they would all come back in the same condition: heroic and vindicated, or disgraced. The servicing mission was a resounding success on nearly all counts. The long and demanding spacewalks were among the most unproblematic ever done, and over several days press and media coverage of the mission was more intense and more favorable than perhaps at any time since Apollo. Live television of the repair activities was carried uninterrupted on many cable television channels, and scenes of the astronauts working on the telescope dominated the evening newscasts. Astronaut Story Musgrave recalled being stopped by people while doing his Christmas shopping shortly after the mission: "They were bleary-eyed from staying up all night. We were better off than they were, . . . they were very excited because they had lived it vicariously."[4]

Tellingly, the crew and ground managers were joined by beaming agency officials and politicians in postmission press conferences and events. When the telescope

---

2. These trends are taken from a letter, Mack to volume authors, November 20, 1994, p. 2.
3. James H. Capshew and Karen A. Rader, "Big Science: Price to the Present," *Osiris* 7 (1992): 4 and 5.
4. Musgrave, answering question at the Crew Post Mission Press Conference, January 4, 1994.

resumed operations and the repairs were fully tested in January 1994, its performance was far better than anticipated, and the end-users were extremely pleased. To what degree the Hubble Space Telescope was restored to its original planned performance is a difficult judgment, and to some extent depends on the criteria one uses. It was, however, dramatically improved. Over the following year, it produced a string of significant results that would not have been possible in its original state.[5]

(2) *Users and developers of complex technology:* The complexity of the process and alienation of the users from the developers is amply visible in the history of the space telescope itself. This aspect is documented and illustrated in published scholarly history, government audits by such units as the General Accounting Office, congressional hearings, and memoirs. From the first organized campaigns by astronomers in the late 1960s, through feasibility studies in the 1970s, development in the 1980s, and operations in the 1990s, the Hubble Space Telescope has been sustained by a complex and protean coalition of diverse parties. If astronomers are the end users (but in no way the only beneficiaries) of the space telescope, then in part the history of its development is one of astronomers struggling to prevent their interests from being submerged and overrun by all the others. The need for the servicing mission at all arose because the users (astronomers) failed to protect fully their interests and allowed the developers (engineers and managers) to produce what *looked* like a space telescope but did not *function* as a space telescope should. This illustrates big science as what James Capshew and Karen Rader have called "at once a broadly diffused mode of cognition and a concentrated form of organized labor."[6]

Not that the astronomers did not try—they tried mightily over the decades, singly and in groups. But the scope of what Robert Smith has called "the biggest kind of big science," big engineering, and big management was overwhelming. In my view, the system that was building the space telescope was bigger than any putative system builder (in the sense of historian Thomas Hughes) could ever hope to manage. Nor should we assume there was bad faith or villainous subversion. The users (astronomers) were not the only, nor necessarily the most important reasons for building it at all. Many other interests (engineering, commerce, politics) wanted a space telescope, and the astronomers only got one because these other interests wanted it. These other interests, however, did not require that the telescope function in quite the same way as the astronomers wished. These other interests did not willfully ignore or subvert the astronomers' requirements; mostly they were just insensitive to the astronomers and far more powerful. Their needs for a successfully functioning space telescope were far more relaxed than

---

5. The industry trade journal *Aviation Week & Space Technology* 140 (24 January 1994): 20 and 21 noted in bestowing its "Laurels for Space/Missiles," Thoughtful memoirs by scientists and managers involved include Robert Bless, "Space Science: What's Wrong at NASA," *Issues in Science and Technology* 5 (Winter 1988): 67–82 and Eric Chaisson, *The Hubble Wars: Astrophysics Meets Astropolitics in the Two Billion Dollar Struggle Over the Hubble Space Telescope* (New York, NY: Harper Collins, 1994).

6. Capshew and Rader, "Big Science," pp. 3, 10. The definitive history of the Hubble Space Telescope program itself is Robert W. Smith, with contributions by Paul A. Hanle, Robert Kargon, and Joseph N. Tatarewicz, *The Space Telescope: A Study of NASA, Science, Technology, and Politics,* paperback edition with a new Afterword (New York, NY: Cambridge University Press, 1993). "Not since Apollo has a [*sic*] such a cohesive space mission team been formed from so many disparate quarters within NASA, the aerospace industry and the academic world, yielding such thorough success. And the team did it on time and within budget." A thoughtful, but early and incomplete assessment of the repair results is given by Eric J. Chaisson, *The Hubble Wars,* Afterword.

the needs of the astronomers. Even though all the different interests assembled around a single big machine, it was an instrumentality with many meanings and many criteria for success. One manager put the issue succinctly: "If the agency did nothing more than take one unaberrated picture from the Wide Field Camera in 1993, it would have been declared a success; so why add any risk to [the servicing mission by fixing other problems]? Well, . . . we promised a functional telescope, not a stunt."[7]

The servicing mission was successful in part because the users asserted themselves and interjected themselves forcefully into the process. The idea for the optical fix came from the user advocates. Implementing the fixes was possible because the other interests found their salvation through the needs of the users. If the development of the telescope illustrates the alienation of the users in big science and the dispersion of interests, its repair provides an illustration of how big science can work beautifully when interests merge. The servicing mission evolved from a diffused set of activities with many parties operating somewhat independently into a tightly focused and exquisitely choreographed group effort. It was well funded, had the rapt attention of management, and a working level *espirit de corps* that transcended institutional and other loyalties.[8]

(3) *Attitudes towards funding:* Some ill-chosen hyperboles aside, the Hubble Space Telescope's benefits were as pure as the driven snow. It is hard to quantify the value of knowing better the nature, history, and origin of the universe in order to compare it to the cost of the telescope. The general public and its representatives might not comprehend fully the subtleties of astrophysics and cosmology, but they did expect at least entertainment and edification for the investment. While the several problems of the telescope affected all the instruments, not just the cameras, it was not fuzzy spectra or muddy digital data that became the public scandal, but rather "blurred vision." The myriad of misleading and horribly muddled similes and analogies used to convey the problems and remedies all came down to the easily and instantly understood pictures and clarity. Even Senator Barbara Mikulski, herself one of the most knowledgeable and savvy advocates, referred to her outrage at the "cataract" and her elation at the successful "contact lens." Hence, the Hubble servicing mission

---

7. Robert W. Smith, "The Biggest Kind of Big Science: Astronomers and the Space Telescope," in *Big Science: the Growth of Large Scale Research*, Peter Galison and Bruce Hevly, eds. (Stanford, CA: Stanford University Press, 1992); Hughes summarized his concept of system builders in Thomas P. Hughes, "The Evolution of Large Technological Systems," in *The Social Construction of Technological Systems: New Directions in the Sociology and History of Technology*, eds. Wiebe E. Bijker, Thomas P. Hughes, and Trevor Pinch, (Cambridge, MA: MIT Press, 1987), pp. 51–82, which contains references to his other writings on the subject. Robert Smith and the author have discussed this and other theoretical perspectives from the social studies of science and technology as applied to the Space Telescope in "Counting on Invention: Devices and Black Boxes in Very Big Science," *Osiris* (New Series) 9 (1994): 101. On big science as instrument and instrumentality see Capshew and Rader, pp. 8, 23. Andrew Butrica in his contribution to this volume notes the dependence of the scientific users on the technology of planetary exploration, and especially the users' location at the end of a long string of interlocking communications processes. The final quote is from Joseph Rothenberg Interview, 8 August 1995, 1/A: 572. Unless otherwise identified, all interviews are by the author and are identified by date, tape number/side, and approximate tape counter number.

8. Many people were impressed by the cooperation and team spirit evident throughout the several years of planning. Astronaut Story Musgrave recalled the horizontal integration was "seamless" like "Apollo" (Musgrave Interview, March 23, 1995, 1/B: 080). On coalitions and multiple interests in science and technology see Bruno Latour, *Science in Action. How to Follow Scientists and Engineers Through Society* (Cambridge, MA: Harvard University Press, 1987) and Smith and Tatarewicz, "Counting on Invention."

was in some sense a redemption, because the defects in the extremely expensive instrument were immediately graspable by the public, who could also easily see the improvement after the repair. They could also easily grasp the magnitude, if not the specific dollar amount, of its cost. As one of the astronomers involved in the program teased in a 1984 lecture:

"That *is* a nice topic," said Alice, "I will put the project in its historical perspective and draw analogies with other great scientific projects."

"Nonsense!" said Humpty Dumpty, "The answer is very short. Space Telescope will revolutionize mankind's understanding of the Universe because it cost a billion dollars! That's all. They wouldn't have spent the money in the first place if it hadn't been so, now would they?"[9]

Great Expectations, Bitter Disappointments

Initially, maintenance and refurbishment missions (including even return of the entire spacecraft for ground overhaul and then reflight) were part of the Space Telescope's routine scenario. The first official NASA telescope planning phase began around 1971, just as the Space Shuttle was being defined. With austere times ahead after Apollo, the Space Telescope and the Space Shuttle soon found common cause. An orbital telescope of that size and cost could not be justified unless it could operate for years or decades, and to that end the Space Shuttle promised routine access for repair and upgrade. Lacking the Space Station, deleted from planning due to cost, the shuttle needed a place to go and useful work to do that could not be accomplished by expendable boosters. To that end, the Space Telescope and Space Shuttle pair became an exemplar of a new and cost-effective way of doing Earth orbital science. Initially projected at about two-and-a-half-year intervals over the fifteen-year life of the Telescope, such service calls were expected to be unproblematic and routine. Indeed, the frequency of access to the spacecraft, predicated on the presumed regular and routine shuttle traffic, was to have removed urgency. If something could not be accomplished on one call, it could wait until the next. It also meant that expensive pre-flight design and testing could be relaxed somewhat, secure in the knowledge that the spacecraft would be accessible for servicing. This notion of routine maintenance remained with the Telescope through its planning in the 1970s, its new start approval in 1977, and through some traumatic budgetary times in 1980. However, several unanticipated circumstances emerged to cast the spotlight on maintaining the Hubble Space Telescope and to raise the stakes.[10]

First, early in 1983 NASA Administrator James Beggs was startled to be told, on the very eve of release of the President's budget for fiscal year 1984, that the program was once again significantly over budget and behind schedule. The full extent of the crisis emerged over the following few months, and provoked painful congressional hearings,

---

9. Malcolm Longair, *Alice and the Space Telescope* (Baltimore, MD: Johns Hopkins University Press, 1989), p. 5; Longair delivered the lecture on which this book is based in June 1984, at which time the cost of the Hubble Space Telescope was estimated at about a billion dollars.
10. Smith, *The Space Telescope*, Chapters 2-3. "Maintenance and Refurbishment (M&R)" was the official designation during development. The first "Hubble Space Telescope Servicing Mission (HST SM-1)" was in planning even before launch and deployment. A lot of words and bile were expended behind the scenes over such banished terms as "repair," "fix," and "rescue," because these terms implied a spacecraft that was "broken." See Chaisson, *The Hubble Wars*, for examples.

shakeups of the program and project management, and serious loss of political capital. By the time the new launch target date was in place, and the new budget understood, the program was operating under "last chance" understandings. It could not go back to the well again, and the program remained under scrutiny from within and without NASA. One salutory effect of the reorganization was that some orbital replaceable units, deleted earlier in the program for budgetary reasons, were reaffirmed, making the Telescope more easily maintainable than it otherwise might have been.[11]

Second, after recovering from this setback, the spacecraft was waiting in storage and testing at Lockheed's Sunnyvale, California, plant when the Space Shuttle *Challenger* accident in January 1986 halted all shuttle missions for an indefinite time. The Telescope program had to keep its "marching army" idling but consuming money during the resulting hiatus, contend with more conservative rules for shuttle bay payloads, and worry about the potential effect on the spacecraft of the unplanned-for extended storage. Moreover, the more conservative shuttle schedule, and even tighter budgets that emerged in the post-*Challenger* era, meant fewer and less-frequent future maintenance opportunities. The program was forced to absorb the additional costs without ability to obtain more money from without. The maintenance and refurbishment budget was an easy target, and money was shifted by adjusting the maintenance schedule and that for future replacement instruments. Some worried about how the spacecraft and its components would fare, not having been designed to reside in storage for so long. Others worried that new instruments would not be ready in time to replace inevitably aging and failing ones.[12]

Third, the Hubble Space Telescope became the centerpiece for recovering from *Challenger* after effects. It was the largest and most complex scientific payload to be delivered to orbit, except perhaps the Galileo Jupiter probe. Unlike Galileo, which would take several years to reach its destination and return data, the Hubble Space Telescope would start providing results in only a few days or weeks of deployment. As launch neared, the press and media coverage was enormous.

When the Space Shuttle *Discovery* roared from the launch pad on April 24, 1990, occupying nearly all of the payload bay was an enormously complex spacecraft. Rather than the usual spacecraft with instruments attached, the Hubble Space Telescope was an immense collection of instruments enshrouded by a spacecraft and various appendages. At the heart, a cylindrical Optical Telescope Assembly held a 2.4 meter diameter primary mirror which would first receive the light from astronomical objects, reflect it forward to a 0.31 meter secondary mirror, which would then send the light back through a central hole in the primary mirror. Just behind the primary mirror five scientific instruments and three fine guidance sensors, themselves large and complex assemblages of optics, motors, and electronics, would share the bundle of precisely focused light. Surrounding the Optical Telescope Assembly, the Support Systems Module contained dozens of electronic and mechanical black boxes to operate the ensemble, and sprouted two deployable antennas and two large solar arrays to generate electricity.

---

11. Smith, *The Space Telescope*, Chapters 8-9.
12. Smith, *The Space Telescope*, Chapter 9; Chaisson, *The Hubble Wars*, Chapter 1. While the $6 million per month seems like a lot for storage, the spacecraft had to be kept in a mammoth clean room with active air conditioning and filtering systems operating constantly, some nitrogen purging of areas of the spacecraft, and some of its systems powered and operating. This was not "dead storage," but much more akin to a patient in intensive care. By the time the spacecraft was launched, it had spent almost a third of its intended design life suspended vertically in a gravity and atmospheric environment for which it was not designed, and with many of its systems powered up and running. Rothenberg, however, believes that this period provided opportunity for testing and debugging that ultimately was beneficial; Rothenberg Interview, August 8, 1995, 1/B: 160.

Designed to be serviceable in orbit, the Hubble Space Telescope was the most mechanic-friendly spacecraft ever flown. Dozens of bright yellow hand holds and sockets for the astronauts' portable foot restraints were strategically located around the spacecraft. Inside the multitude of hinged doors, all of the instruments and many of the black boxes could be removed and inserted by a space-suited astronaut using only a few tools. If motors or actuators failed, the solar arrays could be rolled up (like a window shade) or other appendages deployed or stowed using a ratchet wrench. Some of the deployment crew, notably astronauts Bruce McCandless and Kathy Sullivan, had practiced various emergency procedures for years. McCandless had worked, off and on, for twenty years helping to develop the maintainable features of the Telescope and the special tools required. They had spent many hours in Lockheed's clean room observing and working on the flight spacecraft itself, and in the water tanks at Marshall and Johnson working with various mockups.[13]

On April 25, astronaut and astronomer Steven Hawley grappled the spacecraft with the shuttle's robot arm and eased the 12-ton railroad tank-car sized vehicle out of the payload bay. Still attached to the arm, a carefully orchestrated sequence of commands instructed the spacecraft to deploy its antennas and solar arrays. There was a tense period when one of the solar arrays got stuck, and McCandless and Sullivan donned their spacesuits, waiting inside *Discovery's* airlock, ready to go outside if necessary. Ground controllers freed the array, however, and in mid-afternoon only one orbit later than planned, Hawley released the Telescope. Within ten minutes the freely flying spacecraft had locked onto the Sun, and later maneuvered itself to point the delicate optics, protected by the still closed aperture door at the front, away from the dangerous sunlight. As *Discovery* backed away, ground controllers began the complex sequence of commands to "wake up" the various dormant systems of the Telescope and prepare it for its fifteen-year mission. McCandless and Sullivan doffed their spacesuits, and with the rest of the crew continued *Discovery's* flight, the toolboxes never having been opened. Neither saw the Telescope again, and both left NASA shortly after the mission. Sullivan became chief scientist of the National Oceanic and Atmospheric Administration. McCandless retired from NASA, leaving the maintenance of the Telescope to other astronauts who might never have seen it in person—or so he thought.[14]

The euphoria of the successful deployment was mixed with mild concern as the spacecraft encountered some initial problems. These had nothing to do with the mirror, but rather involved a variety of glitches with the communications antennas and the control systems. The spacecraft somewhat spastically and repeatedly shut itself down into several so-called "safemodes" in response to the motions of various appendages, and had to be coaxed back into operation. When, a few days after deployment the fine guidance sensors attempted to lock onto stars, correctible errors were discovered in the programming but a more persistent oscillation prevented the telescope from keeping itself pointed with the required accuracy.[15]

---

13. Bruce McCandless, Oral History Interview January 8, 1986; Kathy Sullivan, Oral History Interview, January 9, 1986. These interviews are in the Space Telescope History Project collection, National Air and Space Museum, Smithsonian Institution.
14. Chaisson, *The Hubble Wars*, Ch. 1.
15. Chaisson, *The Hubble Wars*, Ch. 2; Smith, *The Space Telescope*, Afterword.

While engineers had forewarned before launch that such a complex spacecraft and ground system was bound to take some time to achieve stable operations, these initial problems earned the spacecraft some bad press that, in the words of a *Washington Post* story, "has turned it into fodder for stand-up comics and prompted some citizens to murmur that it may be a $2.1 billion lemon." On May 15, 1990, late-night comedian David Letterman offered the "Top Ten Hubble Telescope Excuses:"[16]

10. *The guy at Sears promised it would work fine.*
9. *Some kids on Earth must be fooling around with a garage door opener.*
8. *There's a little doohickey rubbing against the part that looks kind of like a cowboy hat.*
7. *See if you can think straight after 12 days of drinking Tang.*
6. *Bum with squeegee smeared lens at red light.*
5. *Blueprints drawn up by that "Hey Vern!" guy.*
4. *Those damn raccoons!*
3. *Shouldn't have used G.E. components.*
2. *Ran out of quarters.*
1. *Race of super-evolved galactic beings are screwing with us.*

After stabilizing, but not solving these and other operational problems and taking the first test image a month after deployment, there was a brief period of almost giddy elation. By mid-June, however, some scientists had become more and more worried about subtle characteristics of the star images. This concern turned to near despair as scientists and engineers realized the reason for the Telescope's inability to focus. The heart of the instrument, either the primary or the secondary mirror, or both (it would later turn out to be the primary) had been ground and polished over several years a decade earlier nearly perfectly to the wrong specifications. That this error had gone undetected over the years of testing was something most people within the program found too startling to believe. Just convincing them that the problem lay with the optics was difficult-it took nearly a month, from late May to late June. Scientists and managers responded to the suggestion with a mixture of bewilderment and outright denial. Once the analyses had finally become too compelling to ignore, NASA Associate Administrator for Space Science Lennard Fisk was told, and he responded that it might be space science's equivalent of the *Challenger* accident. A NASA press conference announcing the conclusion on June 27 showed a panel of somber faces, and the official making the first announcement literally stuttered over the words, "spherical aberration." Outside the program and NASA, the news was received with less grace, and more outrage and even ridicule. Congressional representatives were furious, influential Senator and space enthusiast Barbara Mikulski, calling the Telescope a "technoturkey," expressed outrage over Hubble's "cataract." Comedians and editorial cartoonists, from Herblock to Gary Larson's Far Side, could hardly believe their good fortune. Even filmmakers were quick to seize on the opportunity, and before the year was out: "An opening scene in the comedy film *Naked Gun 2 1/2* features a dark lounge with a depressing atmosphere, downbeat music, and walls lined with pictures showing historically horrible disasters. There, between a picture of the Hindenburg and a half-sunk Titanic, is the Hubble Space Telescope."[17]

---

16. Sawyer, Kathy, "Hubble Expected to Yield First Images Today; Problems Encountered by Telescope are not Unusual Among Spacecraft, Scientists Say," *Washington Post*, 20 May 1990, p. A-6.

17. Faye Flam, "NASA Stakes its Reputation on Fix for Hubble Space Telescope," *Science* 259 (February 12, 1993): 887–89; Smith, *The Space Telescope*, pp. 414–15; Chaisson, *The Hubble Wars*, Ch. 3.

Fix it or Write it off?

Initially, many feared that NASA or Congress would finally decide to write off the loss.[18] The first servicing mission, already planned for 1993, then became much more than a simple scheduled service call: It became the only chance to save the program and the spacecraft from either euthanasia or perhaps resignation to living with its diminished performance. Scientists and managers organized themselves along several strategic lines of work. While fending off the various congressional reviews, NASA Headquarters appointed a "Hubble Space Telescope Optical Systems Failure Review Board," chaired by Jet Propulsion Laboratory director Lew Allen. The Allen Committee, as it came to be called, began work in July and within a month had concluded the trouble lay with the main mirror. Investigating the records and the hardware that remained from the fabrication and testing of the mirror in 1980-1981, the Allen Committee found that a simple error in the test setup had skewed the measurement checks on the mirror, and that the computer-controlled polishing machine had dutifully shaped the mirror to the wrong curvature. Discordant test results at the time had been ignored under the schedule and cost pressures, and independent tests were not done. Eventually, the Justice Department and Hughes Danbury Optical Systems (Hughes had earlier bought the portion of the Perkin Elmer Corporation that had built the Optical Telescope Assembly) settled out of court.[19] While the Allen Committee was working to determine the cause of the spherical aberration, an HST Strategy Panel at the Space Telescope Science Institute under astronomers Holland Ford and Robert A. Brown began work to determine options for recovering from the mirror problem. The Science Institute had been established long before launch to be the focus for the academic community who wished to use the Telescope. With Goddard managing the mission and controlling the Telescope, the Science Institute would receive proposals from astronomers, manage peer review, and then work with the selected astronomers to schedule and obtain the desired observations. Institute staff included scientists who specialized in calibrating and understanding the Telescope's instruments, and engineers who specialized in merging thousands of approved observation requests into an efficient observing schedule. Under the leadership of the assertive and persistent Riccardo Giacconi, the Science Institute had made itself the watchdog for the scientists, at times much to the consternation of NASA managers. They interpreted their charge broadly, and since well before launch had been proactive without much regard for stepping on government toes.[20]

---

18. Hubble Space Telescope and the Space Shuttle Problems: Hearing Before the Subcommittee on Science, Technology, and Space of the Committee on Commerce, Science, and Transportation, United States Senate, One Hundred First Congress, Second Session . . . July 10, 1990 (Washington, DC: GPO, 1990); Hubble Space Telescope: Hearing Before the Subcommittee on Space of the Committee on Science, Space, and Technology, U.S. House of Representatives, One Hundred Third Congress, First Session, November 16, 1993. (Washington, DC: GPO, 1993.).

19. Lew Allen, Roger Angel, John D. Mangus, George A. Rodney, Robert R. Shannon, and Charles P. Spoelhof, *The Hubble Space Telescope Optical Systems Failure Report*, NASA TM-103443, (Washington, DC: NASA, 1990); Robert S. Capers and Eric Lipton, "Hubble Error: Time, Money, and Millionths of an Inch," *Hartford Courant* (March 31–April 3,1991), reprinted in abbreviated form in *Academy of Management Executive* 7 (1993): 41–57. I am grateful to the *Hartford Courant* for supplying a reprint of the original series. See also "Hubble Board Expertise Covers Astronomy, Reconnaissance, Quality," *Aviation Week & Space Technology* 133 (July 9, 1990): 18.

20. Smith, *The Space Telescope*, Chapters 6 and 9; Chaisson, *The Hubble Wars*, p. 200 ff.; "Space Telescope Institute: Inside the Black Box. A New Director Promises to Focus on Service," *Science* 260 (June 18, 1993): 1716.

At Goddard, the newly appointed Associate Director for Flight Projects for the HST, Joe Rothenberg, and the Project Scientist, Al Boggess, met in late July to lay out a strategic recovery plan. First, they would do whatever science could be done with the Telescope in its current state; second, they would try to fix the Telescope's problems at the 1993 maintenance opportunity; finally, they would concentrate on extending the wavelength coverage into the infrared on subsequent scheduled servicing missions that would install replacement instruments. Rothenberg was careful to establish close working relationships with the relevant Headquarters managers and with the Science Institute.[21]

Planners were very lucky that the figure of the mirror was so precisely and uniformly in error. Concurrent with the Allen Committee and the Strategy Panel, a "Hubble Independent Optical Review Panel" under Duncan Moore, University of Rochester Institute of Optics Director, worked to determine the precise prescription for the aberrated mirror, based on records and artifacts at the manufacturer and data taken by the Telescope instruments. The 2.4 meter wide main mirror was too flat by about 2 micrometers, or 1/40 the thickness of a human hair. That meant that an optical element with the reverse prescription could correct much of the aberration. It had already become clear that future replacements for the existing scientific instruments, scheduled to be inserted every few years beginning with the 1993 servicing mission, could incorporate internal optics that would reverse much of the aberration of the Telescope's mirror. However, the existing five instruments and the fine guidance sensors (used by the astrometry team as a sixth virtual instrument) would remain severely compromised for many years unless some other solution were found. In particular, the European Space Agency's Faint Object Camera had no follow-on in the plans, and it was scientifically as well as politically important to find some way to address the European concerns. The primary mirror defect had two serious consequences. First, because the light rays were not precisely brought to a single focus, the images would lack resolution and the other instruments could not pick out individual objects in crowded fields or very small features of extended objects. Second, because the light was diffused the Telescope's instruments could not reach the planned limits of faintness. These two desiderata, clarity and faintness, were precisely the reasons for putting a Telescope of this size above the atmosphere in the first place. While the Telescope was still capable of providing valuable data and addressing much of its observing program, it was compromised in the areas for which it was supposed to be uniquely suited. The existence of a very serious flaw that was nonetheless amenable to compensation was the foundation for an expanded servicing mission.

The Hubble Space Telescope Strategy Panel included many distinguished and experienced astronomers and engineers, as well as retired astronaut Bruce McCandless to advise on the on-orbit feasibility of various proposals. Co-Chair Robert Brown had served as Space Telescope Project Scientist for several years before launch. The Panel considered a wide variety of schemes, including: mechanically or thermally slightly deforming the main mirror; overcoating the main mirror to alter its shape; installing full-aperture glass or gas-filled corrective optics at the front of the telescope; replacing the secondary mirror.

---

21. Joseph Rothenberg, Oral History Interview August 10, 1995, 1/A: 229, 432 and September 8, 1995, 2/B: 000; September 8, 1995, 2/B: 040. Rothenberg had worked in industry on space telescope concepts during the 1970s, and had managed much of the ground data and control system development for several years prior to HST's launch. Boggess was an alumnus of several space astronomy missions at Goddard, including the long-lived and resilient International Ultraviolet Explorer.

These and other proposals were all found wanting, or downright dangerous (to the Telescope or to the astronauts) to various degrees. The Panel also considered how to incorporate changes into the planned replacement instruments scheduled to be installed every few years starting with the 1993 servicing mission.[22]

Bits and pieces of the solution lay near to hand, but would require innovative modifications. As a precaution, work had begun at Jet Propulsion Laboratory in the early 1980s on a replacement for the main imaging camera, the Wide Field/Planetary Camera. Its internal optics would be modified by adding small mirrors figured to reverse much of the spherical aberration introduced by the main mirror.[23] Early in the HST Strategy Panel discussions, optical expert Murk Bottema, of Ball Aerospace, suggested using mirrors similar to those in the replacement Wide Field-Planetary Camera II to adjust the incoming light for the other, axial scientific instruments. The problem was how to deliver such mirrors to the Telescope and insert them precisely into the light bundle behind the main mirror. The solution to this problem, due to electrical engineer James Crocker of the Space Telescope Science Institute, is so remarkable as to seem apocryphal. One evening, during the Strategy Panel's meeting at the Space Telescope European Coordinating Facility in Garching, German, Crocker stepped into the shower in his hotel room. The European-style fixtures included a showerhead on an arrangment of adjustable rods. While manipulating the shower, Crocker realized that similar articulated arms bearing Bottema's mirrors could be extended into the light bundle from within a replacement axial instrument by remote control: "I could see Murk Bottema's mirrors on the shower head."[24]

In the early 1980s, work had begun on a device called STAR, Space Telescope Axial Replacement, an empty stand-in for one of the four axial scientific instruments, just in case one might have to be removed with no new instrument available for insertion. STAR was designed to be mechanically, thermally, and in other ways benign, so that the complex spacecraft would not "notice" the missing instrument. With Bottema's mirrors and Crocker's mechanical arrangement packed inside, STAR would become COSTAR, Corrective Optics Space Telescope Axial Replacement. Once installed as if it were a replacement axial instrument, COSTAR would deploy the tiny mirrors, each figured to intercept and then reverse the spherical aberration for that portion of incoming light directed to various apertures of the remaining three axial instruments. The three remaining axial instruments had a total of five precisely-placed entrance apertures, and the optical design required two mirrors for each. Thus, ten mirrors had to be inserted into the light path in such a way as to intercept the light, correct it, and direct it into the apertures, all

---

22. R. A. Brown and H. C. Ford, eds., *Report of the HST Strategy Panel: A Strategy for Recovery; Results of a Special Study, August–October 1990*, (Baltimore, MD: Space Telescope Science Institute, 1991); "Hubble Scientists Urge NASA to Consider Early Repair of ESA Faint Object Camera," *Aviation Week & Space Technology* 133 (October 15, 1990): 25.

23. K. Leschly, D. Allestad, and L. Herrell. "The Second Generation Wide-Field/Planetary Camera of the Hubble Space Telescope." *Journal of the British Interplanetary Society* 44 (October 1991): 477. The first camera could switch between the "wide field" and the narrower "planetary" camera modes, hence the slash ("/") in its name. The replacement camera gave up this capability in exchange for the corrective optics, putting the two resolutions into the same image frame, and came to be designated officially the "Wide Field/Planetary Camera II." I have followed this usage.

24. Quoted in Douglas Birch, "Hang on, Hubble; Help is on the Way," *Baltimore Sun Magazine* (March 14, 1993): 17–18; Aguirre, *Rescue Mission in Space*, 9:30; Crocker is perhaps the first scientist to re-enact his moment of discovery for a television documentary by getting into a shower, seemingly clad only in a towel. The four axial scientific instruments, the size and shape of a phone booth, are stacked like four sticks of butter in the very aft end of the telescope; the radial instruments (three Fine Guidance Sensors and the Wide Field/Planetary Camera) are each roughly the size of a baby grand piano and are arrayed radially around and just behind the primary mirror.

the while precisely shadowing the apertures from the flood of the remaining aberrated light. COSTAR was an exceptionally complex and delicate system of 5,300 parts including mirrors, mechanical components, and electronics all controllable from the ground. Some of the coin-sized mirrors were an optician's nightmare to figure, their shapes being "anamorphic fourth-order aspheres on toroidal blanks," painstakingly hand-made by Tinsley Optics in California. COSTAR would "fix" the European Faint Object Camera, the Goddard High Resolution Spectrograph, and the Faint Object Spectrograph, and the replacement Wide Field-Planetary Camera would restore the primary imaging camera abilities. Astrometry and pointing control would have to live with the diminished performance of the Fine Guidance Sensors, and the High Speed Photometer, the least problematic but the least-used instrument, would have to give up its spot for COSTAR. Having settled on the COSTAR approach, Ball Aerospace began design and proposal work in November, and was awarded a contract in January 1991. Ball had built the Goddard High Resolution Spectrograph for the telescope, and was designing two second generation scientific instruments for future servicing missions.[25]

The first servicing mission now included replacing two scientific instruments and it would have to remedy a host of other emerging problems, some of which were becoming very serious. At the end of 1990, engineers operating the telescope had devised software changes to try to counter the oscillations of the solar arrays, which would respond sharply to passage between shade and the Sun that occurred twice each 90-minute orbit. The thermally-induced oscillations slowly damped out, but took longer than half an orbit to do so, when another passage would start them all over again. There was limited available memory in the onboard computers, and the new control laws to handle the solar array oscillations consumed a good portion of it. Solar array replacement looked as though it would have to join the other maintenance tasks, although the final decision to replace the European-provided solar arrays was not made until much later. After a spirited debate, they even edged out the optical fixes to assume highest priority.[26]

The Telescope continued to be temperamental, and NASA finally decided in late 1990 that the engineering commissioning phase, called Orbital Verification, was complete—or that they had gotten the spacecraft operating about as well as they could and regular science operations should wait no longer. The Marshall Space Flight Center team who had been in residence since deployment returned to Huntsville, leaving Goddard Space Flight Center to operate the spacecraft and the Space Telescope Science Institute to continue its work coordinating observing proposals, scheduling objects to be observed, calibrating the instruments, and archiving the data. Between the degraded focus, solar array oscillations, the fine guidance sensors acting temperamentally, and a variety of other sporadic problems, all of the observing plans were continually in flux. Various observing programs, planned with the full capability in mind, had to be reassessed and sometimes deferred. The operational problems had further made scheduling difficult, both by degrading observations and also taking up scheduled observing time to resolve.[27]

---

25. Bernadette C. Stechman, "Just Wait and See!" *Challenge* (Ball Corporation) 6 (1993): 10–13; "Our Stars Behind Costar," *Ball Line Quarterly* (Ball Corporation) 49 (First Quarter, 1994): 4–11; Ball Aerospace and Communications Group, *Corrective Optics Space Telescope Axial Replacement*, Information Packet (Boulder, Colorado: 1990); "NASA Awards Contract to Complete COSTAR, Designed to Fix Hubble Telescope Optics," *Aviation Week & Space Technology* 98 (October 21, 1991). I am grateful to John Hetlinger and Charles Scaglia, Ball Aerospace, for providing background materials on COSTAR; "Corrective Optics to Star in Drama to Fix Telescope," *Aviation Week & Space Technology* (May 24, 1993): 44–45.
26. Rothenberg Interview, August 8, 1995, 1/B: 103.
27. Rothenberg Interview, August 10, 1995, 1/A: 113. The October 1 date for formally turning over full control of the telescope from Marshall to Goddard had been agreed to in August.

During 1991, the teams settled into doing the observing programs not preempted by the hardware problems, while developing COSTAR and the other major hardware for the servicing mission. Ball Aerospace began assembling hardware for COSTAR in July, receiving the first flight optics at the end of the year. As the servicing mission planning group considered how to include the new tasks (they had been meeting formally since August 1988), they found themselves under pressure to include still more. In May, a memory unit in the spacecraft's main computer failed, sending the Telescope into the deepest safemode available, nearly the equivalent of a coma. In June, a second of six gyroscopes failed, leaving only one spare since three were required for the spacecraft to determine its position and attitude accurately enough to conduct observations. In July, the Goddard High Resolution Spectrograph developed a problem in its power supply, eliminating half of its capability. By the end of the year, it was clear that the first servicing mission was going to be much more ambitious than anyone had expected. It would be more of an overhaul than a repair. Cost estimates are difficult, both for the servicing mission and for the Telescope as a whole, but the repair cost a significantly large fraction of the Telescope's initial cost. Had it been an automobile, it might well have been declared a total loss.[28]

Planning a Service Call

The close cooperation required of so many parts of NASA, combined with the high stakes and growing anxiety evident in Washington, encouraged a number of institutional clashes that had to be overcome. To the managers and astronauts at the Johnson Space Center, the servicing mission at first appeared as one among others, and it never occurred to them to treat it any differently. Confident and proud of their system for planning and executing Space Shuttle missions, with a host of other missions at various stages of execution, and already interacting with various institutional "customers" for those missions, Johnson was slow to change its approach. As Headquarters scrutiny increased, the complexity of the mission grew, and the need for more subtle interaction with other elements of the program became more apparent, Johnson managers began to feel the pressure.[29]

The result was that planning for the mission at Johnson accelerated much earlier than was their typical procedure. For "customer driven" missions, a multidisciplinary and multi-institutional Payload Operations Working Group would spend one or more years developing the outlines of a mission. Actual flight crews and ground controllers were seldom selected earlier than a year before launch. In the Spring of 1992, some twenty months before launch, Story Musgrave was the first crew member named. Musgrave, an Apollo-era veteran, polymath, and extremely accomplished in extravehicular activity, would be the Payload Commander, foreman for the repair crew and ultimately responsible for the Telescope. Musgrave had started working on making satellites maintainable and serviceable in orbit in 1976, and had been the astronaut most involved in developing the space suits and other equipment. Eleven astronauts began sixteen sessions in Marshall's Neutral

---

28. Typical estimates for the repair mission are $500 million, and for the Telescope's initial construction $1.5 billion, making the repair about one-third the "value" of the Telescope. However, the figures change substantially if one includes the transportation costs of a Shuttle mission or the costs of mission operations and data analysis.

29. Capshew and Rader, "Big Science," p. 16, note the "importance of local context, including geographical and cultural factors" in big science. On the independence of NASA Centers see Howard E. McCurdy, *Inside NASA: the Changing Culture of the American Space Program* (Baltimore, MD: Johns Hopkins University Press, 1992). On the urging from Goddard planners for early assignment of the flight crew see Rothenberg Interview, September 8, 1995 2/B: 320.

Buoyancy Simulator, working with mockups in simulated underwater microgravity to obtain rough approximations of the times required to do various maintenance tasks. All the while, new black box failures on the spacecraft put pressure on the roster of chores to be accomplished.[30]

At about the same time, J. Milton Heflin was chosen lead flight director. With almost thirty years experience at Johnson, Heflin had been involved in ocean recovery of the Apollo astronauts, and had been in mission control since the very first shuttle landing tests in 1977. With ten years and twenty shuttle missions under his belt as a flight director, he was one of the most experienced and seasoned. As Heflin got organized, he was also supporting several other missions as well. He, like most at JSC, recognized that the Hubble Servicing Mission was challenging and needed more attention than some other missions. He would soon realize, however, just how much more attention and resources it would require. Initially, the core of his mission control team included payloads officer Jeff Hanley, robot arm engineer Sal Ferrera, and extravehicular activity experts Jim Thornton and Susan B. Rainwater.[31]

Meanwhile, events in Washington transpired that would have decisive and far-reaching effects on the agency and on the Hubble servicing mission. In early 1992, after escalating discord with the White House, Richard Truly was effectively fired as NASA Administrator. Truly, a career astronaut, was admired and respected throughout the agency but perhaps nowhere more than at Johnson. By March, a new, non-NASA face appeared and was confirmed on April 1. Not perhaps since James Webb had there been a NASA Administrator who was more of an outsider, more inscrutable, or more difficult to adjust to, than Daniel Goldin. Having worked in mostly classified space programs at TRW, Goldin was expected to bring a no-nonsense industrial agility and accountability to the agency. The Bush-Quayle administration had decided that Truly was too closely identified with the old guard, and were determined to bring a fresh approach to NASA. While Goldin championed what he considered to be NASA's strengths, and went out of his way to praise the achievements of NASA, he was determined to bring change to the agency. To some old hands Goldin seemed impulsive, abrasive, and fearfully insensitive to the agency's core traditions and values. To Johnson, anyone who replaced the beloved Richard Truly, under duress no less, boded ill. Stories of strife in "Code A," the office of the Administrator, were rife.[32]

---

30. Rothenberg Interview, August 8, 1995, 2/A: 010.

31. Heflin Interview, March 17, 1995, 1/A: 020, 1/B: 075; it is against the prevailing ethos at Johnson to single out a particular mission as more important than others, and to characterize some people as better than others. However, it is clear from interviews and conversations that while all missions are equal, Hubble was more equal than many others. Accomplished people were eager to be a part of it, and supervisors were concerned to assign their best talent to the mission. On flight assignments, particularly the astronaut mandate of never asking for a particular assignment, see Henry S. F. Cooper, Jr., *Before Liftoff: the Making of a Space Shuttle Crew* (Baltimore, MD: Johns Hopkins University Press, 1987).

32. John Logsdon details elsewhere in this volume Truly's role in the two-and-a-half year return to flight after the 1986 *Challenger* accident, a contribution that earned Truly a Collier Trophy. On Goldin's appointment as NASA Administrator and his management style see: Kathy Sawyer, "The Man on the Moon; NASA Chief Dan Goldin and a Little Chaos Just Might Save the Space Program," *Washington Post* (July 20, 1994): B-1; Theresa M. Foley, "Mr. Goldin Goes to Washington," *Air and Space* 10 (April–May 1995): 36–43. Early in his career Goldin had worked at the NASA Lewis Research Center, but had been with TRW for 25 years prior to his appointment.

The strong signals of change grew more potent when, in May, Marine Major General Jeremiah W. "Jed" Pearson III replaced astronaut William B. Lenoir as Associate Administrator for the Office of Space Flight ("Code M"). Pearson had been deputy commander of Marine forces in Operation Desert Storm before serving briefly at Marine Headquarters, and he now commanded the very heart and soul of NASA—the Space Shuttle program. Goldin continued to replace many Headquarters and Center officials, while the old hands at the agency went through the anxiety of organizational change.[33]

In mid-May Goldin traveled to Johnson to observe a Space Shuttle mission first-hand. The STS-49 *Endeavour* crew's mission was to capture and repair the *Intelsat VI* communications satellite and evaluate Space Station construction techniques. *Intelsat VI*'s apogee kick motor had failed two years earlier, leaving it stranded in a low orbit. The crew was to approach the spinning satellite, insert a special tool into the central motor chamber about which it turned, and fire a grappling device that would seize the chamber, allowing them to slow it down. Once stabilized, they would repair it and then send it on its way to geostationary orbit.

Problems developed almost immediately. The four and one half ton satellite bounced away each time Pierre Thuot, in foot restraints at the end of the robot arm, tried to thrust the capture bar into position. Worse, it began to wobble, and fearing that further attempts might be too dangerous they decided to quit for the day and revisit their options. The seven million dollars capture bar had worked during numerous ground simulations. Something was wrong. Overnight, the crew worked with ground planners and astronauts in simulators to devise a bold contingency plan. The next day, after the capture bar again failed to work, for the first time in history, three astronauts were outside their spacecraft at once. Thuot, Richard Hieb, and Thomas Akers, "the gang of three," stationed themselves in a circle in the payload bay, their feet anchored in restraints, while pilot Dan Brandenstein eased the shuttle toward the spinning satellite. They used their gloved hands to reach out, grab, brake *Intelsat VI* to a stop, and then lower it into its repair fixture.[34]

While the crew repaired the satellite and sent it on its way to geosynchronous orbit, they were hailed as "space wizards." Goldin praised the bold move as a "return to [the] can-do NASA of old," and editorials and commentaries gushed about the drama of the satellite rescue and its proof of the usefulness of astronauts. However, outside the limelight, it "set NASA managers scrambling to rethink their training methods and assumptions about handling large masses in orbit." The flight and ground crews had saved the mission, but they had taken what some considered to be undue, perhaps even foolhardy risks. Caught between deeply engrained and conflicting values, Johnson had been torn between conservatism and safety, on the one hand, and the driving desire to complete the mission, on the other. They had chosen deliberately on the ground and in space to be bold and complete the task. Tom Akers and Kathy Thornton ran into further problems assembling prototype Space Station elements in the payload bay later in the mission. This, too, did not go as it had in simulations and training. It shook Johnson's planning, training, and simulation groups to the core.[35]

---

33. Sawyer, Kathy, "The Federal Page: New NASA Chief Appoints Four Senior Officials," *Washington Post*, April 29, 1992, p. A-21.

34. William Triplett, "Reality Check," *Air and Space* (November 1993): 35-42; Kathy Sawyer, "Can-do NASA is Upbeat, but Dramatic Satellite Rescue Raises Questions," *Washington Post* (May 15, 1992): A-9.

35. Sawyer, "Can-do NASA is Upbeat..."; Story Musgrave Interview, March 23, 1995 1/B: 000; K. T. Thornton Interview, April 3, 1995; Rothenberg Interview, September 8, 1995, 3/A: 010. "Endeavour's flight advances station, Intelsat objectives," *Aviation Week & Space Technology* 136 (June 1, 1992): 43; "Mission Control saved Intelsat rescue from software, checklist problems," *ibid.* 136 (May 25, 1992): 78; "Endeavour's Intelsat rescue set EVA, rendezvous records," *ibid.*, 136 (May 18, 1992): 22.

Worse, while Goldin praised publicly his charges for the daring satellite rescue, he is remembered by others to have commented on "those cowboys!" Barely six weeks into his tenure as NASA Administrator, this was not getting off on the right foot. While Johnson quietly looked inward to reassess their training methods, Goldin appointed a high-level group task force on satellite rescue and repair to study how NASA should handle future such situations. Implications for the Space Station and the Hubble servicing mission were not lost on anyone. Center Directors and Headquarters Associate Administrators decided they, too, needed to track the servicing mission more closely. By the end of July, four new review teams were looking at the Hubble servicing mission, and many more would follow.[36]

At the end of August, sixteen months before launch, three astronauts were named to join Story Musgrave as the extravehicular activity contingent for the servicing mission, making it the most experienced and seasoned crew ever: Tom Akers, Kathy Thornton, and Jeff Hoffman. A lot of work was scheduled in Marshall's Neutral Buoyancy Simulator and Johnson's Weightless Environmental Training Facility to refine further the timelines for the servicing mission, and "EVAs of opportunity" were inserted into upcoming shuttle missions wherever possible to gain more experience. In January 1993, on STS-54, Mario Runco and Greg Harbaugh of the *Endeavour* crew improvised the first of these by carrying one another around the payload bay in a "mass handling" exercise and tested some new tools. Several other crews practiced aspects of the repair mission in orbit, spacewalks being added at every possible opportunity.

Meanwhile, the Telescope continued to be a problem child. In September 1992, the Faint Object Camera's power supply developed problems eliminating half of that instrument's capabilities. A third gyroscope failed, leaving the Telescope running on its minimum complement of three. If another gyroscope were to fail, the spacecraft would be safe but unable to collect any scientific data. A second flight computer memory unit failed. Most of these failures added relatively easy individual tasks to the servicing mission, but the timelines were already overbooked and confidence in them somewhat shaken by the new conservatism. They also caused a scramble among spacecraft engineers to determine which problems could be repaired at all, and of those which had highest priority.[37]

Goddard and the Science Institute, resigned to operating what had become a positively cantankerous spacecraft, became quite adept at dealing with sudden hiccups and replanning observations again and again. The Goddard controllers and scientists had been operating astronomical spacecraft for thirty years. In particular, they could draw on their experience and institutional memory of the remarkable International Ultraviolet Explorer (IUE), which had survived well beyond its design life. They had even learned to operate IUE on a single gyroscope. It was befitting that in the IUE control room sat a stuffed toy "Energizer Bunny." Beginning at the end of the summer of 1991, after the

---

36. Sawyer, "Astronauts Call Rescue of Satellite 'worth doing'; Benefits Said to Outweigh Spacewalks Dangers, Costs," *Washington Post* (May 30, 1992): A-3. Joining the Headquarters Office of Space Flight's "HST Servicing Mission Review Team" under former astronaut General Thomas Stafford, in June the Space Shuttle Program Office at Johnson chartered an "HST Review Team" under Richard Fitts. In July, the "HST SM-1 Program Review Team" (Michael Greenfield Committee) was chartered by the Headquarters Administrator's office; former astronaut Joseph Allen reviewed the extravehicular activity aspects of the mission for the Headquarters Office of Space Science. Astronaut John Young was commissioned to review the plans as well. In July, the HST Servicing Mission Critical Design Review (CDR) was held.

37. The baseline for the servicing mission was three days of spacewalks. Not until April 1993, only eight months before launch, would the number be raised officially to five. In part this represented conservatism of planners trying to be absolutely certain that such an ambitious schedule was necessary. Flight Director Milt Heflin recalled it as a bit of a poker game, not wanting to give in to additional spacewalks unless they were shown to be absolutely necessary, adding, "I can guarantee you that prior to nine months we knew we were headed to five EVAs." Heflin Interview, March 23, 1995, 1/B: 040 & 330; see also Rothenberg Interview, August 8, 1995, 2/A: 140.

shock and denial had run its course, the astronomers and operators had resolved to make the most of the capabilities they had. The first science results were published in *Astrophysical Journal Letters* at the end of 1991. By the end of 1992, astronomers had accumulated a respectable suite of results, some of which surprised even themselves. The Telescope was performing at a level somewhat better than Earth-based telescopes, but far below its expected capabilities.[38]

In December 1992, a year before launch and at about the time it would have been usual to first assign a crew, Richard Covey (Commander), Ken Bowersox (Pilot), and Claude Nicollier (a European Space Agency astronaut and expert on the robot arm) were named to complete the STS-61 flight crew. With these three, the servicing mission continued to enjoy the most experienced and seasoned astronauts available, now with sixteen previous shuttle flights among them. Also in December, a new face at Johnson was named to a position that had not been used since Apollo: Randy Brinkley was to occupy the newly created position of Mission Director.[39]

Brinkley and his position were controversial from the first at Johnson, even though the post had been recommended by former Astronaut Thomas Stafford's review of the servicing mission. "Badged Headquarters," Brinkley reported to Washington but was located at and carried on his work at Johnson. The term carried special significance at Johnson, originally created as the Manned Spacecraft Center in the early 1960s. The Space Task Group, a small band of space enthusiasts among aeronauticists at the Virginia Langley Research Center, had virtually invented human space flight at NASA. As Apollo grew into a behemoth, Headquarters decided to create a new Center on some politically favorable bayside land south of Houston. The space pioneers from Langley had little choice but to leave the Hampton, Virginia, area that many of them loved, for to resist moving to Houston meant to be out of the action. They built the space center from the ground up, developing all the camaraderie and social structures of pioneers, and absorbing much of the rugged, self-reliant culture of Texas. They developed and flew Mercury, Gemini, Apollo, Skylab, and the Space Shuttle. They invented astronaut training, mission control, and spacewalking. At JSC, as they referred to their institution, they operated a veritable space university, with several associated colleges where they trained recruits in how to operate in space and support space flight from the ground. Through Apollo and into the shuttle era there had been a constant tension between Headquarters and JSC, a continually and dynamically negotiated balance of independence and subordination. From Webb onward, Headquarters administrators had placed personnel at JSC who reported not to the Center but to Headquarters.[40]

Brinkley had been at Johnson for several months as a special assistant to Headquarters' space flight chief Jed Pearson before being appointed to the servicing mission. A Marine aviator for twenty-five years, he had served with Pearson during Desert Storm and then later

---

38. Chaisson, *The Hubble Wars*; "Early Results from the Hubble Space Telescope," *Scientific American* 266 (June 1992): 44-46; Faye Flam, "NASA PR: Hype or Public Education?," *Science* 260 (June 4, 1993):1416-18.

39. NASA News Release 92-218, December 8, 1992; Rothenberg Interview, August 8, 1995: 2/A: 020.

40. Ronald L. Newman and Randy H. Brinkley, *STS-61 Mission Director's Post-Mission Report* (Houston, TX: Johnson Space Center, January 1995), pp. 31–32, contains background on creating the post; I am grateful to Messrs. Brinkley and Newman for providing a copy of this detailed and lengthy document. The attitudes toward Brinkley's reception at Johnson are drawn from several oral history interviews, including Brinkley (16 March 1995), referenced in this article, as well as conversations with other Johnson personnel. On the Center's culture, see Henry C. Dethloff, "*Suddenly Tomorrow Came. . . . A History of the Johnson Space Center*", (Washington, DC: NASA SP-4307, 1993). On Headquarters and Johnson relationships, see Arnold S. Levine, *Managing NASA in the Apollo Era*, (Washington, DC: NASA SP-4102, 1982) and W. Henry Lambright, *Powering Apollo, James E. Webb of NASA* (Baltimore, MD: Johns Hopkins University Press, 1995). As John Logsdon notes in his contribution to this volume, as part of the post-*Challenger* management reforms, lead responsibility for the shuttle was taken away from Johnson and returned to Headquarters.

spent a brief time with McDonnell Douglas. Before taking on the servicing mission, Brinkley had studied the shuttle program and tried to get oriented, absorbing some aspects of the JSC culture in the process. He had been impressed by JSC's technical competence, but no less by its independence and his own difficulty in "break[ing] into that subculture." Brinkley researched the Mission Director concept and tried to figure out how to implement it, seeking the advice of many old hands. His charter was broad, to ensure the success of the servicing mission, and largely undefined: "I knew I had the responsibility," he recalled, "somehow I had to grow my own authority." This was a broad mission order commonly used in the Marine Corps, similar to others Brinkley had received there, and in keeping with Pearson's style. Brinkley assembled a small staff, representative of the various kinds of inside expertise he would need.[41]

He had also, early on, been impressed by how important this mission was to many beyond JSC. From meetings with Pearson, Goldin, and congressional representatives (especially the plain-speaking Barbara Mikulski), the message he received was: "this was a make or break mission for NASA . . . Brinkley, don't screw this up, the future of NASA lies in the balance." To gain insight into the progress of the many phases of the mission, he turned to using the independent review groups that had already been set up by others, and created some of his own. The many reviews, some of them uncoordinated and motivated by growing upper-level anxiety over the mission, put a great deal of pressure on the working troops. If Brinkley added to this pressure somewhat by commissioning still more reviews and assessments, he also drew gratitude from the workers by managing and coordinating the reviews and serving as somewhat of a lightning rod. "Answering the mail from Headquarters," while initially not seen as much of a genuine contribution to the mission, eventually was seen as a valuable activity that allowed the people working on various aspects of the mission to function with minimal diversion. The various review recommendations and attention also gave the Hubble servicing mission team a good deal of clout in getting the resources they needed on a priority basis, clout which Brinkley was not afraid to exercise from time to time. This also caused some dismay among the crews of other missions competing for flight controllers, simulation and training time, and attention from the various technical service group.[42]

In January 1993, Administrator Goldin appointed the most formal and highest level review committee of all. Tapping Joseph F. Shea, an Apollo manager, and several senior aerospace executives and experts, he formed "The Task Force on the Hubble Space Telescope Servicing Mission." After several formal meetings and numerous briefings at all the sites where the mission was being planned, the panel in May called the repair mission "achievable" but recommended continued close management attention, strongly endorsing the Mission Director concept. They were concerned about the "escalating nature of the mission," and the "instability" of the plans and on-orbit schedule, based on a "worrisome" trend of equipment failures on the spacecraft.[43]

---

41. Brinkley Interview, March 16, 1995, 1/A: 055; March 22, 1995, 1/B: 060. Among others, Brinkley consulted astronauts Tom Stafford and John Young, both of whom had flown an Apollo mission.

42. Brinkley Interview, March 16, 1995, 1/A: 138; March 22, 1995, 1/B: 000, and 300ff; Heflin Interview, March 17, 1995, 1/A: 300 and March 23, 1995, 1/B: 620. Mission Commander Covey told a reporter, "For this mission we were basically able to get what we needed," but that might not be the case for all missions, he added, Covey, answering question at Crew Post Mission Press Conference, 4 January 1994. Several JSC managers independently admitted to initially resisting the mission director concept and later appreciating Brinkley's contribution. See also Cooper, *Before Liftoff*.

43. Joseph F. Shea, et al., *Report of the Task Force on the Hubble Space Telescope Servicing Mission*, "Shea Panel Report", (Washington, DC: NASA, May 21, 1993), p. 8; James R. Asker, "Hubble: Risky, but 'Achievable,'" *Aviation week & Space Technology* (May 24, 1993): 40–44.

Confluence

The servicing mission itself lay at the confluence of several streams of work involving NASA Headquarters, Jet Propulsion Laboratory, four NASA Centers, the Space Telescope Science Institute, and a half-dozen contractors. During the summer of 1993, these streams would truly converge, and any residual turbulence had to be dealt with. Choreographing the mission was easily as challenging as any other aspect. Before and after capture, myriad instruments and systems on the spacecraft would have to be turned off systematically, the aperture door closed, antennas retracted, and solar arrays rolled up. The shuttle crew would have to grapple the spacecraft with the robot arm, and gingerly berth it into the receiving fixture of the payload bay where electrical connections could supply "life support" during the repairs. The Goddard controllers would have to watch the health of the spacecraft, putting some systems into standby and turning others off as the repair crews began to remove connectors and components, and then turning the systems back on afterward to ensure the "aliveness" of the new parts. As various access doors on the Telescope were opened, the shuttle's attitude had to keep stray sunlight from entering the Telescope while still maintaining lock on the tracking and data relay satellites.

The "EVAs of opportunity" inserted into the shuttle schedule during 1992-1993 revealed crucial information that probably was decisive in the success of the servicing mission. For various reasons the areas of the Telescope being worked on had to be kept out of sunlight. If subjected to sunlight, for example, the black insulation that suppressed interior reflections might "outgas," exuding contaminants that might later deposit onto the optical surfaces. Solar heat might cause insulation to de-bond or expand. Over most of the extravehicular activity experience since Gemini, the problem had been keeping astronauts cool, since they were almost always in sunlight or reflected earthshine, and so the suits and gloves were very efficient at cooling. Mission designers, not taking crew temperatures into consideration, designed a trajectory optimum for the Telescope's and orbiter's needs. This oversight revealed itself dramatically and at a most inopportune time.

In May 1993, Story Musgrave began a series of human thermal vacuum tests, similar to those done by many astronauts since Apollo. After four hours in the airlock, breathing oxygen in his suit to rid his blood of nitrogen, he finally entered the vacuum chamber for an experience he described as, "the world's worst hell. That's the toughest day that you are ever going to have as an astronaut." Inside the black chamber, pumped down to the deepest vacuum possible, and dragging his counterbalanced 480-pound inflated suit "like a plough horse," he started several hours of tool fit checks. His job was to go through all possible combinations of tools—sockets, ratchets, extensions—and fasteners to see whether they would fit and behave at a hundred and seventy degrees below zero. Even Musgrave himself, a medical doctor, did not realize what was happening as he squeezed the tools harder to get them to snap together and apart.[44] It was, he says, the "insidiousness of going from pain to injury." Working in an inflated glove, and feeling numb anyway, he worked on for hours, occasionally pulling his fingers out of the gloves and up, the way a person in a parka might pull his hands into the sleeves, and occasionally trying to warm them in another part of the chamber that simulated sunlight temperatures. After finishing the fit checks, hours of decompressing, and emerging from the airlock, the metal on his suit was still too cold for anyone to approach. As he struggled to get his gloves off, and they dropped away, the damage was evident: severe frostbite, tissue death, in eight fingers.[45]

---

44. Musgrave Interview, March 23, 1995, 1/A: 000.
45. Musgrave Interview, March 23, 1995, 1/A: 620ff; Johnson Space Center, Crew and Thermal Systems Division, *Hubble Space Telescope Servicing Mission Manned Thermal Vacuum Test Report, Part III*, CTSD-SH-756, JSC-37856, Houston, TX: September 1993.

"It was an essential thing which had to happen . . . it redefined the entire EVA world," he said. John Young later noted that astronauts had "complained of cold EVA temperatures for years," and Musgrave recalled cold hands on his very first shuttle spacewalk on STS-6 in April 1983. But previous shuttle astronauts had only been in the shade for portions of their spacewalks, and the Hubble mission would require them to work for extended periods away from sun- or earthshine. With only seven months to go before the mission, Musgrave was flown to be treated by frostbite experts at the University of Alaska, mission planners turned to reconsider extravehicular activity equipment and procedures, and high-level managers fretted. Musgrave recovered fully in time for the mission, and was even in the water tank within days of the injury. But the incident produced "a whole new attitude"—quite literally.[46]

John McCune, an experienced "pointer" who specialized in designing orbiter attitudes for missions, suggested flying an upcoming mission in the orbiter attitude planned for the servicing mission to learn about the thermal environment. Astronauts on STS-57 in June found that the payload bay was indeed far too cold for extended work. The mission design engineers went back to their computers to determine a series of orbiter attitudes that would keep the payload bay temperatures manageable, while keeping the direct sunlight out of the Telescope bays, accommodating the orbiter's needs, yet still conserving already tight maneuvering propellant. Others developed warmer overgloves and other techniques, and revisited the idea of storing the tools inside the warmer (but already full) orbiter cabin rather than in the payload bay toolboxes. Had the cold hands problem emerged on-orbit, in the middle of the repairs, it could have been very serious, even disastrous. As it was, planners had to pull out a thread tightly interwoven throughout the warp and woof of a complex and interlocking mission. It took considerable cooperation between Goddard and Johnson to work out a new flight plan that would accommodate the new thermal requirements.[47]

Over the summer, as headquarters was preoccupied with President Clinton's decision concerning the Space Station, Goddard engineers and astronomers tested and evaluated the instruments and other components in preparation for shipping them, in August, to the Kennedy Space Center. As the instruments and other components flowed through Goddard, they were subjected to multiple, independent, and rigorous testing to cross-check the results and avoid the kind of error that had befallen the mirror. At Goddard and the Science Institute, planners refined the complex and interlocking sequences of instructions that would have to be sent to the spacecraft to prepare each component for replacement and test its successor. Houston was occupied with preparing for joint integrated simulations of the mission. A total of seven would be held, August through November, and represented the most complex total exercise of any mission plan in the history of the shuttle program. With various astronauts in the Neutral Buoyancy Simulator at Marshall, ground controllers at mission control in Houston, and various supporting engineers and scientists at Goddard, the Science Institute, and contractor sites around the world, they would rehearse various parts of the mission as well as numerous "failures" concocted by the simulation supervisors.

One recommendation from several of the review teams concerned increasing the fidelity of the water tank simulations. The robot arm at Marshall's Neutral Buoyancy Simulator was crude, and planning began on a higher-fidelity version. Johnson's tank was too shallow to accomodate an arm. Also, to allow longer and more realistic simulations plans were made to install a nitrogen-oxygen ("nitrox") breathing mixture, which would allow simulations of entire six-hour spacewalks, rather than having to break them up into smaller portions.[48] Here Brinkley ran into center parochialism, as the upgrades to the

---

46. Musgrave Interview, March 23, 1995, 1/A: 260, 580; John Young Memorandum ACS-94-05, Appendix G of *Mission Director's Report*.
47. Heflin Interview, March 23, 1995, 1/B: 414ff.
48. Jay Apt, Oral History Interview, March 22, 1995.

Marshall tank were seen by some as competing with proposals to upgrade the much smaller and more-limited Johnson Weightless Environmental Training Facility to a full Neutral Buoyancy Laboratory, conveniently available to Houston flight crews. After paperwork "got lost" along the way, the changes were finally implemented barely in time for the October simulations.[49]

The flight crew took advantage of the long water tank simulations to hone their bodies and their spirits as well. They would have only one chance on-orbit, and so prepared for that as if it were the Olympics. "We did things at Marshall like we were going to do upstairs . . . at seven a.m. we would be in there and we would brief what we were going to do, and then we would position our tools . . . then we'd get in the suit, and then we would go work, we'd do six or seven hours there, we would get out, we would debrief how everything went, we'd capture our lessons [learned]. Then we went off to the gym, . . . we would put in an hour and a half or two, . . . yeah, hard work . . . then we would come back from there . . . and start talking about and reviewing the next day's activities . . . and then it's ten p.m. and at 7 a.m. we're back at Marshall and the whole day repeats, and we did this day after day after day without missing a day for three weeks. Now, you wonder why it worked?"[50]

Brinkley "wasn't worried about the flight team," but was worried about upper level management's readiness for real-time decision making. Brinkley found that previous mission simulations had effectively concentrated on exercising decision-making among the crew, flight controllers, and the many "back room" technical groups that advise them. Generally, decisions that were too serious to be made within mission control had relied on a Mission Operations Director console position, the occupant of which would be the Flight Director's interface to the upper management. During the joint integrated simulations, a real-time mission management team comprising administrators from the highest levels of the agency participated and were faced with contrived situations that pitted the "safety of the orbiter versus the survival of the Hubble." Associate Administrators for Space Science (Wes Huntress), Space Flight (Jed Pearson), and Center Directors (John Klineberg, Goddard) worked with the team to practice how they would confront such situations during the flight.[51]

In addition to the water tank simulations (738 hours), several other areas of ground training enjoyed renewed emphasis. Since the viscosity of the water tended to make handling massive objects less realistic, an air-bearing floor simulator was used to gain more realistic experience in the dynamics. Twenty hours of manned thermal vacuum tests were done. Computer graphics virtual reality simulators were developed, and were used to research positions for the crew and robot arm to use on the repairs. In the Manipulator Development Facility, Claude Nicollier and Ken Bowersox worked with a realistic robot arm, hoisting full-scale helium-filled balloons in the shape of Hubble and of a space-suited astronaut (nicknamed "Gumby").[52]

---

49. Brinkley Interview, March 16, 1995, 1/A: 230; Jay Apt, Oral History Interview, March 22, 1995.
50. Musgrave Interview, March 22, 1995, 1/B: 380.
51. Brinkley Interview, March 16, 1995, 1/A: 300; March 22, 1995: 360; Cooper, *Before Liftoff*; Heflin Interview, March 23, 1995, 1/B: 290.
52. Opalko, Jane. "Virtual Reality Helps Out on Hubble Repair." *Odyssey* 3 (November 1994): 19; Claude Nicollier, Oral History Interview by Joseph N. Tatarewicz, March 17, 1995.

At the High Fidelity Mechanical Simulator at Goddard, the crew practiced replacing components using highly realistic models and even flight hardware. Goddard had built the simulator, which reproduced the aft portions of the telescope, where the instruments fit, and the equipment bays that held the various electronics components, so that replacement and new components could be checked accurately on the ground. In addition to being responsible for the Hubble Space Telescope, Goddard had worked on satellite servicing concepts since the early 1970s, developing tools and techniques used on other earth-orbiting spacecraft. Rothenberg, Frank Ceppolina, and others had been sensitized to the minute detail required in testing and training simulators by many experiences, among which was the repair of the Solar Maximum mission spacecraft on STS-41C in 1984. There, astronauts had been temporarily halted by a small piece of insulation that had sagged out of place. Goddard also maintained a separate electrical simulator for testing the numerous data and control connections. Most of the components for the servicing mission would travel to the launch site via Goddard's clean room, where they would be tested and where the crew would have a chance to work with real hardware. "Goddard was just magnificent in knowing how to get a crew ready . . . they were much more than just the customer . . . they were EVA trainers, they swam with us all the time . . . ," Musgrave recalled.[53]

The astronauts also went to various other locations around the country and in Europe to see and handle the actual flight hardware. They even went to the Smithsonian's National Air and Space Museum, riding a cherry picker crane after-hours to inspect the old Structural Dynamic Test Vehicle. The only full-scale version of the telescope to have been built in a program that tried to save money by building only one full-fledged flight spacecraft and no prototypes, the vehicle had served as a wire-form on which the 25 miles of flight wiring had been laid out. Not realistic in many respects, the carefully restored and preserved artifact nonetheless retained the many simulated connectors and black boxes where the cabling now in space had once been meticulously strung. Unlike the more aesthetic models built later to represent the flight spacecraft, this vehicle did not have a covering of multilayer insulation to get in the way and obscure details.[54]

The Hubble Space Telescope program had eschewed building prototypes, and so there was no single place the crew and engineers could go to see a truly accurate and completely realistic version of the spacecraft that orbited more than three hundred miles up. There had arisen by necessity a panoply of mockups, models, and simulators, supplemented by thousands of photographs and video tapes taken of the flight spacecraft before it left the ground. Each was accurate in some important respects, yet each was dangerously misleading in most other respects. To get the whole picture, it all had to be synthesized in the mind. No one experience was adequate, and all had defects. The astronaut would have to pick out mentally the appropriate parts of each training exercise and suppress the rest. Musgrave described what it was like:

> *"In your imagination you extract from the manned thermal vacuum testing the appropriate parts, so when you're in the water doing those marvelous things with those gloves, you have to say, 'this is not the way its gonna be.' You have to go back to the manned thermal vac, and think what my real gloves were like there, at flight temperatures . . . only in the head does the entire mission exist. . . . You cannot go dumb in the*

---

53. Musgrave Interview, March 22, 1995, 1/B: 280; Rothenberg and Frank Ceppolina had worked on the Solar Maximum Mission repair planning 1982-1983; the experience influenced the Hubble servicing mission in many ways. See Rothenberg Interview August 10, 1995, 1/A: 229, 1/B: 520, 2/B: 300 and September 8, 1995 2/B: 159.

54. The Structural and Dynamic Test Vehicle is documented in the Artifact Files of the Department of Space History, National Air and Space Museum, Smithsonian Institution, Washington, DC, and in associated documents and oral history interviews of the Space Telescope History Project.

> water tank and say, 'this is the way it's gonna be,' . . . you mentally pull from the water experience, you pull from the air bearing floor, . . . you pull from the vacuum chambers in your real suit, you pull from JPL, from Ball, from the clean room at Goddard, you pull from the previous zero-g experience, you pull from your previous EVA experience. Now you sit down and in your imagination you go through five days of work . . . just like a ballet, where every finger every toe, I knew where every tool was throughout five days. So I built, just like the score of a symphony, I had the whole five days in my head. . . . every single motion, every translation of the body, every worksite. . . . The reason Hubble worked was, number one, Hubble was incredibly friendly to being serviced by an EVA crewperson, and we were able to attack all the details in all of those environments and build a mission."[55]

Such an ambitious and expensive servicing mission might never have been seriously considered had not much more than the Hubble Space Telescope been riding on it. The Space Station had long been NASA's choice for the next logical step beyond the shuttle. As the Hubble Space Telescope's problems and their potential solutions emerged, and with the new awareness of the difficulty of spacewalking under certain circumstances, planning work on space station assembly revealed that it would require an unprecedented amount of extravehicular activity. With only the restrictive volume and weight capacity of the shuttle, assembling the Space Station seemed to call for spacewalks in a nearly implausible number, duration, and complexity and variety of tasks. Review committees questioned seriously whether such a scenario was reasonable. A successful Hubble servicing mission would provide dramatic proof that the Space Station assembly could be done. While the reviews of the servicing plans were going on in the summer of 1993, President Clinton was making an important and extremely contested decision concerning the future form of the Space Station. The Space Station "Freedom" of the Bush administration was replaced by Space Station "Alpha," a less-ambitious and less-costly version. That a space station at all emerged from the presidential decision process was cause for rejoicing, but it was now tied more firmly to the success of the Hubble servicing mission.[56]

Later in the summer, several other unfortunate events served to tarnish further NASA's reputation and raise the stakes. A weather resources satellite, *NOAA-13* failed shortly after launch in early August. The Galileo spacecraft, enroute to Jupiter after a perils-of-Pauline life of its own, had been unable to deploy its high-gain antenna, and attempts to free it seemed doomed, threatening the viability of its mission. The next shuttle mission, STS-51, was delayed from its late July launch date because of concerns about a particularly active annual Perseid meteor shower. When *Discovery* finally flew in mid-September, the crew tested Hubble tools on-orbit during a full seven-hour spacewalk September 16, 1993.[57]

---

55. Musgrave Interview, March 22, 1995, 1/A: 600; the importance of the telescope's having been designed and built to be serviced is cited by many as the most important factor for the mission's success.

56. The National Research Council had reviewed Space Station Freedom assembly plans and worried about the extravehicular activity as early as November 1988. In January 1991, the Presidential Advisory Committee on the Future of the U.S. Space Program (Augustine Committee) raised EVA concerns. Musgrave thought that this was a valid point, and that the use of the Hubble Servicing Mission as a proof-of-capability for space station assembly was fair. Musgrave Interview, March 22, 1995, 1/B: 160.

57. James T. McKenna, "Discovery Crew Tests Hubble Repair Tools," *Aviation Week & Space Technology*, 139 (September 20, 1993): 36–37.

The most serious, and mysterious, mishap of those remarkable few weeks, however, was the *Mars Observer* spacecraft which, on the verge of entering Mars orbit suddenly went silent. Again the cries of critics rose to a din and NASA suffered the barbs of pundits. On September 6, 1993, NASA was honored with its second Dave Letterman Top Ten list.[58]

In addition, 1993 was a frustrating year for the shuttle program. By mid-August, four launch countdowns had proceeded to within twenty seconds of ignition and had to be scrubbed for various reasons. Two had actually ignited the main engines and then shutdown just a split second before the solid rocket boosters were to ignite—the point of no return in a launch. In October, the *Landsat-6* satellite's kick motor failed after launch, leaving the Earth resources satellite stranded and useless. Some of these untoward events were not even within NASA's resposibility or control, and others were only remotely related to the Hubble mission, if at all, yet the servicing mission was acquiring a significance far beyond just repairing the Telescope. While upper-level managers worried about the larger picture, engineers and astronauts tried to maintain their focus on the details that would make or break the mission. As Musgrave recalled: "Did the pressure come to me? No. It's out there, and I know it's out there. But I am gonna go do my job, apart from the pressure and apart from the outside world. I'm the ballerina, and I know the opera company may be resting on my shoulders. OK, it's tough, but I'm gonna go do what I've got to do. It doesn't matter that the opera company is resting on my shoulders, or isn't. It's my art, and I'm going to perfect it to the best of my ability, and it's me and it's my art. . . . Yes, there's pressure, but it's not external pressure, it's internal."[59]

By December, as launch approached, nearly every press story on the mission declared it to be do-or-die. It was a mission of superlatives: more spacewalks, by more astronauts, for more total time than ever-before attempted; one-chance to rendezvous; no second chance to rendezvous once the telescope was released at the end of the repairs. The complexity and high stakes of the mission were reflected in the unprecedented series of reviews, so numerous and extensive that some feared the mission was being reviewed to death, and that the preparation required for these reviews was taking resources away from the mission itself. By the November 17, 1993, Flight Readiness Review, 195 formal recommendations had been made by twelve review teams, of which only twenty-seven remained to be closed.[60]

When the priorities for various repairs were put alongside the order in which certain tasks had to be done, it was not possible to do the various tasks one after the other using 6-hour spacewalks. In priority order, the tasks for full mission success were: solar arrays; two gyroscopes; Wide Field-Planetary Camera; COSTAR; magnetometer; and Solar Array Drive Electronics. Minimum success would be three reliable gyroscopes and an operational WF/PC II or COSTAR. Secondary objectives would include a fix for the Goddard High Resolution Spectrograph, a 386 coprocessor to replace the failed computer memory, a second magnetometer, and gyroscope control electronics. The final timeline scheduled these tasks to get the highest priority items done as soon as feasible, but not in strict priority

---

58. Top Ten NASA Excuses For Losing The Mars Space Probe; 10. "Mars probe? What Mars probe?" 9. Forgot to use The Club; 8. Those lying weasels at Radio Shack; 7. Too much Tang 6. Made by G.E.; 5. Them Martians musta shot it down with a ray gun; 4. Heh, heh, heh. . . Our space probe sucks– heh, heh, heh; 3. At least we didn't blow all our money on some dork screwing around with a car phone; 2. Remember Watergate? Well, Nixon's up to his old tricks again!; 1. Space monkeys.

59. Harwood, William, "Space Shuttle Launch is Aborted 3 Seconds Before Liftoff," *Washington Post*, August 13, 1993, p. A-9; Musgrave Interview, March 22, 1995, 1/B: 180.

60. Mission Director's Report, p. 73; James R. Asker, "Rehearsal Intense for Hubble Flight," *Aviation Week & Space Technology* 139 (November 29, 1993):51–53. The Shuttle program's return to flight after the Challenger accident took place under the scrutiny of numerous review committees, as John Logsdon recounts in his contribution to this volume. The Hubble Servicing Mission committees, however, had a greater proportion of NASA insiders as members, and were almost exclusively appointed from within NASA. In this Goldin and others were proactive, and succeeded in exercising some degree of internal control.

order. Each day's work had to leave the Telescope and the payload bay in a condition that, should the worst emergency arise, the crew could release the Telescope quickly, close the payload bay doors, and head home.[61]

The Servicing Mission: "A Ballet of Bodies and Three Hundred Tools"

On December 2 at 04:27 EST, following a one-day delay for weather, *Discovery* finally roared away from the Kennedy Space Center in an uneventful launch. After nearly two days of catching up with Hubble, and periodically firing thrusters to slow down the catch-up rate, Pilot Ken Bowersox reported the most-often used words to describe the Telescope: "Houston, it's *really* big!"[62] Claude Nicollier grappled the Telescope, and gingerly berthed it into the Flight Support Station where the shuttle's power would substitute for that from the solar arrays. The tiltable turntable would be used to orient the Telescope work areas so that Nicollier and Bowersox, operating the robot arm from the aft flight deck, would have the appropriate portions of the Telescope facing them.

On flight day four, December 4 at 10:46 p.m. EST, "the odd couple" of Musgrave and Hoffman began the first spacewalk. For five days one or another pair would exit the airlock at around the same time each evening and spend between six and eight hours in the payload bay. This was the real test of all their training, as Musgrave put it: "I like the heat of the kitchen; I live in it, I've lived in it for decades, and I thrive on it. And so, it gets me up–to where I've got to be. Goin' out the door for the first time, I was incredibly interested, *concerned* about, have we nailed it in terms of our imaginative process, are the simulators right, are the mockups right, did we approach this job right?"[63] After setting up the work site, Musgrave and Hoffman replaced the gyroscope packages inside the aft shroud of the Telescope, and gyroscope electronics and fuses in the modular bays around the "waist" of the spacecraft. They ran into some problems closing the aft shroud doors, but otherwise all went according to plan. "*Endeavour*, Houston; not to get you spun up, but we've got six good gyros on the telescope," Capsule communicator Greg Harbaugh punned to the crew after the aliveness test spun the new gyros up to operating speed.[64]

On flight day five, Akers and Thornton removed the solar arrays. One array had refused to roll up into its cassette, a situation that had been the subject of one of the sim-

---

61. Two gyroscopes are contained in a package called a Rate Sensing Unit; the magnetometers, at the top of the telescope, sensed the orientation of the Earth's magnetic field and provided an additional attitude reference. For the astronaut's perspective on the plans see: Jeffrey A.Hoffman, "How We'll Fix the Hubble Space Telescope," *Sky and Telescope* 86 (November 1993): 23–29 and Hoffman, "Reflections on Hubble's Repair," *ibid.* 89 (February 1995): 26. Heflin Interview, March 23, 1995, 1/B: 470. On the development of the mission success criteria see Rothenberg Interview, August 8, 1995, 2/A: 060, and especially 110.
62. Ken Bowersox, quoted in Aguirre, *Rescue Mission in Space*, 20:00; Jim Crocker recalled his first encounter with the spacecraft in the Lockheed clean room: "It's so big, this is really big," *ibid.* 20:00; unless otherwise documented, mission events and times are taken from Johnson Space Center, Flight Engineering Office, *STS-61 Space Shuttle Mission Report*, NSTS-08288 (February 1994), Johnson Space Center, Mission Operations Directorate, Systems Division, *STS-61 EVA Debriefing*, JSC-26584 (March 1994); and *STS-61 Mission Director's Post-Mission Report*. The quote in the title of this section is from Story Musgrave Interview, March 22, 1995, 1/A: 440.
63. Story Musgrave, quoted in Aguirre, *Rescue Mission in Space*. They were called the "odd couple" because they were assigned to the odd-numbered spacewalks.
64. Harbaugh, quoted in Aguirre, *Rescue Mission in Space*, 26:00; Heflin Interview, March 17, and March 23, 1995. The balky aft shroud door was a small problem, but one that had important implications. Worried about applying too much force and distorting the vital and irreplaceable door, ground engineers caucused extensively about how to get it closed and latched, while Musgrave wanted to use a strap to pull the door into alignment. Heflin finally decided to let the crew on-site do it their way, to the mild consternation of some ground engineers. Heflin Interview, March 23, 1995, 1/B: 350.

ulations. Working during orbital night, when the array would not be generating electricity, Thornton removed the huge and cumbersome wing, holding it while Nicollier moved her into position with the robot arm. At sunrise, as Thornton let go and precisely spread her arms to ten and two o'clock, the array stood rock-steady for a moment, and as thrusters fired *Endeavour* began to back away from the shiny gold "bird." It rotated slowly, and then as a thruster plume hit the array it bent as if a bird or a pterodactyl flapping its wings. The replacement solar arrays went on without a hitch.[65]

On flight day six, Musgrave and Hoffman removed the Wide Field/Planetary Camera, stationed it temporarily on a fixture, and prepared to install the replacement. Musgrave likened it to moving a baby grand piano, holding it by the keyboard, and trying to insert it into a gigantic dresser like a drawer. At the far end, however, to be inserted right into the heart of the Telescope just behind the primary mirror, was the camera's all-important "pickoff mirror." Musgrave had to remove the protective mirror cover, at which time the optics would be exposed. "That is the most powerful camera in the world, it sees the furthest out there in space and time, if you touch that mirror, however you touch it, that will be on every single image that comes down from the world's most powerful camera." The camera went in without incident, and Musgrave and Hoffman moved to the very top of the Telescope to install new magnetometers.[66]

On flight day seven, mission control played the traditional wake-up music, appropriately the popular song, "I can see clearly, now." Working in the aft shroud, Akers and Thornton removed the High Speed Photometer and then installed COSTAR. Crocker got a message from a friend shortly after COSTAR was installed, "Congratulations . . . there's a train leaving Baltimore tomorrow night at 8:30; if this thing doesn't work, be under it." All they could tell from the aliveness test was that the instrument was ready, but the mirror arms would not be deployed until several days after the servicing mission. In one of the equipment bays, they installed the 386 coprocessor to augment the flight computer's failing memory. Initially the new coprocessor seemed not to work, providing one of the very few serious problems in the mission. Planners considered adding another spacewalk to remove the new coprocessor and simply replace the entire computer, but it turned out that the problem was on the ground in processing the telemetry.[67]

Flight day eight began with a boost from *Endeavour*'s engines to get the ensemble to the highest orbit possible, 320 nautical miles. The last scheduled spacewalk was challenging, for it involved mostly equipment that had not been designed to be replaced or repaired. Musgrave and Hoffman installed a special cable to restore the power supply in the Goddard High Resolution Spectrograph, replaced insulation on the magnetometers, and replaced the electronics box for controlling the solar arrays. When the rolled up solar arrays did not extend, the crew had to crank them out manually. After that, the blankets extended from their cassettes normally.[68]

On flight day nine, December 9 at 11:43 p.m. EST, the Telescope was transferred to its own internal power. At 5:27 the next morning, Nicollier released the spacecraft and the

---

65. Thornton Interview, April 3, 1995.
66. Story Musgrave, quoted in Aguirre, *Rescue Mission in Space*, 34:00.
67. Aguirre, *Rescue Mission in Space*, 36:00; James Crocker, quoted in *ibid.*, 41:30. Heflin Interview, March 23, 1995,1/B:440.
68. Due to the remnants of the atmosphere present at this orbit, the Telescope slowly loses altitude over the years and requires a periodic reboost. The Solar Array Drive Electronics, one of the components changed from EVA-friendly during the cost-cutting decisions of 1980, required plugging in the types of "D-shell" connectors found in the back of home computers, and installing tiny screws. The final minor troublesome event of the mission occured when one of the screws got away, and Nicollier masterfully drove Hoffman on the end of the arm in hot pursuit, Hoffman snatching up the screw at the last moment.

shuttle slowly backed away. While the controllers at Goddard began the carefully-orchestrated sequence of reactivating the various Telescope systems, the crew enjoyed a day off, and then prepared the shuttle for return. They landed at the Kennedy Space Center on December 13 at 25 minutes past midnight. The most remarkable thing about the mission was that nearly everything had gone according to plan. The few glitches that had occurred were minor, and relatively easily solved. *Sky and Telescope* magazine, in a relatively brief post-mission article, simply referred its readers to the preview published the preceding month, since things had gone so well. The telemetry signs were encouraging, but it was a tense Christmas season while everyone waited to see whether the new instruments would really work as well as they seemed to be functioning.

Postlude—Redemption

"It's fixed beyond our wildest expectations," Program Scientist Ed Weiler beamed at a mid-January press conference. Barely five weeks after the servicing mission, after numerous engineering checks, the Telescope had exercised both the new Wide Field-Planetary Camera and COSTAR with the Faint Object Camera. Jim Crocker said the performance was "as perfect as engineering can achieve and the laws of physics will allow."[69] The newly repaired Telescope "overshadowed everything else at the American Astronomical Society (AAS) Meeting" in January 1994.[70] The astronomical end-users of the technology were delighted. Eventually even skeptics had to admit that the Telescope was making good on its most extravagant claims. The dramatic images released only hinted at what was to come. Over the following months, as the rest of the mirror-arms of COSTAR were deployed and the various instruments focused and calibrated, the Telescope really got down to work.

In May astronomers announced the first generally convincing evidence for the existence of black holes, an object at the center of the nearby galaxy M-87 with a mass two-three billion times that of our Sun compressed into the size of our solar system. This led the *Washington Post* to editorialize that the "trials and tribulations" of the Hubble Space Telescope were, in the end, "worthwhile."[71]

In June, a team led by former Space Telescope Program Scientist Bob O'Dell announced that they had obtained images of protoplanetary disks around young stars in the Orion Nebula, and found the process of planetary formation going on around 56 of the 110 stars observed. These rotating disks of gas and dust had been inferred from other orbiting observatories' data and in one case even imaged, but not in the detail or in the numbers reported from Hubble.[72]

---

69. Edward J. Weiler, Program Scientist, quoted in James R. Asker, "Scientists elated by images from refurbished Hubble," *Aviation Week & Space Technology* 140 (January 17, 1994): 24; Science Overview Briefing, January 13, 1994.
70. John Travis, "Hubble Repair and More Wins Astronomers' Acclaim," *Science* 263 (January 28, 1994): 467-68.
71. "Editorial: The Depths of Space," *Washington Post* (June 2, 1994): p. A-22. Evidence for black holes has been announced periodically for many years, even by astronomers using the unrepaired Hubble Space Telescope. Such announcements were, and continue to be controversial. However, this "discovery" was, if not the genuine article, the most substantial and convincing evidence yet found. H. C. Ford, et al., "Narrowband HST Images of M87: evidence for a disk of ionized gas around a massive black hole," *Astrophysical Journal* 435 (1994): L27-L30.
72. C. R. O'Dell and Zheng Wen. "Postrefurbishment Mission Hubble Space Telescope Images of the Core of the Orion Nebula: Proplyds, Herbig-Haro Objects, and Measurements of a Circumstellar Disk," *Astrophysical Journal* 436 (November 20, 1994).

In mid-July the fragments of Comet Shoemaker-Levy 9 hit Jupiter, and Hubble was ready. Day after day, for more than a week, images of the giant planet from Hubble showed the comet fragments as they went in, and the dusky spots where they had disturbed the giant planet's atmosphere.[73]

In October, the astronomers were startled and the public bewildered by the first results from one of the projects that had been of prime importance since conception of the Telescope. Observations seemed to suggest an age for the universe of eight to twelve billion years, dramatically downward from the fifteen to eighteen billion years previously estimated. Paradoxically, this age was younger than some estimates of the ages of certain stars.[74]

By November, confidence in the Telescope's abilities was so high that its *failure* to find stellar objects was considered a major discovery. For decades astronomers and cosmologists had been increasingly uneasy about the so-called "missing mass," some 90 percent of the matter expected to be in the universe that nevertheless does not show up in surveys. When two separate teams using Hubble Space Telescope observations failed to see anywhere near the expected number of red dwarf stars or other objects in certain fields, it deepened the mystery. The number of conventional places to look for the "missing mass" was dwindling, forcing theorists toward more exotic locales.[75]

In December, three teams using the Telescope revealed that they had obtained images of galaxies from very early in the universe, perhaps only one tenth of the total time elapsed since the big bang. Surprisingly, the primeval galaxies were found to be of a variety of complex shapes rather than uniformly simple, clustered rather than evenly distributed, and to harbor apparently very old stars. At such an early age of the universe, they had expected to see more uniformity and less structure, and were somewhat at a loss to understand how so much evolution could have taken place in such a short time.[76]

These and many other results flowed in a steady stream from the teams using the Hubble Space Telescope. Satisfied customers? Ecstatically so. The telescope carried a lot of baggage. But even if one takes into consideration the various ways in which the performance of the repaired telescope fell short of its original planned performance, it still was enormously useful. It had graduated from making-do to normal operations.

The political success of the mission can be read in who decided to appear on the dais at the press conference announcing the first corrected images from Hubble on January 13. Headquarters Acting Associate Administrator for Public Affairs Jeffrey Vincent, in a classic Freudian Slip, welcomed everyone to the "Goddard Space Flight Senator-ah, Center." He had good reason to be nervous. Instead of scientists and engineers, the first panel to speak included Administrator Dan Goldin, White House Science Adviser John Gibbons, and Senator Barbara Mikulski (D-MD), head of the Senate Appropriations Committee in charge of NASA's budget and in whose district was Goddard, as well as the

---

73. "HST Science Highlights: Comets Smash into Jupiter," *Space Telescope Science Institute Newsletter* 11 (December 1994): 1–5.

74. W. L. Freedman, et al, "The Hubble Space Telescope Extragalactic Distance Scale Key Project, I. The Discovery of Cepheids and a New Distance to M81," *Astrophysical Journal* 427 (1994): 628-55; S.M.G. Hughes, "The Hubble Space Telescope Extragalactic Distance Scale Key Project, II. Photometry of WFC Images of M81," *ibid*. 428 (1994): 143–56. Popular accounts emphasized the vigorous debates and tried to make sense of the numerous and conflicting assumptions that underlay such calculations. Among others, see "When Did the Universe Begin?" *Time* 145 (March 6, 1995): 76-82.

75. "Hubble Rules Out a Leading Explanation for Dark Matter," Space Telescope Science Institute Press Release PR-94-41, November 15, 1994.

76. A. Dressler, et al., "The Morphology of Distant Cluster Galaxies. I. HST Observations of CL0934+4713," *Astrophysical Journal* 430 (1994): 107–20; Mark Dickinson, et al., "Galaxies at High Redshift," *Space Telescope Science Institute Newsletter* 11/2 (December 1994): 6 and 7; "Hubble Identifies Primeval Galaxies, Uncovers New Clues to the Universe's Evolution," Space Telescope Science Institute Press Release PR-94-52, December 6, 1994.

Space Telescope Science Institute. Goldin kept breaking a grin throughout his wide-ranging introduction that praised the servicing mission and likened it to great missions of exploration throughout history. Gibbons, referring erroneously to the formerly "astigmatic" Telescope, took the occasion to praise the vision of the Clinton-Gore administration as well. The remarks of all were rhetorical, ceremonial, political, and tinged with minor scientific and technical errors. Senator Mikulski got right to the point: "I chair the subcommittee that financed the manufacture of the most significant contact lens in American history, the fix on the Hubble Space Telescope, and then bankrolled this extraordinary space HMO that went out and gave Hubble Telescope a new contact lens, and I am happy to announce today, that after its launch now in 1990 and some of its earlier disappointments, [raising her voice] the trouble with Hubble is over!" Suddenly, she became a scientific briefer as she proudly held up two images taken by the Faint Object Camera before and after COSTAR and explained in detail how much better the performance was. "This shows what COSTAR can do, and, Mr. Goldin, I'm going to ask you to hold that, because there's more to come," drawing laughter from the assembled reporters as she turned the NASA Administrator into her chart-holder. Holding up an earlier Wide Field/Planetary Camera image, she said it "looks like the way you would look at a road map New Year's eve," and then pointed out the much clearer image from the camera's successor. "I believe these pictures are tangible evidence that not only has Hubble been fixed, but NASA is well on its way to fix that culture that created some of these problems . . . this was a high stakes repair for Hubble. . . ."[77]

The success can also be read in what happened to the people who led the mission. Of the Collier awardees, two managers went on to significant positions in the Space Station program, the most important effort of the agency since it represented the future. Randy Brinkley became Space Station Program Manager at Johnson, and Milt Heflin began working on integrating extravehicular activity in the Space Station assembly. Brewster Shaw continued to lead Space Shuttle activities at Johnson. Joe Rothenberg eventually became Director of Goddard. The flight crew members went on a long public relations tour that took them to the White House, Congress, Europe, the hottest late-night talk shows, and even an appearance in an episode of ABC's hit comedy series 'Home Improvement.' There they played a bit of tape from one of the spacewalks that included two of the male astronauts in the cargo bay of the shuttle doing the primitive grunt popularized by Tim "The Tool Man" Allen. As Story Musgrave reflected, ". . . it was something so basic and primitive about it, humans and their tools, and the drama of whether it was all going to get done."[78]

Commander Richard Covey retired from NASA and joined Unisys Space Systems to head their simulation and training activity. Shortly after the mission he said, "This would be a great mission to end my astronaut career on, and it would be hard for me as a commander to look to another one that would bring as much reward and joy. . . ."[79] Characteristically, the rest of the astronauts returned from their publicity tours to move on to the next assignment. Characteristically, when asked about their roles in the mission they would praise the team, and point to somebody else as responsible for the success.

The experience and lessons learned from the mission were carried far and wide by the many people who worked on it, and it became a paradigm for emulation. Brinkley, as he moved to head the Space Station planning, deliberately sought out the people and expertise from the Hubble Servicing Mission to incorporate them into Space Station. Even 18

---

77. Hubble Space Telescope First Corrected Image Press Conference, Goddard Space Flight Center, January 13, 1994.
78. Story Musgrave, quoted in Aguirre, *Rescue Mission in Space*, 31:30.
79. Covey, answering question the Crew Post Flight Press Conference, January 4, 1994.

months later, on June 25, 1995, after the first docking of the shuttle with the Russian space station, *Mir*, Will Trafton, NASA Headquarters Space Station Director said: "We're estimating now some six hundred plus hours," [of spacewalks to assemble the Space Station, with the Russians doing an additional 200 hours.] "We look at EVA as a resource . . . it's not just another way for a bad event to happen. Hubble has taught us a lot, we're using the Hubble crew and the Hubble experiences to look at the EVA work that's required to assemble space station, and we're pretty happy about where we are." From the Hubble servicing mission came confidence that more than a hundred days of Hubble-type work could be done in assembling the station.[80]

The Hubble Space Telescope, amid the many mundane reasons for its existence, represented at least in part a transcendent and pure purpose: to explore and try to understand the deepest mysteries of the cosmos. It was also in serious trouble, yet not so serious it could not be saved. NASA, an agency that also represents, at least in part, transcendent and lofty goals, was also in trouble, and its fate tied to that of the Telescope. This brought out the best in people, in the samaritans who extended themselves far beyond their job descriptions to come to the aid of the machine and the idea. They worked long hours, pushed themselves, studied, became innovative and clever. They even were able to forget for a time business, Headquarters, Center, and divisional boundaries, and turf, a remarkable achievement in such institutions. Because people believed in the worth of the mission, and because they also feared in their very bones that failure would bring the most dire of consequences, they gave the mission attention and resources. From Congress and the White House all the way down to small divisions and work groups, they put in money, time, and attention. The anxiety of executives looking over their shoulders might have made more work for the executors, but that anxiety also gave them clout to override ordinary bureaucratic barriers. Amid all the other missions, tasks, and priorities swirling about, they put this one on top for a while. Thus the Hubble servicing mission represents the kind of infrequent and special kind of push that people and organizations do from time to time. "Not since Apollo," was a phrase that many people used to describe how they felt working on this mission. Like Apollo, or the Olympics, however, such a special conjunction of will, spirit, and effort probably cannot be sustained in the ordinary course of things. It was, quite literally, an achievement of focus.[81]

Like Apollo too, however successful the mission, it could not by itself create a rising tide for NASA. "That's a lot of baggage to carry with you," Covey said in answer to a question at the post-mission press conference, "we had a task to do, and that was to fix the Hubble, . . . and we knew that the best thing we could do was to do that job very well, . . . and that was all we were thinking about during the course of the mission. . . . We hope that someone else is able to translate that into NASA doing well."[82]

Significant historical movements were afoot that were not controllable, indeed, hardly predictable. *Aviation Week* warned, shortly after the mission: "in political and public relations terms, Mission 61 probably presented more risks than 'up-side' potential. Failure could have been disasterous. But few space policy insiders expect success to give NASA a big, lasting 'bounce'."[83] Even while truly spectacular results were flowing in from the Telescope, and the Clinton administration reaffirmed its faith in the Space Station, the

---

80. Will Trafton, answering reporter's question, post Mir-Docking Press Conference, Johnson Space Center, June 29, 1995.

81. Rothenberg noted, "If you take over a problem . . . that's the number one or two priority to the organization that you're in, you're going to be successful, because people are not going to get in your way, and people are going to rally around it." Rothenberg Interview, August 8, 1995: 1/A: 510.

82. Covey, answering question at the Crew Post Flight Press Conference, January 4, 1994.

83. James R. Asker, "So far, Hubble checks out well," *Aviation Week & Space Technology* 140 (January 3, 1994): 47.

NASA budget was in trouble. As an agency, NASA had been selected by the administration to be a showcase for "reinventing government," and from the White House came directive after directive to reduce its budget. Later in 1994, as a new Republican majority was elected to Congress, even more pressure was brought to bear on NASA. Many of the congressional representatives who had been friends to the agency and who had been impressed by the Hubble servicing mission found themselves in the minority party, having been replaced by other representatives who were either hostile or indifferent. Most ironically, the space science budgets were squeezed ever more tightly, Hubble's included. Scientists feared that the very mission that saved their observatory had lent credibility to the Space Station, which threatened to devour it.

Had the mission not been attempted, the Hubble Space Telescope would have been a constant, orbiting reminder of failure, even while producing very good science. With the successful repair came redemption-but only redemption. The agency's credibility, in this and many other areas, was saved but not boosted. Hubble was no longer an albatross around the neck of NASA, and it appeared the agency could indeed do what it said it would do. But what the public, the politicians, and others *wanted* NASA to do remained as it has for most of the history of the agency: uncertain, fickle, and contested.

About the Contributors

John D. Anderson, Jr., is a member of the Department of Aeronautical Engineering at the University of Maryland, College Park. He is the author of several books and articles including, *Fundamentals of Aerodynamics* (McGraw-Hill Series Aeronautical and Aerospace Engineering, 1991); *Computational Fluid Dynamics: The Basics With Applications* (McGraw-Hill Series in Mechanical Engineering, 1995); and *A History of Aerodynamics and Its Impact on Flying Machines* (Cambridge University Press, 1997).

Mark D. Bowles is vice president of History Enterprises, Inc. and is finishing his Ph.D. dissertation in the program of the history of technology, science, and medicine at Case Western Reserve University. His research focuses on an analysis of information overload as a multi-disciplinary problem from 1945 to the present. He can be reached at mdb@HistoryEnterprises.com.

Glenn E. Bugos is head of The Prologue Group, a company specializing in the study of the history of high technology. He has worked on supersonic flight, having written a history of the development of the McDonnell F-4 Phantom II, *Engineering the F-4 Phantom II: Parts into Systems* (Naval Institute Press, 1996).

Andrew J. Butrica, a graduate of the doctoral program in the history of science and technology at Iowa State University, is a research historian and author of numerous articles and papers on the history of electricity and electrical engineering in the United States and France and the history of science and technology in nineteenth-century France. He is the author of a corporate history, *Out of Thin Air: A History of Air Products and Chemicals, Inc., 1940-1990*, published by Praeger in 1990; *To Seen the Unseen: A History of Planetary Radar Astronomy*, published as SP-4218 in the NASA History Series in 1996; and editor of *Beyond the Ionosphere: Fifty Years of Satellite Communication* (NASA SP-4217, 1997).

Steven T. Corneliussen, a longtime student of NACA history, is an editor and writer at Jefferson Lab, a national particle physics laboratory near NASA Langley Research Center, Hampton, Virginia.

Virginia P. Dawson is founder and president of History Enterprises, Inc., located in Cleveland, Ohio. The company specializes in producing business and institutional histories in a variety of print and electronic formats and organizing archives. She is author of *Engines and Innovation: Lewis Laboratory and American Propulsion Technology* (NASA-SP-4306).

Henry C. Dethloff is a member of the History Department at Texas A&M University. He is the author of numerous books and articles including, *"Suddenly Tomorrow Came . . . ": A History of the Johnson Space Center, 1957–1990* (NASA SP-4307, 1993); *Texas A&M University: A Pictorial History, 1876–1996* (Texas A&M University Press, 1996); and several other works.

Donald C. Elder teaches history at Eastern New Mexico University, Portales. He has published several articles on space flight and his book, *Out from the Behind the Eightball: A History of Project Echo*, appeared in 1995 from Univelt, Inc., in the American Astronautical Society History Series.

ABOUT THE CONTRIBUTORS

Michael H. Gorn is an independent scholar of aerospace technology. He has written *The Universal Man: Theodor von Kármán's Life in Aeronautics* (Smithsonian Institution Press, 1993), and is presently working on a biography of Hugh L. Dryden, NASA deputy administrator in the 1960s, and a history of flight research.

James R. Hansen is a member of the Department of History, Auburn University, Auburn, Alabama. He has written two volumes on the history of the Langley Research Center. The first of these is *Engineer in Charge: A History of the Langley Aeronautical Laboratory, 1917–1958* (NASA SP-4305, 1987). The second is *The Spaceflight Revolution at NASA Langley: From the Sputnik Crisis to the Lunar Landings* (NASA SP-4308, 1995). He has also published *From the Ground Up: The Autobiography of an Aeronautical Engineer* (Smithsonian Institution Press, 1988), with Fred E. Weick, and several articles.

W.D. Kay is a political scientist at Northeastern University in Boston, Massachusetts. He is the author of *Can Democracy Fly in Space? The Challenge of Revitalizing the U.S. Space Program* (Praeger, 1995).

W. Henry Lambright is on the faculty of The Maxwell School of Citizenship and Public Affairs, Syracuse University. He has published widely on science and technology policy, including, *Governing Science and Technology* (Oxford University Press, 1976); *Shooting Down the Nuclear Plane* (Bobbs-Merrill, 1976); *Technology Transfer to Cities* (Westview Press, 1979); and *Presidential Management of Science and Technology: The Johnson Presidency* (University of Texas Press, 1985). He has just published *Powering Apollo: James E. Webb of NASA* (Johns Hopkins University Press, 1995), an administrative biography of NASA Administrator James E. Webb who served between 1961 and 1968.

John M. Logsdon is Director of both the Center for International Science and Technology Policy and the Space Policy Institute of George Washington University's Elliott School of International Affairs, where he is also Professor of Political Science and International Affairs. He holds a B.S. in physics from Xavier University and a Ph.D. in political science from New York University. He has been at George Washington University since 1970, and previously taught at The Catholic University of America. Dr. Logsdon's research interests include space policy, the history of the U.S. space program, the structure and process of government decision-making for research and development programs, and international science and technology policy. He is author of *The Decision to Go to the Moon: Project Apollo and the National Interest* (MIT Press, 1970), general editor of *Exploring the Unknown: Selected Documents in the History of the U.S. Civil Space Program*, three volumes to date, and has written numerous articles and reports on space policy and science and technology policy.

Pamela E. Mack is associate professor of history at Clemson University, Clemson, South Carolina. A Ph.D. in the history of technology from the University of Pennsylvania, Philadelphia, she is the author of the seminal study, *Viewing the Earth: The Social Construction of the Landsat Satellite System* (MIT Press, 1990), and has written several articles in aerospace history.

Anne Millbrooke is an historian at Montana State University, Bozeman. Formerly, she served as archivist and historian with United Technologies, Inc.

Joseph N. Tatarewicz teaches history at the University of Maryland, Baltimore County and has a substantial record of publication in history of science and technology, specializing in space science and planetary exploration. He is the author of *Space Technology and Planetary Astronomy* (Indiana University Press, 1990) and is the author of the forthcoming history entitled *Exploring the Solar System: The History of Planetary Geosciences Since Galileo*, which will be published by the Johns Hopkins University Press.

Lane E. Wallace is an independent aviation writer. She is the author of *Airborne Trailblazer: Two Decades with NASA Langley's Boeing 737 Flying Laboratory* (NASA SP-4216, 1994), *Flights of Discovery: 50 Years of the Dryden Flight Research Center* (NASA SP-4309, 1996), and numerous articles.

Jannelle Warren-Findley is a member of the Department of History at Arizona State University, Tempe. She is the co-editor of *Exploring the Unknown: Selected Documents in the History of the U.S. Civil Space Program, Volume I, Organizing for Exploration* (NASA SP-4407, 1995).

Index

A

A.B. Chance Company, 223
Academy of Sciences, Vienna, 65
Ackeret, Jakob, 82, 107, 127
Adams, Mike, pilot, 156
Advanced Orbiting Solar Observatory, AOSO, 217
Advanced Research Projects Agency, ARPA, 172, 307
Advanced Turboprop Project, ATP, 333, 342, 343
Aero Digest, 17, 18, 19, 20, 111
Aerojet-General, corporation, 285, 307
Africa, 187
Agriculture, U.S. Department of, 237, 238, 239, 242
Aircraft Energy Efficiency, ACEE, 328, 329, 337
Aircraft Engine Research Laboratory, NACA, 55
AiResearch Manufacturing Company, 41
Air Force, U.S., 123, 135, 145, 146, 148, 150, 151, 153, 154, 155, 157, 160, 161, 163, 164, 172, 173, 176, 197, 199, 200, 281, 286, 304, 305, 306, 307, 308, 309, 311, 312, 318
Air Force Space and Missile Systems Organization, SAMSO, 312
Air Mail Service, U.S., 8
Air Transport Association of America, ATAA, 46, 57
Air Transport Command, ATC, 51
Akers, Thomas, astronaut, 380, 381, 390, 391
Akins, David S., 216n13
Akridge, Max, 278n3, 281n14,15
Aldrich, Arnold, 358, 359, 362
Aldrin, Edwin "Buzz" E., Jr., 193, 208, 311
Alexander, Charles C., 167n12, 168n13,18, 176n52, 179n65, 180n69, 182n77, 184n85, 185n91, 187n97, 189n103,10, 190n110, 214n5
Allen, Joseph P., astronaut, 299, 314, 315, 317
Allen, H. Julian, 120, 127, 128, 129
Allen, Lew, 374
Allen, Tim "The Tool Man," 394
Allestad, D., 376n23
Allison Gas Turbine Division, General Motors, 323
American Airlines, 31
American Society of Mechanical Engineers, ASME, 50
Ames Aeronautical Laboratory, 35, 37, 58, 120, 127, 279
Ames Flight Research Branch, 35
Ames, Joseph S., 13, 14, 15, 16, 36, 71, 78, 106, 111, 112, 124
Ames, Milton, 101n29
Ames Research Center, ARC, 255, 256, 323
Anders, William A., 193, 206, 281
Anderson, Senator Clinton, 203, 260
Anderson, John D., Jr., 60, 65, 78, 82n33, 84n36, 104n39, 122n84, 141n20, 265
Anita, experimental spider, 230
Apollo-Applications Program, AAP, 216, 217, 219, 220, 221, 242, 311
Apollo spacecraft, 156, 157, 163, 169, 193, 194, 195, 196, 197, 198, 199, 200, 201, 202,, 203, 204, 205, 206, 207, 208, 209, 210, 211, 213, 216, 219, 220, 221, 224, 232, 239, 253, 257, 262, 277, 278, 280, 281, 286, 287, 288, 292, 293, 297, 299, 301, 306, 313, 328, 331, 347, 354, 358, 359, 365, 367, 370, 378, 381, 384, 395
Apollo Telescope Mount, ATM, 217, 218, 223, 226, 227, 229
Apollo: The Race to the Moon, 103

Apt, Jay, 385
Arabella, experimental spider, 230
Arabian, Don, 223n31
Ardan, Michel, 302, 303
Armstrong, Neil A., astronaut, 193, 206, 208
Army Air Forces, AAF, 40, 41, 42, 43, 50n46, 51, 54, 59, 92, 93
Army Air Corps, USAAC, U.S., 35, 38, 39, 85
Army Air Service, 13
Army Corps of Engineers, U.S., 237
Army, U.S., 6, 86, 113, 170, 173, 175, 176, 214, 304, 307
Arnold, Henry "Hap" H., 85
Asbury, Scott C., 99n26
Asker, James R., 395n83
Astronaut Maneuvering Unit, AMU, 305
Astrophysical Journal Letters, 382
Atlantis, space shuttle, 296, 313
Atlas, 168, 189
Atomic Energy Commission, AEC, 263
Augenstein, Bruno, 247n42
Aurora 7, spacecraft, 189
Automatically Stabilized Maneuvering Unit, ASMU, 312
Aviation Week magazine, 128, 349, 358, 395

B

B-1 aircraft, 147
B-9 aircraft, 17
B-10 aircraft, 17
B-17 aircraft, 49, 115
B-24 aircraft, 49, 55
B-25 aircraft, 55
B-29 aircraft, 59, 87, 137
B-52 aircraft, 152, 157
B-58 aircraft, 147
Baals, Donald D., 8n14, 97n16, 98, 106, 130, 145
Badgley, Peter C., 237n9, 240
Baker, Bobby, 203
Ball Aerospace, 376, 377, 378
Bangert, Linda S., 99n26
Bare, E. Ann, 99n26
Basalla, George, 3n5
Bateman, H., 106n42
Battelle/Columbus laboratories, 294
Bay of Pigs, 196
Bean, Alan L., 228, 229, 230
Beaumont, Texas, 297
Becker, John V., 22, 78, 91, 93, 97n18, 98, 99n22, 101, 102, 107n46, 108, 109, 113, 115, 116, 117, 121, 122n86, 123n88, 125, 129, 130, 131n120, 133, 140, 150, 155
Beggs, James, 347, 352, 370
Beisel, Rex, 23n64
Belew, Leland F., 220, 223n30, 224n32, 226n34, 226n38, 230n43,231, 232n49
Bell Aerosystems, 307, 311
Bell Aircraft Corporation, 86, 87, 160, 304, 308
Bell, Lawrence D., 86, 101, 310n31
Bell X-1 aircraft, 59, 60, 61, 62, 63, 64, 85, 87, 88, 89, 91, 94, 105, 126, 131, 135, 142, 147

Bellanca aircraft, 1
Benesch, Jay, 91n1
Bennett, William A., 51
Benson, Charles D., 213, 216n11, 217n17, 221n26, 222n8, 224n33, 226n35, 229n40, 231n47, 232n48
Berkopec, Frank D., 333n40,41,42
B.F. Goodrich Rubber Company, 32, 34, 37, 55, 183, 304, 308
B.F. Goodyear company, 304
Biermann, Arnold E., 22
Bijker, Wiebe E., 339
Billings, Richard N., 189n109
Bilstein, Robert E., 151n9, 201n18, 21, 207n24, 294n50
Bingham, Senator Hiram, 14
Bioletti, Carlton, 22
Birch, Douglas, 376n24
Bird, John D., 311
Biskind, Peter, 182n76
Bobbitt, Percy J., 99n26, 127, 129
Boeing corporation, 196, 202, 216, 262, 284, 323, 329, 340, 356
Boelter, Llewellyn Michael Kraus, 41, 42
Boggess, Al, 375
Bohr, Niels, 142
Boisjoly, Roger, 355
Bollendonk, Walter W., 300, 301, 318
Bombers, 1928 to 1980s, U.S., 17
Bond, Peter, 184n86, 186n96, 187n98, 191n118
Bonney, Walter T., 103n36, 136n2
Boothman, Flt.Lt. John N.,71
Borman, Frank, 193, 203, 206, 281
Boston, Ronald G., 150
Bottema, Murk, 376
Boushey, A.H., 278
Bowersox, Ken, astronaut, 382, 386, 390
Brand, Vance D., astronaut, 319
Brandenstein, Dan, astronaut, 380
Braslow, Albert L., 91, 128n111
Braun, Wernher von, 197, 205, 215, 216, 217, 260, 282, 286
Brevoort, Maurice J., 23
Briggs, Lyman J., 69, 70, 75, 106n42, 108, 115, 119
Brinkley, Randy, astronaut, 382, 383, 394
British Advisory Committee for Aeronautics, 66
British National Physical Laboratory, NPL, 17, 86
Bromberg, Joan Lisa, 274
Brooks, Courtney G., 221n24, 222n27, 230n42,44
Brooks, Senator Overton, 278
Brown Field, Chula vista, California, 340
Brown, Robert A., astronomer, 374, 375, 376n22
Browne, J.A., 31
Bryan, G. H., 66
Buck Rogers, 302
Burden, William, 56
Bureau of Aeronautics, BuAer, 6, 10, 35, 50, 160
Bureau of the Budget, 240, 241, 246, 247
Burgess, W. Sterling, 5n7, 271n81
Busemann, Adolf, 138, 142
Bush, President George, 346, 347, 379, 388

Butler, Bryan, 273
Butler, H. Scott, 91n1
Butrica, Andrew J., 265n58
Byrd, Commander Richard E., 6
Byrne, John V., 249
Byrne, Robert W., 127
Byrnes, Mark E., 177

C

C-46 aircraft, 44, 47, 48, 51, 52,54
C-47 aircraft, 50
C-54 aircraft, 37, 50
C-74 aircraft, 50
C-99 aircraft, 53
Caldwell, Frank, 66, 67, 68, 69, 75, 107, 108
California Institute of Technology, CIT, 267, 273
Callon, Michel, 194
Cambridge University Press, 3n5
Canadian National Research Council, 37, 46
Canberra, Australia, 273
Cape Canaveral, Florida, 170, 176, 184, 185, 190
Cape Canaveral Launch Operations, See Kennedy Space Center, 289
Capone, Francis J., 91, 99n26
Capshew, James H., 140n13, 166n9, 274, 275n91, 323n6, 367n3, 378n29
Carothers, Wallace H., 324
Carpenter, Commander M. Scott, 165, 189
Carr, Gerald P., 213, 230, 231, 232
Carroll, Thomas, 31n5,
Carroll, F.O., 54
Carter, President Jimmy, 248, 249, 266
Casani, J.R., 266n61
Case Western Reserve University, 174
Centaur, spacecraft, 360
Center for Radar Astronomy, Stanford University, 265
Ceppolina, Frank, 387
Cernan, Eugene A., astronaut, 310
Chaffee, Roger, 219
Challenger, space shuttle, 296, 313, 316, 317, 319, 342, 345, 347, 348, 349, 351, 352, 353, 354, 355, 356, 357, 358, 359, 361, 362, 363, 370, 373
Chambers, Washington I., 6
Chapman, Richard LeRoy, 238
Chew, W.L., 99n25
Chinese National Airways, 53
Christensen, Julien M., 305n14
Chrysler Corporation, 176
Churchill, Winston, 169
Cicero Field, 8
Civil Aeronautics Authority, CAA, 46, 57, 58
Civil Aeronautics Board, CAB, 55
Civil War, 190
Clark, John E., 235
Clay, William C., 31
Clift, J.R., 218
Clinton, President William "Bill," 385, 388, 394, 395

Clousing, Lawrence A., 34-44 passim
Cochran, Jackie, 56
Coffin, J.G., 106n42
Cockroft, Sir John, 91
Cohen, Aaron, 287, 288
Cohn, Benedict, 123
Cold War, 147, 167, 169, 170, 176, 179, 185, 186, 194, 209, 210, 301, 303, 318, 323, 325, 329
Cold Weather Test Station, 39
Colley, Russell, 304
Collier Trophy, 1929, 1, 3, 6, 14, 21, 25, 29, 50, 55, 56, 89, 90, 91, 93, 95, 97, 98, 111, 120, 135, 147, 149, 150, 151, 164, 165, 166, 167, 176, 191, 193, 200, 204, 205, 207, 208, 210, 213, 233, 235, 244, 251, 274, 277, 297, 301, 306, 313, 318, 323, 324, 325, 342, 345, 364, 365, 366, 394
Collins, Michael, 193, 208,, 297n56, 306n15, 310
Collins, Robert, 337
Columbia, space shuttle, 296, 297, 298, 299, 313
Colvocoresses, Alden P., 241n27
Commerce, Department of, 31, 249, 250
Compton, William David, 207, 213n1, 216n11, 217n17, 221n26, 222n28, 224n33, 226n35, 229n40, 231n47, 232n48, 311n35
Congress, U.S., 92, 113, 117, 162, 177, 186, 189, 194, 196, 197, 199, 200, 202, 203, 205, 210, 214, 216, 217, 219, 220, 235, 238, 240, 249, 262, 280, 284, 286, 289, 296, 307, 318, 323, 326, 328, 345, 350, 358, 364, 374, 394, 395, 396
Conner, P.K., 246
Conrad, Charles "Pete," 222, 223, 224, 226, 227
Constant, Edward W., 95, 97n18 , 111, 114, 115, 117, 118
Convair Corporation, 144, 145, 146, 147, 148, 171, 176, 281
Cook, Earl, 326
Cooke, H. Lester, 190n114
Coons, D. Owen, 310n30
Cooper, Leroy Gordon, Jr., 165, 168, 189, 190
Copp, Martin, 120
Corliss, William R., 8n14, 97n16, 145
Cornell University, 15
Corrective Optics Space Telescope Axial Replacement , COSTAR, 376, 377, 389, 391, 392, 394
Covault, Craig, 315n48, 316n52, 318n65
Covey, Richard O., astronaut, 346, 381, 394, 395
Cowling, NACA, 1-27
Cox, Catherine Bly, 103, 205n23
Crabill, Norman L., 91
Crippen, Robert L., astronaut, 297, 317, 349, 358, 359
Crocco, Arturo, 78
Crocker, James, 376
Croft, Thomas A., 265
Crossfield, Scott, pilot, 155, 161, 162
Cubbage, James M., Jr., 99n25
Cummings, C.M., 35
Curtis, Robert I., 120
Curtiss Aeroplane & Motor Company, 29
Curtiss C-46 aircraft, 29
Curtiss, Glenn H., 3; 5n7
Curtiss Hawk AT-5A aircraft, 2, 4, 13, 20
Curtiss-Wright Corporation, 127

405

D

D-558-1 aircraft, 105
Dana, William H., 151, 163
David Clark Co., 160, 304, 308
Davies, J.K., 260n30, 263n49
da Vinci, Leonardo, 105
Dawson, Virginia P., 58n69
DC-3 aircraft, 26, 32, 55, 115
DC-4 aircraft, 55
DC-6 aircraft, 50, 51, 55, 56
Debus, Kurt H., 282
Defense Advanced Research Projects Agency, DARPA, 318
Defense, U.S. Department of, DOD, 157, 162, 168, 199, 200, 216, 250, 282, 284, 305, 306, 307, 318, 328, 360
DeFrance, Smith, 35, 38, 41n30, 49n42, 50, 51, 53,54, 58
De-icing, NACA, 29-58 passim
de Laplace, Pierre Simon Marquis, 62
Denver Research Institute, 144n29,30,32
Desert Storm, war, 250
Dethloff, Henry C., 223n29, 277n2, 280n12, 287n38
DiGregorio, Barry E., 307n20
Discovery, space shuttle, 296, 299, 313, 317, 319, 345, 357, 361, 362, 363, 364, 371, 388
Disher, John, 232
Distinguished Service Medal, 185
Doenhoff, Albert E. von, 120, 128
Doig, Jameson W., 194n5
Donaldson, Coleman duP., 130
Donlon, Charles J., 295
Dornberger, Dr. Walter R., 279
Douglas Aircraft Company, 49, 127, 196, 216, 217, 222
Douglas, G.P., 68
Draley, Eugene C., 99, 136
Dressler, A., 393n76
Dryden, Hugh, 59, 62, 63, 69, 70, 75, 93, 94, 99, 102, 107, 108, 112n59, 116, 119, 121, 122, 123, 125, 130, 149, 174, 195, 197, 215, 216
Dryden Flight Research Center, 323, 335
Dr. Zarkov, fictional character, 302
Dual-Purpose Maneuvering Unit, DMU, 311
Duluth Junior College, 29
DuPont corporation, 324
Durand, William F., 10, 106, 109n51, 116n66

E

Earth, 176, 183, 186, 187, 189, 191, 196, 206, 208, 209, 214, 216, 217, 229, 230, 232, 235, 237, 242, 244, 248, 250, 251, 253, 258, 260, 267, 269, 270, 272,, 274, 276, 277, 279, 280, 302, 303, 316, 362, 370, 381
Earth Observation Satellite Company, 249
Earth Observing System, EOS, 194
Earth Orbit Rendezvous, EOR, 198
Earth Resources Survey Committee, 241
Earth Resources Surveys, ERS, 240
Earth Resources Technology Satellite, ERTS, 238
Easterbrook, Gregg, 151, 163n43
Edson, Lee, 110n54
Edison, Thomas, 3, 21, 102, 341

Edwards Air Force Base, California, 305
Einstein, Albert, 142
Eisenbeis, Kathleen M., 235n2, 239n18, 249n49,52, 250n54
Eisenhower, President Dwight D., 166n10, 171, 172, 173, 177, 180, 186, 196, 251
Electronics Research Center, Cambridge, Massachusetts, 198, 257
Ellis, Linda S., 340
Endeavour, space shuttle, 367, 380, 381, 390, 391
Energy Efficient Engine (E3) (need superscript 3)
Energy, U.S. Department of, 325
Energy Research and Development Administration, ERDA, 325
Engle, Joe, astronaut, 277
Enos, chimp, 187
Enterprise, space shuttle, 296, 313
Eriche, Krafft A., 279
Erickson, Paul C., 308n22,23
Eshleman, Von R., 265
European Space Agency, ESA, 375, 382
Everhart, Joel L., 99n26, 127, 129
Executive Branch, U.S. Government, 240
Explorer 1, 172
Extravehicular Activity, 384, 385, 387, 395
Extravehicular Life Support System, ELSS, 310
Extravehicular Mobility Unit, EMU, 315
Ezell, Linda Neuman, 281, 284n22,23, 288, 290n43, 292n44

F

F-102, aircraft, 144, 145, 146, 147, 148
F-102A, aircraft, 145, 146, 147
F-105, aircraft, 147
F-106, aircraft, 147
F-107A, aircraft, 154
F-108A, aircraft, 154
Facey, John R., 342
Faget, Dr. Maxime "Max" A., 175, 176, 187, 284
Faint Object Spectrograph, 377
Fairchild Industries, 290
Faith 7, spacecraft, 168, 189
Fales, Elisha, 66, 67, 68, 69, 75, 106, 107, 108, 109, 110
Farren, W.S., 123
Federal Aviation Administration, FAA, 328
Ferguson, Eugene S., 139n12, 141n18
Ferrera, Sal, 379
Ferri, Antonio, 127, 130
Feynman, Richard, 353
Finger, Harold, 202
Fink, Daniel J., 235
Fisk, Lennard, 373
Fjedlbo, Gunnar, 265
Flam, Faye, 373
Flash Gordon, fictional character, 302
Fletcher, James C., 261, 262, 284, 286, 328, 352, 354, 355, 358, 359, 361
Flight Safety Foundation, 29
Flandro, Gary A., 254n13Fokker trimotor, 20, 21
Fong, Louis B.C., 143

Foot-Controlled Maneuvering Unit, FCMU, 311
Forbes magazine, 338
Ford, Gerald, 233, 355
Ford, H.C., 376n22
Ford, Holland, astronomer, 374
Ford trimotor, 20
Foreign Agriculture Service, 242
Foss, R.L., 83
Foundation, Alessandro Volta, 78
France, 264
Freedman, W.L., 393n74
Freedom 7, spacecraft, 185, 186
Freitag, Robert F., 282n18,19
Friedman, Herbert, Naval Research Lab, 258, 259, 261
Friendship 7, spacecraft, 187
Frost, Robert, poet, 319
Frutkin, Arnold W., 244
Full-Scale Tunnel, FST, 97

G

Gagarin, Major Yuri Alekseyevich, 184, 185, 186, 196
Galileo spacecraft, 271, 360, 371, 388
Gardner, Dale A., 315, 317
Garn, Senator Jake, 347
Garriott, Owen K., 228, 229, 230
Garrison, Peter, 1n1
Garrell, Edgar S., 38n25
Gemini, spacecraft, 156, 157, 197, 199, 200, 201, 210, 211, 277, 280, 287, 297, 299, 301, 305, 306, 308, 309, 310, 311, 312, 313, 314, 382, 384
General Accounting Office, GAO, 337, 368
General Dynamics, 144, 281, 284
General Electric, GE, 43, 108, 160, 235, 241, 262, 323, 328, 336, 337, 339, 340, 342
General Motors, 127
General Thayer, fictional character, 303
Genome Project, 194
Geological Survey, U.S., USGS, 237, 238
George Washington University, 21
Georgia Tech, 297, 349
German Reich, 169
Germany, 169
Gibbons, John, White House, 393, 394
Gibson, Edward, 230, 231, 232
Gilruth, Robert R., 87, 168, 193, 205, 211, 217n18, 222, 282, 287
Glauert, Herman, 70, 75
Glenn, Lt.Col. John H., Jr., 165, 181, 184, 185, 187, 189, 347
Glenn L. Martin Company, 17, 40, 169
Glennan, T. Keith, 114, 166n10, 168, 173, 174, 175, 197, 214, 215
Goddard High Resolution Spectrograph, 377, 378, 389, 391
Goddard, Dr. Robert H., Memorial Trophy, 251
Goddard Space Flight Center, NASA GSFC, 240, 241, 256, 265, 317, 375, 377, 381, 383, 385, 387, 392, 393, 394
Goethert, Bernhard H., 99, 109, 110, 112, 120, 126, 129
Goetzmann, William H., 177
Goldberg, Leo, 232
Goldin, Daniel, 379, 380, 381, 383, 393, 394

Goldwater, Senator Barry, 328, 329
Gordon, Richard F., astronaut, 310
Gore, Albert, Vice President, 394
Gottingen University, Germany, 62
Grace Episcopal Church, Yorktown, Virginia, 89
Graham, William, 347, 351
Grand Tour, planetary exploration, 251, 253, 254, 255, 256, 257, 258, 259, 260, 261, 262, 263, 264, 269, 271, 272, 275
Gray, C.G., 16, 53
Gray, Edward G., 213
Gray, George W., 31, 37, 106n41, 107n47
Great Britain, 264
Great Depression, 26
Great Society, 200, 210, 211, 220, 253
Grimwood, James M., 167n12, 168n13, 176n52, 179n65, 180n69, 182n77, 184n85, 185n91, 187n97, 189n103,108, 190n110, 201n20, 214n5
Grissom, Virgil "Gus" I., 165, 181, 185, 186, 219
Greene, Howard E., 37
Greenwood, John T., 97n17
Gregg, David, 45n37
Grossman, Arie, 273
Grumman Aircraft Engineering Corporation, 145, 146, 284, 290
Gubitz, Myron B., 150
Guggenheim Aeronautical Laboratory, California Institute of Technology, 62, 99
Gulfstream corporation, 323, 340
Gunston, Bill, 145
Gwynn-Jones, Terry, 6

H

Hacker, Barton, 10n19, 201n20
Hager, Roy D., 323n7,8, 329n31
Haggerty, James J., 340n67
Hale, Edward Everrett, 213
Hale, N. Wayne, 277
Hallion, Richard P., 60n2, 86n40, 88, 92n2, 92n3, 94n11, 114, 122n82, 150n5, 154n16, 156n25, 157n27
Hamblin, Dora Jane, 187n100
Hamilton Aero Manufacturing Company, 22
Hamilton, Martha M., 321n1, 336n52
Hamilton Standard, 44, 46, 315, 321, 323, 330, 333, 339, 340
Hampton, Virginia, 382
Hand-Held Maneuvering Unit, HHMU, 305, 310
Handler, Philip, National Academy of Sciences, 258
Hanley, Jeff, 379
Hansen, James R., 5n9, 8n14, n16, 10-13, 15, 20, 21, 23, 32, 50, 61, 80n32, 85n38, 87, 91n1, 92, 94n12, 95n14, 99n22, 101, 102n32, 103, 104, 110n55, 111, 114, 120, 125, 130, 135n1, 137n4,5, 139n12, 140n16, 141n19, 145, 324
Harbaugh, Greg, astronaut, 381, 390
Hardy, J.K., 43, 53
Hargrove, Erwin C., 194n5
Hart, Terry J., astronaut, 317
Hartinger, James V., 318
Hartman, Edwin P., 29, 49
Hartsfield, Henry W., astronaut, 319
Harwood, William, 389n59

Hauck, Frederick "Rick" H., astronaut, 346
Hawks, Frank, 13
Hawley, Steven, astronaut, 372
Hay, T. Park, 32
Heacock, Raymond L., 265n55
Hearth, Donald P., 254
Heflin, J. Milton, 379, 385n47, 394
Heinemann, E.H., 49
Heinlein, Robert A., 303
Heisenberg, Werner, 142
Hemsley, Richard T., 309
Herblock, editorial cartoonist, 373
Herrell, L., 376n23
Hewes, Donald E., 311
Hibbard, Hall L., 35
Hieb, Richard, astronaut, 380
High Altitude Observatory, HAO, 229
Hiller, Fairchild, 89
Hilmers, David C., astronaut, 346
Hilton, W.F., 86, 121n77, 124
Hindler, Dr. Ernest, 229
Hoffman, E. L., 5n7
Hoffman, Jeff, astronaut, 381, 390, 391
Holmes, D. Brainerd, 199, 200
Hoover, President Herbert, 13, 15, 21, 102
Hounshell, David, 324
House of Representatives, U.S., 173
Houston, Robert S., 150, 151, 154n17, 157, 160n32,33, 163, 202, 230, 287
Houston, Texas, 196, 210, 229, 230, 278, 281, 359, 382, 385
Hsue-shen, Tsien, 125
Huber, William C., 306n15
Hubble Space Telescope, HST, 194, 365, 366, 367, 368, 369, 370, 371, 372, 373, 374, 375, 376, 378, 379, 381, 383, 386, 387, 388, 390, 391, 392, 393, 394, 395, 396
Hughes Aircraft Company, 204, 262
Hughes Danbury Optical Systems, 374
Hughes, Thomas P., 140n13, 141n19, 142, 194, 249, 324, 341, 368
Hugoniot, Pierre, 65
Huguenard, E., 115, 116, 119
Hull, G.F., 69, 108, 119
Humphrey, Hubert H., Vice President, U.S., 195, 203
Hunley, J.D., 91n1, 166n10
Hunsaker, Jerome C., 6, 103n36, 111, 112n59
Hunter, Wilson H., 55
Huntress, Wesley T., 386
Hurd, Peter, 190, 191
Hurricane Gilbert, 364
Hyland, Lawrence A., 193, 203
Hypersonic Weapon and Research and Development System, HYWARDS, 306

I

Ice Research Base, IRB, 51
Inconel X, 152, 160
Institute of the Aeronautical Sciences, 37
Integral Launch and Reentry Vehicle, ILRV, 281

Intelsat, satellite, 380
Interior, U.S. Department of, 237, 238, 239, 241, 244, 249
International Aeronautical Federation, 213
International Council of Scientific Unions, 171
International Geophysical Year, IGY, 171, 177
International Latex Corporation, ILC, 315
International Nickel Company, 160
International Ultraviolet Explorer, IUE, 381
Io, satellite, 275
Iron Curtain, 169
Irvin, James B., 193, 211

J

J-57 aircraft, 144
J-65 aircraft, 146
Jacobs, Eastman, 71, 75, 78, 79, 109n51, 110, 120, 121
Jakab, Peter, 10n21
Janos, Leo, 60
Jenkins, Dennis R., 151n7, 161n36
Jet Propulsion Laboratory, JPL, 169, 172, 203, 204, 214, 253, 254, 255, 256, 259, 260, 261, 262, 263, 264, 265, 268, 269n76, 272, 273, 275, 374, 376, 383
JetStar aircraft, 335
Johns Hopkins University, 14, 69, 71, 107
Johnson, Clarence L., 35
Johnson, Harold A., 42, 306n15
Johnson, Kelly, 73
Johnson, President Lyndon B., U.S., 165, 172, 173, 186, 196, 200, 201, 202, 203, 206, 207, 217, 280n11, 347, 348
Johnson Space Center, JSC, 223, 227, 230, 241, 242, 278, 295, 350, 358, 359, 360, 372, 378, 379, 380, 381, 383, 385, 386, 394
Johnson, Thomas H., 151
Johnston, S. Paul, 13
Johnsville, Pennsylvania, 156, 184
Jones, Alun R., 32, 34, 35, 40, 41, 43, 49, 50, 51, 52,53, 57, 58n68
Jones, Lloyd S., 17
Journal of the Aeronautical Sciences, 118, 119
Journal of Geophysical Research, 268
Journal of the Royal Aeronautical Society, 17
Jupiter, missile, 168, 172
Jupiter, planet, 251, 253, 254, 256, 259, 260, 261, 262, 263, 264, 266, 267, 268, 270, 271, 272, 275, 276
Justice, U.S. Department of, 374

K

Kahoutek, comet, 231
Kann, Clifton F. Von, 326n20, 327n22, 329n33
Kansas City Junior College, 29
Karman, Theodore von, 59, 62, 84, 110, 125, 294n48
Karth, Joseph E., 240
Kay, W.D., 151, 162
KC-135 aircraft, 309
Kennedy, President John F., U.S., 150, 165, 167, 174, 177, 184, 185, 186, 189, 193, 196, 198, 199, 200, 205, 207, 208, 211, 257, 307, 318
Kennedy, Robert, 206

411

Kennedy Space Center, John F., KSC, 197, 202, 219, 222, 228, 289, 291, 296, 348, 358, 359, 363, 385, 390, 391, 392
Kelly, G.F., 310
Kelly, Thomas C., 137n8
Kephart, J.F., 309n25
Kerwin, Joseph P., astronaut, 222, 223, 224, 226, 349
Kevles, Daniel J., 91n1, 114
Khrushchev, Nikita, Soviet Premier, 184
Kill Devil Hills, North Carolina, 60
Killian, James R., 173
Kimball, Leo B., 48
King, Martin Luther, 206
Kingsbury, James, 351, 352
Kingsford-Smith, Sir Charles, 6
Kinzler, Jack, 223
Kirk, Robert L., 318
Kitt Peak National Observatory, 232
Kitty Hawk, North Carolina, 168, 190
Klineberg, John, 329, 337, 386
Kloeppel, Peter, 91n1
Knight, William "Pete" J., 155
Knutson, Roy K., 279
Kochendorfer, Fred D., 270n78
Kohlhase, C.E., 266n61
Kohrs, Richard, 359
Korean War, 170
Kotcher, Ezra, 84, 85, 86
Kraft, Christopher C., Jr., 183, 287
Kraemer, Robert S., 256
Kramer, James, 328
Kuchemann, Dietrich, 141
Kuettner, Joachim P., 182
Kuhn, Thomas S., 142n23

L

Ladd Field, 39
Lake Okeechobee, Florida, 185
Lambright, W. Henry, 174, 198, 200n17, 202n22, 250n55, 348, 348n8
Land Remote Sensing Policy Act, 250
Landsat, satellite, 235, 237, 238, 239, 240, 242, 243, 244, 245, 246, 247, 248, 249, 250, 277, 389
Langley Memorial Aeronautical Laboratory, LMAL, 1-32 passim, 71, 72, 75, 78, 79, 80, 82, 85, 86, 89, 91, 92, 94, 95, 97, 98, 101, 102, 105, 106, 107, 108, 109, 110, 112, 113, 118, 119, 120, 124, 125, 126, 127, 129, 130, 131, 132, 133, 136, 153, 168, 219
Langley Research Center, 13, 135, 136, 137, 138, 139, 140, 144, 145, 146, 174, 175, 282, 286, 311, 312, 323, 382
Lannan, John, 259
Large-Scale Advanced Prop-fan, LAP, 338
Larson, Gary, author, 373
Larson, Karl O., 39
Launius, Roger D., 168n17, 169n22,23, 170n23,24,25, 171n28,29,30,31, 173n37, 177n54, 185n90, 189n107, 307
Law, John, 194
Lawrance, Charles W., 3n6, 5n7, 6, 8
Lawrence, L.J., 359n61
Layman, Richard T., 62, 77, 91n1
Lee, Frederick B., 208

LeMay, Maj. Gen. Curtis E., 93, 94, 121
Lenoir, William B., 380
Leschly, K., 376n23
Lesley, Everett P., 10
Leslie, Stuart, historian, 251, 252n6
Letterman, David, comedian, 373, 389
Levenson, Thomas, 133
Levine, Alan J., 170, 171, 172, 174, 179, 197n11
Levine, Arthur L., 256n17
Levintan, R.M., 333
Levy, Gerald S., 265
Lewis, Eugene, 210n27
Lewis, George W., 10- 22 passim, 36, 51, 57, 58n68, 78, 85, 102, 107, 111, 112, 114, 116, 117, 118, 125
Lewis Laboratory, 45, 174, 215
Lewis Research Center, 321, 322, 323, 324, 325, 326, 328, 329, 330, 331, 333, 335, 337, 340, 342, 343 passim
Liberty Bell 7, 186
Liepmann, Hans W., 69
Life, 167, 184, 187
Limerick, Patricia Nelson, 178, 179
Lindal, Gunnar, 265
Lindsay, Nathan, astronaut, 349
Ling-Temco-Vought (LTV), 307, 308, 309, 311
Lilly, Howard, 105
Lindbergh, Charles A., 2, 6, 15, 100
Lindsey, W.F., 79n30, 82n34, 121
Littell, Robert E., 79n30, 82n34, 121
Little, B.H., Jr., 99n25
Lockheed 12A, aircraft, 29, 34, 35, 36, 38, 40, 46
Lockheed Air Express, 2, 13, 284, 323
Lockheed Propulsion Company, 29, 36, 37, 83, 280, 281, 285, 290, 294, 304, 329, 371
Loening, Grover, 3n6, 5n7
Loftin, Laurence K., Jr., 11, 115, 116, 123n89, 132
Loftus, Joseph, 289n41, 294n50
Logsdon, John M., 151, 157, 173, 196n6, 213n4, 237n5, 379n32
Long, James E., 254n11
Loughborough, Dwight L., 37
Lounge, John M., 346
Lousma, Jack R., 228, 229, 230
Lovell, James A., 193, 206, 281, 311
Low, George, 205, 206, 221, 259, 261n38, 286, 336n54
Luce, Henry, 179, 238n14
Lucky Lindy, 167
Luke Skywalker, fictitious character, 321
Luna II, III, Soviet spacecraft, 280
Lunar Flying Unit, LFU, 311
Lunar Orbit Rendezvous, LOR, 198, 199
Lundin, Bruce, 325

M

Mach, Ernst, 65, 128, 142
Machell, R.M., 310n31
Mack, Pamela, 194, 236n4, 237n7,8, 238n10,14, 239n17, 240n21,24, 241n28, 242n30, 244n36, 245n38, 246n40, 246n40, 249n47
Manhattan Project, 140, 169, 173, 186, 194

Manned Maneuvering Unit, MMU, 299, 300, 301, 305, 306, 309, 311, 312, 313, 314, 315, 317, 318, 319
Manned Space Flight, Office of, OMSF, 197, 198, 199, 202, 216
Manned Space Flight Experiments Board, MSFEB, 217
Manned Spacecraft Center, MSC, 196, 197, 205, 216, 219, 278, 281, 284, 285, 286, 287, 289, 290, 293, 306, 311, 312, 382
Manufacturing Engineering Laboratory, 218
Mariner, spacecraft, 251, 256, 261, 262, 263, 264, 266, 267, 268, 269, 271, 272
Mark II, III, IV, pressure suits, 304
Markusen, Ann, 169n20
Mars, planet, 201, 251, 253, 273, 277, 389
Marshall Space Flight Center, MSFC, 197, 217, 218, 220, 222, 223, 256, 278, 280, 281, 285, 286, 287, 289, 290, 348, 349, 351, 352, 355, 356, 358, 359, 372, 377, 378, 381, 385, 386
Martin Marietta Corporation, 262, 277, 282, 284, 300, 301, 312, 313, 314, 315, 316, 317, 318
Martin, Russell, 299n1
Massachusetts Institute of Technology, MIT, 6, 16, 113, 196
Matthews, Charles W., 219
Mayer, John, 293
Maywood, Illinois, 8
McAvoy, William H., 11-44 passim
McBrien, R.L., 36, 40
McCandless, Bruce, astronaut, 300, 301, 302, 312, 314, 315, 316, 318, 319, 372, 375
McCook Field, 66, 75, 106
McCune, John, 385
McCurdy, Howard E., 103, 104, 173, 175, 197n10, 201n21, 233n52
McDivitt, James A., astronaut, 310
McDonald, Allan J., 354, 355, 356, 357
McDonnell Aircraft Corporation, 176, 222, 306, 308
McDonnell Douglas corporation, 281, 284, 290, 294, 323, 329, 383
McDougall, Walter A., 92n3, 114n63, 171, 172n33, 173, 200n16 , 237n6
McDowell, Jonathan, 149n1, 150n4
McElroy, John H., 235n2, 240n21, 249n53, 250n59
McFarland, Stephen L., 3n4
McHugh, James G., 22
McKay, Jack, pilot, 156
McKee, Daniel D., 306n16
McKenna, James T., 388n57
McKenna, Paul J., 333
McLarren, Robert, 56
McNamara, Robert, Secretary of Defense, 186, 199, 200
McQuay, Inc., 43
Mead, Senator James M., 92
Memorandum of Understanding, MOU, 153, 157
Mercanti, Enrico P., 244
Mercury, Project, 156, 157, 163, 164, 165, 166, 167, 168, 169, 173, 174, 175, 176, 177, 179, 180, 181, 182, 183, 184, 185, 186, 189, 190, 191, 197, 199, 200, 201, 211, 214n5, 215, 231, 273, 277, 279, 280, 287, 299, 304, 305, 306, 382
Mercury-Atlas, 187, 188
Mercury Seven, 191
Messner, Julian, 150
Meyer, Theodor, 65
Michener, James, 100, 103
Mikkelson, Daniel, 321, 337
Mikulski, Senator Barbara, 369, 373, 383, 393, 394
Miller, Elton W., 17, 18, 19, 21
Millikan, Clark, 99, 102

Mills, Charles T.L., 99n26, 271n80
Mississippi Test Facility, see NSTL, 296
Mitchell, John, 120
Mitchell, Royce, 354, 357
Mitz, Milton A., 264n53,54, 267n65,68
Modular Maneuvering Unit, MMU, 305, 306, 313, 314, 315, 316
Moffett, Admiral William A., 6n11, 7
Moffett Field, Navy, 35
Mojave Dessert, California, 59
Mondale, Walter, 286
Moon, 176, 186, 197, 199, 200, 201, 202, 204, 205, 206, 208, 209, 210, 211, 215, 216, 251, 253, 278, 280, 281, 302, 303, 311, 331
Mooneyhan, D.W., 246
Moore, Duncan, 375
Moore, Wendell F., 307
Morris, Owen, 277,289
Morrow Board, 18
Morrow, Dwight, 18
Morrow, T.F., Chrysler Corporation, 279
Moss, Senator Frank, 328
Mueller, George, 200, 202, 205, 216, 217, 218, 219, 220, 221, 278, 281, 285
Muenger, Elizabeth A., 108
Muhleman, Duane O., 273
Munk, Max M., 7, 18, 110, 121
Muroc Dry Lake, 59
Murray, Bruce, 251n2, 263n50
Murray, Charles, 103, 205n23
Musgrave, Story, astronaut, 367, 369, 378, 381, 384, 385, 386n50, 387, 388n55, 389, 390, 391, 394
Muskie, Senator Edmund, 286
Myers, Dale D., 282, 285, 289n41, 295

N

National Academy of Sciences, NAS, 253
National Advisory Committee for Aeronautics, NACA, 1-163 passim, 250 , 279n9, 280, 324, 325, 326, 342
National Aeronautic Association, NAA, 3-15 passim, 147, 208, 251, 318, 342, 365n1
National Aeronautics and Space Administration, NASA, 164-396 passim
National Aeronautics and Space Council, NASC, 307
National Air and Space Museum, Smithsonian, 176, 387
National Bureau of Standards, NBS, 46, 48, 62, 75, 125
National Carbon corporation, 304
National Geographic, 108
National Oceanic and Atmospheric Administration, NOAA, 242, 249
National Research Council, NRC, 352, 355, 356n43, 357, 361, 363
National Science Foundation, NSF, 276, 355
National Security Council, NSC, 171
National Space Club, 251
National Space Testing Laboratory, NSTL, formerly Mississippi Test Facility, 296
Nature, 268
Naugle, John E., 262, 263n46, 268n73
Naval Aircraft Factory, 46
Naval Air Development Center, 156
Naval Research Laboratory, 37, 38, 44, 169, 171
Navy, U.S., 6, 8n15, 10, 35, 42, 86, 113, 123, 146, 153, 154, 156, 157, 171, 172, 173, 304, 318
Navy-Curtiss (NC), 6

Nazis, 303
Neel, Carr, 35, 41, 51
Neely, Frederick J., 3n6
Nelson, George D., astronaut, 315, 317
Neptune, planet, 251, 256, 259, 260, 263, 269, 271, 273, 274, 275, 276
NERVA, engine, 260
New Concord, Ohio, 189
New Deal, 2
Newell, Homer, 232, 254n10, 255, 256, 265n57
New York, New York, 13, 189
New York Times, 117, 347, 349, 350, 351, 352, 354n35, 355, 356n47, 357n52, 359n62, 360n66, 362n76
Newkirk, Ertel, 221n24, 222n27, 227n37, 230n42,44
Newton, Isaac, 62, 141
Nicollier, Claude, ESA astronaut, 382, 386, 390, 391
Nike-Zeus, satellite, 172
Nixon, President Richard M., 184, 207, 209, 221, 253, 256, 257, 261, 281, 286
Noa, astronaut recovery ship, 189
Nobel Prize, 91, 98
Nored, Donald, 324, 326n21, 330, 333n40,1,42, 339, 340n66, 343
Nordberg, William, 239
North, J.D., 17
North American Aviation Corporation, 152, 154, 160, 196, 202, 203, 205, 279, 280, 285
North American Rockwell corporation, 262, 281, 284, 312
Northrop corporation, 127
Northwest airlines, 31
Norton, F.H., 106n42
Nowlan, Philip Francis, 302

O

O-47 aircraft, 35
Oberth, Hermann, 213
O'Connor, Bryan, astronaut, 362
O'Dell, C.R., 392n72
Office of Management and Budget, OMB, 162, 163, 235, 239, 247, 259, 261, 262, 263
Opalko, Jane, 386n52
Ordway, Frederick I., III, 216n10, 232n51
Organization of Petroleum Exporting Countries, OPEC, 326

P

P-38 aircraft, 73, 83, 84, 122, 123, 124
P-39 aircraft, 122
P-40 aircraft, 122
P-47 aircraft, 115, 122
P-51 aircraft, 59, 87, 105, 137
P-59 aircraft, 123
P-80, aircraft, 123
Pacific ocean, 6, 189, 207, 227, 230
Paine, Thomas, 205, 207, 220, 258, 284
Panama Canal, 289
Paterson, New Jersey, 6
PBY-2 boat, 35
PBY-5 aircraft, 49
PB2Y-3 aircraft, 49

PB4Y-1 boat, 49
Pearl Harbor, 117, 119, 122, 125, 171, 177
Pearson, Jeremiah "Jed" W., 380, 382, 386
Pecora, W.T., 238n14
Pendegraft, J.B., 218
Peenemunde, Germany, 177
Pennsylvania-Central Airlines, 31, 53
Petersen, Commander Forrest S., 150, 159
Pettengill, Gordon H., 262
Philadelphia Navy Yard, 6
Phillips, General Samuel, 200, 202, 203, 205, 358
Physics Today, 122
Pickering, William H., 263n46, 268
Pindzola, M., 99n25
Pinkel, Benjamin, 22
Pinkel, Irving, 58
Pinson, Jay D., 83n35
Pioneer, spacecraft, 251, 266, 268
Piper Pawnee aircraft, 8n16
Pittsburgh Plate Glass Co., PPG, 40
Pluto, planet, 251, 254, 259
PN-10 boat, 6, 7
Pogue, William R., 213, 230, 231, 232
Polluck, Andrew, 340
Pope, Alan, 126
Post, Wiley, aviator, 303, 304
Prandtl, Ludwig, 62, 65, 70, 75, 105, 106, 110, 112
Pratt & Whitney, 17, 22,23, 24, 46, 144, 290, 323, 328, 339
President's Aircraft Board, 18
President's Commission on the Space Shuttle Challenger, see Rogers Commission
President's Science Advisory Committee, PSAC, 173
Prim, J.W., 310n31
Principia, 62
Propeller Research Tunnel, PRT, 7-26 passim, 97, 107, 119, 128
Proxmire, Senator William , 286
Puckett, Allen E., 69
Puddy, Donald , 223
Pyne, Stephen J., 178, 179

Q

Quayle, Vice President Dan, 379
Queen Isabella, 190
Quest for Performance: The Evolution of Modern Aircraft, 115

R

Rader, Karen A., 140n13, 166n9, 274, 275n91, 323n6, 367n3, 368, 378n29
Rainwater, Susan B., 379
Ranger, spacecraft, 251, 263
Rankine, William John, 64, 65
Rasool, S. Ichtiaque, 251n4, 261n40, 263, 266
RCA company, 249
Reaction Motors, 160, 161, 163
Reagan, President Ronald, 249, 297, 318, 348, 353, 354, 363

Redstone Arsenal, Alabama, 169, 170, 173, 175
Reed, Sylvanus, 3n6, 5n7
Reedy, George, 172
Rees, Eberhard, 282, 287
Reid, Elliott G., 12
Reid, Henry J.E., 18, 21, 24, 78
Remote Maneuvering Unit, RMU, 307, 311
Remote Manipulator System, RMS, 366
Rensberger, Boyce, 348n11
Rensselaer Polytechnic Institute, RPI, 12
Research Airplane Committee, 154
Reynolds, Osborne, 128
Richardson, Holden "Dick" C. , 6, 7, 8
Riemann, G.F. Bernhard, mathematician, 64
Riolo, Robert, 91n1
Robert J. Collier Trophy, see Collier Trophy
Robie, Bill, 3n6, 147n42, 149n2, 165n5, 167n11, 235n1, 251n1, 313n41
Robinson, Russell G., 131
Rocket Propulsion Laboratory, 305, 308
Rocket Research Corporation, 311
Rockwell International, 290, 295, 296, 313, 314
Rodert, Lewis August, 29-58 passim, 101, 325
Rogers Commission, 348, 349, 350, 351, 352, 353, 354, 355, 356, 358, 359, 360, 362, 363
Rogers, William P., Secretary of State, 348
Rohr Industries, 323
Rohrbach, Carl, 321
Roland, Alex, 5n9, 17, 18, 20, 21, 56, 93, 95, 97n18, 103, 104, 111, 114, 117, 125, 133, 313n40
Rollin, Vernon G., 23n63
Roosevelt, President Franklin D., 15
Rose Garden, White House, 165, 167
Rosholt, Robert L., 157
Rothenberg, Joseph, 375, 377, 379n30, 387, 394, 395n81
Roush, Paul A., 37
Rover, 208
Royal Academy of Science, Rome, 78
Royal Aeronautical Establishment, see Royal Aircraft Establishment
Royal Aircraft Establishment, RAE, 43, 66
Rubashkin, David, 253n7, 257n22,23
Runco, Mario, astronaut, 381
Russia, 169, 177
Rutherford, Lord Ernest, 91n1

S

Sagan, Carl, 266
Sandy, Mary 340
Saturn, planet, 251, 256, 259, 261, 262, 263, 264, 266, 267, 268, 269, 270, 271, 272, 275, 276
Saturn V, launch vehicle, 161, 196, 205, 219, 220, 222, 223, 230, 280
Sawyer, Kathy, 373n16
SB2D-1 aircraft, 50
S.6B aircraft, 71, 73
Schey, Oscar W., 20, 23
Schirra, Walter M., Jr., 165, 189
Schjeldahl, G.T., Company, 223
Schmitt, Senator Harrison, 182, 209, 248

Schneider, William C., 221, 222, 232, 233
Schriever, Bernard A., 307
Schulman, Robert, 190
Schultz, David C., 312
Schurmeier, Harris "Bud" M., 261, 264
Schweickart, Russell "Rusty" Louis, 224, 226
Science, 268
Science Institute, 374, 381, 383, 385, 394
Science News, 259
Scientific American, 115
Scott, David R., 193, 211
Seamans, Robert C., Jr., 151, 197, 202, 217, 219, 238n14
Self-test and repair computer, STAR, 260
Sentinel, newspaper, 351
Serling, Robert J., 321
Shapley, Willis H., 247n42
Shea, Joseph F., 202, 216, 383
Shepard, Commander Alan B., Jr., 165, 181, 184, 185, 186, 231
Shipman, Harry L., 299
Shuttle Avionics Integration Laboratory, SAIL, 287
Shuttle Mission Simulator, SMS, 287
Sievers, Keith, 337n55
Sigma 7, spacecraft, 189
Sikorsky, Igor, 56
Silk, J.M., 308n21
Silverstein, Abe, 112n60
Skoglund, Victor J., 48
Skolnikoff, Eugene, 247n42
Sky and Telescope magazine, 391, 392
Skylab, spacecraft, 213, 214, 215, 221, 222, 223, 224, 225, 227, 231, 232, 233, 277, 295, 299, 301, 311, 312, 313, 382
Slade, Martin A., 273
Slay, Alton, 361
Slayton, Donald "Deke" K., 165, 182, 222, 228, 312n36
Smith, Bradford, 263, 264
Smith, Herschel H., 7
Smith, Senator Margaret Chase, 203
Smith, Michael L., 178, 179, 180
Smith, Richard K., 97, 108, 113, 125n95
Smith, Robert W., 260, 368, 369n7
Smithsonian Institution, 6
Society of Automotive Engineers, 13
Solar Aircraft Company, 41
Solar Maximum satellite, 315, 316, 317, 319, 387
Solid Rocket Motor, SRM, 348, 349, 351, 352, 354, 355, 356, 357, 358, 361, 362
Soule, Hartley A., 111n56
Soviet Union, U.S.S.R., Russia, 144, 156, 169, 177, 180, 184, 185, 186, 196, 199, 201, 205, 210, 214, 257, 307, 323, 329
Spaatz, General Carl, 56
Space, 103
Space Science Board, 253, 259, 261, 262, 271
Space Telescope Axial Replacement, STAR, 376
Space Transportation System, STS, 313, 358, 359, 363, 364, 365
Spencer, Jim, 91n1
Sperry, Elmer, 3n6, 5n7

Sperry Gyroscope, 160
Sperry Messenger, 6, 8
Sputnik, U.S.S.R. satellite, 156, 163, 171, Kaputnik172, 177, 179, 180, 214, 306, 307, 325
Stack, John, 62, 63, 71, 73, 75, 77, 79, 80, 81, 82, 84, 85, 86, 89, 91, 92, 93, 94, 95, 97, 98, 99, 100, 101, 109, 110, 113, 116, 117, 118, 119, 120, 121, 122, 123n90 , 124, 125, 126, 127, 128, 130, 132, 133, 135, 136, 137, 142
Stafford, Thomas, 382
Stainforth, Flt.Lt. George H., 73
Stanford University, 10, 106, 265
Stapleton, Geoffrey, 91n1
Staudenmaier, John M., 3n5
Stechman, Bernadette C., 377n25
Steelman, Donald, 221
Stennis, John C., 286
Stever, H. Guyford, 355, 356, 357
Stewart, Homer Joe, 255
Stewart, Robert L., astronaut, 302, 315, 316
Stewart-Warner Corporation, 41, 51, 53
Stinson aircraft, 2
Stolp, P.C., 333
Stone, Dr. Edward C., 251, 252, 264n52, 266n63, 267, 268, 270, 274
Stoney, William, 250n58
Strack, William C., 326n21
Stradivari, Antonio, 133
Strategic Arms Reduction Talks, START, 318
Strategic Defense Initiative, SDI, 318
Strategic Arms Limitation Talks, SALT, 318
Stuhlinger, Ernst, 216, 232n51
Sullivan, Kathy, astronaut, 372
Summerlin, Lee B., 217
Sun, 272
Sunspot II, MSFC vacuum , 218
Super Guppy aircraft, 222
Supersonic Transport Plane, 286
Surveyor, spacecraft, 201, 204, 210, 251
Sustaining University Program, SUP, 196, 200, 210, 211
Sutter, Joseph, 356
Sweden, 264
Swenson, Loyd S., Jr., 167n12, 168n13, 176n52, 179n65, 180n69, 182n77, 183n79, 184n85, 185n91, 187n97, 189n103,108 , 190n110, 214n5
Symington, Senator Stuart, 286
Systeme Probatoire d'Observation de la Terre, SPOT, French spacecraft, 248

T

Tatarewicz, Joseph N., 265n57, 275
Taylor, Captain David W., 6
Taylor, Michael J.H., 146n39
Teague, Olin E., 282, 286
Theodorsen, Theodore, 23, 24, 25, 31, 78, 110, 112n60, 121
The Right Stuff, 103
Thermoelectric Outer Planets Spacecraft, TOPS, 260, 261, 262, 269
Thiokol Corporation, 277, 285, 313, 348, 354, 355, 356, 357, 358
Thomas, Albert, Texas Rep, 196
Thomas, John, 352, 354, 355, 356, 357
Thompson, Floyd L., 8n15, 219, 220

Thompson, John R., 349
Thompson, Robert F., 284, 286, 293
Thornton, Jim, 379
Thornton, Kathy, astronaut, 380, 381, 390, 391
Thuot, Pierre, astronaut, 380
Tichenor, Frank, 17, 18, 19, 20, 111n57
Ticonderoga, aircraft, 26
Time-Life, corporation, 179, 180, 363n86
Tindall, Howard "Bill" W., 293
Titan, missile, 273, 306, 360
Titov, Gherman, 187
Tomlinson, D.W., U.S.N.R., 38n21, 39
Townend, Hubert C., 17
Townes, Charles H., 261n38, 262
Trafton, Will, 395
Trane Company, 43
Transcontinental airlines, 39
Transonic Wind Tunnel, NASA Langley, 96
Transportation, U.S. Department of, 328
Travis, John, 392n70
Tribus, Lt. Myron, 49, 51, 53, 54
Trimble, William F., 6n11,12
Triplett, William, 380n34
Triton, celestial satellite, 276
Truly, Richard, 231, 277, 345, 347, 348, 349, 350, 351, 354, 355, 356, 357, 358, 359, 360, 361, 362, 363, 364, 379
Truman, President Harry S., 89, 101, 197
TRW, 265, 379
Tsiolkovsky, Konstantin, 213
Turbojet Revolution, 115
TWA, airlines, 31
Tycho Brahe, observatory, 252
Tyler, G. Leonard, 265

U

Udall, Stewart L., Secretary of the Interior, 238
Ulysses, spacecraft, 360
Unducted Fan, UDF, 339, 342
Unisys Space Systems corporation, 394
United Aircraft and Transport Corporation, 23
United Airlines, 31, 32, 38, 40, 51, 336, 337
United Nations, 184, 189
United States, U.S., 2, 16, 25, 35, 37, 85, 91, 115, 144, 146, 151, 156, 163, 164, 165, 168, 169, 172, 177, 178, 179, 182, 183, 184, 187, 190, 196, 199, 200, 201, 214, 228, 235, 237, 248, 264, 304, 307, 318, 319, 326, 328, 329, 338, 341, 345, 349, 365
United Technology Center, 285
University of Alaska, 385
University of California, Berkeley, 71, 84
University of Illinois, 8
University of Michigan, 12, 16
University of Minnesota, 29, 39
University of Prague, Germany, 65
University of Rochester, 375
University of Virginia, 124

Uranus, planet, 251, 256, 259, 260, 263, 269, 271, 272, 275, 276
U.S. News and World Report, 181
U.S. Rubber corporation, 304
U.S.S. New Orleans, ship, 230, 232

V

Valentine, E. Floyd, 22
Van Allen, James, 172
Vandenberg Air Force Base, California, 360
van Hoften, James D., 315, 317
Vanguard, Project, 171, 172
Variable Density Tunnel, VDT, 7, 18, 23, 71, 80, 86, 97, 108, 110, 112, 128, 133
Vaughan, Diane, sociologist, 103, 104n37
Velcro, material, 299
Venturi, G.B., 138
Venus, planet, 253, 262, 270, 273
Verne, Jules, 115, 301, 302
Vickery, Hugh, 321
Victory, John F., 56, 100, 101
Vietnam war, 200, 209, 211, 219, 220, 253, 286
Viking I spacecraft, 169, 251, 257, 259, 261, 263
Vincent, Jeffrey, 393
Vincenti, Walter G. 5n8, 10n18, 20, 23, 61, 91n1, 102, 112, 120, 127, 128, 129, 313
Virden, Ralph, pilot, 83
VJ Day, 55
Very Large Array, VLA, 273
Vostok 1, Soviet spacecraft, 184, 186, 189
Voyager spacecraft, 251, 252, 253, 263, 266, 267, 268, 269, 270, 271, 272, 273, 274, 275
Vrabel, Deborah, 323n7, 329n31
Vultee, Gerry, 13

W

Waff, Craig B., 253n7, 261n36, 262n41
Wainwright, 187n99
Waldrop, M. Mitchell, 247n43
Walker, Dr. Joseph A., pilot, 150, 151, 155, 157
Wallace, Hayes D., 141
Wallops Island, Virginia, 87, 137, 139
Walton, Ernest T.S., 91
Ward, John William, historian, 100, 102, 122, 129
Ward, Vernon G., 94, 98, 99, 132
Warner, Edward Pearson, 13, 17, 18, 19, 20, 38n23, 122, 123n88
Warwick, James W., 265n55
Washington Evening Star, 259
Washington Post, 100, 118, 126, 321, 354n32, 356n45, 357n50, 360n69, 361n71,74, 373, 392
Washington Navy Yard, Washington, DC, 6, 173
Washington Times, 321, 340
Wasp, engine, 24
Watson, James, 274
Webb, James E., 174, 177, 186, 190, 191, 193, 195, 196, 197, 198, 199, 200, 201, 202, 205, 206, 207, 210, 217, 219, 220, 347, 348, 379, 382
Weick, Fred, 8-26 passim, 324
Weiler, Edward J., 392

Weinberg, Alvin, 367
Weinberger, Caspar W., 261n40
Weisner, Jerome, 199
Weitz, Paul J., 222, 223, 224, 226
Wells, H.G., 301, 302
Wen, Zheng, 392n72
West Germany, 264
Western Airlines, 39
Western Union, 316
Westervelt, Captain George C., 6
Westfall, Catherine, 91
Whirlwind Engine, 12, 13
Whirlwind J-5 aircraft, 6, 12,13
Whitcomb, Richard, 123, 133,135, 136, 137, 138, 139, 140, 142, 144, 145, 146, 147, 148, 325
White, Edward H., astronaut, 219, 301, 306n15, 310
White, Major Robert M., 150
White House, 13, 15, 16, 165, 194, 200, 209, 258, 261, 308, 345, 347, 352, 379, 393, 394, 395, 396
White Sands Proving Ground, New Mexico, 169, 190
Whitlow, John B., 326n21
Whitsett, Maj. Charles E., Jr., 300, 301, 312, 313n37, 314, 318
Wide Field-Planetary Camera, 377
Wikete, J.E., 333n43
Will Rogers, humorist, 304
Williams, Walter, 349
Wilson, E.B., 106n42
Winnie Mae aircraft, 304
Winter Flight Laboratory, 39
Wolfe, Tom, *The Right Stuff*, 103, 179, 184,
Wood, Donald, 8n14, 12, 21n55,22
Wood, Gordon E., 265
Wood, R. McK., 68
Woods, Robert J., 86
Woods Hole, Massachusetts, 253, 254, 258, 259, 261, 262
Woody Woodpecker, cartoon character, 303
Worcester Polytechnic Institute, 21, 136
Worden, Lt.Col. Alfred M., 193, 211
World War I, 62, 66
World War II, 11, 16, 25, 37, 59, 65, 85, 91, 93, 102, 111, 129, 135, 138, 140, 142, 167, 169, 173, 175, 177, 182, 195, 214, 325
Wright Aeronautical Corporation, 6
Wright Apache, aircraft, 12
Wright brothers, 3, 10, 16, 54, 65, 89, 95, 118, 122, 126, 168
Wright Brothers Memorial Trophy, 101
Wright Corporation, 6
Wright Field, 35, 38, 39, 41, 43m, 46, 49, 50, 84, 85, 145, 304
Wright Flyer, 66
Wright, Orville, 3n6, 5n7, 100, 105
Wright-Patterson Air Force Base, Ohio, 106, 305
Wright, Ray H., 94, 98, 99, 102, 107n46, 109, 110, 116n66, 120, 122, 125, 126, 127, 129, 130, 131, 132
Wright Whirlwind aircraft, 7
Wright, Wilbur, 105
Wyatt, DeMarquis, 215

X

X-15 aircraft, 149, 151, 152, 153, 154, 155, 156, 157, 160, 161, 162, 163, 164, 174, 304
X-20A aircraft, 155
XBM-1 bomber, 34
XLR-11 rocket engine, 151, 155, 161
XLR-99 rocket engine, 151, 152, 155, 160, 161, 162, 163
XP-79 rocket, 87
XTB2D-1 aircraft, 50

Y

Yackey Aircraft Company, 8
Yackey Transports aircraft, 10
Yeager, Captain Charles E. "Chuck," pilot, 59, 60, 88, 89, 91, 94, 101, 126
Young, Alfred W., 20
Young, James O., 88
Young, John, astronaut, 277, 297, 297, 359, 385
Young, Wayne,   295

Z

Zahm, Albert F., 106n42, 109, 10 , 111

The NASA History Series

Reference Works, NASA SP-4000

Grimwood, James M. *Project Mercury: A Chronology* (NASA SP-4001, 1963)

Grimwood, James M., and Hacker, Barton C., with Vorzimmer, Peter J. *Project Gemini Technology and Operations: A Chronology* (NASA SP-4002, 1969)

Link, Mae Mills. *Space Medicine in Project Mercury* (NASA SP-4003, 1965)

Astronautics and Aeronautics, 1963: Chronology of Science, Technology, and Policy (NASA SP-4004, 1964)

Astronautics and Aeronautics, 1964: Chronology of Science, Technology, and Policy (NASA SP-4005, 1965)

Astronautics and Aeronautics, 1965: Chronology of Science, Technology, and Policy (NASA SP-4006, 1966)

Astronautics and Aeronautics, 1966: Chronology of Science, Technology, and Policy (NASA SP-4007, 1967)

Astronautics and Aeronautics, 1967: Chronology of Science, Technology, and Policy (NASA SP-4008, 1968)

Ertel, Ivan D., and Morse, Mary Louise. *The Apollo Spacecraft: A Chronology, Volume I, Through November 7, 1962* (NASA SP-4009, 1969)

Morse, Mary Louise, and Bays, Jean Kernahan. *The Apollo Spacecraft: A Chronology, Volume II, November 8, 1962–September 30, 1964* (NASA SP-4009, 1973)

Brooks, Courtney G., and Ertel, Ivan D. *The Apollo Spacecraft: A Chronology, Volume III, October 1, 1964–January 20, 1966* (NASA SP-4009, 1973)

Ertel, Ivan D., and Newkirk, Roland W., with Brooks, Courtney G. *The Apollo Spacecraft: A Chronology, Volume IV, January 21, 1966–July 13, 1974* (NASA SP-4009, 1978)

Astronautics and Aeronautics, 1968: Chronology of Science, Technology, and Policy (NASA SP-4010, 1969)

Newkirk, Roland W., and Ertel, Ivan D., with Brooks, Courtney G. *Skylab: A Chronology* (NASA SP-4011, 1977)

Van Nimmen, Jane, and Bruno, Leonard C., with Rosholt, Robert L. *NASA Historical Data Book, Vol. I: NASA Resources, 1958–1968* (NASA SP-4012, 1976, rep. ed. 1988)

Ezell, Linda Neuman. *NASA Historical Data Book, Vol II: Programs and Projects, 1958–1968* (NASA SP-4012, 1988)

Ezell, Linda Neuman. *NASA Historical Data Book, Vol. III: Programs and Projects, 1969–1978* (NASA SP-4012, 1988)

Astronautics and Aeronautics, 1969: Chronology of Science, Technology, and Policy (NASA SP-4014, 1970)

Astronautics and Aeronautics, 1970: Chronology of Science, Technology, and Policy (NASA SP-4015, 1972)

Astronautics and Aeronautics, 1971: Chronology of Science, Technology, and Policy (NASA SP-4016, 1972)

Astronautics and Aeronautics, 1972: Chronology of Science, Technology, and Policy (NASA SP-4017, 1974)

Astronautics and Aeronautics, 1973: Chronology of Science, Technology, and Policy (NASA SP-4018, 1975)

Astronautics and Aeronautics, 1974: Chronology of Science, Technology, and Policy (NASA SP-4019, 1977)

Astronautics and Aeronautics, 1975: Chronology of Science, Technology, and Policy (NASA SP-4020, 1979)

Astronautics and Aeronautics, 1976: Chronology of Science, Technology, and Policy (NASA SP-4021, 1984)

Astronautics and Aeronautics, 1977: Chronology of Science, Technology, and Policy (NASA SP-4022, 1986)

Astronautics and Aeronautics, 1978: Chronology of Science, Technology, and Policy (NASA SP-4023, 1986)

Astronautics and Aeronautics, 1979–1984: Chronology of Science, Technology, and Policy (NASA SP-4024, 1988)

Astronautics and Aeronautics, 1985: Chronology of Science, Technology, and Policy (NASA SP-4025, 1990)

Gawdiak, Ihor Y. Compiler. *NASA Historical Data Book, Vol. IV: NASA Resources, 1969–1978* (NASA SP-4012, 1994)

Noordung, Hermann. *The Problem of Space Travel: The Rocket Motor.* In Ernst Stuhlinger and J.D. Hunley, with Jennifer Garland, editors (NASA SP-4026, 1995)

Astronautics and Aeronautics, 1986–1990: Chronology of Science, Technology, and Policy (NASA SP-4027, 1997)

Management Histories, NASA SP-4100

Rosholt, Robert L. *An Administrative History of NASA, 1958–1963* (NASA SP-4101, 1966)

Levine, Arnold S. M*anaging NASA in the Apollo Era* (NASA SP-4102, 1982)

Roland, Alex. *Model Research: The National Advisory Committee for Aeronautics, 1915–1958* (NASA SP-4103, 1985)

Fries, Sylvia D. *NASA Engineers and the Age of Apollo* (NASA SP-4104, 1992)

Glennan, T. Keith. *The Birth of NASA: The Diary of T. Keith Glennan,* edited by J.D. Hunley (NASA SP-4105, 1993)

Seamans, Robert C., Jr. *Aiming at Targets: The Autobiography of Robert C. Seamans, Jr.* (NASA SP-4106, 1996)

Project Histories, NASA SP-4200

Swenson, Loyd S., Jr., Grimwood, James M., and Alexander, Charles C. *This New Ocean: A History of Project Mercury* (NASA SP-4201, 1966)

Green, Constance McL., and Lomask, Milton. *Vanguard: A History* (NASA SP-4202, 1970; rep. ed. Smithsonian Institution Press, 1971)

Hacker, Barton C., and Grimwood, James M. *On Shoulders of Titans: A History of Project Gemini* (NASA SP-4203, 1977)

Benson, Charles D. and Faherty, William Barnaby. *Moonport: A History of Apollo Launch Facilities and Operations* (NASA SP-4204, 1978)

Brooks, Courtney G., Grimwood, James M., and Swenson, Loyd S., Jr. *Chariots for Apollo: A History of Manned Lunar Spacecraft* (NASA SP-4205, 1979)

Bilstein, Roger E. *Stages to Saturn: A Technological History of the Apollo/Saturn Launch Vehicles* (NASA SP-4206, 1980)

Compton, W. David, and Benson, Charles D. *Living and Working in Space: A History of Skylab* (NASA SP-4208, 1983)

Ezell, Edward Clinton, and Ezell, Linda Neuman. *The Partnership: A History of the Apollo-Soyuz Test Project* (NASA SP-4209, 1978)

Hall, R. Cargill. *Lunar Impact: A History of Project Ranger* (NASA SP-4210, 1977)

Newell, Homer E. *Beyond the Atmosphere: Early Years of Space Science* (NASA SP-4211, 1980)

Ezell, Edward Clinton, and Ezell, Linda Neuman. *On Mars: Exploration of the Red Planet, 1958–1978* (NASA SP-4212, 1984)

Pitts, John A. *The Human Factor: Biomedicine in the Manned Space Program to 1980* (NASA SP-4213, 1985)

Compton, W. David. *Where No Man Has Gone Before: A History of Apollo Lunar Exploration Missions* (NASA SP-4214, 1989)

Naugle, John E. *First Among Equals: The Selection of NASA Space Science Experiments* (NASA SP-4215, 1991)

Wallace, Lane E. *Airborne Trailblazer: Two Decades with NASA Langley's Boeing 737 Flying Laboratory* (NASA SP-4216, 1994)

Butrica, Andrew J. Editor. *Beyond the Ionosphere: Fifty Years of Space Communication* (NASA SP-4217, 1997)

Butrica, Andrew J. *To See the Unseen: A History of Planetary Radar Astronomy* (NASA SP-4218, 1996)

Center Histories, NASA SP-4300

Rosenthal, Alfred. *Venture into Space: Early Years of Goddard Space Flight Center* (NASA SP-4301, 1985)

Hartman, Edwin, P. *Adventures in Research: A History of Ames Research Center, 1940–1965* (NASA SP-4302, 1970)

Hallion, Richard P. *On the Frontier: Flight Research at Dryden, 1946–1981* (NASA SP-4303, 1984)

Muenger, Elizabeth A. *Searching the Horizon: A History of Ames Research Center, 1940–1976* (NASA SP-4304, 1985)

Hansen, James R. *Engineer in Charge: A History of the Langley Aeronautical Laboratory, 1917–1958* (NASA SP-4305, 1987)

Dawson, Virginia P. *Engines and Innovation: Lewis Laboratory and American Propulsion Technology* (NASA SP-4306, 1991)

Dethloff, Henry C. *"Suddenly Tomorrow Came . . .": A History of the Johnson Space Center, 1957–1990* (NASA SP-4307, 1993)

Hansen, James R. *Spaceflight Revolution: NASA Langley Research Center From Sputnik to Apollo* (NASA SP-4308, 1995)

Wallace, Lane E. F*lights of Discovery: 50 Years at the NASA Dryden Flight Research Center.* (NASA SP-4309, 1996).

Herring, Mark R. *Way Station to Space: A History of the John C. Stennis Space Center* (NASA SP-4310, 1997).

Wallace, Harold D., Jr. *Wallops Station and the Creation of an American Space Program* (NASA SP-4311, 1997).

General Histories, NASA SP-4400

Corliss, William R. *NASA Sounding Rockets, 1958–1968: A Historical Summary* (NASA SP-4401, 1971)

Wells, Helen T., Whiteley, Susan H., and Karegeannes, Carrie. *Origins of NASA Names* (NASA SP-4402, 1976)

Anderson, Frank W., Jr. *Orders of Magnitude: A History of NACA and NASA, 1915–1980* (NASA SP-4403, 1981)

Sloop, John L. *Liquid Hydrogen as a Propulsion Fuel, 1945–1959* (NASA SP-4404, 1978)

Roland, Alex. *A Spacefaring People: Perspectives on Early Spaceflight* (NASA SP-4405, 1985)

Bilstein, Roger E. *Orders of Magnitude: A History of the NACA and NASA, 1915–1990* (NASA SP-4406, 1989)

Logsdon, John M., General Editor. With Lear, Linda J., Warren-Findley, Jannelle, Williamson, Ray A., and Day, Dwayne A., *Exploring the Unknown: Selected Documents in the History of the U.S. Civil Space Program, Volume I: Organizing for Exploration* (NASA SP-4407, 1995)

Logsdon, John M., General Editor. With Day, Dwayne A., and Launius, Roger D., *Exploring the Unknown: Selected Documents in the History of the U.S. Civil Space Program, Volume II: External Relationships* (NASA SP-4407, 1996)

Logsdon, John M., General Editor. With Launius, Roger D., Onkst, David H., and Garber, Stephen J. *Exploring the Unknown: Selected Documents in the History of the U.S. Civil Space Program, Volume III: Using Space* (NASA SP-4407, 1998).

ISBN 0-16-049640-3